Environmental Micropaleontology

The Application of Microfossils to Environmental Geology

TOPICS IN GEOBIOLOGY

Current volumes in this series

Environmental Micropaleontology

The Application of Microfossils to Environmental Geology

Edited by

Ronald E. Martin

University of Delaware
Newark, Delaware

Kluwer Academic / Plenum Publishers
New York, Boston, Dordrecht, London, Moscow

Library of Congress Cataloging-in-Publication Data

Environmental micropaleontology: the application of microfossils to environmental geology/edited by Ronald E. Martin
p. cm. — (Topics in geobiology; v. 15)
Includes bibliographical references and index.
ISBN 0-306-46232-X
1. Micropaleontology. 2. Environmental geology. I. Martin, Ronald E. II. Series

QE719 .E58 2000
560—dc21

00-029630

ISBN: 0-306-46232-X

233 Spring Street, New York, N.Y. 10013

http://www.wkap.nl/

10 9 8 7 6 5 4 3 2 1

A C.I.P. record for this book is available from the Library of Congress.

Contributors

Elisabeth Alve Department of Geology, University of Oslo, Blindern, N-0316 Oslo, Norway

V. M. Bresler Institute for Nature Conservation Research, George S. Wise Faculty of Life Sciences, Tel Aviv University, Ramat Aviv 69978, Israel.

Raymond A. Christopher Department of Geological Sciences, Clemson University, Clemson, South Carolina 29634

Rodolfo Coccioni Istituto di Geologia dell' Università, Campus Scientifico, Località Crocicchia, I-61029, Urbino, Italy

Barrie Dale Department of Geology, University of Oslo, Blindern, N-0316 Oslo, Norway

Jean-Pierre Debenay Laboratoire de Géologie, Equipe EA 2644, Faculté des Sciences, Université d'Angers, 49045, Angers Cedex, and Laboratoire d'Etude des Bio-Indicateurs Marins (LÉBIM), Ker Chalon, 85350 Ile d'Yeu, France

Sushil S. Dixit Paleoecological Environmental Assessment and Research Laboratory (PEARL), Department of Biology, Queen's University, Kingston, Ontario K7L 3N6, Canada

Stephen H. Eagar Research School of Earth Sciences, Victoria University of Wellington, Wellington, New Zealand

Michael T. Ebrahim School of Earth Sciences, Victoria University of Wellington, Wellington, New Zealand

Donald W. Engelhardt Earth Sciences Resources Institute, University of South Carolina, Columbia, South Carolina 29801

Emmanuelle Geslin Laboratoire de Géologie, Equipe EA 2644, Faculté des Sciences, Université d'Angers, 49045, Angers Cedex, and Laboratoire d'Etude des Bio-Indicateurs Marins (LÉBIM), Ker Chalon, 85350 Ile d'Yeu, France

Jean-Jacques Guillou Laboratoire de Géologie, Equipe EA 2644, Faculté des Sciences, Université de Nantes, 44072 Nantes Cedex 3, France

Pamela Hallock Department of Marine Science, University of South Florida, St. Petersburg, Florida 33701

Malcolm B. Hart Department of Geological Sciences and Plymouth Environmental Research Center, University of Plymouth, Plymouth PL4 8AA, United Kingdom

Scott P. Hippensteel Department of Geology, University of Delaware, Newark, Delaware 19716–2544

Avraham Honigstein Ministry for National Infrastructures, Oil and Gas Section, Jerusalem 94387, Israel

Scott E. Ishman U.S. Geological Survey, 926A National Center, Reston, Virginia 20192. Present address: Department of Geology, Southern Illinois University, Carbondale, Illinois 62901

Arun Kumar Department of Earth Sciences and Ottawa-Carleton Geoscience Centre, College of Natural Sciences, Carleton University, Ottawa K1S 5B6, Canada

Maurice Lesourd CNRS and SCME, Faculté de Médecine, Rue Haute Reculee, 49045 Angers Cedex, France

Joyce Lucas-Clark Clark Geological Services, Fremont, California 94536

Ronald E. Martin Department of Geology, University of Delaware, Newark, Delaware 19716–2544

John William Murray School of Ocean and Earth Science, Southampton Oceanography Centre, Southampton SO14 3ZH, England

Reuven Ortal Nature Reserves Authority, Jerusalem 94467, Israel

R. Timothy Patterson Department of Earth Sciences and Ottawa-Carleton Geoscience Centre, College of Natural Sciences, Carleton University, Ottawa K1S 5B6, Canada

Fabrice Redois Laboratoire de Géologie, Equipe EA 2644, Faculté des Sciences, Université d'Angers, 49045, Angers Cedex, and Laboratoire d'Etude des Bio-Indicateurs Marins (LÉBIM), Ker Chalon, 85350 Ile d'Yeu, France

Amnon Rosenfeld Geological Survey of Israel, Jerusalem 95501, Israel

Eugenij I. Schornikov Institute of Marine Biology, Far Eastern Branch, Russian Academy of Sciences, Vladivostok 690041, Russia

John P. Smol Paleoecological Environmental Assessment and Research Laboratory (PEARL), Department of Biology, Queen's University, Kingston, Ontario K7L 3N6, Canada

Veronique Stouff Laboratoire de Géologie, Equipe EA 2644, Faculté des Sciences, Université d'Angers, 49045, Angers Cedex, and Laboratoire d'Etudes des Bio-Indicateurs Marins (LÉBIM), Ker Chalon, 85350 Ile d'Yeu, France

G. J. van der Zwaan Institute of Earth Sciences, University of Utrecht, 3508 TA, Utrecht, and Department of Ecology/Biogeology, Faculty of Biology, University of Nijmegen, 6525 ED Nijmegen, The Netherlands

Robert S. Van Pelt Bechtel Savannah River Inc., Environmental Restoration Engineering & Technology, Savannah River Site, Aiken, South Carolina 29808

V. V. Yanko-Hombach Institute for Nature Conservation Research, George S. Wise Faculty of Life Sciences, Tel Aviv University, Ramat Aviv 69978, Israel. Present address: Avalon Institute for Applied Sciences, Winnipeg, Manitoba R3P 2G9, Canada

Foreword

Micropaleontology—the study of microscopic fossils—has had a long and honorable history as an applied science. For much of the twentieth century, micropaleontologists, most with geological backgrounds, have been involved in solving geological problems utilizing a two-pronged approach—biostratigraphy and paleoenvironmental reconstruction. As a result of their efforts, rock strata throughout the Phanerozoic have been dated and correlated, sedimentary basins have been reconstructed, and environments of deposition have been inferred. Millions of barrels of petroleum have been located and/or extracted from the Earth owing, in large part, to the efforts of applied micropaleontologists.

The above is an accurate but now limited characterization of the use of microfossils, as this book, *Environmental Micropaleontology*, so eloquently makes clear. Modern micropaleontologists are as likely to be trained in the biological as in the geological sciences. As our studies of the Earth have diversified, so a much greater variety of geological problems needs to be addressed. Furthermore, micropaleontological approaches are now being utilized not only to study past events but also to solve present-day environmental problems. And these are not minor problems, of little import to humankind. Many chapters in this book, written by experts in the field, address the difficult but immediately relevant topics of environmental quality, pollution, and remediation.

What makes this book so useful, however, is its breadth of coverage. Thirty-two authors, from ten countries, use six different groups of organisms to deal with environmental issues, such as entrophication, heavy metal pollution, storm frequency, and coral reef vitality, in a wide range of settings in Europe, North America, the Mid-East, and the South Pacific, from rivers and lakes, through marshes and lagoons, to atolls and reefs. Thus, readers will be able to find much that is relevant to their own particular interests.

At the core of this volume is the use of microfossils as indicators of environmental disturbance, with important implications for ecosystem conservation and management. What sets microfossils apart from the classical environmental change/pollution indicators such as polychaetes and oligochaetes is the presence of fossilizable "hardparts," whether they be composed of calcium carbonate, silica, sporopollenin or agglutinated grains of sediment. The high preservation potential that the presence of hardparts endows means that microfossils can provide a uniquely historical perspective to environmental disturbance including indication of predisturbance conditions, the onset, development, and apogee of natural or anthropogenic environmental change, followed by recovery to predisturbance conditions or a new environmental state.

This book then, although it deals with microscopic life from the past and the present, is a book about the future. Without doubt, microfossils will continue to be used for classical biostratigraphy, paleoenvironmental reconstruction, and the exploration for oil and gas. But increasingly, the various kinds of environmental problems that are the subject of this volume will be addressed routinely using microfossils—i.e., the members of the meiofauna with preservable hardparts. If this book had been written 50 years ago, it would have been a pamphlet. By 2050, I would predict the need for a multivolume set to deal adequately with this burgeoning field of scientific endeavor.

Stephen J. Culver

Department of Geology
East Carolina University
Greenville, North Carolina
January 2000

Preface

Microfossils are ideally suited to environmental studies because their short generation times allow them to respond quickly to environmental change. A 1995 issue of the *Journal of Foraminiferal Research* (D. B. Scott and J. H. Lipps, editors) emphasized the application of micropaleontology to modern environmental problems. These studies, as well as others published elsewhere, included basic stratigraphic and paleoecological investigations: the correlation of alluvial deposits at seismic hazard sites in the San Francisco Bay area; the correlation of discontinuous waterbearing units of alluvial origin in the San Francisco Bay area that serve as aquifers for municipal and agricultural water supplies; the alignment of the "Chunnel" that now connects England and France as it was being excavated from both sides of the English Channel; and the deciphering of the earthquake history of active margins of the Pacific Northwest.*

It is now time to assess the progress that has been made and delineate future directions of research in the field of environmental micropaleontology. This book is intended primarily for workers in both academia and industry. The taxa studied are mainly foraminifera, but include arcellaceans, diatoms, dinoflagellates, and ostracodes; the papers themselves range from reviews of applications of particular taxa to specific case studies. Some papers present updated baseline data that can be used to evaluate anthropogenic impacts (Chapters 1 and 2), whereas others are concerned with anthropogenic impacts recorded in surface and near-surface samples (Chapters 3, 4, 6, 7, and 8). Some studies have detected subtle, but nevertheless significant, anthropogenic impacts in often *seemingly* healthy environments over time spans of only decades (Chapters 5, 11, 12, and 16) or centuries (Chapters 13, 14, and 17), and, in some cases, documented the response of communities to subsequent remediation efforts. Two papers deal with cellular defense mechanisms against toxic substances and the occurrence of abnormal morphologies in response to pollution (Chapters 9 and 10, respectively; see also Chapter 5); two others with the use of classical and high-resolution biostratigraphy to correlate aquifers (Chapter 18) and site the Thames River Barrier (Chapter 19), respectively; and one with the use of taphonomy (the science of fossil preservation) to evaluate the frequency of storms and their contribution to coastal sedimentation in the Holocene record (Chapter 15).

* For complete references see R. E. Martin, *One Long Experiment: Scale and Process in Earth History*. New York: Columbia University Press, 1998.

The diversity of locations and research of both contributors and reviewers indicates how widespread the application of micropaleontology to environmental problems has become. My sincere thanks for the thoughtful efforts of many external reviewers: (1) Richard N. Benson (Delaware Geological Survey, Newark, Delaware); (2) Joan Bernhard (Department of Environmental Health Sciences and Marine Science Program, University of South Carolina); (3) Richard Brugam (Department of Biological Studies, Southern Illinois University); (4) John Cann (School of Engineering, Applied Geology, University of South Australia); (5) Don Charles (Phycology Section, Patrick Center for Environmental Research, Philadelphia Academy of Sciences); (6) Dan Charman (Department of Geological Sciences, University of Plymouth); (7) Tom Cronin (U.S. Geological Survey, Reston, Virginia); (8) Lucy Edwards (U.S. Geological Survey, Reston, Virginia); (9) Roland Ferry (Water Management Division, Coastal Programs Section, U.S. Environmental Protection Agency, Atlanta, Georgia); (10) Lev Fishelson (Tel Aviv University, Tel Aviv, Israel); (11) Richard Forester (U.S. Geological Survey, Denver, Colorado); (12) Norman Frederiksen (U.S. Geological Survey, Reston, Virginia); (13) Paul Gayes (Center for Marine and Wetland Studies, Coastal Carolina University, Conway, South Carolina); (14) Jean-Pierre Guilbault (BRAQ-Stratigraphie, Montréal, Canada); (15) Lukas Hottinger (Geologisch-Palontologisches Institut der Universitat Basel, Switzerland); (16) Frans Jorissen (Departement de Géologie et Océanographie, Université de Bordeaux I, France); (17) Michael Kaminski, Department of Geological Sciences, University College London); (18) Hiroshi Kitazato (Department of Life and Earth Sciences, Shizuoka University, Japan); (19) Rosalie Maddocks (University of Houston); Franco Medioli (Department of Earth Sciences, Dalhousie University); and (20) Peter Noel-Webb (Department of Geological Sciences, Ohio State University). In addition, a number of contributors to the book were kind enough to review one or more manuscripts: Elisabeth Alve, Jean-Pierre Debenay, Stephen Eagar, Scott Hippensteel, Scott Ishman, Pamela Hallock, Tim Patterson, Amnon Rosenfeld, and Valentina Yanko-Hombach.

My thanks especially to Elisabeth Alve and Valentina Yanko-Hombach for their help in identifying potential contributors, and to Mr. Ken Howell of Plenum Press for his invitation to organize it. Cheryl Doherty kindly assisted with minor editorial revisions of final manuscripts.

Ron Martin

Newark, Delaware
January 2000

Contents

Part II • Water Quality in Modern Marine, Marginal Marine, and Freshwater Environments

Chapter 13 • Dinoflagellate Cysts as Indicators of Cultural Eutrophication and Industrial Pollution in Coastal Sediments

Barrie Dale

Chapter 14 • Environmental Stratigraphy: A Case Study Reconstructing Bottom Water Oxygen Conditions in Frierfjord, Norway, over the Past Five Centuries

Elisabeth Alve

Chapter 15 • Foraminifera of Storm-Generated Washover Fans: Implications for Determining Storm Frequency in Relation to Sediment Supply and Barrier Island Evolution, Folly Island, South Carolina

Scott P. Hippensteel and Ronald E. Martin

Chapter 16 • Benthic Foraminiferal Distributions in South Florida: Analogues to Historical Changes

Scott E. Ishman

Introduction

For many decades, applied paleontology has concentrated on finding energy resources. Despite the extensive, but almost exclusive, use of micropaleontology in petroleum exploration during the last half-century, foraminifera were used much earlier to date strata in a water well near Vienna, Austria in 1877. Further studies followed in the United States, but it was primarily J. A. Udden of Augustana College (Illinois), who, in 1911, began to use microfossils to correlate water wells in the United States. Udden would later forsake academe to become head of the newly organized Bureau of Economic Geology of Texas, where he shifted application of microfossils from water to petroleum. Bandy *et al.* (1964*a*,*b*; 1965*a*,*b*) were among only a handful of workers during the succeeding years who examined the response of foraminiferal populations to environmental disturbance (large inputs of sewage in shallow marine waters). Bandy *et al.*'s studies lay dormant, however, probably because the environmental movement had not yet fully developed and because of the heavy emphasis of applied micropaleontology on petroleum exploration (Bandy's results may also have been confounded by the effects of low salinity; see Debenay *et al.*, Ch. 2).

A second factor affecting the development of environmental micropaleontology is that the fossil record has long been viewed as incomplete—and therefore flawed—especially since the writings of Lyell and Darwin. Recently, however, biologists and paleontologists have begun to realize that this view is itself extreme and that the fossil and subfossil records hold tremendous potential for the evaluation of environmental disturbance, both natural and anthropogenic.

Besides erosion, the process of time-averaging—the mixing of hardparts of different generations and habitats before final burial—has been viewed as another insurmountable obstacle of the fossil record. Time-averaging of fossil and subfossil assemblages results because rates of sedimentation are too slow to prevent mixing of ecological signals into accumulations of longer duration and lower temporal resolution. Nevertheless, time-averaging has come to be viewed in a positive light (Bambach, 1998; see Martin, in press, for review). Short-term population (high-frequency) phenomena are indeed often lost because of time-averaging, but that is advantageous if one is interested in longer-term processes over decades or centuries. Time-averaged assemblages are more likely to be representative of long-term environmental conditions and community dynamics because the dominance of a particular set of environmental parameters will increase with time while comparatively short-term (and perhaps unrepresentative) fluctuations ("noise") are damped or completely filtered out. High loss rates in death assemblages mainly apply to the ecologically most transient parts of communities: thus, some death assemblages appear comparable to the results of repeated biological surveys that document changes in community species composition and diversity over several decades or more, including sudden phenomena that might be missed by short-term sampling regimes (Kidwell and Bosence, 1991; Behrensmeyer and Chapman, 1993; Kidwell and Flessa, 1995).

Besides surface assemblages, another prime target of paleontological research is the state of the environment prior to anthropogenic influence, such as deforestation and coastal eutrophication, which may be subtle but far-reaching (e.g., Vitousek *et al.*, 1997). Studies of the prehistoric record would prove invaluable because they would consider processes—including the effects of *natural* disturbance—that might be missed by short-term ecological studies; the organization and resilience of biological communities; and the occurrence of alternative community states in response to environmental disturbance (Kidwell and Bosence, 1991; Kidwell and Flessa, 1995). Such studies would also provide data on the potential influence of anthropogenic disturbance on actualistic studies of modern environments that are applied to ancient settings.

These sorts of studies are not just academic: they hold important implications for ecosystem conservation and management (as well as ecological and evolutionary theory). Haynes (1985), for instance, concluded that increased subadult mortality of elephants in preanthropogenic death assemblages might be indicative of natural drought and not anthropogenic activities. Bell (1992) suggested that rising nutrient levels may have contributed to outbreaks of the Crown-of-Thorns starfish, *Acanthaster planci*, on the Great Barrier Reef; the starfish then attack the corals and decimate the reefs. When the starfish die and disarticulate, they contribute significant numbers of skeletal ossicles to the sediment. One way to test for the occurrence of similar outbreaks in the past (*prior* to anthropogenic nutrient input), then, is to examine the sedimentary record. In order to further clarify the situation, Greenstein *et al.* (1995) conducted field experiments that simulated outbreaks and mortality of the starfish. Hippensteel and Martin (1999) used offshore microfossils (foraminifera) to identify storm layers in marsh deposits, and they are currently using artificial "storm layers" of glass beads to determine rates of deposition and bioturbation in coastal marshes of Delaware and South Carolina. How would such studies impact legislation affecting the management, preservation, or development of African savannah, coral reefs, and wetlands, respectively?

Micropaleontologists will no doubt continue to be employed in petroleum exploration, especially if the price of oil rises as dramatically as predicted early in the next century (Kerr, 1998). Nevertheless, micropaleontology seems to have come full circle since Udden's time (Martin, 1991, 1995, 1998) and is embarking on a new avenue of research: *environmental* micropaleontology.

References

Bambach, R. K., 1998, Musings of an optimist: The study of past life should be on the brink of a great future, *Palaios* **13**:511–513.

Bandy, O. L., Ingle, J. C., and Resig, J. M., 1964*a*, Foraminiferal trends, Laguna Beach outfall area, California, *Limnol. Oceanogr.* **9**:112–123.

Bandy, O. L., Ingle, J. C., and Resig, J. M., 1964*b*, Foraminifera, Los Angeles County outfall area, California, *Limnol. Oceanogr.* **9**:124–137.

Bandy, O. L., Ingle, J. C., and Resig, J. M., 1965*a*, Foraminiferal trends, Hyperion outfall, California, *Limnol. Oceanogr.* **10**:314–332.

Bandy, O. L., Ingle, J. C., and Resig, J. M., 1965*b*, Modification of foraminiferal distribution by the Orange County outfall, California, *Ocean Sci. Eng.* **1965**:55–76.

Behrensmeyer, A. K., and Chapman, R. E., 1993. Models and simulations of time-averaging in terrestrial vertebrate accumulations, in: *Taphonomic Approaches to Time Resolution in Fossil Assemblages* (S. M. Kidwell and A. K. Behrensmeyer, eds.), Paleontological Society Short Courses in Paleontology No. 6, pp. 125–149.

Bell, P. R. F., 1992. Eutrophication and coral reefs: Some examples in the Great Barrier Reef Lagoon, *Water Res.* **26**:553–568.

Greenstein, B. J., Pandolfi, J. M., and Moran, P. J., 1995. Taphonomy of crown-of-thorns starfish: Implications for recognizing ancient population outbreaks, *Coral Reefs* **14**:91–97.

Haynes, G., 1985. Age profiles in elephant and mammoth bone assemblages, *Quater. Res.* **24**:333–345.

Hippensteel, S. P., and Martin, R. E., 1999, Foraminifera as an indicator of overwash deposits, barrier island sediment supply, and barrier island evolution: Folly Beach, South Carolina, *Palaeogeogr. Palaeoclim. Palaeoecol.* **149**:115–125.

Kerr, R. A., 1998, The next oil crisis looms large—and perhaps close, *Science* **281**:1128–1131.

Kidwell, S. M., and Bosence, D. W. J., 1991, Taphonomy and time-averaging of marine shelly faunas, in: *Taphonomy: Releasing the Data Locked in the Fossil Record* (P. A. Allison and D. E. G. Briggs, eds.), Topics in Geobiology, Vol. 9, Plenum Press, New York, pp. 116–209.

Kidwell, S. M., and Flessa, K. W., 1995, The quality of the fossil record: Populations, species, and communities, *Ann. Rev. Ecol. Systema.* **26**:269–299.

Martin, R. E., 1991, Beyond biostratigraphy: Micropaleontology in Transition?, *Palaios* **66**:437–438.

Martin, R. E., 1995, The once and future profession of micropaleontology, *J. Foram. Res.* **25**:372–373.

Martin, R. E., 1998. *One Long Experiment: Scale and Process in Earth History*, Columbia University Press, New York, 262 pp.

Martin, R. E., In press. *Taphonomy: A Process Approach*, Cambridge, Cambridge University Press.

Vitousek, P. M., Mooney, H. A., Lubchenco, J., and Melillo, J. M., 1997, Human domination of Earth's ecosystems, *Science* **277**:494–499.

I

Baseline Studies of Foraminifera

Chapter 1

When Does Environmental Variability Become Environmental Change?

The Proxy Record of Benthic Foraminifera

JOHN WILLIAM MURRAY

1. INTRODUCTION

The aim of ecological studies is to establish the relationship between the biota (e.g., community structure of populations of living organisms including standing crop, species abundance, and species diversity) and the attributes of the environment (physical, chemical, and biological). Such studies may be *spatial*

JOHN WILLIAM MURRAY • School of Ocean and Earth Science, Southampton Oceanography Centre, Southampton SO14 3ZH, England.

Environmental Micropaleontology, Volume 15 of *Topics in Geobiology*, edited by Ronald E. Martin. Kluwer Academic/Plenum Publishers, New York, 2000.

involving a suite of samples collected over a geographic area during a very short time interval (days), or *temporal*, where samples are collected from one (or more) sites over an extended period of time (ideally several decades). Spatial studies give a snapshot over a broad area, whereas temporal studies give a near-continuous record of a very small area.

Both approaches yield data with a lot of "noise" that may obscure the overall patterns not only within the community structure but also in its relationship to the environment. On a short timescale, conditions may appear to be unchanging, and this has given rise to the concept of ecological stability (for a critique see Grimm, 1996). In reality, there may be changes which operate at such a slow rate that they are not obvious. These have been described as the *invisible present* (Magnusson, 1995). However, such slow changes become apparent when time-series studies are carried out over period of years or decades as, for instance, the 50-year-long Russell cycle observed off southern England (Cushing and Dickson, 1976). Also, there may be time lags of longer than 1 year between a disturbance and the ecological response, although the duration will depend on the frequency of reproduction of the organisms involved (this is commonly only a few months in the case of foraminifera). Time lags may possibly lead to misinterpretation of cause and effect.

Those organisms that have preservable hardparts have the potential to leave a fossil record in stratigraphic sequences. Such records embrace both space *and* time. Geologists are used to interpreting the fossil record to show changes of environment through time. They are also aware that change is normal in the history of the Earth. The principal difference between the ecological and geological approaches to change is the timescale: ecologists are normally concerned with changes on much shorter timescales than geologists.

However, it is possible to combine the two approaches by carrying out high-resolution studies of faunal change (as a proxy for environmental change) on cores of sediment containing well-preserved microfossils, where it is possible to date individual sediment horizons using radiometric methods. The advantage of this approach is that it provides a means to reconstruct past environments from times prior to the documentation of ecological data (e.g., over the past few hundred years) and prior to the onset of significant human interference. This is particularly important in those instances where recent anthropogenic disturbance has caused environmental change, for it is then possible to determine what conditions were like before disturbance and thereby establish its full effects. Geologists can also provide insight into the effects of environmental change on early human activity [see, e.g., Reinhardt *et al.*, 1994; Stanley and Warne, 1997, on the relationship between Holocene sea-level changes and the early utilization of deltas; see also in this volume Patterson and Kumar, Ch. 11; Dixit and Smol, Ch. 12; Dale, Ch. 13; Alve, Ch. 14; Hippensteel and Martin, Ch. 15; Ishman, Ch. 16; van der Zwaan, Ch. 17].

Thus, there are three possible approaches to investigating environmental change using organisms as proxies: comparison of spatial surveys from different time periods, time-series studies of selected sampling areas, and high-resolution studies of cores.

2. The Concept of Environmental Change

Over the last decade or so there has been popular concern about changing climate. This stems partly from fear that the climate may cool to give another ice age or that it may warm through the accumulation of so-called greenhouse gases. But the fear is mainly that man's activities are adversely affecting both the climate and the natural environment as a consequence of pollution. Biologists and geologists are now actively investigating the contemporary effects of man's activity on living organisms and environments.

In spite of the growing literature on the subject of environmental change, and especially its potential effects on humans, it is very hard to find a definition of what constitutes an environmental change. It is first necessary to consider variability in the environment. Variability is fluctuation about a mean and takes place on different timescales: diurnal or semidiurnal; seasonal; and cycles over a period of years/decades. Such variability can be considered as short-term change. True environmental change requires that the variability occur about a mean, which itself has changed through time. But there may still be uncertainty in distinguishing variability from change. For instance, there may be progressive change in variability either on a single trend or with small cycles superimposed on larger cycles (Fig. 1). It is probably best to assume that there is a continuum from short-term variability to longer-term change but the problem remains of where to draw the boundary between variability and change.

It is also customary for the layman to think of environmental change solely in terms of chemical or physical parameters, but biological changes also affect the environment. For instance, the sudden blooming of an organism has consequences for other organisms (e.g., as competition for space and nutrients and as a potential source of food). Similarly, the arrival of a new species or the local extinction of a species has potential consequences for the rest of the biota. It is also a commonly held view that change must be bad but that is not necessarily true. The addition, e.g., of small amounts of nutrients to a system may be beneficial and the breakdown of a barrier between a lagoon and the open sea may improve water circulation with consequent advantages for at least part of the biota.

3. Foraminifera as Proxies of Change

Benthic foraminifera are abundantly present in modern and ancient marine sediments. There are numerous case studies that have demonstrated their utility in documenting present environments (summarized in Murray,

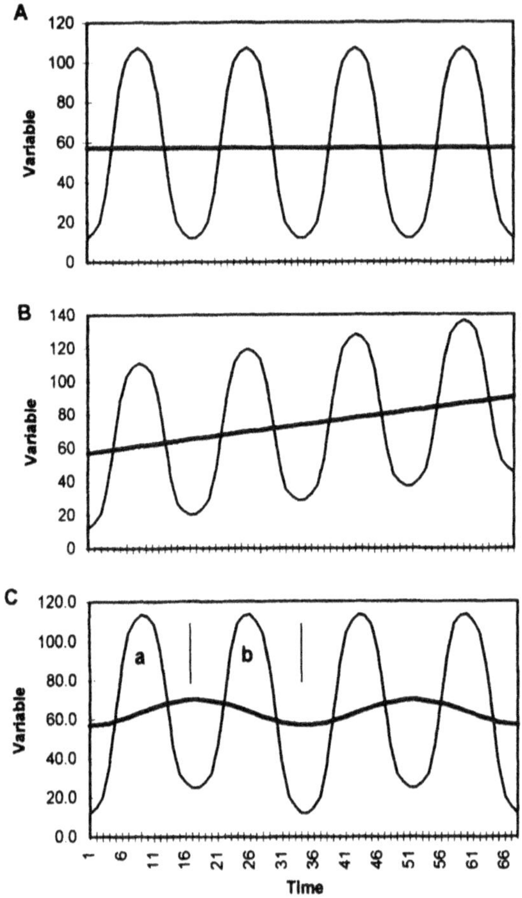

FIGURE 1. Variability and change: (A) variability about a mean with no change; (B) variability about a mean with a progressive change; (C) variability about a mean that has longer-term cyclic variability; segments a and b may *appear* to indicate change.

1991) and a few have documented changes over recent decades. Furthermore, although much remains to be learned about the ecology of benthic foraminifera and the taphonomic processes that affect their preservation in the fossil record, they are nevertheless the best meiofaunal group for giving a proxy record of past marine and marginal marine environments. This is not only because they are abundant but, unlike other meiofaunal organisms such as nematodes and tiny crustaceans, they have a test (shell) that is capable of preservation in the subsurface sedimentary record.

Because of space limitations, it has been necessary to be selective in the choice of examples from the literature but, wherever possible, reference has also been made to reviews that give more comprehensive lists of references (e.g., Murray, 1991, on ecology; Alve, 1995*a*,*b*, on pollution; Alve, 1999, on colonization). For time-series studies, those used are based on living assemblages (stained with Rose Bengal) and collected over a period of 2 years or more.

4. Methods

4.1. Sampling Methodology

A prerequisite for ecological studies is that sampling must be carried out very carefully. Although benthic foraminifera are known to live infaunally to a depth of several centimeters in most environments, the majority live within the top 1 cm and it is imperative that there be no loss of this layer during the sampling process. Many authors study the >63-μm-size fraction but others use >40, >125, or even >250 μm. When comparisons are being made between two or more studies or among different times of sampling, care must be taken to base this on the same size fractions; otherwise errors will arise (Schröder *et al.*, 1987).

4.2. Methods of Data Analysis Used in this Review

The attributes of foraminifera that are related to ecology include univariate measures (where the data for the whole sample are expressed as a single figure such as species diversity or standing crop of the whole assemblage) or multivariate measures (where similarities are computed between two or more samples as in, e.g., cluster analysis, principal component analysis, or multidimensional scaling). It is important to note that while stability or a change in a univariate measure in one area may be of ecological significance, two completely unrelated areas may have the same univariate value. For example, two areas may have exactly the same range of species diversity or standing crop yet not have a single species in common. By contrast, multivariate methods utilize all the assemblage data.

One-way ANOVA (analysis of variance), covariance, and correlation (covariance of two datasets divided by the product of their standard deviations; values close to 1 indicate a close relationship) have been calculated for published data using Excel 5.0. Nonmetric multidimensional scaling (MDS) plotted in two dimensions is used here to recognize groups of samples that are similar (lie close to one another on an MDS plot) and to distinguish samples that are dissimilar (lie in different parts of the MDS plot). The calculations have been carried out with the PRIMER package (Clarke and Warwick, 1994) using the Bray–Curtis coefficient of similarity (the most widely accepted measure of similarity used in ecological studies) and with no transformation of the data.

5. Modern Examples of Environmental Variability and Change

Examples of variability and change can be divided into three groups: short-term, seasonal, repetitive variability; long-term variability/change; non-cyclic change.

5.1. Short-Term, Seasonal, Repetitive Variability

In order to define the range of short-term variability of the environment, it is necessary to have time-series data on both the ecological parameters and the organisms.

5.1.1. Variability in Standing Crop

There are relatively few studies in which foraminiferal samples from the same locality have been taken monthly over a period of 1 or more years. However, these provide important data on the variability of the standing crop both during a year and from one year to another (Table 1). The intertidal and very shallow water examples, with the exception of Bahrain, are all from temperate settings. The annual ranges of variability are large in each area and the data contain much "noise." One way of overcoming this for any one station is to make comparisons between years based on the mean monthly values plus or minus their standard deviations. This shows that there is generally some overlap from one year to the next. However, analysis of variance (single-factor ANOVA; 5% confidence; Table 2) shows that only the Exe data are statistically similar for the two whole-year datasets, whereas the pairs of years for Bahrain, Bottsand, and Puerto Deseado are significantly different, as are those for the deeper sites in the Baltic Sea. The data for Hamana Lake, Japan, are based on a comparison of the four common sampling months over 2 years and have a high degree of similarity.

Much of the noise in the monthly records can probably be attributed to patchiness in distribution patterns. Of the examples discussed above, a survey of patchiness has been carried out only at Puerto Deseado. This revealed great variation both on 10-cm and 1-m sample spacing (Boltovskoy and Lena, 1969*b*). For example, in their area 1, with 23 samples spaced at 1-m intervals, the observed range was 61–280 individuals per 9 cm^2 or 27 cm^3 (mean 148, SD 74). Thus, in this area it is remarkable that any pattern can be observed from one month to the next when only a single sample was taken each month, as in the surveys by Boltovskoy (1964) and Boltovskoy and Lena (1969*a*). Time-series data also reveal whether there is any cyclicity in standing crop. In 1965 in Bottsand Lagoon (Fig. 2*a*), there was a spring low followed by a summer–winter peak but the latter did not occur in 1966 (Lutze, 1968). Covariance correlation is low (0.17), indicating little similarity, and is not significant. However, the two main species showed different patterns: *Elphidium williamsoni* (*Cribroelphidium articulatum* of Lutze) had a pattern closely resembling that of the whole standing crop but *Miliammina fusca*

TABLE 1. Standing Crop Data for Studies of Two or More Years Duration

Area	Year	Standing crop—10 cm³: Range	Standing crop—10 cm³: Monthly mean	Standing crop—10 cm³: Monthly SD	Number of samples	Source
Bottsand	1965	30–280	135	70	12	Lutze, 1968
Germany	1966	20–65	61	54	12	Lutze, 1968
	1967	65–200	89	54	6	Lutze, 1968
Baltic	1973–1974	23.5 m 47–303	141	64	22	Wefer, 1976
Germany	1974–1975	23.5 m 36–121	83	46	14	Wefer, 1976
	1973–1974	27 m 58–364	198	101	21	Wefer, 1976
	1974–1975	27 m 201–1067	536	345	14	Wefer, 1976
Exe	1979[a]	7–65	28	20	12	Murray, 1983
England	1980[a]	9–46	26	13	12	Murray, 1983
	1981[a]	9–32	19	9	7	Murray, 1983
Deseado	1961[b]	2–13	8	4	8	Boltovskoy, 1964
Argentina	1962[b]	1–36	11	11	12	Boltovskoy, 1964
	1963[b]	14–44	36	9	4	Boltovskoy, 1964
Deseado	1964	26–128	61	26	11	Boltovskoy and Lena, 1969*a*
Argentina	1965	12–29	21	6	12	Boltovskoy and Lena, 1969*a*
	1966	35–51	43	11	4	Boltovskoy and Lena, 1969*a*
Hamana	1987–1988	13–493	166	199	5	Matsushita and Kitazato, 1990[c]
Japan	1988–1989	9–971	285	285	12	Matsushita and Kitazato, 1990[c]
Bahrain	1991	2–65	18	17	11	Basson and Murray, 1995
	1992	11–81	39	27	12	Basson and Murray, 1995
	1993	40–42	41	2	2	Basson and Murray, 1995

[a]*Nonion depressulus.*
[b]*Elphidium macellum.*
[c]Original data supplied by Kitazato.

showed much the same range of variability throughout the period without any obvious cyclicity. In the Exe estuary (Fig. 2b), there were lows for *Nonion depressulus* in June–August in all 3 years sampled (Murray, 1983); covariance correlation is 0.39 and not significant. The two studies of Puerto Deseado (Fig. 2c, d) produced conflicting results. From 1961 to 1963 there were peaks in the standing crop of *Elphidium macellum* from November to May (southern summer) and lows in the winter (Boltovskoy, 1964). However, in the continuation of the survey, where the whole assemblage was studied, the peaks of

TABLE 2. Analysis of Variance (Single-Factor ANOVA, 5% Confidence) of Standing Crop Data

		Source of variation					
Site	Period	Sum of squares	Degrees of freedom	Mean of squares	F test	Confidence value	Critical value of F
Bahrain	Between years	3604.93	1	3604.93	7.65	0.01	4.30
	Within years	10364.93	22	471.13			
	Total	13969.86	23				
Exe	Between years	29.38	1	29.38	0.11	0.75	4.30
	Within years	6082.97	22	276.50			
	Total	6112.36	23				
Bottsand	Between years	47259.38	1	47259.38	16.88	0.00	4.30
	Within years	61589.58	22	2799.53			
	Total	108848.96	23				
Puerto Deseado	Between years	7210.67	1	7210.67	16.62	0.00	4.30
	Within years	9542.67	22	433.76			
	Total	16753.33	23				
Baltic Sea 23.5 m	Between years	29313.06	1	29313.06	8.688	0.00575	4.13
	Within years	114708.8	34	3373.789			
	Total	144021.9	35				
Baltic Sea 27 m	Between years	960041.7	1	960041.7	18.1	0.00016	4.1393
	Within years	1750064	33	53032.25			
	Total	2710106	34				
Hamana Lake, Japan	Between years	1770.125	1	1770.125	0.054	0.82397	5.98745
	Within years	196694.8	6	32782.46			
	Total	198464.9	7				

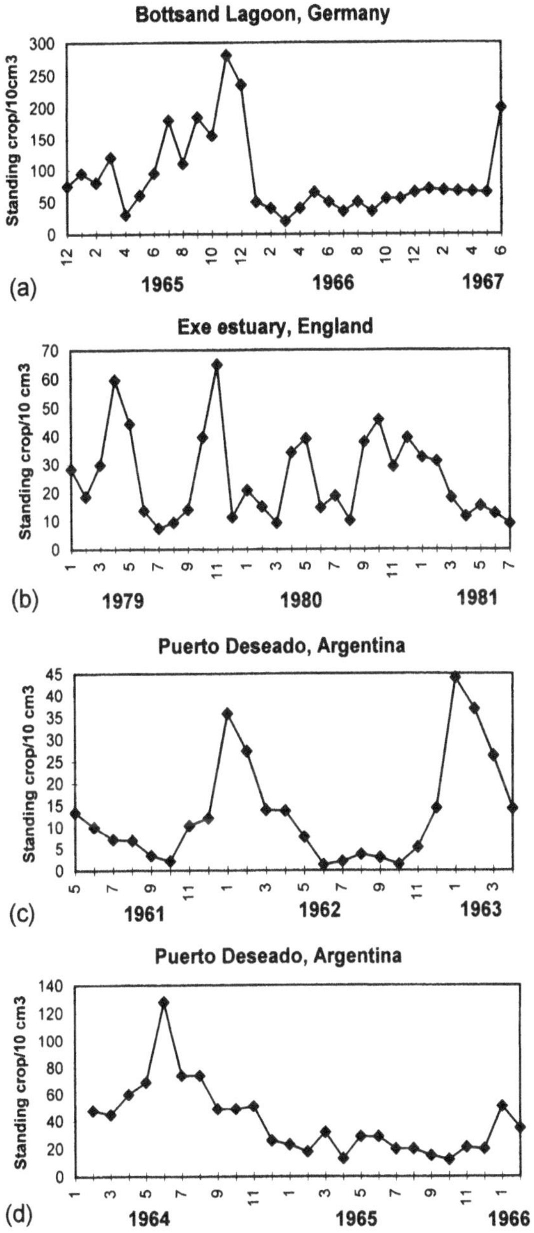

FIGURE 2. Time-series data on standing crop: (a) Bottsand Lagoon, Germany (data from Lutze, 1968); (b) Exe estuary, England (data from Murray, 1983); (c, d) Puerto Deseado, Argentina (data from Boltovskoy 1964; Boltovskoy and Lena, 1969*a*).

standing crop were during April–November 1964 (southern winter), but the pattern was not repeated in 1965 and, consequently, the covariance correlation was very low (−0.12). Boltovskoy and Lena (1969*a*) were unable to account for these differences.

The standing crop of individual species in the intertidal zone in Bahrain also shows different patterns from one year to the next (Basson and Murray, 1995). Whereas the numbers of *Ammonia beccarii* remained much the same throughout the study period, *Elphidium advenum* increased about fourfold in 1992–1993, while *Brizalina pacifica* and *Nonion* sp. suddenly increased 30 to 100 times in August 1992 and remained high thereafter. There was a parallel increase in diversity. This poses the question: does this reflect natural variability or was it an environmental change? At present the cause of these abundance increases remains unknown since the only environmental variables measured were temperature and salinity.

The Lim channel is an arm of the Adriatic Sea in Istria. Daniels (1970) made a transect of stations from the open sea (35 m) to the landward end (5 m) and sampled over a 14-month period. The three seaward stations had the highest standing crops (range 321–1227 per 10 cm^2; means 714, 679, 800; SD 272, 302, 304, respectively), the innermost two stations had the lowest values (range 104–722 per 10 cm^2 means 353, 320; SD 101, 200), while the central stations had intermediate values.

Until recently, it was commonly believed that the deep sea is a very stable environment with virtually no environmental variability. However, this is now known to be incorrect. Bottom photography has shown that there is often quite a considerable amount of bioturbation creating microhabitats that provide scope for patchiness in meiofaunal distribution and abundance. Furthermore there are benthic storms (Kaminski, 1985) and disturbance from turbidity currents (Vénec-Peyré and Le Calvez, 1986). But perhaps the most important discovery is that the deep sea experiences seasonality. This comes as pulses of food (termed phytodetritus) derived from the phytoplankton (Gooday and Lambshead, 1989).

Since almost all deep-sea life is ultimately dependent on food from plankton, the arrival of an increased quantity of phytodetritus triggers a biological response, particularly among bacteria and protozoa. Gooday (1993) has shown that there are certain foraminifera that opportunistically exploit phytodetritus (so-called "phytodetritus species"; mainly *Epistominella exigua* and *Alabaminella weddellensis*). They have a small test size (generally less than 200 and 150 μm, respectively) and live throughout the year in the top 1 cm of sediment. When the seasonal pulse of phytodetritus arrives on the ocean floor they colonize it, reproduce, and increase in number. For instance, at 1350 m in the Porcupine Seabight west of Ireland, phytodetritus species were present in the sediment in April 1982 and July 1982 (means 20 and 110 per 3.5 cm^3, respectively), and during July, when there was a layer of phytodetritus on top of the sediment, there were a further 166 within the phytodetritus (data recalculated from Gooday and Lambshead, 1989). At an abyssal site in the northeast Atlantic (depth of 4550 m) the mean abundance of phytodetritus species in April 1988 in two sediment cores taken when no phytodetritus was present was 14 per 3.5 cm^3, whereas in July 1986 following a phytodetritus pulse, the mean of three sediment cores was 22 per 3.5 cm^3 and 183 for the phytodetritus layer (Gooday and Turley, 1990, plus supplementary data from A. Gooday).

At a permanent research station in Sagami Bay, Japan, situated beneath the oxygen minimum zone in Pacific deep water at 1450 m water depth, the temperature, salinity, and bottom water oxygen concentrations are stable throughout the year. However, the standing crop of the benthic foraminifera varied from 200 to 2800 per 10 cm^2 over a 4-year period. The low values occurred during the summer (200–500 per 10 cm^2) and the high values in the spring (1500–2000 per 10 cm^2) when there was an input of phytodetritus to the seafloor (Ohga and Kitazato, 1997). Furthermore, the foraminifera migrated vertically in the sediment as the redox boundary shallowed following the arrival of the phytodetritus. There were juvenile individuals of the shallow infaunal species *Textularia kattegatensis* and *Bolivina pacifica* in spring, and this suggests that they reproduced following the arrival of phytodetritus.

The San Pedro Basin on the U.S. Californian borderland has a sill depth of 715 m. Box cores collected at different seasons showed that there was essentially the same group of species making up the >150-μm assemblage although the rank order varied slightly from one season to another. The depth distribution within the sediment remained constant throughout the sample period (Silva *et al.*, 1996).

Monthly sampling of four stations over 1 year in the Clyde Sea, Scotland, showed variations in standing crop from 0–200 individuals per 10 cm^3 on coarse sand at 8 m to 800–3600 per 10 cm^3 on muddy sediment at 38–46 m (Hannah and Rogerson, 1997; samples prepared by elutriation so individuals smaller than 63 μm are included; Hannah, pers. comm., February 1998). From analysis by ANOVA the authors concluded that season had no significant effect on the size of the standing crop.

To summarize, all the above studies reveal considerable short-term variability including changes from one year to the next, which mask any underlying cyclicity. Also, most time-series are too short to detect secular changes.

5.1.2. Variability in Assemblages

For the Exe estuary, MDS plots of the individual years 1979 and 1980 and the data for both years combined show a broad scatter with no order from one month to the next. This can be interpreted as confirming that there is essentially one assemblage showing considerable variability through time. On the other hand, MDS plots of the Bahrain data reveal two groups that are separated with no order within either of them (Fig. 3). One group includes February 1991–July 1992 (dark field) and the other August 1992–February 1993 (light field). Thus, the faunal change (increase in *Brizalina* and *Nonion*) has clearly affected the whole assemblage.

5.1.3. Variability in the Dominant Species

The rank order of species may be constant or change from month to month. Examples of constancy include the Exe estuary (*N. depressulus*, Murray, 1983), Puerto Deseado (*Buliminella elegantissima*, Boltovskoy and

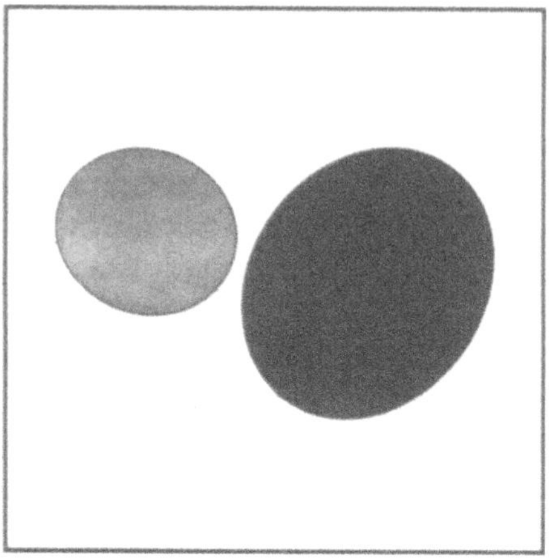

FIGURE 3. MDS plot of Bahrain standing crop time-series data with no transformation and using the Bray–Curtis similarity index: stress = 0.04; dark field for months 1–18, light field for months 19–25 (Month 1 = February) (data from Basson and Murray, 1995).

Lena, 1969*a*), the Plym estuary (*Haynesina germanica* except for 1 month, Castignetti, 1996), and Bahrain (average of replicates, *A. beccarii* except for 2 months out of 26, Basson and Murray, 1995). Examples of changing dominance include the Bottsand Lagoon, the Lim channel, and the southern North Sea.

In Bottsand Lagoon, *Elphidium williamsoni* (=*Cribrononion articulatum* of Lutze, 1968) was dominant in 1965 but throughout most of 1966 the dominant form was *Miliammina fusca*. The differences between the two species is that the standing crop of *M. fusca* had much the same variability throughout the 2 years, whereas *E. williamsoni* was much more abundant in 1965 than in 1966. In the Lim channel, Istria, the seaward stations were dominated by *Nonionella opima* but, with passage toward the landward end, the dominance became more variable (Daniels, 1970).

In some areas of the southern North Sea (Murray, 1992), the same living species was dominant throughout the year (dominant dead species given in parentheses): station 2 at 25 m, *Elphidium excavatum* (*E. excavatum*), and stations 5 and 4 at 63 and 81 m water depth respectively, *Stainforthia fusiformis* (station 5, *S. fusiformis*, *E. excavatum*, *Eggerelloides scaber* depending on the month; stations 4, *S. fusiformis*). However, station 3 at 52 m had dominant *Epistominella vitrea* in April/May 1988, *E. scaber* in September 1988 and May/June 1989, and *S. fusiformis* in December 1988 (*E. excavatum*). It can be seen that these variations in the living assemblages were not always recorded in the preserved dead assemblages.

5.1.4. Species Diversity

There are few published data on species diversity and unless the census data sets are published it is not possible to calculate the values. In the Lim channel (survey conducted over 14 months, Daniels, 1970), species diversity measured by the alpha index showed some variability but no obvious pattern within a single station (seaward, station 39, range 12–17, mean 13.5, SD 1.7; landward, station 1, range 9–14, mean 11.8, SD 1.8). In the Bahrain survey, the information function [$H(S)$] had a mean value of 0.5 from February 1991 to July 1992 and, in association with the faunal changes, it rose to a mean of 1.4 from August 1992 until February 1993 (Basson and Murray, 1995).

5.1.5. Summary

From the time-series studies carried out so far, it can be seen that there is considerable variability from one season to the next; the pattern of variability is not necessarily repeated precisely from year to year; the variability affects the standing crop, the dominant species, and the species diversity. These datasets provide a fundamental baseline from which to evaluate long-term environmental change.

5.2. Long-Term Change

Examples of long-term change are both one-way and cyclical: progressive environmental changes (one-way); pollution (one-way); and anoxia and flushing of basins (cyclical). In one-way changes there may be differences in the dominant species (previously rare species replacing formerly common ones) and changes in the rate of production and accumulation of tests. This is quite different from the cyclic change from anoxia, which kills off the fauna, to oxygenation, which allows recolonization initially by opportunistic species.

5.2.1. Progressive Environmental Changes (One-Way)

One-way progressive environmental changes could result from an increase or decrease in a variable over a long period of time. There is some evidence that the deep Skagerrak Basin in the northeast North Sea has been the site of progressive faunal change over the past 60 years and especially over the last three decades. In 1937 a detailed survey of the benthic foraminifera focused mainly on the taxonomy (Höglund, 1947), but data were also gathered on the total (live plus dead) assemblages. In 1992/1993, the Norwegian Geological Survey and the University of Bergen, Norway, took a series of cores from the same area that Höglund had sampled. Since Höglund did not distinguish live from dead, the only possible comparison was between the total assemblage results from the upper 2 cm of the new cores with Höglund's total assemblage data (also from the top 2 cm) from the same localities.

Alve and Murray (1995) were able to show that since 1937:

1. There has been an almost threefold increase in the total abundance of tests in the deep Skagerrak Basin (Fig. 4).

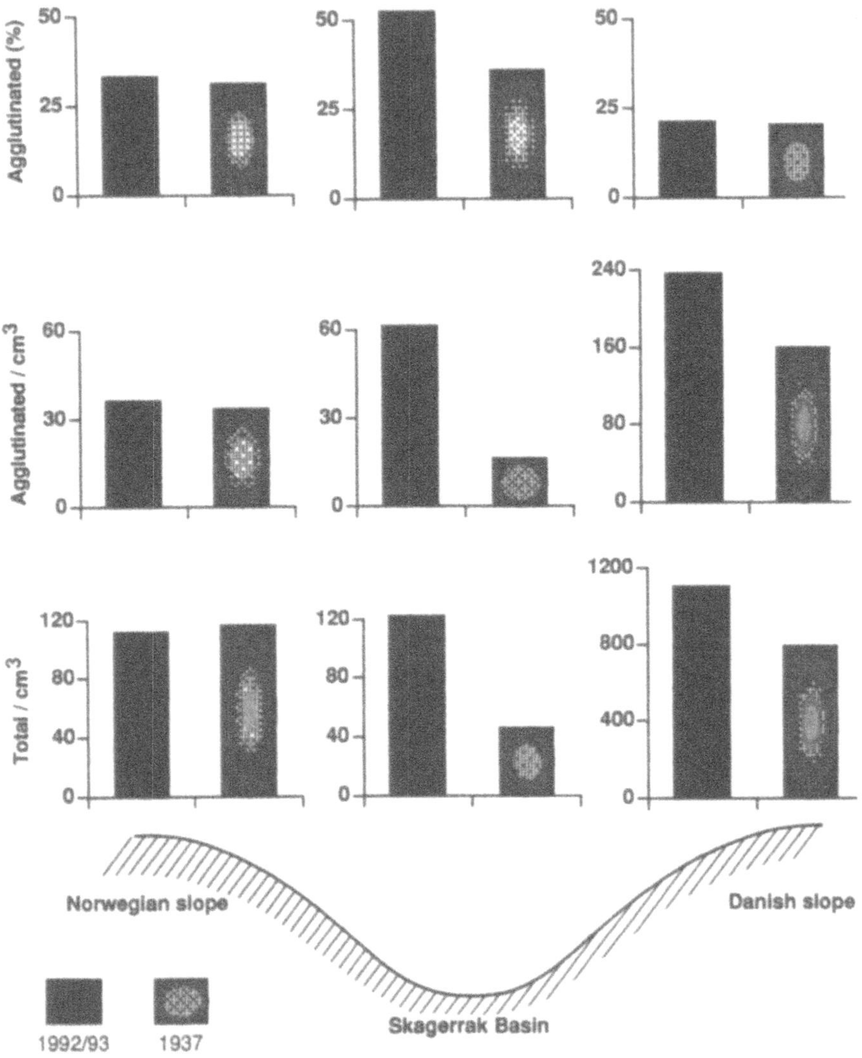

FIGURE 4. Comparison of average relative and absolute abundance of agglutinated tests and total abundance of tests between samples collected in 1937 and 1992/1993 in the Skagerrak (reprinted from Alve and Murray, 1995, with permission from Elsevier Science).

2. The 1937 basin fauna was dominanted by calcareous species (*Cassidulina laevigata*, *Pullenia osloensis*), whereas now it is dominated by agglutinated forms (*Haplophragmoides bradyi*).
3. *Trochamminopsis pusillus* and *Saccammina* spp. were not recorded in 1937 but now they are common and characteristic basin taxa so they must have migrated into the area.
4. On the Norwegian slope the dominant species in 1937 was *C. laevigata*

but now it is *P. osloensis*; since 1937, the absolute abundance of *C. laevigata* has halved, whereas that of *P. osloensis* has tripled.
5. On the Danish slope there has been a dramatic increase in *Stainforthia fusiformis* since 1937 and a significant reduction in both the absolute and relative abundance of *C. laevigata* and *Hyalinea balthica*.

Comparative studies of the living and dead assemblages in the surface sediments have demonstrated that dissolution of calcareous tests is now active in the deep basin so that the dead assemblages are dominated by agglutinated taxa (Alve and Murray, 1997).

In a related study, Alve (1996) studied downcore distributions. Subsurface levels were dated using ^{210}Pb so it is possible to relate the changes in faunal composition to a timescale. Alve presented faunal assemblage data and calculated the accumulation rates of tests for successive time increments. Core 56 from the deep basin (652 m) spans the time period from around 1870 to 1993 and the seafloor at the time Höglund collected his samples is now represented by the 16 to 18-cm interval (1935–1943). To provide an objective analysis of the whole dataset, multivariate analysis has been undertaken. Both cluster analysis and MDS ordination produce two well-defined groups of samples, pre- and post-1971 (Fig. 5), supporting Alve's conclusion that there has been a change upcore, particularly since about 1970.

There was a major change in the accumulation rates of agglutinated tests at around 1970–1971 (Fig. 6: 1935–1970 mean 92 tests/10 cm^2/year; 1971—1993: mean 205 tests/10 cm^2/year). However, the overall accumulation rate decreased during this time (1935–1970: mean 419 tests/10 cm^2/year; 1971—1993: mean 325 tests/10 cm^2/year). Whereas agglutinated tests formed 22% of the accumulating assemblages in 1933–1970, they increased to 63% in 1971–

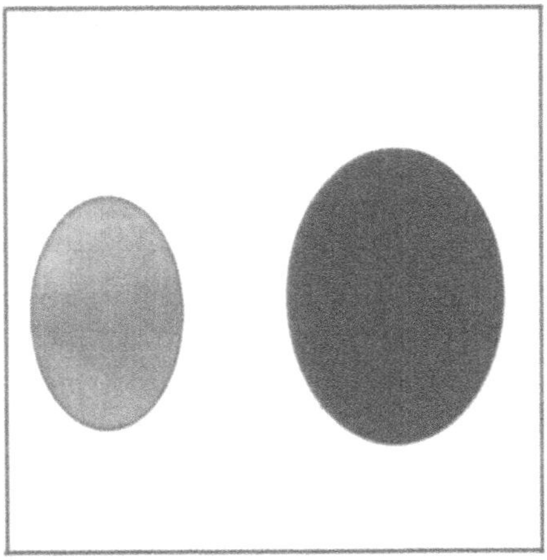

FIGURE 5. MDS plot of Skagerrak core 56 data with no transformation and using the Bray–Curtis similarity index: stress = 0.03; dark field for samples from 0–0.5 to 5–6 cm; light field for samples from 8–10 to 16–18 cm (data from Alve, 1996).

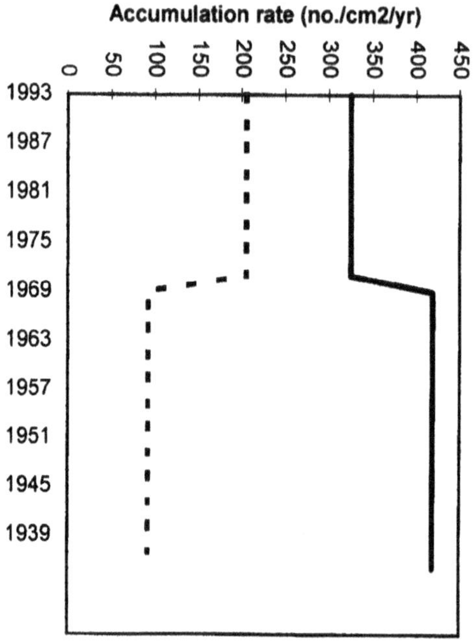

FIGURE 6. Mean accumulation rate of tests in core 56 from the Skagerrak: solid line = whole assemblage; broken line = agglutinated component (data from Alve, 1996).

1993. These accumulation rates could be interpreted as evidence that there was:

1. An upcore increase in the production of agglutinated tests.
2. A downcore increase in destruction of agglutinated tests.
3. An upcore decrease in the production of calcareous tests.
4. An upcore increase in the dissolution of calcareous tests.

Alve considered that although there may have been some destruction of agglutinated tests, overall their preservation is good. The preserved agglutinated fauna at a core depth equivalent to 1937 is very similar to that recorded by Höglund. Furthermore, the upward increase is due in part to the reintroduction of species (*Trochamminopsis pusillus* and *Saccammina* spp.) that were not recorded by Höglund at these sites but are occasionally present in pre-1940 sediments in core 56. A similar upward abundance increase was recorded by Haas (1997) from cores taken between 245 and 450 m. The increase in the dissolution of calcareous tests already noted by Alve and Murray (1995) is confirmed by the core 56 data. *Pullenia osloensis* was one of the commonest living species in the 1991–1993 survey, but rare in surface dead assemblage, although very abundant in pre-1970 assemblages. The following conclusions can be drawn: no amount of dissolution of calcareous tests from the post-1970 samples could lead to an *increase* in the absolute numbers of agglutinated tests or in their rate of accumulation; the upward increase in agglutinated tests could arise if there was a downward increase in the destruction of these tests but the arguments presented above make this unlikely; if there was no downcore increase in the destruction of agglutinated tests, it follows that there must also have been a decline in the production of calcareous tests in the post-1970 period.

How might these results be interpreted? Are there parallel changes in the macrofauna? Biologists report that there has been an increase in the supply of organic matter and this has led to a recent increase in the biomass and standing crop of the macrofauna (Rosenberg *et al.*, 1987; Josefson, 1990), but they do not have the same historical database that has been worked out for the foraminifera. Nevertheless, long-term changes in the North Sea benthos have been attributed to anthropogenic rather than natural causes since there have been no significant changes in water temperature or salinity, but there has been a marked increase in eutrophication, pollution, and fisheries impact on the seafloor (Kröncke, 1995). Plankton biomass increased by three to four times and its effects on the benthic biomass on muddy sediments in the German Bight peaked in the 1970s. Furthermore, in a comparable study of the foraminifera from the northeastern rim of the Skagerrak and from the Kattegat, Moodley *et al.* (1993) considered that there was a *decrease* in foraminiferal densities due to eutrophication, which favored the macrofauna. Alve and Murray (1995) and Alve (1996) concluded that the faunal changes (especially the increased production of agglutinated tests in the deep part of the basin) are likely to be linked to an environmental change (an increase in nutrients and the flux of organic matter) that might be natural but is more likely to have been caused by anthropogenic activity. The changes in the Skagerrak parallel those described as being due to anthropogenic causes in adjacent areas of the North Sea and the Kattegat.

On a smaller timescale, prior to 1991, *Amphistegina gibbosa* was abundant and healthy in the shallow waters of the Florida Keys but in 1991 these forms started to undergo symbiont-loss. This has led to a slowdown in reproduction and an increase in shell breakage. This trend has continued and as a consequence the numbers of individuals have fallen dramatically. The cause is as yet undetermined but may be related to increased solar radiation (Williams *et al.*, 1997).

5.2.2. Pollution (One-Way)

Where there is a buildup of pollutant concentration through time, such an environment provides an example of change in one direction (although if the input of pollutants declines, the prepollution fauna may be reestablished). The effects of pollution can be monitored by studying the consequent faunal changes through time, although in some cases there is no baseline study of the area prior to the onset of pollution.

In a review of the effects of pollution on foraminifera, Alve (1995*a*) noted that the effects of pollution are complex but that a few generalizations can be made: in cases of severe organic pollution, an abiotic zone or an area with extremely reduced abundance is present in the immediate vicinity of point sources; increased abundance (hypertrophic) relative to the natural background level is found close or at some distance from the outfall; there is a reduced number of species close to the outfall, compared to the surrounding seafloor and, in some cases, increased compound diversity (i.e., measures such as the information function, which take into account species proportions) at some distance from the outfall; there is modification of the original assemblage composition. Some tolerant or opportunistic species benefit from organic pollution and reduced competition and predation if some taxa are reduced or excluded. These consequences of pollution are all long-term changes beyond the range of normal environmental variability.

With increased awareness of the need to minimize pollution, many countries have introduced legislation to reduce the discharge of pollutants. Former badly polluted areas are now less polluted and foraminifera can be used to monitor this recovery (see Alve, 1995*a*,*b*, Ch. 14 in this volume; and below).

Where baseline studies have not been carried out prior to the onset of pollution or where it is desired to study the historical development of pollution (and recovery), the analysis of core data that extend back into the pre-pollution time period is invaluable. A schematic diagram shows the possible sequence of events in a heavily organically polluted area (Fig. 7). Alve (1995*a*) divided the foraminifera into "natural" N (i.e, the normal prepollution fauna), "transitional" T, and "opportunistic" O groups, according to their tolerance to pollution disturbance. When the hypertrophic conditions are reached, species diversity falls, faunal dominance increases, and the number of tests reaches a peak; then abiotic conditions develop when pollution is at a maximum. The order of events is reversed as the area recovers following the cessation of pollution. In such studies, particular attention must be paid to any taphonomic loss of information. Good examples of this approach using the preserved foraminiferal record are the reconstructions of past bottom water oxygen conditions in Drammensfjord, Norway (discussed below) and Frierfjord, Norway (Alve, Ch. 14, in this volume). This is a very powerful technique for documenting environmental change.

5.2.3. Flushing of Basins (Cyclical)

Certain enclosed basins have restricted communication with the open sea owing to the absence of tidal influence, salinity contrasts, and usually the presence of a sill, which restricts the ingress of more-saline deep waters. Under these conditions, the deep waters in the basin become depleted in oxygen because renewal is too slow to compensate for the biological and chemical consumption. Classic examples are fjords, the Baltic Sea, and basins on the California borderland.

Periodic flushing has been documented in Drammensfjord, southern Norway (Alve, 1991). The combination of a well-developed halocline and a sill at 10 m restricts deep-water exchange to the inner basin of the fjord, which takes place only every 3 to 5 years. Although from about 1500 to 900 years ago the basin was flushed and oxygenated and had calcareous foraminiferal

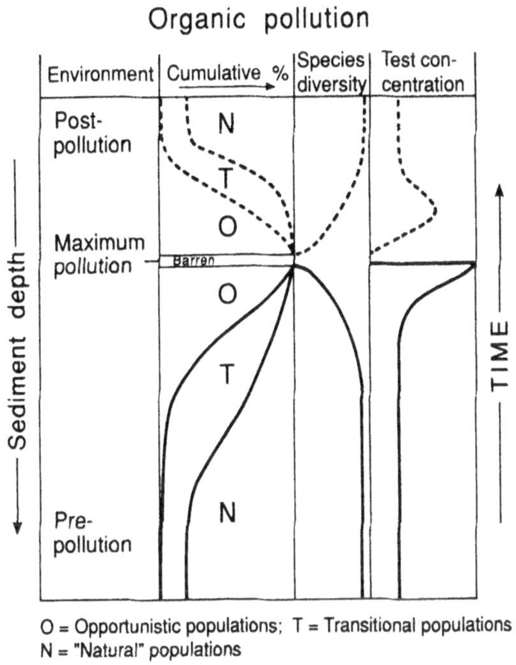

FIGURE 7. Schematic diagram of a core penetrating organically contaminated deposits. From the base up the generalized faunal changes show the response to increasing pollution and extinction at the pollution maximum. The dotted lines above show the possible path of recovery following improvement in the environment. See text for further discussion (reproduced with permission from Alve, 1996).

assemblages, from 900 until about 170 years ago, changes in climatic conditions increased the inflow of freshwater, reduced exchange with the open sea, and led to assemblages that were dominantly agglutinated. Beginning about 170 years ago there have been additional pressures from organic, oxygen-consuming pollution, which caused anoxia.

From about 1830 to 1970 the basin received waste from pulp mills and sewage, but after this the pulp mills closed and the input of sewage was reduced so the environment started to recover. In 1982 anoxia was present at water depths greater than 30–35 m but between 1984 and 1988 the redox boundary was lowered to deeper than 50–55 m. Alve (1995*b*) took samples in 1984 and 1988 to document these improvements. She found that:

1. Living forms were present on substrates only from 0 to 30 m water depth in 1984 but by 1988 they extended down to 50 m.
2. The sediments that had previously been anoxic for more than 5 years did not become suitable for foraminiferal colonization until more than 1 year after they had been exposed to oxic bottom waters (a good example of a lag effect).

3. In areas at 10–35 m (bathed by the transitional water layer), the standing crop doubled between 1984 and 1988 as conditions improved.
4. The first important colonizer of the formerly anoxic sediments, *Stainforthia fusiformis*, is a highly opportunistic species (Alve, 1994); it is naturally small and thin-walled and its juveniles are easily transported.
5. Although there was a diversity increase as the fauna became established, diversity remained low, as would be expected in a marginal marine environment.

The Santa Barbara Basin off California has a maximum depth of 625 m and a sill depth of around 475 m. Episodic ventilation leads to oxygen depletion of bottom waters below the sill depth in between periods of flushing. When oxygen levels in the deep basin are less than 23 $\mu M\ O_2$ ($\sim$0.5 ml O_2/liter), macrofaunal predators are excluded and even much of the meiofauna is affected. Nevertheless, certain foraminifera tolerate these low oxygen levels, and standing crop values are very high (up to 1197 individuals per cm^3, Bernhard and Reimers, 1991; up to 2176 individuals per cm^3, Bernhard *et al.*, 1997). Some foraminifera survive short periods of anoxia but following weeks/months of anoxia, the foraminiferal fauna is entirely wiped out (Bernhard and Reimers, 1991). There is a sequential tolerance to progressively lower oxygen concentrations: (lowest tolerance) *Uvigerina juncea*, *Suggrundina ekkisi*, *Loxostomum pseudobeyrichi*, *Bolivina argentea*, *Trochammina pacifica*, *Bolivina seminuda*, *Buliminella tenuata*, *Chilostomella ovoidea*, *Spiroplectammina earlandi*, and *Nonionella stella* (highest tolerance; dominant when oxygen levels fell to $<2\ \mu M\ O_2$; Bernhard *et al.*, 1997). It should be noted, however, that all these species also live in environments having an ample supply of oxygen and that there are no species confined entirely to very-low-oxygen settings.

5.2.4. Summary

A progressive change, such as that described from the Skagerrak, is in some ways similar to a buildup of pollution in that the change is in one direction. The consequences are that there may be a change in the dominant species (rarer species replacing formerly common ones). There may also be changes in the rate of production and accumulation of tests. This is quite different from the cyclicity that goes from anoxia, which kills off the fauna, to oxygenation, which allows recolonization initially by opportunistic species.

5.3. Rapid Noncyclic Change

Rapid noncyclic changes may be single events (like digging a new harbor channel and thereby changing the water circulation pattern, or the introduction of a new species) or they may be irregular in occurrence (volcanic eruptions, earthquakes).

5.3.1. Change in Water Circulation

The lagoons along the coast of the United Arab Emirates were largely unaffected by human activity until the late 1960s. The foraminiferal assemblages of the Abu Dhabi lagoon were studied in 1965 (Murray, 1970*a*) and again in 1969 (Murray, 1970*b*). In 1965 the tidal flow into the lagoon was restricted by the tidal oolith delta. Seagrass in the lagoon was sparse. By 1969 the channel across the oolith delta had been dredged deeper and a new channel at the landward end of Abu Dhabi island allowed water exchange with the adjacent lagoon. The tidal range within the Abu Dhabi lagoon seemed to be greater, the seagrass was healthy and luxurious, and the fivefold increase in human population was presumably matched by a similar increase in organic matter as the sewage was discharged into the sea.

In the 1965 survey, living foraminifera were found to be confined largely to plant substrates (especially seagrasses) with few on the sediment substrates but no quantitative measurements were made. By 1969 the numbers of live forms must have increased because it was possible to obtain standing crop data from all the subenvironments. These changes, together with the greater amount of seagrass, may have been the consequence of improved water circulation with the open gulf, which reduced the hypersalinity, and higher levels of organic matter (from sewage).

The total assemblages collected in 1964–1965 from the Miramichi River estuary, Canada, were used to define two biofacies: river with dominant *Miliammina fusca* and bay with dominant *Elphidium* spp. (Bartlett, 1966; Tapley, 1969). A decade later the area was sampled again and three assemblage zones defined: upper estuarine (=river) with *M. fusca*, transition with *Ammotium cassis*, and open bay (=bay) with *Elphidium excavatum-clavatum* (Scott *et al.*, 1977). The transition zone fauna occurs in an area that was previously occupied by a calcareous assemblage, and this change was attributed to alterations in water circulation owing to changes in the barrier islands and increased river discharge.

5.3.2. Introduction of New Species

The reintroduction of two species into the Skagerrak Basin between 1937 and around 1970 has already been mentioned in connection with inferred environmental changes. The mechanism of introduction is unknown but is presumably natural. The introduction of *Ammotium cassis* into the Baltic Sea probably took place between 1936 and 1952 (Lutze, 1965), and this species is now common beneath the transition water layer in the southern part of the sea.

The introduction of *Trochammina hadai* into San Francisco Bay is thought to have taken place during the 1980s, and it is possible, though not proven, that it is a result of man's activity. McGann and Sloan (1996) report that it already forms between 8 and 56% of core top total assemblages, suggesting that it is a strong competitor, which may have important long-term consequences for the meiofaunas of the area.

5.3.3. Volcanic Eruptions

Volcanic eruptions occurred on Deception Island, Antarctica, in 1967–1970, and these killed off much of the fauna. Samples were taken each summer from 1971–1975 to investigate recolonization of the area. The aim was to make census studies based on counts of 300 individuals per sample (Finger and Lipps, 1981). Within the caldera (Port Foster), only one of the samples yielded 300 living individuals and many were barren, but by 1973 nine of the eleven samples had a rich fauna. The 1971 assemblage was dominated by *Trochammina malovensis*, *Miliammina arenacea*, *Globocassidulina crassa* and *Nonionella bradii* but by 1975 the dominant forms were *M. arenacea*, *Stainforthia fusiformis*, and *N. bradii*. On the flank slope of the volcano it was not until 1974 that the faunas had become rich enough to yield 300 stained individuals per sample. The dominant species was *G. crassa*. Both within the caldera and on the flanks, the living forms were patchy in occurrence even in 1975, and it is doubtful whether the fauna was fully reestablished.

The preeruption fauna collected in 1927 and documented by Earland (1934) was dominated by *M. arenacea*. It is interesting that, once again, *S. fusiformis* was an early recolonizing species even though it is not normally a component of the assemblages (no calcareous forms were recorded by Earland). Finger and Lipps concluded that the posteruption assemblages in the caldera are sufficiently different from those recognized by Earland in 1934 to suggest that the foraminiferal fauna "may have undergone significant reorganization in this century."

A more recent example is the 1991 eruption of Mount Pinatubo in the South China Sea. The deep seafloor downwind at the time of the eruption became blanketed in volcanic ash that buried the fauna to a depth of 2–6 cm. Hess and Kuhnt (1996) sampled the area 3 years after the eruption. The dead assemblages from beneath the volcanic ash are dominated by tubular agglutinated foraminifera such as *Saccorhiza ramosa* and *Rhabdammina abyssorum*. The living assemblages from the thin layer of mixed hemipelagic sediment and volcanic ash above the ashfall deposits are dominated by *Reophax dentaliniformis* and *Quinqueloculina seminula* (Fig. 8, April 1994). The dead assemblage in this surface layer includes numerous small *Textularia* sp. rarely found living. Hess and Kuhnt speculated that *Textularia* sp. were the earliest recolonizers and had been replaced by *Reophax* and *Quinqueloculina* by the time the samples were taken. There was a major decrease in diversity between the pre- and postash assemblages. The presence of *Reophax* as an early recolonizing species conforms with observations made in experiments on recolonization at 3912 m in the Panama Basin of the Pacific (Kaminski *et al.*, 1988). Further sampling in 1996 (Fig. 8, June 1996) showed that suspension feeding tubular agglutinating foraminifera had reestablished themselves (Kuhnt, pers. comm. 1998).

In summary, deposition of ash was a catastrophic event that caused mass mortality of the foraminifera. In each case, there was a new assemblage of

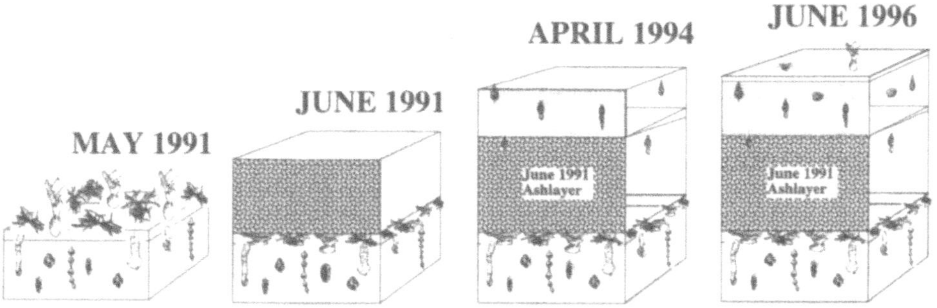

FIGURE 8. Cartoon showing the succession of events from prior to after ashfall in the South China Sea adjacent to Mount Pinatubo (reproduced with permission, Kuhnt, pers. comm., 1998).

opportunistic taxa starting the process of recolonization, and in these two examples even 3–5 years after the eruption the recovery process was clearly only in its early stages.

6. Experiments

Experimental studies can help to determine the causes and responses to change but few have been carried out (Moodley and Hess, 1992; Alve and Bernhard, 1995; Moodley *et al.*, 1997).

The response of foraminifera to declining oxygen levels was tested by Alve and Bernhard (1995). They placed a piece of virtually undisturbed seafloor sediment in each of two tanks. One was maintained as a control with normal oxygenated conditions. In the other the oxygen level was reduced incrementally every month over a period of 4 months to a minimum level of 0.2 ml/liter. At this low oxygen level, virtually all the macrofauna died. However, the effect on the foraminifera was to cause previously infaunal species to move out of the sediment and become epifaunal on polychaete tubes extending above the sediment surface. Following reoxygenation of the water, the foraminifera started to return into the sediment. The dominant species were *Stainforthia fusiformis* and *Bulimina marginata*. *S. fusiformis* is an opportunist and the first species to recolonize previously anoxic sediments in fjords, as noted above. This experiment shows that foraminifera are able to tolerate very low levels of oxygen for extended periods of time and that infaunal organisms may become epifaunal in order to track oxygen and survive.

Moodley and Hess (1992) pointed out that since infaunal foraminifera live close to the redox boundary, it is essential for survival that they are able to tolerate short periods of anoxia (facultative anaerobiosis). This was tested in experiments, and it was reported that those taxa with hardtests survived anoxia for a period of 78 days (the duration of the experiment (Moodley *et al.*, 1997).

7. Rate of Response

The response of organisms to environmental variability/change depends on a number of factors including: whether they have a short or long turnover time in relation to the period of change; whether they are euryhaline/stenohaline, eurythermal/stenothermal; opportunists/generalists/equilibrium, etc.; whether the change is potentially favorable (increase in food; decrease in stress) or unfavorable (increase in stress; development of limiting conditions such as severe oxygen depletion).

It appears that for many benthic foraminifera, the generation time is less than 3 to 4 months. Therefore, there can be a fairly rapid response to an increase in food but there may be a short time lag between the two. For example, Altenbach (1992) reported that *Cribrostomoides subglobosum* from the Norwegian Sea doubled its body mass within 3 days of being given abundant food, although this was mainly due to storage of food in vacuoles. After a further 7 days the vacuoles had been reduced and the body mass had increased by 17% compared with the original weight. In the experiments noted above, foraminifera survived anoxia for 78 days but presumably would ultimately be killed by the absence of oxygen.

In considering the response of organisms to environmental variables, some ecological factors are limiting in the sense that there is a threshold beyond which the effects on the organism are either to inhibit reproduction or, at the extreme, to cause death. Perhaps environmental variability does not matter too much as long as it stays within the nonlimiting range. However, once one parameter approaches the limits of tolerance for any given species, the stress level for that species will increase dramatically.

The term niche is used to describe the place of an organism in the ecosystem. The niche embraces the effects of all the ecological factors, both biotic and abiotic. The *fundamental niche* is where a species could in theory exist and the *realized niche* is where it actually does because, in practice, environmental variables interact. For instance, a species may be able to tolerate low salinity as long as the temperature is close to the optimum. However, if both variables are close to the limits of tolerance, their combined effects might cause death before the true limit of tolerance of either is reached. Because there are so many variables affecting organisms, in practice the realized niche of any given species will differ from one place to another. In other words, different combinations of factors may be limiting distributions in different areas.

Where an environment deteriorates (e.g., by becoming severely oxygen-depleted), it is the more specialized equilibrium species that disappear first. The generalists and opportunists survive the longest. Recovery seems to be a slow process and, in the single example of recovery from anoxia, initiation of colonization took more than 1 year. In all cases, the fauna had not reached former species diversity and composition even after several years. Similarly, recovery from a volcanic ash fall is slow (on a human timescale). For further details of the process of colonization see Alve (1999).

Buzas and Culver (1994) argue that under natural conditions there is a pool of species available to occupy new environments and from this pool more-or-less random subgroups establish themselves in any new space. Therefore, it should not be expected that newly created similar environments will have exactly the same mature fauna or that disturbed environments will reestablish their previous faunas.

8. Is It Possible to Predict the Consequences of Change?

This is a "what if..." question. What will happen if *x* changes by so much? The UN Environment Program commissioned a group of scientists to address the question: what would happen to the coastal areas of the Caribbean if there was a global rise of 1.5°C and a sea-level rise of 20 cm by the year 2025 (Maul, 1993)? One conclusion was that beaches and deltas would be much affected by the sea-level rise but scarcely affected by the temperature change. On the other hand, estuaries and seagrass beds were considered to be equally at risk from both changes. Corals would suffer increased disease with higher temperatures. The important point is that different ecosystems and organisms respond to changes in different ways.

No one has yet attempted to model the effects of a given environmental change on the distribution and abundance of foraminifera, although Culver and Buzas (1995) speculated upon the consequences of global warming. In their opinion, the environments at greatest risk are marshes (including those dominated by mangroves), perhaps more through the associated rise in sea level than the global increase in temperature. However, marsh faunas managed to withstand the rapid rise of sea level following the last glacial maximum ($\sim$125 m) so there is no cause to be too pessimistic about their future. Indeed, the greatest threat to marshes is their destruction through so-called "land-reclamation." Of course, it is easy to see that a 1-m rise in sea level would have serious consequences for a marginal marine environment and little or no effect on those environments deeper than the inner shelf. It would be expected that such a transgressive event would lead to the lateral migration of the marsh microfauna so that former marsh would become covered by unvegetated tidal flats: e.g., in a temperate setting a change from a *Jadammina/Trochammina/Miliammina* marsh fauna to an *Ammonia/Elphidium/Miliammina* mudflat fauna.

It is possible to summarize the main trends seen when there is an environmental change (Fig. 9). If the change is from low to high stress (e.g., increase in pollution, influx of phytodetritus) then dominance will increase,

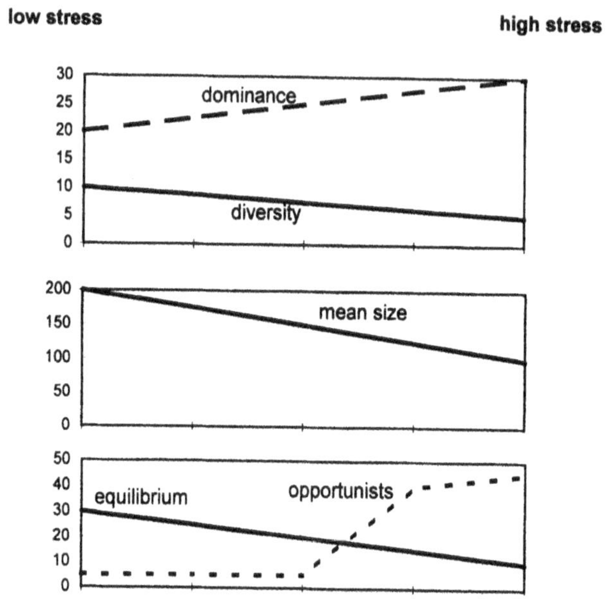

FIGURE 9. Summary of trends shown by assemblages responding to environmental change. The *y*-axis values are arbitrary units. See text for discussion.

whereas diversity and mean size of individuals will decrease. The abundance of equilibrium species will decrease and, when a certain threshold is reached, that of opportunists increase. If the change is from high stress to low stress (e.g., recovery from pollution or from an ashfall) then the trends will be reversed.

9. The Importance of Baseline Studies for Monitoring Environmental Change

In order to monitor contemporary environmental change it is necessary to unravel the signal resulting from natural variability from that caused by the shift in environmental conditions. At the present state of knowledge, it is easy to do this using foraminifera where the change is fast and large (e.g., input of organic pollution, volcanic eruption, major change in estuarine circulation) (see in this volume Debenay *et al.*, Ch. 2; Coccioni, Ch. 3; Ebrahim, Ch. 4; Hallock, Ch. 5; Eagar, Ch. 6; Rosenfeld *et al.*, Ch. 7; Schornikov, Ch. 8).

In order to attempt to document slow environmental change, it is necessary to have a detailed database of the natural variability of both the environment and the fauna (including the species present, dominance, diversity, abundance). This entails gathering time-series data over periods of years. Replicate samples should be taken in order to determine patchiness and to provide statistical rigor. This type of sampling program is very time-consuming, and in a research climate where funding is difficult to obtain, it is unlikely to be regarded as frontier science. Yet this is a mistaken view on the part of funding bodies because such baseline studies are essential prerequisites for monitoring change. Ideally, each country should be responsible for maintaining such studies for its own coastal areas (which are those environments most likely to experience change, whether natural or induced by man). There is also some urgency to undertake such studies because progressive environmental change is already under way. Two hypothetical examples show the difference between progressive and rapid change (Fig. 10*a*, *b*, respectively). In each case there is no doubt that changes are taking place.

In order to plan a time-series survey the following procedure is recommended. If the site is intertidal, one should establish a permanent station by placing a marker (e.g., post). If replicates are to be taken, the markers should be used spaced some meters apart. The first requirement is to determine the

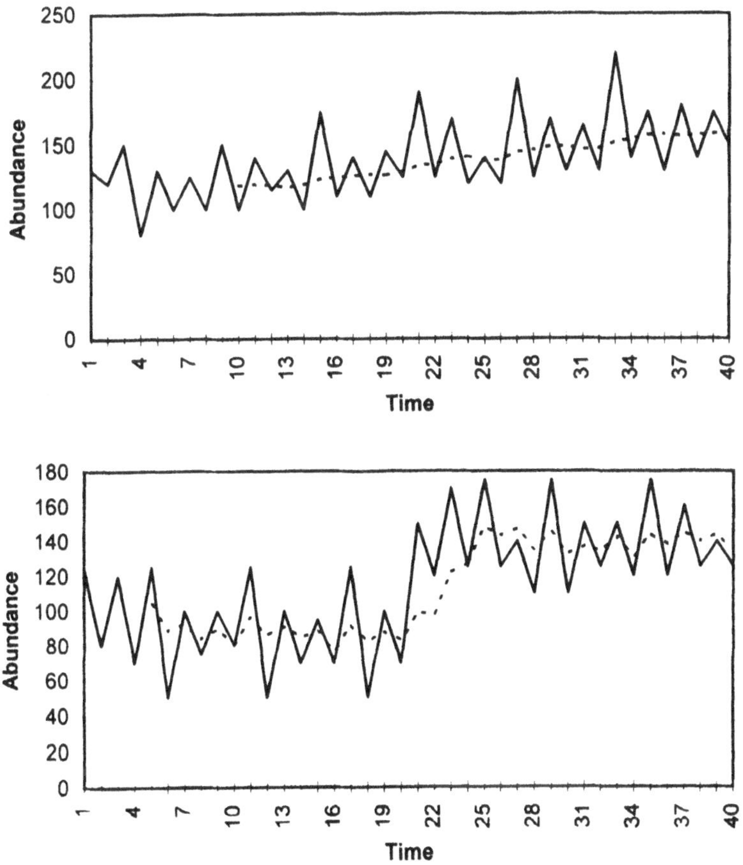

FIGURE 10. Two hypothetical examples of change. Variability in abundance (solid line) and the 10-point moving average (broken line) change from left to right: (a) progressive change, (b) rapid change (time 20/21).

depth below the sediment surface where the majority of individuals live by taking a pilot core and sectioning it horizontally into slices (most living forms are commonly in the 0- to 1-cm interval). Then replicate samples should be taken using a core tube of known area, preferably on a monthly basis. In order to avoid sampling disturbance from month to month, it is necessary to take successive samples from slightly different positions close to the marker and to avoid walking on areas yet to be sampled. The survey should continue for several years. Environmental parameters should be measured at each time of sampling.

10. Summary and Conclusions

Care must be taken to distinguish between an *apparent faunal change*, within the normal range of variability, and true environmental change:

1. With the notable exceptions of catastrophic events such as volcanic eruptions, most environmental changes are progressive, being made up of small incremental shifts.
2. There is a continuum from variability over periods of less than 1 year to conditions that go beyond the range of annual variability over a longer time period.
3. Change is the norm and there are cycles with different periodicities.
4. In general, because of the short generation time, foraminifera respond quickly to seasonal environmental variability.
5. As with the macrofauna, when adverse conditions develop, the first taxa to disappear are the more specialized ones, and the survivors are the generalists with greater tolerance to environmental stress.
6. Similarly, after the destruction of a microfauna, the first colonizers are opportunists, typically very small in size and with short generation times.

The final conclusion is that: through the analysis of changes in abundance of marker species; the introduction of new species or serious loss of previously existing species; and changes in species diversity, dominance, and abundance that extend well outside the established limits of variability, it is possible to document the extent of environmental changes that have taken (or are taking) place.

ACKNOWLEDGMENTS: In 1997 Valentina Yanko kindly invited me to present a plenary lecture at the First International Conference on "Application of Micropaleontology in Environmental Sciences," held in Tel Aviv, Israel, and this paper is a fuller development of that talk. I am grateful to Elisabeth Alve (Oslo) for permission to use Fig. 7 and to Wolfgang Kuhnt (Kiel) for Fig. 8. Very helpful comments on the manuscript were provided by Elisabeth Alve and by Andy Gooday (Southampton). Clarification of data sets have kindly been provided by Andy Gooday, Mary McGann (USA), and Hiroshi Kitazato (Japan). The referees, Jean-Pierre Debenay and Hiroshi Kitazato provided additional helpful comments. I also thank Ron Martin for his encouragement and advice.

References

Altenbach, A. V., 1992, Short term processes and patterns in the foraminiferal response to organic flux rates, *Mar. Micropaleontol.* **19**:119–129.

Alve, E., 1991, Foraminifera, climatic change, and pollution: A study of late Holocene sediments in Drammensfjord, southeast Norway, *Holocene* **1**:243–261.

Alve, E., 1994, Opportunistic features of the foraminifer *Stainforthia fusiformis* (Williamson): evidence from Frierfjord, Norway. *J. Micropalaeontol.* **13**:24.

Alve, E., 1995*a*, Benthic foraminiferal responses to estuarine pollution: A review, *J. Foram. Res.* **25**:190–203.

Alve, E., 1995*b*, Benthic foraminiferal distribution and recolonization of formerly anoxic environments in Drammensfjord, southern Norway, *Mar. Micropaleontol.* **25**:169–186.

Alve, E., 1996, Benthic foraminiferal evidence of environmental change in the Skagerrak over the past six decades, *Bull. Norges Geol. Unders.* **430**:85–93.

Alve, E., 1999, Colonization of new habitats by benthic foraminifera, *Earth-Sci. Rev.* **46**:167–185.

Alve, E., and Bernhard, J. M., 1995, Vertical migratory response of benthic foraminifera to controlled oxygen concentrations in an experimental mesocosm, *Mar. Ecol. Prog. Ser.* **116**:137–151.

Alve, E., and Murray, J. W., 1995, Benthic foraminiferal distribution and abundance changes in Skagerrak surface sediments: 1937 (Höglund) and 1992/1993 data compared, *Mar. Micropaleontol.* **25**:269–288.

Alve, E., and Murray, J. W., 1997, High benthic fertility and taphonomy of foraminifera: A case study of the Skagerrak, North Sea, *Mar. Micropaleontol.* **32**:157–175.

Basson, P. W., and Murray, J. W., 1995, Temporal variations in four species of intertidal foraminifera, Bahrain, Arabian Gulf, *Micropaleontology* **41**:69–76.

Bartlett, G. A., 1966, Distribution and abundance of foraminifera and thecamoebina in Miramichi River and Bay, *Bedford Inst. Oceanogr. Rep.* **66**(2):1–107.

Bernhard, J. M., and Reimers, C. E., 1991, Benthic foraminiferal population fluctuations related to anoxia: Santa Barbara Basin, *Biogeochemistry* **15**:127–149.

Bernhard, J. M., Sen Gupta, B. K., and Borne, P. F., 1997, Benthic foraminiferal proxy to estimate dysoxic bottom-water oxygen concentrations: Santa Barbara Basin, U.S. Pacific continental margin, *J. Foram. Res.* **27**:301–310.

Boltovskoy, E., 1964, Seasonal occurrence of some living foraminifera in Puerto Deseado (Patagonia, Argentina), *J. Cons. Int. l'Expl. Mer* **29**:136–145.

Boltovskoy, E., and Lena, H., 1969*a*, Seasonal occurrences, standing crop and production in benthic foraminifera of Puerto Deseado, *Contr. Cushman Found. Foram. Res.* **20**:87–95.

Boltovskoy, E., and Lena, H., 1969*b*, Microdistribution des foraminifères benthoniques vivants, *Rev. Micropaléontol.* **12**:177–185.

Buzas, M. A., and Culver, S. J., 1994, Species pool and dynamics of marine paleocommunities, *Science* **264**:1439–1441.

Castignetti, P., 1996, A time-series study of foraminiferal assemblages of the Plym estuary, south-west England, *J. Mar. Biol. Assoc., U.K.* **76**:569–578.

Clarke, K. R., and Warwick, R. M., 1994, Changes in marine communities: An approach to statistical analysis and interpretation, Natural Environment Research Council, UK, 144 pp.

Culver, S. J., and Buzas, M. A., 1995, The effects of anthropogenic habitat disturbance, habitat destruction, and global warming on shallow marine benthic foraminifera, *J. Foram. Res.* **25**:204–211.

Cushing, D. H., and Dickson, R. R., 1976, The biological response in the sea to climatic changes, *Adv. Mar. Biol.* **14**:1–122.

Daniels, C. H. von, 1970, Quantitative ökologische Analyse der zeitlichen und rämlichen Verteilung rezenter Foraminiferen im Limski kanal bei Rovinj, *Göttinger Arb. Geol. Paläontol.* **8**:1–109.

Earland, A., 1934, The Falklands sector of the Antarctic (excluding South Georgia), *Discov. Rep.* **10**:1–208.

Finger, K. L., and Lipps, J. H., 1981, Foraminiiferal decimation and repopulation in an active volcanic caldera, Deception Island, Antarctica, *Micropaleontology* **27**:111–139.

Gooday, A. J., 1993, Deep-sea benthic foraminiferal species which exploit phytodetritus: Characteristic features and controls on distribution, *Mar. Micropaleontol.* **22**:187–205.

Gooday, A. J., and Lambshead, P. J. D., 1989, Influence of seasonally deposited phytodetritus on benthic foraminiferal populations in the bathyal northeast Atlantic: The species response, *Mar. Ecol. Prog. Ser.* **58**:53–67.

Gooday, A. J., and Turley, C. M., 1990, Responses by benthic organisms to inputs of organic material to the ocean floor: A review, *Phil. Trans. Roy. Soc. London* **A331**:119–138.

Grimm, V., 1996, A down-to-earth assessment of stability concepts in ecology: dreams, demands and the real problems, *Senckenb. Marit.* **27**:215–226.

Haas, H. C., 1997, Recent and subrecent agglutinated foraminifers in four box cores from the Skagerrak (NE North Sea), in: *Contributions to the Micropaleontology and Paleoceanography of the Northern Atlantic* (H. C. Haas and M. A. Kaminski, eds.), *Grzybowski Foundation Special Publication* **5**:217–226.

Hannah, F., and Rogerson, A., 1997, The temporal and spatial distribution of foraminiferans in marine benthic sediments of the Clyde Sea area, Scotland, *Est. Coast. Shelf Sci.*, **44**:377–383.

Hess, S., and Kuhnt, W., 1996, Deep-sea benthic foraminiferal recolonization of the 1991 Mt. Pinatubo ash layer in the South China Sea, *Mar. Micropaleontol.* **28**:171–197.

Höglund, H., 1947, Foraminifera in the Gullmar Fjord and the Skagerak, *Zool. Bidr. Uppsala* **26**:1–328.

Josefson, A. B., 1990, Increase in benthic biomass in the Skagerrak-Kattegat during the 1970s and 1980s—effects of organic enrichment? *Mar. Ecol. Prog. Ser.* **66**:119–130.

Kaminski, M. A., 1985, Evidence for control of abyssal agglutinated foraminiferal community structure by substrate disturbance: Results from the Hebble area, *Mar. Geol.* **66**:113–131.

Kaminski, M. A., Grassle, J. F., and Whitlatch, R. B., 1988, Life history and recolonization among agglutinated foraminifera in the Panama Basin, *Abh. Geol. Bund., Wien* **41**:229–243.

Kröncke, I., 1995, Long-term changes in North Sea benthos, *Senckenb. Marit.* **26**:73–80.

Lutze, G. F., 1965, Zur Foraminiferen-Fauna der Ostsee, *Meyniana* **15**:17–142.

Lutze, G. F., 1968, Jahresgang der Foraminiferen-Fauna in der Bottsand-Lagune (westliche Ostsee), *Meyniana* **18**:13–30.

Magnusson, J., 1995, The invisible present, in: *Ecological Time Series* (T. M. Powell and J. H. Steele, eds.) Chapman and Hall, New York, pp. 448–464.

Matsushita, S., and Kitazato, H., 1990, Seasonality in the benthic foraminiferal community and the life history of *Trochammina hadai* Uchio in Hamana Lake, Japan, in: *Paleoecology, Biostratigraphy, Paleoceanography and Taxonomy of Agglutinated Foraminifera* (C. Hemleben, M. A. Kaminski, W. Kuhnt, and D. B. Scott, eds.), Kluwer, Dordrecht, 695–715.

Maul, G. A., 1993, *Climatic Change in the Intra-Americas Sea*, Arnold, London.

McGann, M., and Sloan, D., 1996, Recent introduction of the foraminifer *Trochammina hadai* Uchio into San Francisco Bay, California, USA, *Mar. Micropaleontol.* **28**:1–3.

Moodley, L., and Hess, C., 1992, Tolerance of infaunal benthic foraminifera for low and high oxygen concentrations, *Biol. Bull.* **183**:94–98.

Moodley, L., Troelstra, S. R., and van Weering, C. E., 1993, Benthic foraminiferal response to environmental change in the Skagerrak, northeastern North Sea, *Sarsia* **78**:129–139.

Moodley, L., van der Zwaan, G. J., Herman, P. M. J., Kempers, L., van Breugel, P., 1997, Differential response of benthic meiofauna to anoxia with special reference to foraminifera (Protista: Sarcodina), *Mar. Ecol. Prog. Ser.* **158**:151–163.

Murray, J. W., 1970*a*, The Foraminiferida of the Persian Gulf 6: Living forms in the Abu Dhabi area, *J. Nat. Hist.* **4**:55–67.
Murray, J. W., 1970*b*, The foraminifera of the hypersaline Abu Dhabi lagoon, Persian Gulf, *Lethaia* **3**:51–68.
Murray, J. W., 1983, Population dynamics of benthic foraminifera: Results from the Exe estuary, England, *J. Foram. Res.* **13**:1–12.
Murray, J. W., 1991, *Ecology and Palaeoecology of Benthic Foraminifera*, Longman, Harlow.
Murray, J. W., 1992, Distribution and population dynamics of benthic foraminifera from the southern North Sea, *J. Foram. Res.* **22**:114–128.
Ohga, T., and Kitazato, H., 1997, Seasonal changes in bathyal foraminiferal populations in response to the flux of organic matter (Sagami Bay, Japan), *Terra Nova* **9**:33–37.
Reinhardt, E. G., Patterson, R. T., and Schröder-Adams, C. J., 1994, Geoarchaeology of the ancient harbor site of Caesarea Maritima, Israel: evidence from the sedimentology and palaeoecology of benthic foraminifera, *J. Foram. Res.* **24**:37–48.
Rosenberg, R., Gray, J. S., Josefson, A B., and Pearson, T. H., 1987, Petersen's Benthic stations revisited: II. Is the Oslofjord and eastern Skagerrak enriched? *J. Exp. Mar. Biol. Ecol.* **105**:219–251.
Rosenberg, R., Hellman, B., and Lundberg, A., 1996, Benthic macrofaunal community structure in the Norwegian Trench, deep Skagerrak, *J. Sea Res.* **35**:181–188.
Schröder, C. J., Scott, D. B., and Medioli, F. S., 1987, Can smaller benthic foraminifera be ignored in paleoenvironmental analyses? *J. Foram. Res.* **17**:101–105.
Scott, D. B., Medioli, F. S., and Schafer, C. T., 1977, Temporal changes in foraminiferal distributions in Miramichi River estuary, New Brunswick, *Canadian J. Earth Sci.* **14**:1566–1587.
Silva, K. A., Corliss, B. H., Rathburn, A. E., and Thunell, R. C., 1996, Seasonality of living benthic foraminifera from the San Pedro Basin, California borderland, *J. Foram. Res.* **26**:71–93.
Stanley, D. J., and Warne, A. G., 1997, Holocene sea-level changes and early human utilization of deltas, *GSA Today* **7**(12):1–7.
Tapley, S., 1969, Foraminiferal analysis of the Miramichi estuary, *Marit. Sed.* **5**:20–29.
Vénec-Peyré, M. T., and Le Calvez, Y., 1986, Foraminifères benthiques et phénomènes de transfert: Importance des études comparatives de la biocènose et de la thanatocènose, *Bull. Mus. Hist. Nat., Paris, ser. 4* **8**:171–184.
Wefer, G., 1976, Umwelt, Produktion und Sedimentation benthischer Foraminiferen in der Westlichen Ostee, *Rept. Sondersforschungsbereich 95, Kiel* **14**:1–103.
Williams, D. E., Hallock, P., Talge, H. K., Harney, J. N., and McRae, G., 1997, Response of *Amphistegina gibbosa* populations in the Florida Keys (USA) to a multi-year stress event (1991–1996), *J. Foram. Res.* **27**:264–269.

Chapter 2

Distribution Trends of Foraminiferal Assemblages in Paralic Environments

A Base for Using Foraminifera as Bioindicators

JEAN-PIERRE DEBENAY, JEAN-JACQUES GUILLOU, FABRICE REDOIS, and EMMANUELLE GESLIN

JEAN-PIERRE DEBENAY • Laboratoire de Géologie, Equipe EA 2644, Faculté des Sciences, Université d'Angers, 49045 Angers Cedex, and Laboratoire d'Étude des Bio-Indicateurs Marins (LÉBIM), Ker Chalon, 85350 Ile d'Yeu, France JEAN-JACQUES GUILLOU • Laboratoire de Géologie, Faculté des Sciences, Université de Nantes, 44072 Nantes Cedex 3, France FABRICE REDOIS and EMMANUELLE GESLIN • Laboratoire de Géologie, Equipe EA 2644, Faculté des Sciences, Université d'Angers, 49045 Angers Cedex, and Laboratoire d'Étude des Bio-Indicateurs Marins (LÉBIM), Ker Chalon, 85350 Ile d'Yeu. France.

Environmental Micropaleontology, Volume 15 of *Topics in Geobiology*, edited by Ronald E. Martin. Kluwer Academic/Plenum Publishers, New York, 2000.

1. Introduction

Increasing pollution by industrial, agricultural, and other anthropogenic chemicals makes it ever more necessary to develop a thorough water management system. Chemical analyses are often used, but they are expensive and can measure only a fraction of the contaminants present at a given time. Such analyses may also require continuous monitoring in rapidly changing environments such as estuaries and coastal lagoons (paralic environments). Moreover, they reveal nothing about the adverse effects of contaminants that are readily taken up into the tissues of resident organisms (Walker and Livingstone, 1992) and that induce biological responses at all levels of biological organization, from the molecular to the ecosystem level (Cairns and McCormick, 1992). Therefore, the need to assess the impact of pollution leads to the study and development of biomarkers and bioindicators that detect the presence of both known and unknown contaminants (McCarty and Shugart, 1990; see also in this volume Geslin *et al.*, Ch. 9; Bresler and Hombach, Ch. 10).

Owing to a poor knowledge of the impact of natural variables, the use of bioindicators merits many more studies. This need is particularly critical in estuaries and coastal lagoons that are, on one hand, subject to tidal and seasonal changes in natural variables and, on the other hand, heavily exposed to a number of chemicals, including industrial pollutants and agricultural pesticides. These environments have great economic potential particularly in tourist areas and fisheries.

The choice of bioindicators depends on the ecosystem (Wilson *et al.*, 1995; Shear, 1996; Yazvenko and Rapport, 1996). In coastal settings, large numbers of foraminifers can be collected in small sample volumes; moreover, tests are often preserved in sediment, allowing comparison between anthropogenic and preanthropogenic conditions. Also, several species are widespread in estuaries and lagoons, allowing comparative studies between these environments, which is impossible with macrobenthos because estuarine species are largely replaced in lagoons by specifically lagoonal species (Barnes, 1989, 1994; Bamber *et al.*, 1992). Biomonitoring based on foraminifers may deal with: (1) individuals affected by morphological abnormalities due to chemical contamination (Alve, 1991; Yanko *et al.*, 1994; see also in this volume Bresler and Yanko, Ch. 10); or (2) species or groups of species indicating pollution impact by their absence, presence, or abundance.

Before using foraminiferal assemblages as pollution indicators in paralic environments, a precise understanding of their response to environmental variables is necessary. For example, the use of foraminifera as pollutant indicators at a sewage outfall (e.g., Bandy *et al.*, 1965; Schafer, 1973; review in Alve, 1995) generally disregards the impact of the freshwater input that is obviously one of the main causes of environmental disturbance. Thus, it is necessary to consider the distribution of foraminifera under natural conditions more carefully (Fig. 1).

The aim of this work is to provide information on the distribution of foraminifers in paralic environments so as to distinguish between the re-

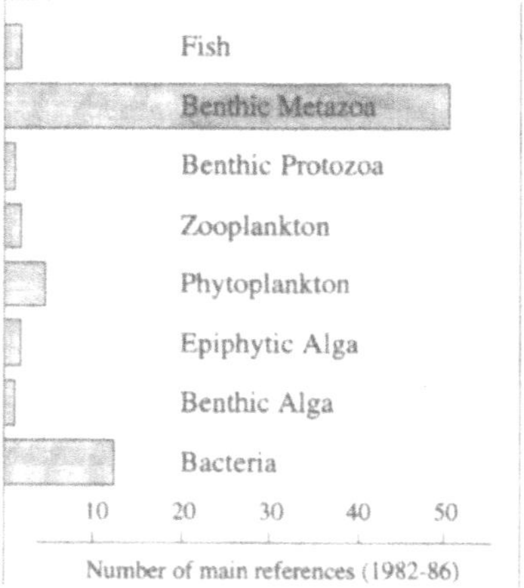

FIGURE 1. Diagram showing the proportion of studies concerning various groups of organisms used in biological methods of water-quality assessment. The 690 references used to draw this diagram were selected with the key word "bioindicator" in three databanks (Pascal, CNRS France; Pollution and <enviroline, U.S.A.) in 1982–1983. Benthic protozoa, including foraminifers, takes last place with benthic algae (after Blandin, 1986).

sponses due to anthropogenic stress and those due to changes in the natural environment. It is based on studies carried out on 1500 samples collected in various paralic environments from 1972 to 1997, supplemented by a compilation of selected works published by other authors in other paralic environments (see also in this volume Murray, Ch. 2).

2. Paralic Environments

The principal characteristic of paralic environments is that they occur at the transition between marine and continental realms—estuaries, coastal lagoons, marshes, and coastal zones subject to high freshwater input. Their ecological characteristics depend on various parameters including (1) the general climatic conditions (e.g., temperature, insolation, evaporation); (2) horizontal salinity distribution, which controls the zonation of flora and fauna in brackish and hypersaline habitats; (3) the vertical mixing intensity of fresh and marine waters (degree of stratification); (4) the hydrodynamic turnover time resulting from freshwater input (river discharge, rain, or groundwater seepage), tidal currents, gravitational circulation, and volume loss by evaporation; (5) the energy of wind-forced currents and waves; (6) the nature of the bottom, from rock to mud; (7) the presence of seaweeds or seagrass meadows; (8) the nature and amount of input of sediment (turbidity), dissolved organic matter, and nutrients; (9) the chemical characteristics of the water such as gas in solution (O_2, CO_2, SH_2...) and pH; (10) the presence of hypoxic episodes in the most restricted areas; and (11) the impact of anthropogenic activities.

The complex interplay among these parameters results in great difficulty in evaluating the average environmental characteristics at a given point. Moreover, mesotidal (tidal range between 2 and 4 m) and macrotidal (tidal range >4 m) environments undergo strong stresses resulting from the drastic changes during each tide cycle, such as penetration of full-strength seawater during flood tide, which can be almost entirely replaced by freshwater during ebb tide; emersion of the intertidal area resulting in changes in temperature, salinity (concentration by evaporation or dilution by rain), pH, and oxygen content of the interstitial waters.

3. Study Areas

Paralic environments were mainly studied on the Atlantic coast of Africa, from Mauritania (latitude about 20°N) to Congo (latitude about 5°S) (Table 1 and Fig. 2). The climatic shift from the north (precipitation less than 200 mm a year) to the south (precipitation over 1500 mm a year) and the morphological characteristics of the studied areas provide a wide range of environmental characteristics (Table 1).

Detailed studies were also carried out in Brazilian lagoons and on the Atlantic coast of France. Other samples were collected in paralic environments in Vietnam, Australia, New Caledonia, and on the Mediterranean coast of France. These areas provide examples of microtidal to macrotidal coasts and of normal (freshwater discharge to the sea) to inverse (penetration of marine water) estuaries.

3.1. Estuaries

A three-year interdisciplinary survey of about 40 stations was carried out in the estuary of the Casamance River (Senegal) from 1984 to 1986 (Debenay and Pagès, 1987). This hypersaline inverse estuary provides a linear model that allows a precise description of the changes in microfaunas from marine to very restricted conditions, 250 km from the sea (Table 1). In Senegal, the normal estuary of the Senegal River and the inverse estuary of the Saloum River were also investigated (Debenay *et al.*, 1987*a*; Debenay, 1990). Samples were also collected in the Zaire estuary and in the small Songolo estuary in Congo (Debenay and Moguedet, 1990) (Table 1).

In Vietnam, samples were collected in the region of Can Gio (Guélorget *et al.*, 1997) and in the tidal channels of Camau peninsula (Bui Thi Luan *et al.*, 1994, and unpublished data) (Table 1).

Three estuaries were also studied on the macrotidal French Atlantic coast: (1) in the Auray River estuary (Gulf of Morbihan), several stations were sampled every month for one year (Redois, 1996; Redois and Debenay, 1996); (2) samples were collected in the Vie estuary in order to follow the changes in the assemblages from the open sea to the marshes (Table 1); (3) the impact of various sources of pollution on foraminiferal assemblages was investigated in the estuary of Adour River (Casamajor and Debenay, 1995).

3.2. Restricted Lagoons

Restricted lagoons are characterized by two or more entrance channels or inlets and a well-defined tidal circulation (Kjerfve, 1986). Several restricted lagoons were investigated along the Atlantic coast of Africa, including small lagoons in the arid region of Cape Timiris in Mauritania (Debenay, 1990) (Table 1), small lagoons scattered along the Senegalese coast, and coastal lagoons in Congo (Debenay and Moguedet, 1990). All these lagoons are shallow, and their waters, partially renewed during each tidal cycle, are slightly hypersaline (Cape Timiris) or hyposaline, but they are not stratified.

Another restricted lagoon was studied on the southern Brazilian coast, the Cananéia-Iguape lagoon (Eichler *et al.*, 1995). It is a deep (up to 12 m) lagoon with a salt wedge due to freshwater inflow from several small coastal rivers (Miyao *et al.*, 1986; Eichler *et al.*, 1995) (Table 1). Data from the lagoon of Venice (Petrucci *et al.*, 1983; Albani *et al.*, 1991) are also considered in this study (Table 1).

3.3. Choked Lagoons

Choked lagoons are characterized by one or more long, narrow entrance channels, long residence time, and dominant wind-forcing (Kjerfve, 1986). Nonstratified choked lagoons were studied in Brazil: the Araruama lagoon (unpublished data); in Benin: Lake Nokoué (Debenay *et al.*, 1993*a*,*b*); and on the French Mediterranean coast: the Étang du Prévost (Favry *et al.*, 1998) and adjacent lagoons (Étang de l'Arnel and Étang de Vic, unpublished data). Lake Nokoué was studied in greater detail as it provides a two-dimensional model in which the hydrodynamics are well known (Table 1).

Stratified choked lagoons were studied in Brazil (Lagoa da Conceição; Debenay *et al.*, 1997, 1998), and on the Ivory Coast (Ebrie lagoon; Debenay *et al.*, 1987*b*) (Table 1).

3.4. Mangrove Swamps and Salt Marshes

Samples were collected in mangrove swamps of Cairns (eastern Australia, unpublished data) and New Caledonia, and on the salt marshes of New Caledonia (Table 1). Mangrove swamps and salt marshes adjacent to African and Brazilian areas studied were also investigated.

Data from the literature were also considered: saltworks of Salins de Giraud (south of France) (Zaninetti, 1982); high marshes of Acheloos and Evinos deltas (northern side of the Gulf of Patras, western coast of Greece)

TABLE 1. Location and General Characteristics of Study Areas and Areas Selected from the Literature for Additional Data[a]

Area	Location and climate	Environment
Estuaries		
1 Senegal (Casamance estuary) (Debenay, 1990)	About 12° 30′N, 15° to 17° W Transition between Soudanian and Guinean climate. Becoming Sahelian about 1970 (Debenay *et al.*, 1994)	Short river. Inverse estuary with mangrove swamps in its lower part; the small drainage basin has been under arid conditions since about 1970
2 Senegal (Senegal estuary) (Debenay *et al.*, 1987*a*)	About 16° N, 17° W Arid Sahelian	Normal estuary; the upper drainage basin is the humid Fouta Djalon region
3 Senegal (Saloum estuary) (Debenay, 1990)	About 14° N, 17° W Arid Sahelian	Inverse estuary with mangrove swamps; the whole drainage basin is under arid conditions
4 Congo (Zaire and Songolo estuaries) (Debenay and Moguedet, 1990)	About 4° 30′ S, 11° 30′ E Tropical	Small and large (Congo River) estuaries
5 Vietnam (Mekong delta) (Bui Thi Luan *et al.*, 1994; Guélorget *et al.*, 1997) and unpublished data	About 10° N, 105° E Tropical	Estuary and tidal channels
6 France (Auray River estuary) (Redois and Debenay, 1996)	About 47° 30′ N, 3° 30′ W Temperate	Restricted bay and estuaries
7 France (Vie estuary, Atlantic) (unpublished data)	About 46° 30′ N, 2°E Temperate	Estuary
8 France (Adour estuary, Atlantic) (Casamajor and Debenay, 1995)	About 4° N, 1° 30′ E Temperate	Estuary
Restricted Lagoons		
9 Mauritania (Cape Timiris) (Debenay, 1990)	About 19° 20′ N, 16° 30′ W Saharian, very arid	Small and shallow lagoons
10 Senegal (coastal lagoons) (Debenay *et al.*, 1987*a*; Debenay, 1990)	About 13° 50′N to 15° N Arid Sahelian	Small and shallow lagoons
11 Congo (coastal lagoons) (Debenay and Moguedet, 1990)	About 4° 30′ S, 11° 30′E Tropical	Small brackish lagoons periodically closed by a sandspit
12 Brazil (Cananéia-Iguape Lagoon—São-Paulo State) (Eichler *et al.*, 1995)	About 24° 30′ to 25° S, 47° 30′ to 48° W Subtropical humid	Restricted lagoon, about 70 km long, parallel to the coastline
13 Italy (Venice lagoon) (Albani *et al.*, 1991; Petrucci *et al.*, 1983)	About 47° N, 12° 30′ E Mediterranean	Restricted lagoon

[a] All unpublished data were studied by the authors. The numbers in the first column are used to locate areas in Fig. 2.

Salinity	Dominant hydrodynamics	Origin of waters
Salinity increasing upward to 172‰, 230 km from the sea during the dry season; annual changes of more than 100‰ in the upper reaches (Pagès and Debenay, 1987)	Penetration of marine water concentrated by evaporation up to 250 km during the dry season; fresh water input into the upper reaches during the rainy season (July–September)	Marine water concentrated by evaporation in most of the estuary; freshwater in the upper reaches during the rainy season
Salinity 35‰ decreasing upward, more rapidly during the rainy season	Microtidal	Mixing between marine and freshwater
Salinity increasing upward	The water deficit results in an inflow of marine water then concentrated by evaporation	Marine water concentrated by evaporation
From more than 30‰ at high tide down to about 1‰ at low tide	Tidal currents and freshwater input	Marine water alternating with freshwater
35‰ in the coastal area, decreasing inward to 15‰ in the study areas during the dry season; no more than 25‰, even in the coastal area during the rainy season	Tidal cycles and freshwater input	Marine water diluted by continental water input
34‰ at high tide near the mouth decreasing upward to 1‰ in the study area	Tidal currents and river discharge	Mixing of marine water and freshwater
34‰ at high tide near the mouth decreasing upward to 7‰ in the study area	Tidal currents and river discharge	Mixing of marine water and freshwater
34‰ at high tide near the mouth decreasing upward to 0‰ in the study area	Tidal currents and river discharge	Mixing of marine water and freshwater, with pollution sources
Slightly hypersaline (38–40‰)	Tidal currents	Marine waters almost entirely renewed at each tide
Slightly hypersaline or slightly hyposaline depending on seasonal changes	Microtidal	Marine waters almost entirely renewed at each tide
From more than 30‰ at high tide down to less than 10‰ depending on the season and on the opening of the lagoons	Tidal currents when the lagoons are open and freshwater input during the rainy season	Marine water partly mixed with freshwater
About 35‰ near the entrance channel down to 0‰ in the tributaries. Inclined salinity stratification (Miyao *et al.*, 1986)	Tidal currents and freshwater input by the tributaries, inclined stratification	Mixing of marine water and freshwater, with penetration of marine water near the bottom
From about 36‰ near the entrances to about 10‰ in the inner parts of the lagoon	Tidal and wind-forced currents	Mixing of marine water and freshwater

(*Cont.*)

TABLE 1. Continued

Area	Location and climate	Environment
Choked Lagoons		
14 Brazil (Araruama lagoon—Rio de Janeiro state) (unpublished data)	22° 50′ to 22° 57′ S, 42° to 42° 25′ W Tropical	Choked lagoon about 50 km long; connection to the sea by a straight channel
15 Benin (Lake Nokoué) (Debenay *et al.*, 1993*a*)	About 6° 30′ N, 2° 30′ E Tropical	Shallow choked lagoon about 160 km² (maximum depth = 3 m)
16 France (Étang du Prévost and Étang de l'Arnel, Mediterranean) (Favry *et al.*, 1998 and unpublished data)	About 43° 30′ N, 3° 55′ E Mediterranean	Choked lagoons; the Étang de l'Arnel is connected to the sea through the Étang du Prévost, they are very shallow (<1 m deep)
17 Brazil (Lagoa de Conceição—Santa Catarina State) (Debenay *et al.*, 1997)	About 25° 30′ S, 48° 30′ W Subtropical humid	Choked lagoon about 13 km long connected to the sea by a 2-km long and 2-m deep channel
18 Ivory Coast (Ebrie Lagoon) (Debenay, 1990)	About 5° 15′ N, 3° 45′–4° 45′ W Tropical	Choked elongated lagoon more than 100 km long parallel to the coastline; connection to the sea by a straight channel
Mangrove Swamps and Salt Marshes		
19 Australia (Cairns area and Lizard Island) (unpublished data)	17° S, 146° E Tropical	Mangrove swamps
20 New Caledonia (unpublished data)	About 21° to 22° S, 164° to 166° 30′ E Tropical with strong trade winds	Mangrove swamps
21 France (Saltworks of Salins de Giraud, Mediterranean) (Zaninetti, 1982)	About 43° 20′ N, 4° 40′ E Mediterranean	Man made salt-works
22 Greece (Western coast, Gulf of Patras) (Scott *et al.*, 1979)	About 38° N, 21° 30′ E Mediterranean	River deltas
23 Canada (Nova Scotia) (Scott and Medioli, 1980)	About 44° to 47° N, 60° to 65° W Temperate cold	Salt marshes
Low Salinity Coastal Zone		
24 Guinea (Los Archipelago) (Debenay *et al.*, 1987)	About 9° 30′ N, 13° 30′ W Guinean tropical humid	Internal shelf and mangrove swamps

Salinity	Dominant hydrodynamics	Origin of waters
Hypersaline (about 70‰), except near the entrance channel	Wind-forced currents and waves; no stratification of the water, even in the deepest areas (>10 m)	Marine water concentrated by evaporation due to strong winds. Weak freshwater input by small rivers
Salinity from 0.2‰ during the rainy season up to 33‰ during the dry season	Discharge of the Oueme River and wind-forced clockwise current preventing stratification (Colleuil, 1984)	Mixing between marine and freshwater with predominance of freshwater during the rainy season
18 to 40‰ depending on seasonal changes around the periphery of the lagoons, 27 to 38‰ near the inlet, 10‰ and lower near the small freshwater streams in the Étang de l'Arnel	Microtidal (wind-forced and tidal currents)	Marine water concentrated by evaporation or diluted by freshwater input on the periphery
Salinity of the surface water from 15‰ in rainy season up to 30‰ in dry season; salinity of the bottom water always over 30‰	Stratification with a strong halocline between 2 m and 4 m	Brackish surface water, saline anoxic bottom water
Always more than 30‰ in the depressions. During the rainy season: 10‰ near the entrance channel lowering to 1‰, 20 km westward and 10 km eastward; during the dry season: 25 and 10‰ in the same areas (Tastet, 1974; Durand and Guiral, 1944)	Tidal current, discharge of Comoe River and stratification	Diluted to very diluted marine water near the surface; anoxic salted water in the depression; pollution near Abidjan
From marine water, down to about 10‰ in the creeks, in Cairns	Tidal currents and freshwater input in the creeks; immersion at low tide	Marine water alternating with brackish water in the lower reaches of the creeks; brackish water upward
Salinity increasing from the channels in the mangrove swamps toward the hypersaline salt marshes	Tidal currents and waves in the channels; evaporation of marine water between spring tides on the marshes	Marine water alternating with brackish water in the channels; water concentrated by evaporation in the salt marshes
Hypersaline: study area from 35 to more than 150‰	Controlled circulation of water in channels	Marine water concentrated by evaporation
12 to 27‰ during rainy period, 50 to 70‰ during dry period	Microtidal	Mixing of marine water and freshwater
	Tidal cycles	Estuarine waters
Salinity as low as 20‰ in all the coastal area (Uschakov, 1970)	Discharge of the Konkoure River	Marine waters diluted by river discharges

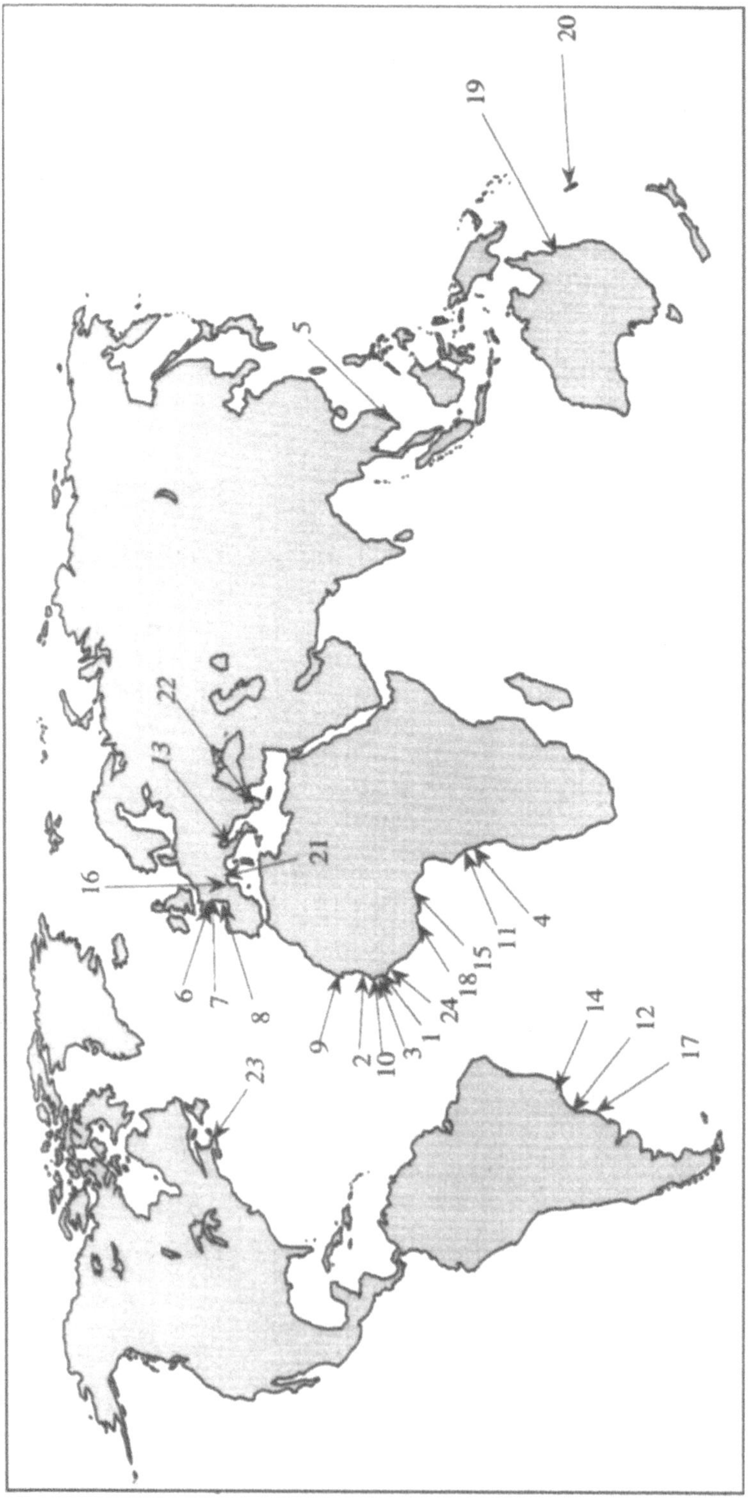

FIGURE 2. Location of study areas (numbers in italics indicate data from the literature). See Table 1 for an explanation of the numbers.

(Scott *et al.*, 1979); Atlantic coast of Canada (Scott and Medioli, 1980; Scott *et al.*, 1981)

3.5. Low-Salinity Coastal Zone

Los Archipelago (Guinea) is located 5 km southwest of Conakry (Table 1). From April to June, tropical waters reach the coastal zone and salinity is about 35‰. Afterward, they are replaced by Guinean waters and the salinity decreases to about 21‰, with exceptionally low salinities (8–9‰) occurring during the rainy season (Uschakov, 1970). The temperature of the water is almost constant, except in January, when cold upwelling waters may reach the region.

4. Materials and Methods

Depending on the study area, samples were collected only in the shallow subtidal zone or from the subtidal zone to the supratidal one as far as the upper marshes. Generally, they were collected during a single field trip. However, in the Casamance River (Senegal) they were collected during a 3-year interdisciplinary survey of about 40 stations from the mouth up to 250 km from the sea. In the Gulf of Morbihan (France), several stations were sampled every month for 1 year.

Samples were collected either by hand at low tide in the shallow areas, with a grab-sampler from a boat, or by diving when the depth was greater than 1 m. In all cases only the uppermost layer of the sediment (about 5 mm) was scraped off and kept in alcohol or in neutralized formaline. In the laboratory, a constant volume of 50 cm^3 of sediment was washed on a 50-μm sieve and the microfauna separated by flotation on carbon tetrachloride. The total assemblage was studied without distinction between empty tests and living individuals. From 100 to 300 specimens were counted and classified according to the Loeblich and Tappan (1988) classification. Salinity was measured at each station with an Atago refractometer. Additional studies of water and sediment were sometimes carried out.

Ozarko *et al.* (1997) consider that sampling of the uppermost layer of the sediment does not give an accurate representation of modern marsh foraminiferal assemblages. This is true for living assemblages, which additionally need a continuous survey owing to seasonal changes. However, inaccurate results may be obtained for ecological zonations based on total faunas collected in a thick surface interval, owing to different sedimentation rates and taphonomic processes in different areas. Thus we consider that the study of the total faunas in a thin surface interval provides less biased ecological information in paralic environments.

The species mentioned in the text are listed in Table 2 according to the nature of the test and grouped into the three suborders to which they belong.

TABLE 2. Species Mentioned in the Text Listed According to the Nature of the Test and Grouped into the Three Suborders to Which They Belong

Textulariina (agglutinated tests)	Rotaliina (calcerous hyaline tests)	Miliolina (calcareous porcelaneous tests)
Ammobaculites exiguus	*Ammonia beccarii*	*Pseudotriloculina* cf. *oblonga*
Ammotium salsum	*Ammonia parkinsoniana*	*Quinqueloculina seminula*
Arenoparrella mexicana	*Ammonia tepida*	*Quinqueloculina trigonula*
Eggerelloides scabrus	*Asterorotalia pulchella*	
Gaudryina exilis	*Bolivina* spp.	
Haplophragmoides wilberti	*Cribroelphidium excavatum*	
Jadammina macrescens	*Cribroelphidium excavatum var. selseyense*	
Jadammina polystoma	*Cribroelphidium gunteri*	
Miliammina earlandi	*Cribroelphidium williamsoni*	
Miliammina fusca	*Discorinopsis aguayoi*	
Miliammina spp.	*Elphidiella* sp.	
Polysaccammina ipohalina	*Elphidium excavatum*	
Polysaccammina hyperhalina	*Elphidium pulvereum*	
Pseudoclavulina sp.	*Ephidium limbatum*	
Pseudothurammina limnetes	*Eponides repandus*	
Scherochorella moniliformis	*Glabratella baccata*	
Siphotrochammina lobata	*Haynesina depressula*	
Tiphotrocha comprimata	*Haynesina germanica*	
Trochammina inflata	*Neoconorbina nitida*	
Trochammina spp.	*Nonion* cf. *commune*	
	"Nonion depressulum"	
	Nonion pauciloculum	
	Pararotalia niponica	
	Pararotalia sp.	
	Pseudononion atlanticum	
	Rosalina spp.	

5. Results

5.1. Estuaries

5.1.1. East Atlantic Tropical Estuaries

In the Casamance estuary, species of the suborder Rotaliina dominate the assemblage up to about 100 km inland, where the salinity reaches about 50‰. The main species are *Ammonia tepida* and *Cribroelphidium gunteri*, which

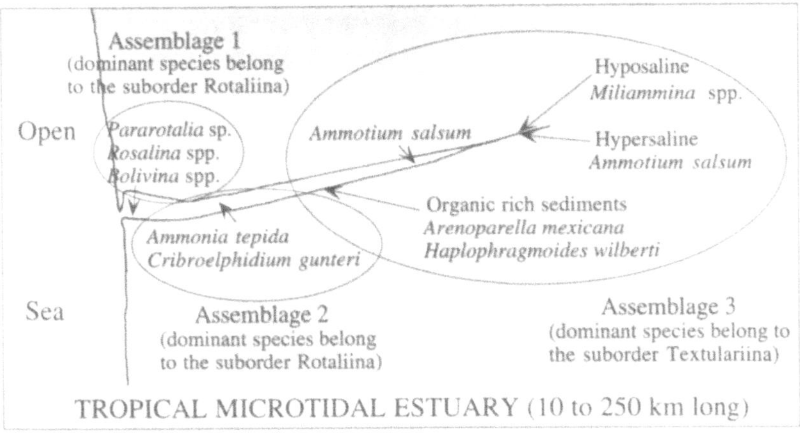

FIGURE 3. Longitudinal distribution of the dominant species in a tropical microtidal estuary (example: the Casamance River—Senegal).

disappear completely about 130 km inland in the dry season, when salinity exceeds 70‰. After the rainy season, they appear again about 170 km from the sea, where salinity remains between 35 and 50‰ for more than 4 months (Debenay and Pagès, 1987). Species related to strong marine influence, such as *Pararotalia* sp., *Bolivina* spp., and *Rosalina* spp. occur up to 20 km from the mouth (Fig. 3). Locally, within the first 100 km, an agglutinated species, *Eggerelloides scabrus*, becomes dominant.

Further, the abundance of the suborder Textulariina (mainly *Ammotium salsum*) increases rapidly, reaching a maximum of more than 1000 individuals per 50 cm^3 of superficial sediment about 180 km from the sea. Farther upstream, the abundance of foraminifers decreases but a small number of tests can be found up to about 230 km from the sea, where salinity is over 100‰ during the dry season and near 0‰ at the end of the rainy season. *Haplophragmoides wilberti* and *Arenoparella mexicana* may be abundant in organic-rich sediments, especially in the lateral salt marshes, where salinity can reach 80‰.

The same trends with the same species are observed in the inverse Saloum estuary and in the normal Senegal River estuary. However, in the latter, the zonation occurs over a shorter distance (about 20 km) owing to freshwater inflow, which limits marine influence. *Miliammina fusca* is present in the upper areas (Debenay *et al.*, 1987*a*). The same trends within a shorter distance were also observed in the small Songolo estuary.

In the Zaire River estuary, where freshwater inflow is strong, the marine influence is limited to the deep central channel. The waters of the lateral shallows are brackish, and foraminiferal assemblages are dominated by *Ammonia tepida*, *A. parkinsoniana*, *Eggerelloides scabrus* and *Polysacammina ipohalina*.

5.1.2. Mekong Delta and Tidal Channels on the Camau Peninsula (Vietnam)

Near the mouth of the channels assemblages under marine influence include *Pararotalia niponica*, *Asterorotalia pulchella*, *Bolivina* spp., and *Elphidiella* sp. *Ammonia tepida* is dominant in areas under moderate marine influence (salinity >20‰). Upstream, the percentage of *Ammotium salsum*, *Ammobaculites exiguus*, *Arenoparrella mexicana*, and *Gaudryina exilis* increases. Near mangrove swamps, *Trochammina inflata*, *Jadammina macrescens*, *Miliammina* spp., and *Haplophragmoides wilberti* dominate (Fig. 4).

5.1.3. East Atlantic Temperate Estuaries (Coast of France)

The upper estuary of the Auray River is dominated by agglutinated forms mainly *Miliammina fusca* in low-salinity areas. Seaward, calcareous forms appear in the order *Haynesina germanica*, *Ammonia tepida*, and near the mouth *Elphidium pulvereum*, *Haynesina depressula*, and *Ammonia beccarii*. This general horizontal freshwater-to-seawater transition may be disturbed by local conditions and, e.g., *Eggerelloides scabrus* or *Quinqueloculina seminula* may become abundant when seagrasses are present (Fig. 5).

In the middle estuary, a vertical water-to-land transition is observed. The lower intertidal zone is dominated by *Ammonia tepida*, which is replaced upward by *Haynesina germanica*. The upper intertidal zone is dominated by *Ammotium salsum*, *Scherochorella moniliformis*, and *Cribroelphidium williamsoni*. The assemblages of the supratidal zone, which are rich in organic matter, are composed of *Trochammina inflata* and *Jadammina macrescens*.

In the Vie estuary, the main species are *Ammonia tepida*, *Haynesina germanica*, and *Cribroelphidium excavatum*. Some species characteristic of

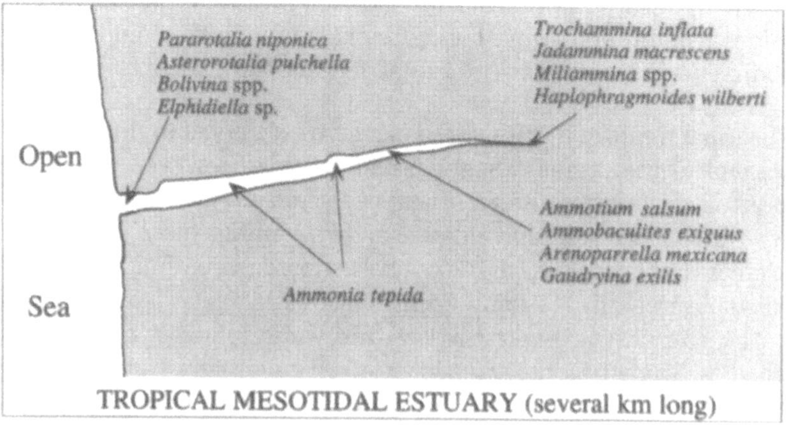

FIGURE 4. Longitudinal distribution of the dominant species in a tropical mesotidal estuary (example: the Mekong Delta—Vietnam).

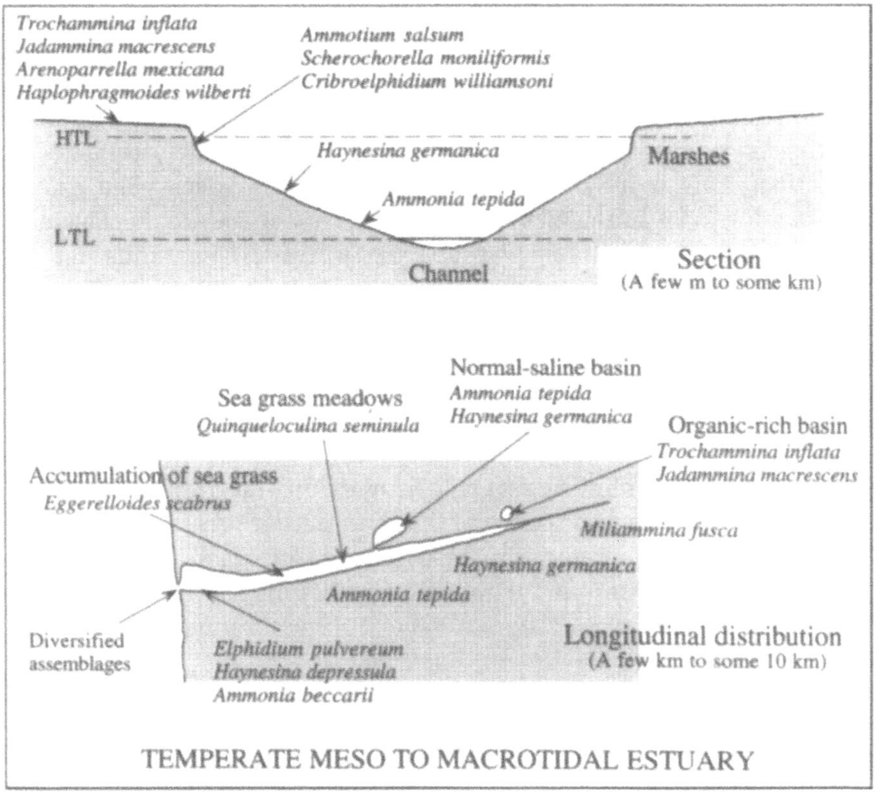

FIGURE 5. Longitudinal and transverse distribution of the dominant species in a temperate meso- to macrotidal estuary (example: the Auray River—France).

marine influence, such as *Bolivina* spp., *Glabratella baccata*, *Neoconorbina nitida*, and *Rosalina* spp. may occur near the mouth. In the tidal channels and in the small basins connected to the estuary, *Jadammina macrescens* and *Trochammina inflata* are dominant, and are sometimes associated with *H. germanica*. These species are associated with *Cribroelphidium williamsoni* in the upper intertidal zone and in the supratidal zone.

In the estuary of the Adour River, the impact of different sources of pollution disturbed the general landward trend. Effluents from the abattoir characterized by the abundance of organic matter (mainly blood) favored an increase in the number of tests. The input of degraded organic matter from a fish farm caused an increase in the size of the tests. Chemical pollution in the harbor, associated with freshwater input from boat washing, has led to the dominance of *Miliammina fusca*, which is usually found in low-salinity areas. These trends are consistent with the impact of pollution on foraminiferal assemblages reported in other estuaries (review in Yanko *et al.* 1999).

5.1.4. Generalization

In all the estuaries studied, the foraminiferal assemblages change landward from highly diverse assemblages, dominated by calcareous forms (suborder Rotaliina) near the open sea, to oligospecific assemblages dominated by agglutinated species (suborder Textulariina), and ultimately a monospecific population. This trend of decreasing diversity, and decreasing calcareous forms passing to agglutinated ones landward, corresponds to the general trend in paralic foraminiferal faunas around the world in hyposaline and highly hypersaline environments (review in Murray, 1973, 1991; Debenay, 1990; Hayward and Hollis, 1994). In normal estuaries, the increasing proportion of agglutinated species landward may be related to decreasing pH with increasing continental runoff that affects calcareous species. However, the same trend exists in *inverse* estuaries where the horizontal zonation is related to increasing salinity landward, reaching extreme values where only high-salinity-tolerant species are able to survive. In meso- to macrotidal estuaries, the horizontal zonation is also related to the short-period stress resulting from the alternation of marine water and freshwater during each tidal cycle in the upper estuarine zone and in the transition zone. This alternation, which leads to dramatic changes in salinity and osmotic pressure, requires increasingly well-developed osmoregulation up the estuary.

Indeed, Hayward and Hollis (1994) show that salinity and exposure during tidal cycles have the strongest influences on the ecological distribution of brackish foraminifera in New Zealand estuaries, and Wang (1992) considers the dominant factors controlling the distribution of living estuarine species to be salinity and pH; Yassini and Jones (1989, 1995) point out direct correlations among salinity, water depth, foraminiferal diversity, and population density. It should be noted that the foraminiferal distribution around sewage outfalls (e.g., Watkins, 1961; Bandy *et al.*, 1964) follows the same trend as the foraminiferal distribution in estuaries, showing that the salinity stress resulting from the freshwater input in these areas is probably the major factor acting on foraminiferal assemblages.

Miliammina fusca, which is often dominant in upper normal estuarine environments or in hyposaline marshes, is commonly associated with thecamoebians. Although De Rijk (1995) did not find any correlation between the frequency distribution of this species and salinity, it is often reported in low-salinity areas (e.g., Murray, 1968, 1973; Boltovskoy and Wright, 1976; Eichler *et al.*, 1995; Redois and Debenay, 1996). The vertical zonation of decreasing percentages of *Ammonia tepida* and increasing percentages of *Haynesina germanica* (both species associated with *Elphidium excavatum*) was also reported in dead assemblages from the Hamble estuary (Alve and Murray, 1994).

5.2. Lagoons

5.2.1. Restricted Lagoons

In the small African lagoons that have been studied, foraminiferal assemblages are dominated by *Ammonia* spp., *Quinqueloculina seminula*, elphidiids (mainly *Cribroelphidium gunteri* and *E. limbatum*), and *Haynesina germanica*. *Ammotium salsum* and *Trochammina inflata* only occur in the most landward areas.

In the Iguape-Cananéia lagoon (Brazil), the foraminiferal distribution is characterized by calcareous marine assemblages near the mouth, including *Pararotalia* sp.; high proportions of *Ammonia* spp. and *Cribroelphidium gunteri* in the outer lagoon; and a gradual increase in agglutinated species (mainly *Ammotium salsum*, *Gaudryina exilis*, and *Miliammina earlandi*) with increasing distance from the sea and/or decreasing depth. *Miliammina. earlandi* is dominant in low-salinity areas. The only exception in this general seawater-to-freshwater transition is a local influence of sewage where *Arenoparrella mexicana* and *Haplophragmoides wilberti*, which we generally observed in organic-rich sediment, have an abundance higher than expected. A peculiar assemblage, composed mainly of *Trochammina inflata* and *Siphotrochammina lobata* lives on mangrove tree roots in the intertidal zone.

5.2.2. Nonstratified Tropical Choked Lagoons

The assemblages of the Araruama lagoon (Brazil) are dominated by three species that often make up more than 80% of the assemblages: *Ammonia tepida*, *Cribroelphidium excavatum* var. *selseyense*, and *Pseudotriloculina* cf. *oblonga* (Debenay *et al.*, unpublished data). Their distribution seems to depend mainly on the input of organic matter with no relation to distance from the sea. The species characteristic of marine influence (e.g., *Pararotalia* sp., *Pseudononion atlanticum*, and *Eponides repandus*) are only present near the mouth of the entrance channel.

Near the entrance channel of Lake Nokoué (Benin), in the area under marine influence (Cotonou channel), foraminiferal assemblages are dominated by *Ammonia tepida*, *A. parkinsoniana*, and *Cribroelphidium gunteri*, with a number of additional species (Debenay *et al.*, 1993*a*). In the main water body, they are dominated by *Ammotium salsum*. *Miliammina earlandi* and *M. fusca* predominate in the low-salinity areas. Samples collected in organic-rich sediments of the So and Ouémé estuaries contain *Arenoparrella mexicana* and *Haplophragmoides wilberti*.

5.2.3. Mediterranean Choked Lagoons

The dominant species in the subtidal area of the Étang du Prévost and the Étang de l'Arnel (France) is *Ammonia tepida*, associated with *Haynesina germanica*, *Quinqueloculina seminula*, and *Cribroelphidium gunteri*. Nearer the mouth, the assemblage is also dominated by *A. tepida* but the number of additional species is greater; at the mouths of freshwater tributaries, the dominant species are *Trochammina inflata*, *Haplophragmoides wilberti*, and *Jadammina macrescens*. *Miliammina fusca* is present in the areas of very low salinity. In the upper salt marsh, *J. macrescens* dominates (unpublished data) (Fig. 6).

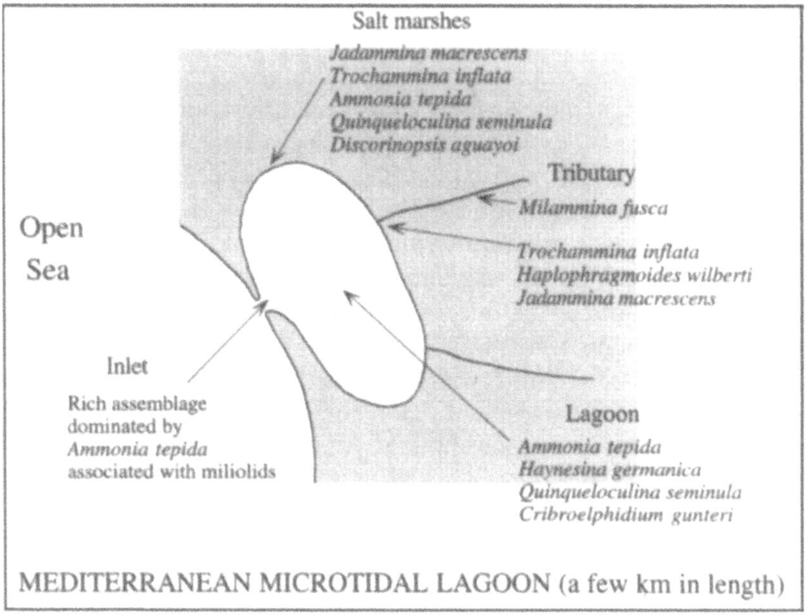

FIGURE 6. Distribution of the dominant species in Mediterranean microtidal choked lagoons.

5.2.4. Stratified Tropical Choked Lagoons

Foraminiferal populations of the Lagoa da Conceição (Brazil) are under two influencing factors:

First, surface water is diluted by rain and continental freshwater input. As a result, the foraminiferal assemblages on the shallow (<1.5 m) sand banks are dominated by agglutinated species, mainly *Ammotium salsum* associated with *Ammobaculites exiguus*, *Miliammina earlandi*, *Pseudoclavulina* sp., and *Gaudryina exilis*. The proportion of *M. earlandi* is high only in one sample collected at the mouth of a small tributary.

Second, the saltier bottom water results from the penetration of marine water at high tide during the dry season. During the rainy season, the stagnation of this water in depressions leads to anoxic conditions, while salinity remains about 34‰. The main foraminiferal taxa in the depressions are *Ammonia tepida* and *Cribroelphidium gunteri* (Fig. 7). However, this microfauna develops only during periods of renewal of bottom water and only empty tests are found during anoxic conditions (Debenay *et al.*, 1997).

In the Ebrie lagoon (the Ivory Coast), the dominant species is *Ammotium salsum*. Near the entrance channel, it is associated with *Ammonia tepida* and with very rare specimens of the Textulariina or Buliminidae (Debenay *et al.*, 1987*b*). Inward, the proportion of these associated species decreases. In the most restricted areas, *A. salsum* is the only remaining foraminifer and is associated with rare thecamoebians. A high proportion of *Miliammina* spp.

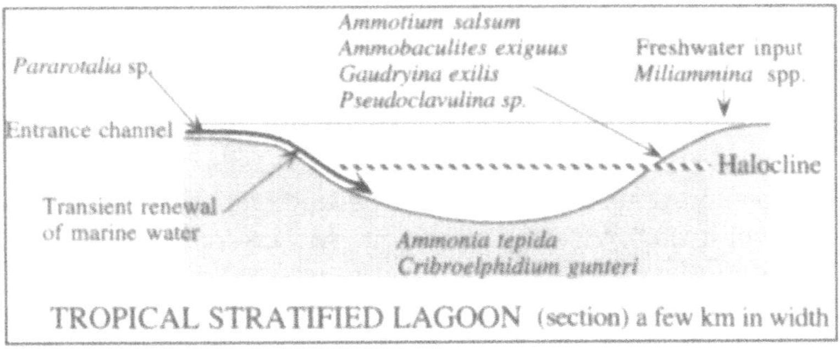

FIGURE 7. Distribution of the dominant species in a tropical stratified lagoon (example: the Lagoa da Conceição—Brazil). Halocline is at the transition between surface water diluted by rain and continental freshwater input and saltier bottom water resulting from the penetration of marine water at high tide during the dry season.

occurs near the Comoé River mouth, where freshwater input is strong. Neither foraminifera nor other microorganisms were found in the deep depressions located near the entrance channel. This absence of microfauna can be related to the anoxic conditions prevailing in these depressions.

5.2.5. Generalization

Foraminiferal distribution in restricted lagoons is similar to foraminiferal distribution in estuaries, owing to the fact that the same marine-to-continental water gradient occurs in both environments. Moderate stratification leads to both horizontal and vertical zonations of foraminifera. In stratified choked lagoons, foraminiferal distribution depends on the stresses resulting from low salinity in the superficial water and from anoxic conditions in the depressions. Even in shallow lagoons, the foraminiferal assemblages are strongly influenced by water stratification. Constant anoxic conditions in the depressions lead to the absence of foraminifers.

In moderately hypersaline oligotrophic homogeneous areas such as the Araruama lagoon, the distribution of the microfauna, with a few dominant species and a great proportion of porcelaneous tests, seems to depend mainly on the nutrient input.

5.3. Mangroves and Salt Marshes

5.3.1. Tropical Mangroves and Salt Marshes

In the external part of the mangroves (*Rhizophora* zone) of the areas studied in New Caledonia and Australia, calcareous species are dominant with a great proportion of *Helenina anderseni* occurring in areas of low salinity.

Quinqueloculina seminula and *Ammonia tepida* are dominant in the *Avicennia* zone. *Jadammina macrescens* and *Trochammina inflata* are dominant in upper marshes, except in small pools where hypersaline water remains at low tide and where *Q. seminula* becomes dominant.

In the areas studied in Brazil, agglutinated species are dominant in well-developed mangroves (*Arenoparrella mexicana*, *Haplophragmoides wilberti*, and *Gaudryina exilis*), whereas calcareous species (mainly *Ammonia tepida* and *Cribroelphidium gunteri*) dominate in recently regenerated mangroves. *Siphotrochammina lobata* is abundant on *Rhizophora* roots (Eichler *et al.*, 1995; Debenay *et al.*, 1996).

5.3.2. Mediterranean Salt Marshes

In the saltworks of Salins de Giraud (south of France), the foraminiferal assemblages are subject to big variations in temperature and salinity. They are also affected by the nature of sedimentary environments (e.g., sand, mud, algal mats). Agglutinated species (mainly *Trochammina inflata* and *Jadammina macrescens*) are favored by muddy sediments with algal mats. On the other hand, in sandy sediments, the calcareous species are dominant (*Ammonia tepida* and "*Nonion depressulum*"; Zaninetti, 1982).

High marsh areas on the northern side of the Gulf of Patras (Greece) are characterized by *Jadammina polystoma*, *Trochammina inflata*, and *Discorinopsis aguayoi* (see Fig. 6). Low marsh and mudflat areas are characterized by a high percentage of *Ammonia "beccarii"* (*A. tepida*?) and *Haynesina depressula* (*Protelphidium*) (Scott *et al.*, 1979).

According to Albani *et al.* (1991), *Ammonia "beccarii"* (*A. tepida*?) constitutes more than 50% of marsh foraminiferal assemblages in the lagoon south of Venice, except in high-stress marginal urban biotopes, where *Nonion pauciloculum* becomes dominant. Petrucci *et al.* (1983) made the distinction between the upper marsh dominated by *Trochammina inflata* associated with *Jadammina macrescens*, *Quinqueloculina seminula*, and *A. "beccarii"*; the middle marsh dominated by *Miliammina fusca* with lesser amounts of *Polysaccammina hyperhalina*, *T. inflata*, *A. "beccarii"* and *Q. seminula*; and the lower marsh dominated by *A. "beccarii."*

5.3.3. Temperate North Atlantic Salt Marshes

On the Atlantic coast of France, the upper intertidal zone is dominated by *Ammotium salsum*, *Scherochorella moniliformis*, and *Cribroelphidium williamsoni*. The assemblages of the supratidal zone, which are rich in organic matter, are composed of *Trochammina inflata* and *Jadammina macrescens*, and sometimes *Arenoparrella mexicana* and *Haplophragmoides wilberti* (see Fig. 5).

In Canada, Scott and his collaborators studied mainly the vertical zonation of the microfauna, with a view to using this zonation for Quaternary sea-level studies. Scott and Medioli (1980), e.g., determined four foraminiferal zones in Nova Scotia: IA (uppermost high marsh) association dominated by *Jadammina macrescens*; IB (high marsh and upper middle marsh) association dominated by *J. macrescens* and *Tiphotrocha comprimata*; IIA (lower middle marsh and upper low marsh) association dominated by *Trochammina inflata* and *Miliammina fusca*; IIB (lower low marsh) association dominated by *M. fusca* and *Ammotium salsum*. The zonation was somewhat different in the tidal systems of Prince Edward Island, where high percentages of *Pseudothurammina limnetes* were found in zone IB and *A. salsum* was absent (Scott *et al.*, 1981).

5.3.4. Generalization

Marsh foraminifers have a widespread distribution (Murray, 1971; Scott, 1976). Their vertical zonation is similar in microtidal and macrotidal paralic environments (Scott *et al.*, 1979) and mainly depends on the frequency and duration of subaerial exposure (e.g., Scott and Medioli, 1980; Jennings and Nelson, 1992; Hayward and Hollis, 1994). However, species living primarily in intertidal conditions such as *Trochammina inflata* and *Jadammina macrescens* can live near extreme high-water levels in the salt marshes (Scott and Medioli, 1980; Patterson, 1990; Jennings and Nelson, 1992; Hayward and Hollis, 1994).

Scott (1977 in Scott *et al.*, 1980) infers that marsh foraminifers are sensitive to small differences in subaerial exposure while they are relatively insensitive to salinity change, as compared with the estuarine forms. This assertion is consistent, for instance, with the distribution of *Ammotium salsum*. This widespread paralic species is a typical marsh species in temperate macrotidal environments but it lives in the subtidal zones of tropical microtidal environments down to 5 m. In both cases, it is subjected to strong variations in salinity.

Species living in temperate marshes such as *Ammotium salsum*, *Arenoparrella mexicana*, and *Haplophragmoides wilberti* occur in subtidal areas of tropical environment. Thus they are not universal indicators of sea level, and the vertical zonation concept proposed by Scott and Medioli (1978, 1980) cannot be used with them. Moreover, this concept was not supported by the observations of De Rijk (1995) in salt marshes.

5.4. Low-Salinity Coastal Zone

The foraminiferal assemblages of Los Archipelago (Guinea) are dominated by *Ammonia tepida*, *A. parkinsoniana*, and *Cribroelphidium gunteri*, characteristic of external paralic environments, and *Quinqueloculina trigonula* and *Nonion* cf. *commune*, usually found on the inner shelf.

5.5. Influence of Selected Parameters

Numerous parameters that are not directly related to the general oceanic-to-continental trend also act on foraminiferal distribution. Among them are the presence or absence of vegetation, the nature of the sediments, the temperature, and the salinity.

1. *The Presence or Absence of Vegetation.* Although Hayward and Hollis (1994) and Wang (1992) consider that vegetation type has little influence on foraminiferal associations, the presence or absence of vegetation is important because less organic production leads to fewer variations in sediment pH, which benefits the calcareous species. On the other hand, some agglutinated species such as *Arenoparrella mexicana* and *Haplophragmoides wilberti* are associated with organic-rich sediment. Other species are directly related to the presence of vegetation, such as *Eggerelloides scabrus*, which may be associated with seagrass in temperate environments (Redois, 1996) and *Siphotrochammina lobata*, attached to mangrove tree roots in tropical environments (Eichler *et al.*, 1995). *Jadammina macrescens* has also been recorded in epiphytic communities (Matera and Lee, 1972).

2. *The Nature of Sediments.* In the saltworks of Salins de Giraud, for instance, foraminiferal distribution seems to be related to sedimentary environments (Zaninetti, 1982). Muddy sediments with algal mats are favorable to agglutinated species (mainly *Trochammina inflata* and *Jadammina macrescens*), whereas the calcareous species (mainly *Ammonia tepida* and "*Nonion depressulum*") are dominant in sandy sediments (Zaninetti, 1982).

In the Araruama lagoon, the abundance of miliolids is probably related to the presence of shell bottom, as assessed by Phleger and Lankford (1957), and as reported by Levy (1970) in some Mediterranean lagoons.

3. *Temperature.* In higher latitudes, the distribution of *Ammonia "beccarii"* (or *A. tepida*) is related to intertidal areas where the summer temperature may be sufficiently high to allow reproduction (Scott *et al.*, 1980). In tropical or Mediterranean environments, this species lives in subtidal environments.

4. *Salinity.* The same euryhaline species are found either in hypersaline or in hyposaline environments, suggesting that osmoregulation is an important parameter and that salinity distribution controls the distribution of at least some of the foraminiferal species. *Ammotium salsum* is one of the most adaptable species. In the upper reaches of the Casamance estuary, e.g., it can resist drastic salinity changes from 50‰ to 100‰ between the rainy and the dry season (Debenay and Pagès, 1987).

6. Discussion

Paralic environments are transitional between oceanic and continental waters. A few highly adaptable and widespread foraminiferal species occupy most of these environments (review in Murray, 1973, 1991). These "paralic species" occur in hyposaline environments as well as in hypersaline ones. They are salinity-tolerant and have been used as salinity indicators in the standard estuarine classifications based on the spatial distribution of salinity (e.g., Nichols, 1974).

The distribution of foraminiferal assemblages in paralic environments under different climatic conditions and with different tidal ranges shows a general marine-to-restricted gradient with high diversity near the mouth, decreasing landward. Some species indicate a strong marine influence, e.g., *Pararotalia* spp. and *Bolivina* spp. in tropical areas, and *Elphidium pulvereum*, *Haynesina depressula*, and *Ammonia beccarii* on the Atlantic coast of France. Other species are characteristic of the outer paralic system (*Ammonia tepida*, *Cribroelphidium excavatum*, *Cribroelphidium gunteri*, *Haynesina germanica*), and others of the inner paralic system (*Ammobaculites exiguus*, *Arenoparrella mexicana*, *Gaudryina exilis*, *Haplophragmoides wilberti*, *Miliammina* spp., *Trochammina* spp., *Siphotrochammina lobata*, and *Ammotium salsum*, which is one of the most adaptable species). However, there are some differences in this general distribution in the various environments (Fig. 8).

In some choked lagoons, particularly in the Mediterranean, and in inverse estuaries located in arid areas where the residence time of the water is long enough for it to take on special characteristics, calcareous species dominate in most cases, but agglutinated species are found in peripheral marshes or in zones of freshwater input where organic matter is abundant and pH is low (Scott *et al.*, 1979; Zaninetti, 1982; Petrucci *et al.*, 1983; Albani *et al.*, 1991). This indicates that elevation (aquatic-to-land transition) and/or freshwater input (seawater-to-freshwater transition) also act on foraminiferal distribution in these environments.

Tidal cycles generate a great diversity of stresses in meso- to macrotidal temperate environments. These stresses disturb the general trends and cause different assemblages to occur. As a result, the composition of benthic foraminiferal assemblages reflects the complex interaction between biotic and abiotic parameters and their multiple changes in space and time.

Nevertheless, foraminiferal dead assemblages give an average picture of the ecologic conditions, despite the great annual or even interannual variability. Thus, they allow a synecological approach to the environment. The bias introduced by postmortem transport and dissolution (e.g., Murray, 1984, 1986; Alve and Murray, 1994) may have been overestimated because it was generally determined on the basis of a comparison between living and dead assemblages at a given time, and did not take into account the great seasonal variability of the living assemblages (Redois, 1996).

Most of the studies dealing with pollutant impact on biological communities used numerical characterizations such as the diversity index and the dominance index. These biocenotic indexes are often used in foraminiferal studies (Murray, 1973). They allow the detection of changes in the assemblages, and it is generally accepted that they can be used as an early warning. However, they do not take into account the role of the species in the community or their potential as bioindicators. On the contrary, methods based

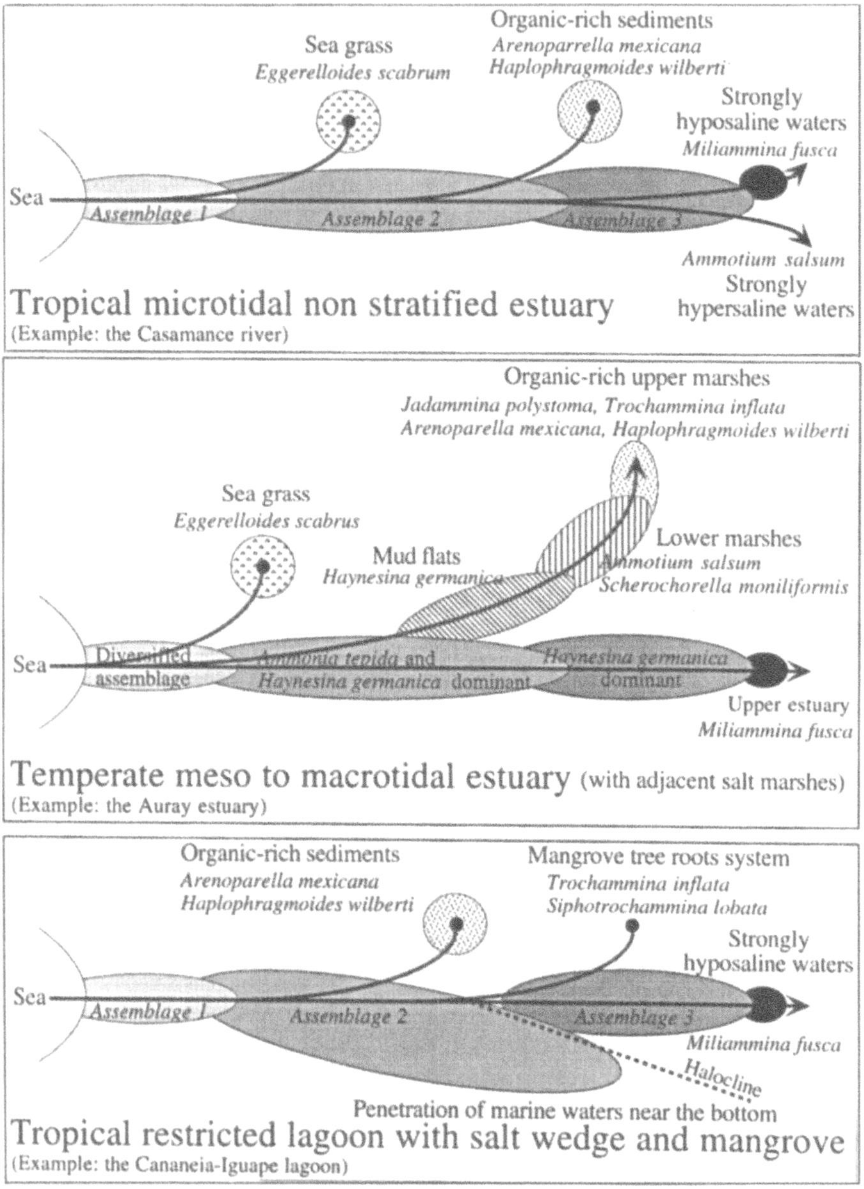

FIGURE 8. Schematic diagram showing the general distribution of foraminiferal assemblages in paralic environments under different climatic conditions and tidal ranges. This distribution shows a general marine-to-restricted gradient but some differences exist between the environments according to their various characteristics.

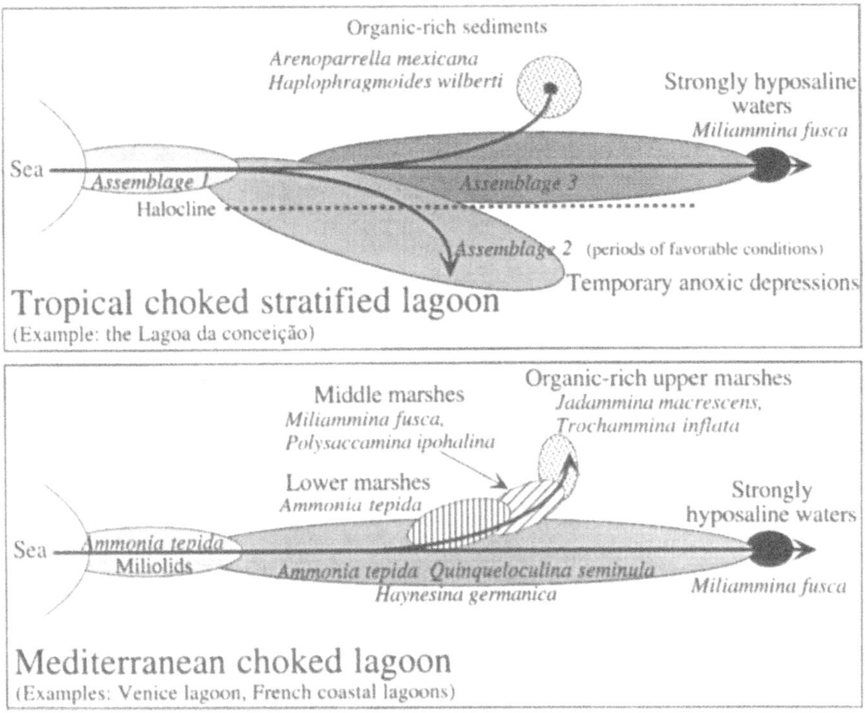

FIGURE 8. Continued

on the presence or absence of bioindicator species give information of greater ecological value (Caquet and Lagadic, 1998). For example, the distribution index "Ic" calculated on the basis of three selected foraminiferal assemblages of well-known ecological significance in tropical environments allows the description of the progressive changes in ecological conditions and may indicate impacted areas by anomalous changes (Debenay, 1990; Debenay *et al.*, 1993*b*).

7. Conclusion

Despite the great diversity of ecological parameters acting in paralic environments, this work shows that foraminiferal assemblages are distributed along several major gradients, such as the horizontal seawater-to-freshwater transition and the vertical water-to-land transition, in relation to natural parameters and to various stresses. Knowledge of these distribution patterns will allow the detection of impacted areas and help to distinguish between responses due to changes in natural variables and those due to anthropogenic stress.

Because foraminifers have a short life cycle, they react quickly and can be used as an early warning indicator (Kramer and Botterweg, 1991). They should be employed as part of an integrated program of pollution monitoring, including chemical analysis of the contaminants. Their uses should include routine long-term surveillance programs, hazard assessment at specific discharge sites, and monitoring of the effectiveness of remedial actions.

However, given what we know today, more field studies and laboratory experiments are necessary before foraminifers can be used alone to make reliable distinctions between natural and anthropogenic environmental stresses or determine the effects of pollutants.

ACKNOWLEDGMENTS: This work was possible thanks to the collaboration of the Centers for Oceanographic Research (CRO) of Abidjan (the Ivory coast) and of Dakar-Thiaroye (Senegal), of the Institut Pédagogique National of Nouakchot (Mauritania), of the EPEEC (Dakar) and of the Geological Department of the University of Dakar for research in Africa. The Brazilian investigations were possible thanks to the CAPES/COFECUB program 178/95 and to the collaboration of the Instituto Oceanográfico of São Paulo. Samples from Vietnam were provided by J. Pagès and Bui Thi Luan and some samples from the Mediterranean were provided by O. Guélorget. We are particularly grateful to R. Arfi, J. Pagès, D. Carité, M. Ba, A. Ly, C. Moreau, M. N. de Cazamajor, B. Robineau, I. Niang-Diop, B. B. Eichler, W. Duleba, and C. Bonetti for providing samples and help in the field.

References

Albani, A., Favero, V. M. and Barbero, R. S., 1991, The distribution and ecological significance of Recent foraminifera in the lagoon south of Venice (Italy), *Rev. Españ. Micropaleontol.* **23**(2):29–45.

Alve, E., 1991, Benthic foraminifera in sediment cores reflecting heavy metal pollution in Sorfjord, Western Norway, *J. Foram. Res.* **21**:1–19.

Alve, E., 1995, Benthic foraminifera response to estuarine pollution: A review, *J. Foram. Res.* **25**:190–203.

Alve, E., and Murray, J. W., 1994, Ecology and taphonomy of benthic foraminifera in temperate mesotidal inlet, *J. Foram. Res.* **24**:18–27.

Bamber, R. N., Batten, S. D., Sheader, M., and Bridgwater, N. D., 1992, On the ecology of brackish water lagoons in Great Britain, *Aquat. Conserv.* **2**:65–94.

Bandy, O. L., Ingle, J. C., and Resig, J. M., 1964, Foraminifera:Los Angeles County outfall area, California, *Limnol. Oceanogr.* **9**:124–137.

Bandy, O. L., Ingle, J. C., and Resig, J. M., 1965, Modifications of foraminiferal distributions by the Orange County outfall, California, *Mar. Technol. Soc., Trans.* **1965**:54–76.

Barnes, R. S. K., 1989, The coastal lagoons of Britain: An overview and conservation appraisal, *Biol. Conserv.* **49**:295–313.

Barnes, R. S. K., 1994, A critical appraisal of the application of Guélorget and Perthuisot's concepts of the paralic ecosystem and confinement to macrotidal Europe, *Est. Coast. Shelf Sci.* **38**:41–48.

Blandin, P., 1986, Bioindicateurs et diagnostic des systèmes écologiques, *Bull. d'Ecol.* **17**:215–307.

Bolstovskoy, E., and Wright, R., 1976, *Recent Foraminifera*, W. Junk, The Hague.

Bui Thi Luan, Debenay, J.-P., and Pagès, J., 1994, Holocene foraminifera in the Camau Peninsula (Southern Viet Nam), *Hydrobiol. Trop.* **27**:23–31.

Cairns, J. Jr, and McCormick, P. V., 1992, Developing an ecosystem based capability for ecological risk assessments, *Environ. Prof.* **14**:186–196.

Caquet, T., and Lagadic, L., 1998, Consequences d'atteintes individuelles précoces sur la dynamique des populations et de la structuration des communautés et des écosystèmes, in: *Utilisation des biomarqueurs pour la surveillance de la qualité de l'environnement* (L. Lagadic, T. Caquet, J.-E. Amiard, and F. Ramade, eds.) Lavoisier Tec Doc, Paris, pp. 265–298.

Cazamajor, M.N. de, and Debenay, J.-P., 1995, Les foraminifères, bio-indicateurs des environnements paraliques: réaction à divers types de pollution dans l'estuaire de l'Adour, in: *Biological Markers of Pollution.* ANPP, Annales Volume, Chinon-France (September 1995), pp. 371–378.

Colleuil, B., 1984, Un modèle d'environnement lagunaire soumis aux conditions du climat équatorial tempéré: le lac Nokoué (République populaire du Bénin), Thesis, University of Bordeaux, France, p. 135.

Debenay, J.-P., 1990, Recent foraminiferal assemblages and their distribution related to environmental stress in the paralic environments of West Africa (Cape Timiris to Ebrie Lagoon), *J. Foram. Res.* **20**:267–282.

Debenay, J.-P., and Pagès, J., 1987, Les peuplements de Foraminifères et de Thécamoebiens de l'estuaire hyperhalin du fleuve Casamance (Sénégal). Leur évolution de février 1984 à novembre 1985, *Rev.Hydrobiol. trop.* **20**:231–254.

Debenay, J.-P., and Moguedet, G., 1990, Les lagunes côtières et l'estuaire de la Songolo, au Congo et l'estuaire du fleuve Congo-Zaire : présentation et description des peuplements de foraminifères benthiques, *Rev. Paléobiol.* **9**:15–26.

Debenay, J.-P., Ba, M., Ly, A., and Sy, I., 1987*a*, Les écosystèmes paraliques du Sénégal. Description, répartition des peuplements de Foraminifères benthiques, *Rev. Paléobiol.* **6**:229—255.

Debenay, J.-P., Arfi, R., and Konate, S., 1987*b*, Foraminifères récents des milieux paraliques des côtes d'Afrique de l'Ouest, *Géol. Méditer.* **14**:5–13.

Debenay, J.-P., Colleuil, B., and Texier, H., 1993*a*, Peuplements de foraminifères du lac Nokoué (Bénin) avant fermeture de la lagune. Comparaison avec d'autres environnements paraliques d'Afrique de l'Ouest, *Rev. Micropaléont.* **36**:225–234.

Debenay, J.-P., Perthuisot, J.-P., and Colleuil, B., 1993*b*, Expression numérique du confinement par les peuplements de foraminifères. Applications aux domaines paraliques d'Afrique de l'Ouest, *C. R. Acad. Sci. (Paris), t. 316, série II* **12**:1823–1830.

Debenay, J.-P., Pagès, J., and Guillou, J.-J., 1994, Transformation of a subtropical river into a hyperhaline estuary: The Casamance River (Sénégal). Paleogeographical implications, *Paleoecol. Paleoclim. Paleogeogr.* **107**:103–119.

Debenay, J.-P., Eichler, B. Beck, Bonetti, C., and Coelho, C., 1996, Foraminifers, biomarkers in the mangrove swamps of Bertioga (São Paulo, Brasil), in: 3° reunião especial da SBPC *Ecosystema costeiros: do conhocimento a gestão.* Anais, Florianopolis, SC, Brasil (May 1996): 448–449.

Debenay, J.-P., Eichler, B. Beck, Guillou, J.-J., Eichler-Coelho, P., Coelho, C., and Porto-Filho, E., 1997, Comportement des peuplements de foraminifères et comparaison avec l'avifaune dans une lagune fortement stratifieé : la Lagoa da conceição (S.C., Brésil), *Rev. Paléobiol.* **16**:55–75.

Debenay, J.-P., Eichler, B. Beck, Duleba, W., Bonetti, C., and Eichler-Coelho, P., 1998, Stratification in coastal lagoons: Its influence on foraminiferal assemblages in two Brazilian lagoons, *Mar. Micropaleontol.* **35**:65–89.

De Rijk, S., 1995, Salinity control on the distribution of salt marsh foraminifera (Great Marshes, Massachussetts), *J. Foram. Res.* **25**:156–166.

Durand, J.-R., and Guiral, D., 1994, Hydroclimat et hydrochimie, in: *Environnements et ressources aquatiques en Côte d'Ivoire, II milieux lagunaires* (J.-R. Durand, P. Dufour, D. Guiral, and S. G. F. Zabi, eds.) pp. 59–90, ORSTOM Éditions, Paris.

Eichler, B. Beck, Debenay, J.-P., Bonetti, C., and Duleba, W., 1995, Répartition des foraminifères benthiques dans la zone Sud-Ouest du système laguno-estuarien d'Iguape-Cananeia (Brésil), *Boll. Inst. Oceanográf. USP*, São Paulo **43**:1–17.

Favry, A., Guélorget, O, Debenay, J.-P., and Perthuisot, J.-P., 1998, Distribution des peuplements de foraminifères actuels dans une lagune mediterranéenne, l'Étang du Prévost, *Vie et Milieu Banyuls-sur-mer* **48**:41–53.

Guélorget, O., Lefebvre, A., Martin, J. L., Favry, A., Debenay, J.-P, Vo Duy, S., Supangat, A., Nguyen Tac, A., and Perthuisot, J.-P., 1997. Organisation hydrologique et biologique de la région estuarienne de Can Gio, Delta du Mekong, Vietnam, in: *Proc. Symp. Franco-Vietnamien sur l'aquaculture en Zone Tropicale*, Na Trang (December 1995), pp. 55–85.

Hayward, B. W., and Hollis, C. J., 1994, Brackish foraminifera in New Zealand: A taxonomic and ecologic review. *Micropaleontology* **40**:185–222.

Jennings, A. E., and Nelson, A. R., 1992, Foraminiferal assemblage zones in Oregon tidal marshes: Relation to marsh floral zones and sea level, *J. Foram. Res.* **22**:13–29.

Kjerfve, B., 1986, Comparative oceanography of coastal lagoons, in: *Estuarine Variability* (D. A. Wolfe, ed.), Academic Press, New York, pp. 63–81.

Kramer, K. J. M., and Botterweg, J., 1991, Aquatic biological early warning system: An overview, in: *Bioindicators and Environmental Management* (D. W. Jeffrey and B. Mudden, eds.), Academic Press, London, pp. 95–126.

Levy, A., 1970, Contribution à l'étude des milieux margino-littoraux: influence des facteurs hydrologiques et sédimentologiques sur les peuplements microfaunistiques (foraminifères). Golfe du Lion. Thèse doctorat d'état Sciences, Paris. 2 vols., p. 396.

Loeblich, A. R. Jr., and Tappan, H., 1988, *Foraminiferal Genera and Their Classification*, 2 vols. Van Nostrand Reinhold, New York.

Matera, N. J., and Lee, J. J., 1972, Environmental factors affecting the standing crop of foraminifera in sublittoral and psammolittoral communities of a Long Island salt marsh, *Mar. Biol.* **14**:89–103.

McCarty, J. F., and Shugart, L. R. (eds.), 1990, *Biomarkers of Environmental Contamination*, Lewis, Boca Raton.

Miyao, S. Y. Nshihara, L., and Sarti, C. C., 1986, Caractersticas Físicas e químicas do sistema estuarino-lagunar de Cananéia-Iguape, *Boll. Inst. Oceanográf. USP*, São Paulo **34**:23–36.

Murray, J. W., 1968, Living foraminifers of lagoons and estuaries, *Micropaleontology* **14**:435–455.

Murray, J. W., 1971, Living foraminiferids of tidal marshes: A review, *J. Foram. Res.* **1**:153–161.

Murray, J. W., 1973, *Distribution and Ecology of Living Benthic Foraminiferids*, Heinemann, London.

Murray, J. W., 1984, Benthic foraminifera: Some relationships between ecological observations and paleoecological interpretations, in: *Benthos 83, 2nd Int. Symp. Benthic Foraminifera* (H. Oertli, ed.), April 1983, pp. 465–469.

Murray, J. W., 1986, Living and dead Holocene foraminifera of Lyme Bay, southern England, *J. Foram. Res.* **16**:347–352.

Murray, J. W., 1991, *Ecology and Palaeoecology of Benthic Foraminifera*, Longman, Harlow.

Nichols, M. M., 1974, Foraminifera in estuarine classification, in: *Coastal Ecological Systems of the United States* (H. T. Odum, B. J. Copeland, and E. A. McMahon, eds.) Conservation Foundation, Washington D.C., pp. 85–103.

Ozarko, D. L., Patterson, R. T., and Williams, H. F. L., 1997, Marsh foraminifera from Nanaimo British Columbia (Canada): Implications of infaunal habitat and taphonomic biasing, *J. Foram. Res.* **27**:51–68.

Pagès J., and Debenay J.-P., 1987, Evolution saisonnière de la salinité de la Casamance (Sénégal). Description et essai de modélisation, *Rev. Hydrobiol. Trop.* **20** :203–217.

Patterson, R. T., 1990, Intertidal benthic foraminiferal biofacies on the Fraser River Delta, British Columbia: Modern distribution and paleoecological importance, *Micropaleontology* **36**:229–244.

Petrucci, F., Medioli, F. S., Scott, D. B., Pianetti, F. A., and Cavazzini, R., 1983, Evaluation of the usefulness of foraminifera as sea level indicators in the Venise lagoon (N. Italy), *Acta Nat. Anteneo Parm.* **19**:63–77.

Phleger, F. B., and Lankford, R. R., 1957, Seasonal occurrences of living benthonic foraminifera in some Texas Bays, *Contr. Cushman Found. Foram. Res.* **8**:93–105.

Redois, F., 1996, Les Foraminifères benthiques actuels, bioindicateurs du milieu marin. Exemples du plateau continental sénégalais et de l'estran du golfe du Morbihan (France). PhD Thesis, University of Angers, France, 2 vol, p. 401.

Redois, F., and Debenay, J.-P., 1996, Influence du confinement sur la répartition des foraminifères benthiques. Exemple de l'estran d'une ria mésotidale de Bretagne méridionale, *Rev. Paléobiol. (Genève)* **15**:243–260.

Schafer, C. T., 1973, Distribution of foraminifera near pollution sources in Chaleur Bay, *Water Air Soil Pollut.* **2**:219–233.

Scott, D. B., 1976, Brackish-water foraminifera from southern California and description of *Polysaccammina ipohalina* n. gen., nsp, *J. Foram. Res.* **6**:312–321.

Scott, D. B., and Medioli, F. S., 1978, Vertical zonations of marsh foraminifera as accurate indicators of former sea levels, *Nature* **272**:528–531.

Scott, D. B., and Medioli, F. S., 1980, Quantitative studies of marsh foraminiferal distributions in Nova Scotia: Implications for sea level studies, *Cushman Found. Foram. Res. Spec. Publ.* **17**:58 pp.

Scott, D. B., Piper, D. J. W., and Panagos, A. G., 1979, Recent salt marsh and intertidal mudflat foraminifera from the western coast of Greece, *Riv. Ital. Paleontol.* **85**:243–266.

Scott, D. B., Schafer, C. T., and Medioli, F. S., 1980, Eastern Canadian estuarine foraminifera: a framework for comparison, *J. Foram. Res.* **10**:205–234.

Scott, D. B., Williamson, M. A., and Duffett, T. E., 1981, Marsh foraminifera of Prince Edward Island: Their recent distribution and application for former sea level studies, *Marit. Sed. Atlan. Geology* **17**:98–129.

Shear, H., 1996, The development and use of indicators to assess the state of ecosystem health in Great Lakes, *Ecosys. Health* **2**:241–258.

Tastet, J.-P., 1974, L'environnement physique du système lagunaire Ébrié, *Univ. Abidjan, Ser. Doc. Dep. Sc. de la Terre* **11**:1–28.

Uschakov, P. V., 1970, Observation sur la faune benthique du littoral guinéen, *Cah. Biol. Mar.* **11**:435–457.

Walker, C. H., and Livingstone, D. R. (eds.), 1992, *Persistent Pollutants in Marine Ecosystems, SETAC Spec. Publ.*, Pergamon Press, Oxford, 272 pp.

Wang, P., 1992, Distribution of foraminifera in estuarine deposits: A comparison between Asia, Europe and Australia, in: *Centenary of Japanese Micropaleontology* (K. Ishizaki and T. Saito, eds.), Terra Scientific Publishing Co, Tokyo, pp. 71–83.

Watkins, J. G., 1961, Foraminiferal ecology around the Orange County, California, ocean sewer outfall, *Micropaleontology* **7**:199–206.

Wilson, R. C. H., Harding, L. E., and Hirvonen, H., 1995, Marine ecosystem monitoring network design, *Ecosys. Health* **1**:222–227.

Yanko, V., Kronfeld, A., and Flexer, A., 1994, The response of benthic foraminifera to various pollution sources: implications for pollution monitoring, *J. Foram. Res.* **24**:1–17.

Yanko, V., Arnold, A., and Parker, W., 1999. The effect of marine pollution on benthic foraminifera, in: *Modern Foraminifera* (Barun Sent Gupta, ed.), Kluwer Academic Publishers, London, pp. 217–235.

Yassini, I., and Jones, B., 1989, Estuarine foraminiferal communities in Lake Illawarra, New South Wales, *Proc. Linn. Soc. N.S.W.* **110**:229–266.

Yassini, I., and Jones, B., 1995, *Foraminiferida and Ostracoda from Estuarine and Shelf Environments on the Southeastern Coast of Australia*, University of Wollongong Press, Wollongong, 484 pp.

Yazvenko, S. B., and Rapport, D. J., 1996, A framework for assessing forest ecosystem health, *Ecosys. Health* **2**:40–51.

Zaninetti, L., 1982, Les foraminifères des marais salants de Salins de Giraud (Sud de la France): Milieu de Vie et transport dans le salin, comparaison avec les microfaunes marines, *Géol. Méditerr.* **9**:447–470.

II

Water Quality in Modern Marine, Marginal Marine, and Freshwater Environments

Chapter 3

Benthic Foraminifera as Bioindicators of Heavy Metal Pollution

A Case Study from the Goro Lagoon (Italy)

RODOLFO COCCIONI

1. Introduction

Foraminifera are extensively used in different fields of earth and environmental sciences by virtue of several factors such as (1) a hard exoskeleton, which records fundamental environmental changes and evolutionary processes of earth history, (2) small size and, consequently, high abundance in small samples, (3) wide distribution over all marine environments, (4) high taxonomic diversity, and (5) very short reproductive cycles (month to year), which make them excellent recorders of environmental changes covering a short time span.

Over the last four decades many studies (e.g., Watkins, 1961; Lidz, 1965; Wright, 1968; Boltovskoy and Boltovskoy, 1968) on benthic foraminiferal

RODOLFO COCCIONI • Istituto di Geologia dell' Università, Campus Scientifico, Località Crocicchia, I-61029 Urbino, Italy.

Environmental Micropaleontology, Volume 15 of *Topics in Geobiology*, edited by Ronald E. Martin. Kluwer Academic/Plenum Publishers, New York, 2000.

assemblages have been carried out from different parts of the world in areas exposed to different kinds of marine pollution (see also Alve, 1995 for a review). Through these studies a considerable effort has been made to develop new methodologies for biological monitoring of different contaminants. In addition, because of the increased knowledge of the biology of foraminifera (see Lee and Anderson, 1991) these studies have shown that benthic foraminifera have great potential as indicators of pollution, providing one of the most sensitive and inexpensive markers of environmental stress in both naturally and anthropogenically stressed aquatic environments.

More recently, several studies (Schafer *et al.*, 1980; Naidu *et al.*, 1985; Ellison *et al.*, 1986; Alve, 1991; Sharifi *et al.*, 1991; Banerji, 1992; Stubbles, 1993; Yanko *et al.*, 1994, 1998; Ashraf, 1997; see also in this volume Patterson and Kumar, Ch. 11) from different environmental settings have focused on benthic foraminiferal response to heavy metal pollution, which has markedly increased over the last few decades and caused deleterious effects on ecosystems. These investigations have documented the fact that such contamination, which might cause pathological processes in the foraminiferal cell, plays an important role in the development of frequent to common occurrence of abnormal (teratological) tests as well as changes in foraminiferal abundance and taxonomic composition, size variation, and structural modification. Accordingly, most of the authors cited above have suggested using the presence of deformed tests of benthic foraminifera as a powerful *in situ* bioindicator of heavy metal pollution; test deformation also provides information about which pollutants have the most deleterious effects on benthic foraminifera and which taxa are mostly affected.

Apart from anthropogenic influences, however, high proportions of abnormal tests may also result from natural environmental stresses resulting either from extreme environmental conditions (e.g., Zaninetti, 1982; Zampi and D'Onofrio, 1984, 1987; Almogi-Labin *et al.*, 1992) or from very rapid changes in ecological parameters (e.g., Seiglie, 1964; Closs and Madeira, 1968; Caralp, 1989). Based on a substantial literature review, Boltovskoy *et al.* (1991) wrote that morphological test variations in benthic foraminifera might be related to several environmental parameters (i.e., temperature, salinity, carbonate solubility, depth, nutrition, substrate, dissolved oxygen, illumination, pollution, water motion, trace elements, and rapid environmental fluctuation) and suggested that normally a combination of these factors is what really affects the test morphologies.

To date, then, it has been difficult to separate anthropogenic from natural sources of stress and to isolate any single specific cause. Moreover, it has been difficult to separate the effects of chemical pollution from those of trace elements because many trace elements can be present in the chemical pollutants. Although certain parameters are measured much more frequently than others, their importance has probably been overemphasized. It is necessary, therefore, before using benthic foraminifera as biomarkers for heavy metal pollution, to establish the real role of heavy metals in controlling their distribution, growth, and test abnormalities.

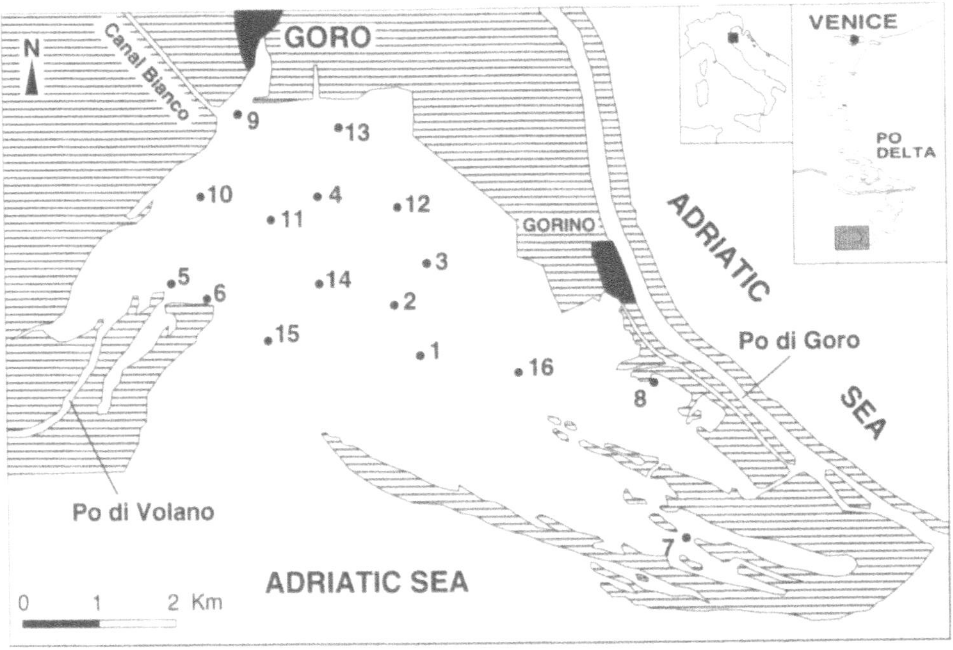

FIGURE 1. Map of Goro Lagoon showing location of sampling stations.

Consequently, benthic foraminiferal distribution and trace element concentration were investigated in surface sediments of several stations from the Goro Lagoon (northern Italy) (Fig. 1), an unexplored study area for foraminiferal analysis where only preliminary studies (Coccioni *et al.*, 1997; Gentiloni Silverj, 1997) have so far been carried out. The study area is particularly suitable for studying the response of benthic foraminifera to trace metal pollution, as it: (1) is subject to strong levels of heavy metal pollution, (2) displays a high proportion of abnormal tests, and (3) was the subject of a multiyear (1988–1991) program that monitored its main physical, chemical, and biological parameters (see Bencivelli and Castaldi, 1991; Bencivelli *et al.*, 1994).

2. Environmental Setting

The Goro Lagoon, with a total area of 26 km^2 and an average depth of 1.5 m, is the southernmost bay in the Po River delta (Fig. 1). The Goro Lagoon is the most important source of income for the local communities, since many of the inhabitants of this area trade locally produced goods such as mussels, clams, edible bivalves, and fry from fish hatcheries. The increased frequency and intensity of environmental pollution over the last few years has caused great damage both to the fisheries and to the lagoon ecosystem, with serious economic consequences (see Bencivelli and Castaldi, 1991; Bencivelli *et al.*, 1994; Vollenweider *et al.*, 1992).

The Goro Lagoon was formed during the last two centuries by the continuous eastward progradation of the southern channel of the Po River delta, the Po di Goro. The river sediments reworked by tides, waves, and longshore currents built up the bar that closed the bay on its southern side. At the beginning of the 20th century, the lagoon began to assume a more defined shape and reached its present topography in the 1960s (Ciabatti, 1967). The inlet of Goro (Sacca di Goro) has been preserved in recent years by several interventions of bar reinforcement and embankment construction, which blocked its natural evolution.

Sediments are predominantly of a muddy type, except for the area in the vicinity of the lagoon's mouth, where predominantly sandy sediments are present (Dal Cin and Pambianchi, 1991). Physical and chemical parameters throughout the lagoon vary greatly owing to the input of freshwater from the Po di Volano and the exchange of water with the Adriatic Sea. Over the years 1988–1991 (see Colombo *et al.*, 1991, 1994) the salinity and temperature ranges were 6.5‰–32.2‰ (23.5‰ on average) and 5°–27.1°C (16.6°C on average) in surface water and 11.7‰–31‰ (26.9‰ on average) and 4.2°–27.5°C (16.6°C on average) in bottom water, respectively. pH varied from 7.3 to 9.1 in surface waters and 7.4 to 9 in bottom waters. Oxygen concentrations in the bottom water were generally above critical levels ranging from 3.7 to 12.4 mg/liter. Average values of both surface and bottom water temperatures did not change throughout the lagoon. Lowered values of both surface and bottom water salinity characterized the eastern side (21.7‰ and 24.2‰ on average, respectively) and western side (20.7‰ and 26.1‰ on average, respectively) sides of the lagoon. The highest values of these parameters occurred toward the mouth of the lagoon (26.0‰ and 28.4‰ on average, respectively). According to these salinities, the Goro Lagoon can be considered mostly as a polyhaline water body.

The Goro Lagoon can currently be considered as an eutrophic, moderately to strongly polluted body of water receiving solid and liquid substances from the hinterland through the Po di Volano (Fig. 1), which drains heavily industrialized and crop-cultivated areas, with a total surface of about 3000 km^2 (Fagioli *et al.*, 1991, 1994; Barbanti *et al.*, 1992; Pugnetti *et al.*, 1992; Rinaldi *et al.*, 1992; Pambianchi *et al.*, 1994). Severe hypoxic and anoxic conditions have occurred in recent years throughout the lagoon: these events are restricted to the summer when the presence of large quantities of organic matter coincides with high temperatures and static hydrodynamic conditions.

3. Material and Methods

This study is based on 47 surface grab sediment time-averaged samples collected within the Goro Lagoon at 16 selected stations (Fig. 1 and Table 1). These samples are time-averaged and therefore "noisy" environmental fluctuations have presumably been damped. At each station two sets of samples (30 g each of wet sediment) were collected three different times over a period of

TABLE 1. Coordinates of Sampling Stations

Station	UTM coordinates
1	0288077, 4965804
2	0287807, 4966540
3	0288123, 4967016
4	0286848, 4967864
5	0285170, 4966480
6	0284866, 4966248
7	0291319, 4963352
8	0290938, 4965461
9	0285697, 4968975
10	0285367, 4967847
11	0286198, 4967555
12	0287765, 4967769
13	0287064, 4968601
14	0286875, 4966790
15	0286018, 4966098
16	0289165, 4965517

6 months to obtain information on a reasonable time interval of maximum values of concentrations and their variations. These sets of samples have been used for foraminiferal, trace element, and organic carbon analyses, respectively. Sampling periods were (a) spring 1997 (mid-April), (b) summer 1997 (mid-July), and (c) autumn 1997 (mid-October). Owing to unfavorable weather conditions, one station was not sampled during the spring cruise.

Following the method described in Walton (1952), one set of samples was treated with buffered Rose Bengal solution immediately after being brought on board ship to distinguish living (i.e., stained) specimens. Despite its well-known limits (see Boltovskoy and Wright, 1976; Walker *et al.*, 1974; Gooday, 1986; Bernhard, 1988; Jorissen *et al.*, 1995), the Rose Bengal method remains the most practical way to quantify living foraminiferal fauna and, according to Lutze and Altenbach (1991), if carefully employed, it leads to 96% correct identifications. In this work, the method proposed by De Stigter *et al.* (1998) was followed to distinguish living (i.e., well-stained) specimens.

All the samples from this set were gently washed through 63- and 125-μm sieves to remove any excess stain and then dried. At least 300 specimens were separated from the fraction >125 μm by hand-picking and counted for faunal analysis. Living and nonliving, normal specimens and deformed specimens were distinguished.

Examples of morphological abnormalities are illustrated by SEM microphotographs in Figs. 2 and 3. All the materials used in this study were deposited in the collection at the Geological Institute, University of Urbino.

In evaluating the correlation between the percentage of morphological abnormalities and heavy metal concentrations, only living specimens were taken into account. In order to document changes in the assemblages and to determine a possible relation with heavy metal pollution, the faunal density

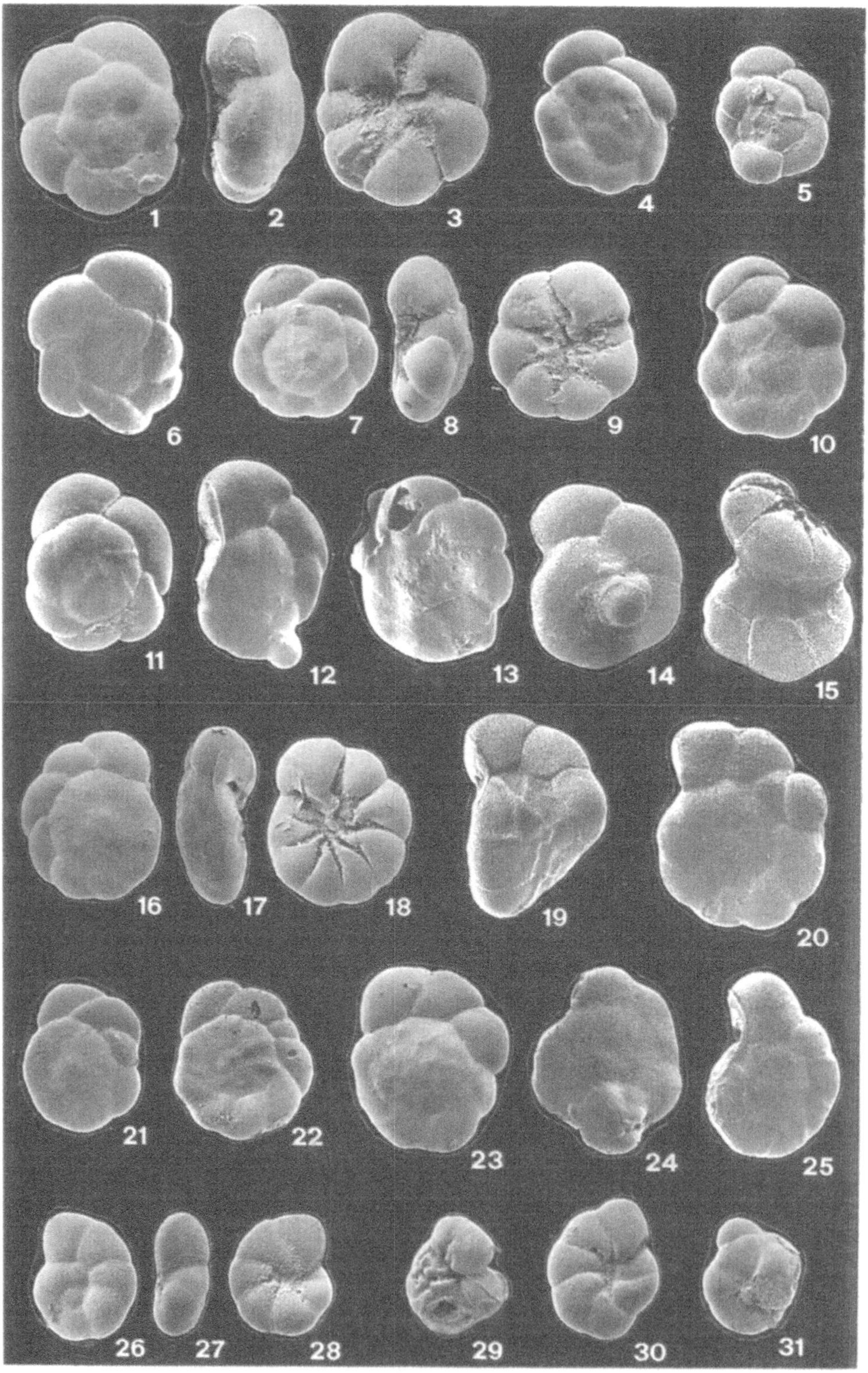
1
2
3
4
5
6
7
8
9
10
11
12
13
14
15
16
17
18
19
20
21
22
23
24
25
26
27
28
29
30
31

from the fraction >125 μm was calculated as the number of living individuals per gram of dry sediment.

The classification of Loeblich and Tappan (1988) was adopted for identification at the supraspecific level. Also, following Poag (1978) and Jorissen (1988), some species (i.e., *Ammonia beccarii, A. parkinsoniana, A. tepida, Cribroelphidium decipiens,* and *Haynesina lidoense)* were designated as formae and three morphotypes were assigned to *A. parkinsoniana* forma *tepida* (see the appendix for the faunal list).

Standard laboratory methods (Franzini *et al.*, 1972, 1975; Leoni and Saitta, 1976; Leoni *et al.*, 1982) were used to determine the concentrations (in ppm) of nineteen trace metals (Sc,V, Cr, Co, Ni, Cu, Zn, Nb, Pb, Hg, Ga, Sr, Ba, Y, Zr, Rb, La, Ce, Th) and three other pollutants (S, As, Br) in selected samples from spring and autumn using a Philips PW 1480 spectrometer. All these trace elements, which include nonmetals, heavy metals, transition and rare-earth metals, and one actinide metal were used to evaluate a possible relation between trace element concentration and the number and type of deformed living specimens. However, in order to compare results with those of previous studies and also to show new findings, only nine trace elements (V, Cr, Co, Ni, Cu, Zn, Hg, Pb, Th) were considered. Total organic carbon (TOC) content in weight (wt.) % was also determined from the selected samples by the Leco combustion method (Leco Industrial Furnace).

4. Results and Discussion

4.1. Foraminiferal Analysis

4.1.1. Structure and Composition of Foraminiferal Assemblages

A total of 97 species (including 5 formae and 3 morphotypes), 95 calcareous and 2 agglutinated, representing 50 genera were identified. The

←

FIGURE 2. SEM photomicrographs showing examples of normal (nondeformed) specimens and morphological abnormalities in benthic foraminifera found in Goro Lagoon (reduced 45% for reproduction). **1–6** *Ammonia parkinsoniana* (d'Orbigny) forma *tepida* (Cushman) morphotype 1, normal specimen (1 spiral view, 2 lateral view, 3 umbilical view, ×75) and individuals showing distorted chamber arrangement (4, ×50 and 6, ×135) and reduced size chamber (5, ×75). **7–15** *Ammonia parkinsoniana* (d'Orbigny) forma *tepida* morphotype 2, normal specimen (7 spiral view, 8 lateral view, 9 umbilical view, ×50) and specimens exhibiting aberrant chamber shape with abnormal additional chamber (10, ×75; 12, ×82.5; 13, ×120), distorted chamber arrangement (11, ×120), abnormally protruding chamber (14, ×290), and Siamese twins (15, ×75). **16–25** *Ammonia parkinsoniana* (d'Orbigny) forma *tepida* morphotype 3, normal specimen (16 spiral view, 17 lateral view, 18 umbilical view, ×50) and individuals showing change in coiling direction (19, ×75), reduced size chamber (20, ×80; 21, ×50; 22, ×50), nondeveloped test (23, ×50), abnormal additional chamber (24, ×75), and aberrant chamber shape (25, ×110). **26–31** *Ammonia perlucida* (Heron-Allen and Earland), normal specimen (26 spiral view, 27 lateral view, 28 umbilical view, ×50) and individuals displaying Siamese twins (29, ×50), nondeveloped test (30, ×100), and abnormal additional chamber (31, ×50).

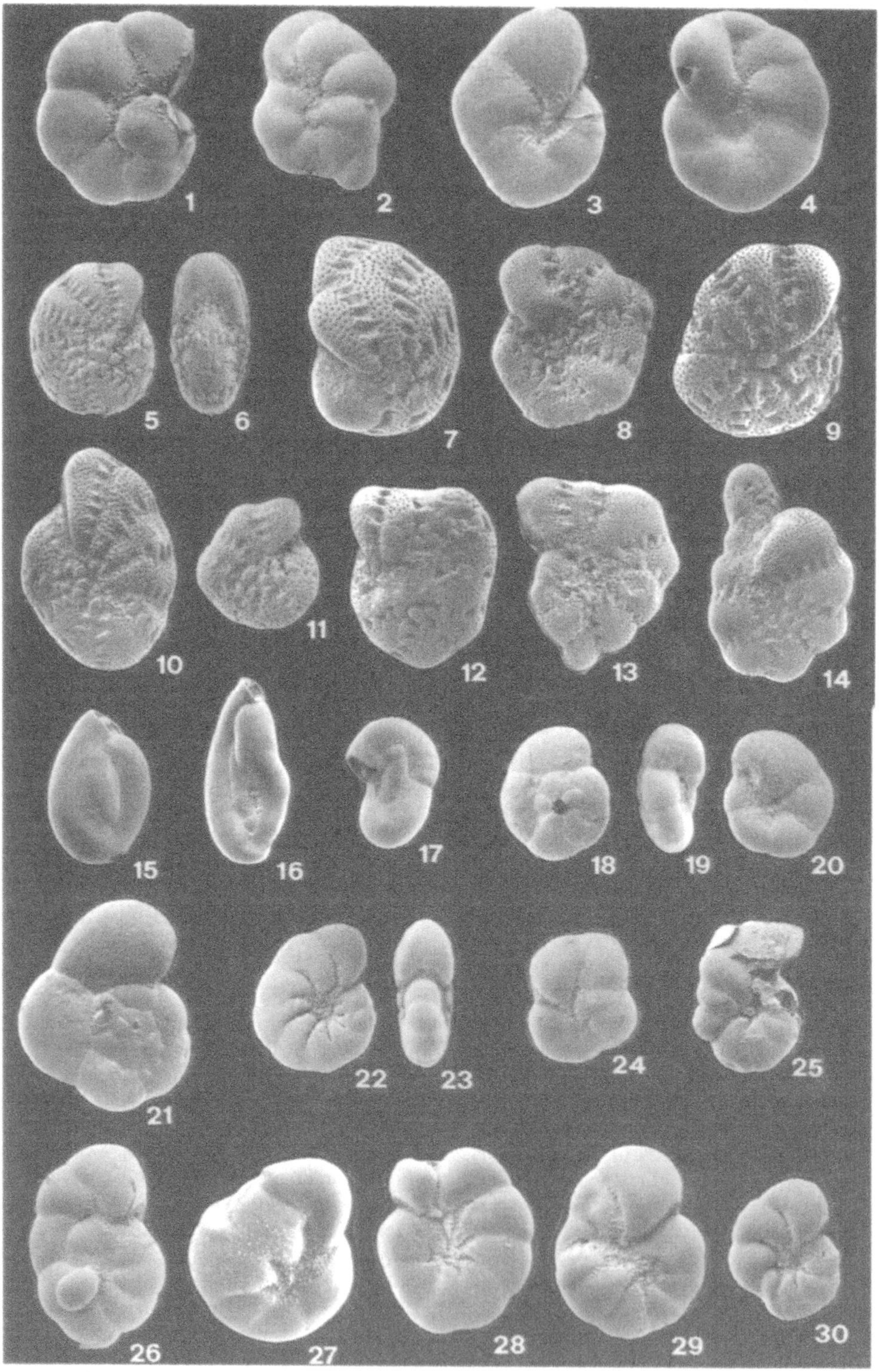
1
2
3
4
5
6
7
8
9
10
11
12
13
14
15
16
17
18
19
20
21
22
23
24
25
26
27
28
29
30

distribution of species, formae and morphotypes recognized in each station over the surveyed time interval is listed in Tables 2–4.

Most of the identified forms occur in very small numbers and are unstained. Accordingly, they are considered as having been transported into and redeposited within the Goro Lagoon.

Only 8 foraminiferal species (including 2 formae and 3 morphotypes) were identified as living. Living assemblages are largely dominated by *A. parkinsoniana* forma *tepida* mph. 2 (34–38%), *H. germanica* (33–35%), *C. gunteri* (11-16%) and, subordinately, by *A. parkinsoniana* forma *tepida* mph. 3 (4-11%) and *Q. seminula* (3–5%), which constitute 98% of the whole assemblages. *A. parkinsoniana* forma *parkinsoniana, A. parkinsoniana* forma *tepida* mph. 1, *T. inflata, A. perlucida, C. laevigata carinata,* and *E. advenum* comprise only low to very low percentages of assemblages (Fig. 4).

These very low-diversity assemblages composed of forms having different modes of life and belonging to different morphotypes are comparable to the *Ammonia beccarii* association characteristic of lagoons along the Mediterranean coast (see Murray, 1991). The dominance of *A. parkinsoniana* forma *tepida* within the assemblages is linked to its preference for low salinity—including polluted—environments, and high numbers of morphotypes 2 and 3 are related to their tolerance for depleted oxygen conditions (Jorissen, 1988).

Living populations of *A. parkinsonia* forma *tepida* consist of very high numbers of megalospheric forms, confirming the stressed conditions of the Goro Lagoon. Megalospheric forms are, in fact, produced during the asexual cycle, which is suggested by Furssenko (1978) to prefer stressed conditions. Zampi and D'Onofrio (1984, 1987) reported a high percentage of megalospheric specimens of *Ammonia beccarii* (corresponding to the forms here determined as *A. parkinsonia* forma *tepida*) from the San Gilla (Sardinia) and eastern Orbetello (Tuscany) lagoons. According to these authors this morphogenetic feature is related to anthropogenic alteration of physical and chemical parameters such as dissolved O_2 and pH. Yanko *et al.* (1994) found

←

FIGURE 3. SEM photomicrographs showing examples of normal (nondeformed) specimens and morphological abnormalities in benthic foraminifera found in Goro Lagoon (reduced 45% for reproduction): **1–4** *Ammonia perlucida* (Heron-Allen and Earland) showing abnormally protruding chamber (1, ×50), abnormal additional chamber (2, ×75), reduced size chamber (3, ×100), and aberrant chamber shape (4, ×50). **5–14** *Cribroelphidium gunteri* (Cole), normal specimen (5, side view and 6, lateral view; ×50) and individuals exhibiting Siamese twins (7, ×100), reduced size chamber with abnormal additional chamber (8, ×75), reduced size chamber (9, ×150; 11, ×75; 12, ×125), reduced size chamber with aberrant chamber shape (13, ×75), and abnormally protruding chamber (14, ×75). **15–17** *Quinqueloculina seminula* (Linnaeus), normal specimen (15, ×75) and individuals with distorted chamber arrangement (16, ×75) and aberrant chamber shape (17, ×75). **18–21** *Trochammina inflata* (Montagu), normal specimen (18 spiral view, 19 lateral view, 20 umbilical view; ×75) and individual with aberrant chamber shape (21, ×100). **22–30** *Haynesina germanica* (Ehrenberg), normal specimen (22, side view and 23, lateral view; ×100) and individuals displaying nondeveloped (24, ×50; 30, ×50), Siamese twins (25, ×50), distorted chamber arrangement with abnormal additional chamber (26, ×75), and reduced size chamber (27, ×150; 28, ×105; 29, ×75).

TABLE 2. Percentages of Living *plus* Nonliving (*T*), Living *plus* Nonliving Abnormal (TD), living (L), and Living Abnormal (LD) Benthic Foraminifera at Each Station in April 1997

Goro Lagoon	Station 1				Station 2				Station 3				Station 4				Station 5				Station 6				Station 8				Station 9			
Species	T	TD	L	LD	T	TD	L	LD	T	TD	L	LD	T	TD	L	LD	T	TD	L	LD	T	TD	L	LD	T	TD	L	LD	T	TD	L	LD
Adelosina partschi																	1,0															
Ammodiscus sp.																	0,3															
Ammonia beccarii forma *beccari*																	3,9				6,3	10,4										
Ammonia park. forma *parkinsoniana*	2,0																27,7	31,3			23,1	29,2	8,3		1,3				0,3			
Ammonia park. forma *tepida mph. 1*	6,6	9,3			3,3	5,0			1,2				3,3	2,6			3,5				1,3	2,1			3,3	4,6			5,3	13,2		
Ammonia park. forma *tepida mps. 2*	38,2	34,9			41,7	47,5			40,4	48,7			31,3	38,5	66,7	60,0	3,9	12,5			15,5	29,2	33,3	50,0	19,9	24,6	36,4	33,3	20,1	22,6	51,7	57,1
Ammonia park. forma *tepida mph. 3*	18,3	27,9			29,3	25,0			41,6	25,6			34,0	15,4			8,7	12,5			12,5	12,5	16,7		22,9	21,5	9,1	33,3	7,2	7,5	3,4	
Ammonia perlucida													0,7				2,6	6,3			4,3	4,2			0,3				1,3	1,9		
Asterigerinata planorbis																																
Brizalina alata																					0,3											
Brizalina catanensis																	1,6				0,3				1,0				0,3			
Brizalina sp.																	0,3															
Bulimina exilis																	0,3															
Bulimina jarvisi																	0,3															
Bulimina marginata					0,3												3,9				0,7											
Cassidulina laevigata carinata	0,3																3,2				1,7				0,3							
Cibicides advenum																																
Cibicides lobatulus																	0,3															
Cibicidoides pachyderma																	3,5				1,3				2,0							
Cribroelphidium gunteri	12,6	25,6			9,0	10,0	40,0		4,8	12,8	40,0		14,3	17,9	6,7		1,6	12,5			6,6	8,3	25,0	50,0	11,3	16,9			10,7	13,2	6,9	14,3
C. poeyanum forma *decipiens*	0,3				0,3												3,9				2,0											
Dentalina leguminiformis																																
Elphidium aculeatum																																
Elphidium advenum																	1,0								0,7							
Elphidium crispum																	3,5				5,6		8,3									
Elphidium macellum																									0,3							
Elphidium sp. 4 Cimerman & Langer																	0,3															
Elphidium sp. 7 Cimerman & Langer	0,3																															
Eponides repandus																	0,3															
Fissurina piriformis																																
Fursenkoina sp. 1																	0,3															
Globobulimina ovula																	0,3															
Gyroidinoides laevigatus																																
Gyroidinoides soldanii	0,3																0,3				1,3				0,3							
Haplophragmoides sp.																									2,7							
Haynesina germanica	13,6	2,3			15,3	12,5	60,0		12,0	12,8	60,0		14,0	25,6	26,7	40,0	2,6	25,0			6,9	4,2	8,3		13,6	15,4	27,3	33,3	50,9	39,6	27,6	28,6
Haynesina granosa forma *lidoense*																	1,3				1,0											

Goro Lagoon	Station 1				Station 2				Station 3				Station 4				Station 5				Station 6				Station 8				Station 9			
Species	T	TD	L	LD	T	TD	L	LD	T	TD	L	LD	T	TD	L	LD	T	TD	L	LD	T	TD	L	LD	T	TD	L	LD	T	TD	L	LD
Lagena aspera																																
Lagena sp.																	0,3															
Lenticulina gibba																					0,3				0,3							
Lenticulina sp. 1																	0,6															
Lenticulina sp. 2																	0,3															
Melonis barleeanum	0,3																1,0				1,0											
Melonis pompilioides																	1,0				0,3											
Neoconorbina orbicularis																																
Nodosaria hispida																													0,3			
Nonion commune	0,3																0,6				0,7											
Nonion fabum																																
Oridorsalis umbonatus stellatus																																
Orthomorphina stainforthi	0,7																															
Orthomorphina tenuicostata																																
Planulina ariminensis																	0,3															
Planulinoides ? sp. 1 Cim & Langer																	0,6				0,3											
Praeglobobulimina affinis																									2,3	1,5						
Praeglobobulimina pupoides																					0,3											
Pseudoeponides falsobeccarii																					0,7				0,3							
Pseudotriloculina laevigata																	0,3															
Pullenia bulloides	0,7																0,6												0,3			
Quinqueloculina badenensis																	0,3															
Quinqueloculina boueana																																
Quinqueloculina limbata	0,7																															
Quinqueloculina seminula	4,0				0,7								2,3				5,1				1,0				2,0	3,1			2,8	1,9	10,3	
Rectuvigerina siphogenerinoides																																
Reussella spinulosa																									0,3							
Siphonina reticulata	0,3																															
Siphonina tubulosa																									0,3							
Sphaeroidina bulloides																	0,6				0,7											
Spiroloculina sp.																	0,3															
Trifarina bradyi																																
Triloculina sp.																	0,3															
Trochammina inflata																					0,3				13,6	12,3	27,3					
Uvigerina peregrina	0,3																6,8				2,3				0,7							
Valvulineria bradyana																	0,3				1,3											
Number of Specimens	301	43	0	0	300	40	5	0	332	39	5	0	300	39	15	5	311	16	0	0	303	48	12	4	301	65	11	3	318	53	29	7

(*Cont.*)

TABLE 2. Continued

Goro Lagoon	Station 10				Station 11				Station 12				Station 13				Station 14				Station 15				Station 16			
Species	T	TD	L	LD	T	TD	L	LD	T	TD	L	LD	T	TD	L	LD	T	TD	L	LD	T	TD	L	LD	T	TD	L	LD
Adelosina partschi																												
Ammodiscus sp.																												
Ammonia beccarii forma *beccari*																												
Ammonia park. forma *parkinsoniana*																	0,3				5,0	3,0						
Ammonia park. forma *tepida mph. 1*	3,9	4,8			5,4	7,3	14,3		3,0						11,1		3,0		12,5		3,6	6,1						
Ammonia park. forma *tepida mps. 2*	16,0	20,6	15,4		21,8	25,5	14,3		44,2	45,9	46,7	40,0	40,3	17,1	55,6		30,6	18,2	25,0		33,1	45,5	12,5		56,2	50,0	42,9	
Ammonia park. forma *tepida mph. 3*	14,7	9,5			8,5	7,3			17,5	13,1	13,3	20,0	25,9	28,6			11,6	12,1			7,3	15,2	12,5		17,8	28,1		
Ammonia perlucida	2,0	1,6			0,3				1,3												1,3							
Asterigerinata planorbis					0,3																0,3							
Brizalina alata																												
Brizalina catanensis																	0,3				2,3							
Brizalina sp.																												
Bulimina exilis																												
Bulimina jarvisi																					0,3							
Bulimina marginata																					0,7							
Cassidulina laevigata carinata																					2,0							
Cibicides advenum	0,3																				3,3							
Cibicides lobatulus																					0,7							
Cibicidoides pachyderma	0,3																				0,7							
Cribroelphidium gunteri	34,9	46,0	15,4		32,9	43,6	28,6		11,6	23,0	13,3		10,4	11,4	5,6		17,6	48,5	25,0		12,3	15,2			14,9	15,6	28,6	
C. poeyanum forma *decipiens*																												
Dentalina leguminiformis																					0,3							
Elphidium aculeatum																					0,3							
Elphidium advenum									0,3																			
Elphidium crispum																												
Elphidium macellum																												
Elphidium sp. 4 Cimerman & Langer																												
Elphidium sp. 7 Cimerman & Langer																												
Eponides repandus																												
Fissurina piriformis					0,3																							
Fursenkoina sp. 1																												
Globobulimina ovula																												
Gyroidinoides laevigatus																					0,7							
Gyroidinoides soldanii																					0,3							
Haplophragmoides sp.																												
Haynesina germanica	26,4	17,5	69,2	100	21,2	16,4	42,9		21,5	18,0	26,7	40,0	20,7	37,1	22,2		32,6	21,2	37,5		16,9	15,2	37,5		9,5	6,3	28,6	
Haynesina granosa forma *lidoense*																												

Goro Lagoon	Station 10				Station 11				Station 12				Station 13				Station 14				Station 15				Station 16			
Species	T	TD	L	LD	T	TD	L	LD	T	TD	L	LD	T	TD	L	LD	T	TD	L	LD	T	TD	L	LD	T	TD	L	LD
Lagena aspera																					0,7							
Lagena sp.																												
Lenticulina gibba																												
Lenticulina sp. 1																												
Lenticulina sp. 2																												
Melonis barleeanum																												
Melonis pompilioides																					0,3							
Neoconorbina orbicularis																					0,7							
Nodosaria hispida																												
Nonion fabum																					1,0							
Oridorsalis umbonatus stellatus																					0,3							
Orthomorphina stainforthi																												
Orthomorphina tenuicostata																					0,3							
Planulina ariminensis					0,3																							
Planulinoides ? sp. 1 Cim & Langer																												
Praeglobobulimina affinis																					0,7							
Praeglobobulimina affinis																					0,3							
Praeglobobulimina pupoides																												
Pseudoeponides falsobeccarii																												
Pseudotriloculina laevigata																												
Pullenia bulloides																												
Quinqueloculina badenensis																												
Quinqueloculina boueana																									0,3			
Quinqueloculina limbata									0,3																			
Quinqueloculina seminula	1,6				8,9				0,3				2,6	5,7	5,6		4,0				2,6		37,5		1,3			
Rectuvigerina siphogenerinoides																					0,3							
Reussella spinulosa																												
Siphonina reticulata																												
Siphonina tubulosa																												
Sphaeroidina bulloides																												
Spiroloculina sp.																												
Trifarina bradyi																					0,3							
Triloculina sp.																												
Trochammina inflata																												
Uvigerina peregrina																												
Valvulineria bradyana																					0,7							
Number of Specimens	307	63	13	1	316	55	7	0	303	61	15	5	347	35	18	0	301	33	8	0	302	33	8	0	315	32	7	0

TABLE 3. Percentages of Living *plus* Nonliving (T), Living *plus* Nonliving Abnormal (TD) Living (L), and Living Abnormal (LD) Benthic Foraminifera at Each Station in July 1997

Goro Lagoon	Station 1				Station 2				Station 3				Station 4				Station 5				Station 6				Station 7				Station 8			
Species	T	TD	L	LD	T	TD	L	LD	T	TD	L	LD	T	TD	L	LD	T	TD	L	LD	T	TD	L	LD	T	TD	L	LD	T	TD	L	LD
Adelosina cliarensis																																
Adelosina partschi																																
Ammonia beccarii forma *beccari*																	1,3															
Ammonia park. forma *parkinsoniana*	3,0	3,4															16,0	27,0	15,4		2,0				2,3							
Ammonia park. forma *tepida mph. 1*	5,3				6,6				4,8	2,7	3,3		1,7				4,0	5,4											0,3			
Ammonia park. forma *tepida mph. 2*	21,3	17,2			40,3	22,2	72,7		41,8	48,6	43,3	60,0	41,1	45,8	34,9	20,0	18,3	24,3	26,9	66,7	29,7	54,7			14,6	12,5	9,3		36,4	41,2	47,6	50,0
Ammonia park. forma *tepida mph. 3*	36,7	27,6			6,2	3,7	9,1		24,4	18,9	26,7	20,0	10,9	8,3	7,9		12,7	18,9	11,5		2,3	1,9			8,3		4,7		27,2	17,6	33,3	25,0
Ammonia perlucida	1,7	10,3			1,6	3,7			0,9				1,7				4,7	5,4											2,0			
Amphicoryna sp.																																
Anomalinoides flinti																	0,3															
Brizalina alata																																
Brizalina catanensis																	2,0															
Bulimina elongata																																
Bulimina fusiformis																									0,3							
Bulimina inflata																	0,3															
Bulimina marginata																	1,0				0,3											
Bulimina minima																	0,3															
Bulimina sp.	0,3																															
Cassidulina laevigata carinata	0,3																2,7	5,4	7,7	33,3	0,3											
Cibicides advenum																	0,3															
Cibicides lobatulus																																
Cibicides refulgens																	0,3															
Cibicidoides pachyderma																	2,0															
Cibicidoides subhaidingerii																	0,3				0,3											
Cribroelphidium gunteri	12,3	27,6			21,3	29,6	9,1	100	6,3	10,8			14,9	16,7	11,1	20,0	8,7	8,1	15,4		23,7	28,3	12,5		55,8	71,9	41,9	57,1	5,6	8,8		
Dentalina leguminiformis																	0,3															
Elphidium aculeatum																	1,0				0,7											
Elphidium advenum																	3,0		3,8													
Elphidium sp. 4 Cimerman & Langer																																
Elphidium sp. 7 Cimerman & Langer																	0,3															

Goro Lagoon	Station 1				Station 2				Station 3				Station 4				Station 5				Station 6				Station 7				Station 8			
Species	T	TD	L	LD	T	TD	L	LD	T	TD	L	LD	T	TD	L	LD	T	TD	L	LD	T	TD	L	LD	T	TD	L	LD	T	TD	L	LD
Eponides repandus																	0,3															
Globobulimina ovula																	0,3															
Gyroidinoides laevigatus																																
Gyroidinoides soldanii																	0,3															
Hanzawaia boueana																	0,3															
Haynesina germanica	17,3	13,8			21,3	40,7	9,1		20,5	18,9	26,7	20,0	28,1	25,0	42,9	40,0	8,0	5,4	15,4		38,7	15,1	87,5		16,9	15,6	44,2	42,9	28,5	32,4	19,0	25,0
Haynesina granosa forma *lidoense*													0,3				1,0															
Heterolepa floridana																	0,3															
Lagena aspera																																
Leticulina sp. 1																	0,3															
Marginulina costata																					0,3											
Melonis barleeanum																	0,3															
Melonis pompilioides																	0,7															
Neoeponides bradyi																	0,3															
Nonion commune					0,3												0,3								0,3							
Nonion fabum																																
Planulina sp.																					0,3											
Praeglobobulimina affinis																	0,3															
Praeglobobulimina pupoides																	0,7															
Pullenia bulloides																									0,3							
Quinqueloculina agglutinans																	0,3															
Quinqueloculina boueana																																
Quinqueloculina seminula	1,3				2,0				1,4				1,3	4,2	3,2	20,0	3,3		3,8		1,3											
Siphonina tubulosa																	0,3															
Triloculina fichteliana																																
Trochammina inflata					0,3																				1,0							
Uvigerina peregrina																	1,7															
Valvulineria bradyana																	0,7															
Number of specimens	300	29	0	0	305	27	11	1	352	37	30	5	302	24	63	5	300	37	26	3	300	53	16	0	301	32	43	7	302	34	21	4

(*Cont.*)

TABLE 3. Continued

Goro Lagoon	Station 9				Station 10				Station 11				Station 12				Station 13				Station 14				Station 15				Station 16			
Species	T	TD	L	LD	T	TD	L	LD	T	TD	L	LD	T	TD	L	LD	T	TD	L	LD	T	TD	L	LD	T	TD	L	LD	T	TD	L	LD
Adelosina cliarensis																									0,3							
Adelosina partschi									0,3																							
Ammonia beccarii forma *beccari*																																
Ammonia park. forma *parkinsoniana*	0,3																								0,3				1,0			
Ammonia park. forma *tepida mph. 1*	7,0		11,9		3,7	2,4			6,8	9,8	7,4		2,5	3,8	2,4	50,0	2,5	7,4			1,7				0,3				4,6	9,7	3,7	
Ammonia park. forma *tepida mph. 2*	15,8	20,7	7,1		41,9	38,1	54,7	50,0	31,1	24,4	51,9	50,0	54,5	34,6	41,5	50,0	54,1	44,4	50,0		25,5	37,5	37,5	50,0	24,9	38,5	22,2		58,6	45,2	25,9	
Ammonia park. forma *tepida mph. 3*	2,8		2,4		1,7		1,9		1,6	4,9	3,7		5,0	15,4	2,4		4,2	14,8	15,0		8,6	9,7	25,0		6,3	3,8			10,2	9,7	14,8	
Ammonia perlucida	1,9				1,0								0,9	3,8			2,8				1,0		12,5		1,0	3,8						
Amphicoryna sp.																																
Anomalinoides flinti																																
Brizalina alata	0,3																															
Brizalina catanensis	0,9																								5,3							
Bulimina elongata																									0,3							
Bulimina fusiformis																																
Bulimina inflata																									0,3							
Bulimina marginata																									2,7							
Bulimina minima																																
Bulimina sp.																																
Cassidulina laevigata carinata	0,6				0,3																				5,6							
Cibicides advenum																									3,7							
Cibicides lobatulus																									1,7							
Cibicides refulgens																																
Cibicidoides pachyderma																									3,3							
Cibicidoides subhaidingerii																																
Cribroelphidium gunteri	25,6	41,4	28,6	33,3	30,9	47,6	17,0	37,5	22,7	53,7	14,8	50,0	13,1	23,1	9,8		14,6	14,8	20,0		42,4	41,7			9,0	23,1	11,1		6,9	25,8	14,8	100
Dentalina leguminiformis																																
Elphidium aculeatum	0,3																									1,3						
Elphidium advenum																									1,0							
Elphidium sp. 4 Cimerman & Langer																													0,3			
Elphidium sp. 7 Cimerman & Langer																									1,0							

Goro Lagoon	Station 9				Station 10				Station 11				Station 12				Station 13				Station 14				Station 15				Station 16			
Species	T	TD	L	LD	T	TD	L	LD	T	TD	L	LD	T	TD	L	LD	T	TD	L	LD	T	TD	L	LD	T	TD	L	LD	T	TD	L	LD
Eponides repandus																									2,3							
Globobulimina ovula																																
Gyroidinoides laevigatus																									3,0							
Gyroidinoides soldanii																									0,3							
Hanzawaia boueana																																
Haynesina germanica	37,7	31,0	33,3	33,3	16,6	9,5	26,4	12,5	26,9	7,3	22,2		22,4	19,2	43,9		19,4	18,5	15,0		19,5	9,7	25,0	50,0	18,6	23,1	66,7		8,9	6,5	7,4	
Haynesina granosa forma *lidoense*	0,3																															
Heterolepa floridana																																
Lagena aspera																									0,3							
Leticulina sp. 1																																
Marginulina costata																																
Melonis barleeanum																																
Melonis pompilioides																									0,3							
Neoeponides bradyi																																
Nonion commune																									1,7							
Nonion fabum																									0,3							
Planulina sp.																																
Praeglobobulimina affinis																																
Praeglobobulimina pupoides																																
Pullenia bulloides																																
Quinqueloculina agglutinans																																
Quinqueloculina boueana																									0,3							
Quinqueloculina seminula	6,0	6,9	16,7	33,3	4,0	2,4			10,4				1,6				2,3				1,3	1,4			3,3				9,2	3,2	33,3	
Siphonina tubulosa																																
Triloculina fichteliana																													0,3			
Trochammina inflata	0,3								0,3																0,3							
Uvigerina peregrina																									0,3							
Valvulineria bradyana																									0,3							
Number of specimens	316	29	42	3	301	42	53	8	309	41	27	2	321	26	41	2	355	27	20	0	302	72	8	2	301	26	9	0	304	31	27	2

TABLE 4. Percentages of Living *plus* Nonliving (T), Living *plus* Nonliving Abnormal (TD), Living (L), and Living Abnormal (LD) Benthic Foraminifera at Each Location in October 1997

Goro Lagoon	Station 1				Station 2				Station 3				Station 4				Station 5				Station 6				Station 7				Station 8			
Species	T	TD	L	LD	T	TD	L	LD	T	TD	L	LD	T	TD	L	LD	T	TD	L	LD	T	TD	L	LD	T	TD	L	LD	T	TD	L	LD
Ammonia park. forma *parkinsoniana*					0,3												7,5	9,8			2,0	4,9										
Ammonia park. forma *tepida mph. 1*	1,0				2,3	2,2	2,3		0,9				8,3	17,0	11,5	20,0	0,6				1,7								8,7	9,5	4,8	
Ammonia park. forma *tepida mph. 2*	55,8	53,8	60,0	80,0	38,7	53,3	54,5	87,5	58,2	33,3	59,1	33,3	39,0	25,5	34,6	40,0	21,1	36,6	11,1		7,4	12,2	4,1		27,3	27,7	12,5		35,0	40,5	38,1	33,3
Ammonia park. forma *tepida mph. 3*	19,1	15,4	25,5		14,9	11,1	11,4	12,5	13,8	11,1	18,2		15,6	10,6	11,5		9,7	7,3	5,6		4,7	2,4	2,0		37,5	19,1			20,7	21,6	33,3	33,3
Ammonia perlucida																	7,8	7,3			0,3				0,7							
Asterigerinata planorbis																	0,6															
Brizalina catanensis																	0,3															
Bulimina inflata																	1,3															
Bulimina marginata																	0,3															
Cassidulina laevigata carinata																	1,3		5,6		0,3											
Cibicides advenum																	0,6															
Cibicides lobatulus																	0,3															
Cibicidoides pachyderma																	1,3															
Cribroelphidium gunteri	6,9	15,4	1,8		14,9	13,3	4,5		6,8	16,7			19,0	31,9	11,5	20,0	11,7	26,8	22,2	66,7	20,1	31,7			28,0	48,9	56,3	83,3	22,0	21,6	9,5	16,7
C. poeyanum forma *decipiens*																	4,5	2,4			0,3											
Dentalina leguminiformis																	0,3															
Elphidium advenum																																
Elphidium crispum																	0,6															
Fursenkoina sp. 2																																
Gyroidinoides laevigatus																	0,3		5,6													
Gyroidinoides soldanii																	1,0															
Haynesina germanica	10,2	11,5	9,1	20,0	27,8	20,0	27,3		19,1	38,9	22,7	66,7	15,6	14,9	30,8	20,0	24,0	9,8	50,0;	33,3	60,9	48,8	91,8	100	6,6	4,3	31,3	16,7	12,0	6,8	14,3	16,7
Lenticulina sp. 3																																
Melonis barleeanum																					0,3											
Melonis pompilioides																	0,3				0,3		2,0									
Neoconorbina orbicularis																																
Nonion commune																																
Nonion fabum																					0,3											
Praeglobobulimina pupoides																																
Pseudoeponides falsobeccarii																	0,3															
Pullenia bulloides																					0,3											
Quinqueloculina lamarckiana																	0,3															
Quinqueloculina seminula	6,9	3,8	3,6		1,0				1,2				2,5				1,3												1,3			
Siphonina reticulata																					0,3											
Sphaeroidina bulloides																	0,6															
Trochammina inflata																					0,7											
Uvigerina peregrina																	0,6															
Valvulineria bradyana																	0,6															
Number of Specimens	303	26	55	5	302	45	44	8	340	18	22	3	326	47	26	5	308	41	18	3	300	41	49	4	304	47	16	6	300	74	21	6

Goro Lagoon	Station 9				Station 10				Station 11				Station 12				Station 13				Station 14				Station 15				Station 16			
Species	T	TD	L	LD	T	TD	L	LD	T	TD	L	LD	T	TD	L	LD	T	TD	L	LD	T	TD	L	LD	T	TD	L	LD	T	TD	L	LD
Ammonia park. forma *parkinsoniana*																									0,3							
Ammonia park. forma *tepida mph. 1*	0,7				1,2				5,6	4,0	5,4		1,2				3,0								4,3	5,0	15,4	33,3				
Ammonia park. forma *tepida mph. 2*	27,5	35,4	19,2	20,0	27,1	52,9	50,0	100	23,8	36,0	32,4	75,0	48,6	42,9	33,3		41,1	48,4	35,0		30,0	17,2	42,9	20,0	38,3	50,0	61,5		59,1	62,5	45,5	50,0
Ammonia park. forma *tepida mph. 3*	2,3	2,1							3,6	4,0	5,4		11,8	4,8	8,3		17,4	22,6	20,0		6,6	10,3	2,9	20,0	5,3	12,5	15,4	66,7	15,3	12,5	13,6	
Ammonia perlucida					0,3	2,9			0,7							0,7																
Asterigerinata planorbis																									0,7							
Brizalina catanensis	0,3																				0,6				2,7							
Bulimina inflata																									0,3							
Bulimina marginata	0,3																								1,3							
Cassidulina laevigata carinata																									2,0							
Cibicides advenum																									0,3	2,5						
Cibicides lobatulus	0,3																								0,3							
Cibicidoides pachyderma	0,3																								2,7							
Cribroelphidium gunteri	36,1	39,6	30,8	20,0	25,8	32,4			25,8	36,0	21,6	25,0	19,3	42,9	41,7	100	7,7	9,7	10,0		34,1	55,2	14,3	40,0	9,3	17,5			6,8	6,3	4,5	25,0
C. poeyanum forma *decipiens*																									0,7	2,5						
Dentalina leguminiformis																																
Elphidium advenum																									1,0							
Elphidium crispum																									1,0							
Fursenkoina sp. 2																									0,7							
Gyroidinoides laevigatus																																
Gyroidinoides soldanii																																
Haynesina germanica	30,1	22,9	50,0	60,0	38,8	8,8	50,0		19,2	20,0	18,9		17,8	9,5	8,3		28,8	19,4	35,0	100	26,9	17,2	37,1	20,0	20,0	10,0			14,0	18,8	27,3	25,0
Lenticulina sp. 3					0.3																											
Melonis barleeanum	0,3																															
Melonis pompilioides																									0,7							
Neoconorbina orbicularis																									0,3							
Nonion commune																									1,7							
Nonion fabum																					0,3											
Praeglobobulimina pupoides	0,3																															
Pseudoeponides falsobeccarii																																
Pullenia bulloides																																
Quinqueloculina lamarckiana																																
Quinqueloculina seminula	1,3				6,5	2,9			21,2		16,2		1,2		8,3		1,3				1,6		2,9		2,7		7,7		4,9		9,1	
Siphonina reticulata																																
Sphaeroidina bulloides																									0,3							
Trochammina inflata																																
Uvigerina peregrina																									1,3							
Valvulineria bradyana																									1,0							
Number of Specimens	302	48	26	5	325	34	20	3	302	25	37	4	331	21	12	2	300	31	20	2	320	29	35	5	300	40	13	3	308	32	22	4

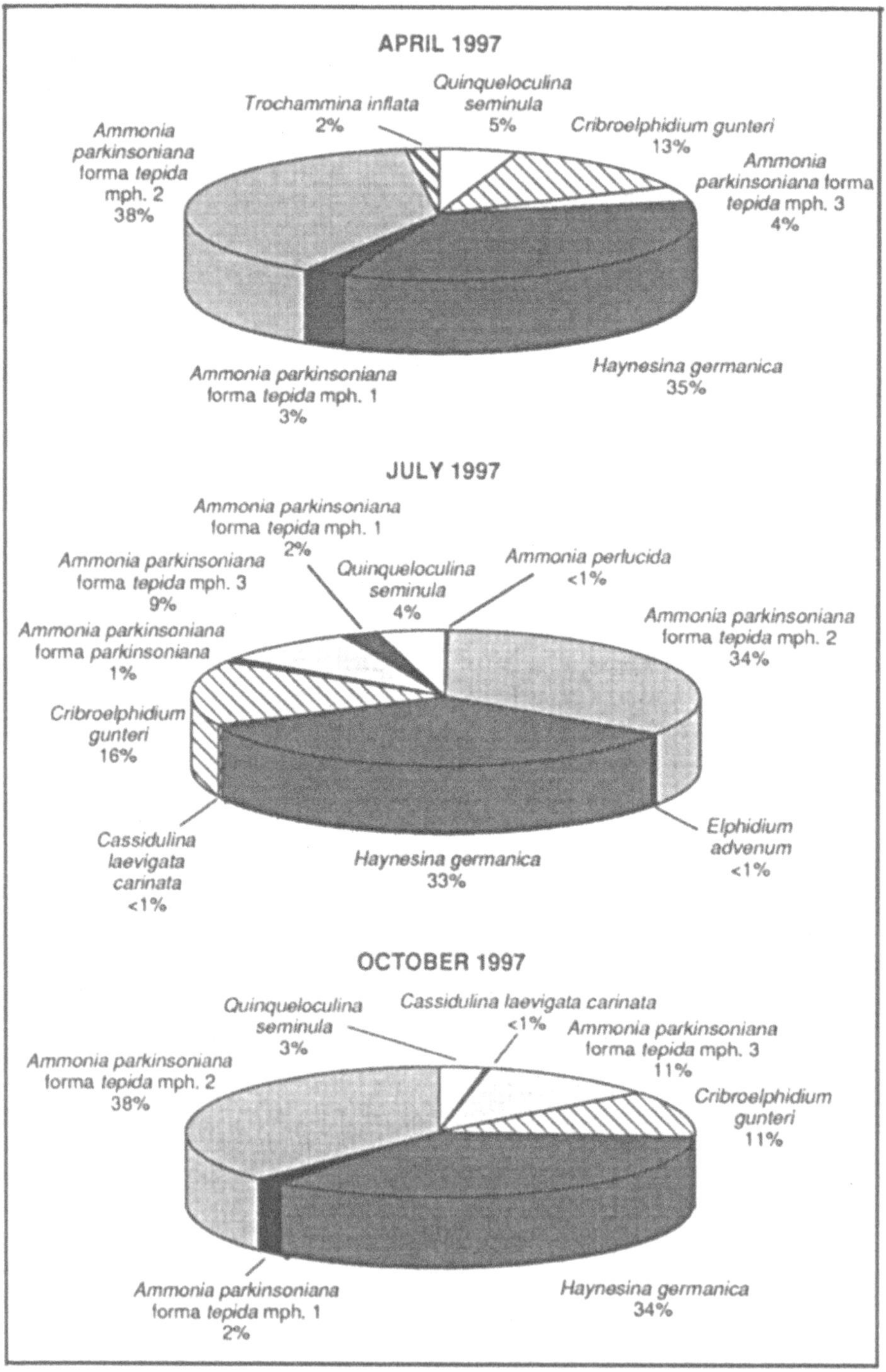

FIGURE 4. Average percentages of species, formae, and morphotypes constituting the living assemblages.

high numbers of megalospheric forms of *A. tepida* in a contaminated site along the Mediterranean coast of Israel where toxic trace metal pollution was prevalent.

Living assemblage diversity is highest in July, when the nutrient flux, according to Rinaldi *et al.* (1992), is more abundant. Faunal density ranges from 0 (stations 1 and 6, April) to 96 (station 6, October) with generally higher values in stations where sediments are mainly muddy, suggesting a close correlation between faunal density and sediment type (see Figs. 6 and 7).

4.1.2. Morphological Abnormalities

The development of abnormalities in test morphology is a common and noticeable feature of benthic foraminifera from the Goro Lagoon (Figs. 2 and 3), which is attributed to pathological morphogenesis rather than to mechanical damage. Test morphological abnormalities have been observed in all the stations over the surveyed time interval. The percentage of deformed specimens, however, varies from station to station (Tables 2–4). Within the living assemblages this percentage reached a maximum value of 37% at station 7 in October.

Throughout the surveyed time interval, only minor fluctuations were observed in the average percentages of deformed specimens within the living assemblage (9–17%). However, higher proportions of abnormal forms were usually found at stations located on the western and eastern sides of the lagoon, where, according to Colombo *et al.* (1991, 1994), water salinity is lowered. This would suggest that there is a correlation between the percentage of deformed tests and salinity. Interestingly, an increase in the proportion of living deformed foraminifera caused by low salinity has recently been reported by Yanko *et al.* (1998) from assemblages along the Mediterranean coast of northern Israel.

Within living assemblages seven forms exhibit morphological abnormalities according to the affected part of the test and the nature of the deformation: *A. parkinsoniana* forma *tepida* mph. 1, *A. parkinsoniana* forma *tepida* mph. 2, *A. parkinsoniana* forma *tepida* mph. 3, *C. laevigata carinata, C. gunteri, H. germanica.* and *Q. seminula* (Fig. 5). Thus these forms display morphological deformities of their test independent of their taxonomic affinities, mode of life, and test morphotype. The living deformed assemblages are largely dominated by *A. parkinsoniana* forma *tepida* mph. 2, *C. gunteri, H. germanica,* and *A. parkinsoniana* forma *tepida* mph. 3, although marked changes in their abundance occurred over the surveyed time interval. The order of these taxa tends to reflect a decreasing degree of their sensitivity to deformation.

Also following Alve (1991), Sharifi *et al.* (1991), Almogi-Labin *et al.* (1992), and Yanko *et al.* (1994, 1998), seven different modes of abnormality in test morphology have been recognized: (1) reduced chambers(s) size, (2) aberrant chamber shape, (3) distorted chamber arrangement or change in coiling, (4) abnormal additional chamber(s), (5) abnormally protruding chamber(s), (6)

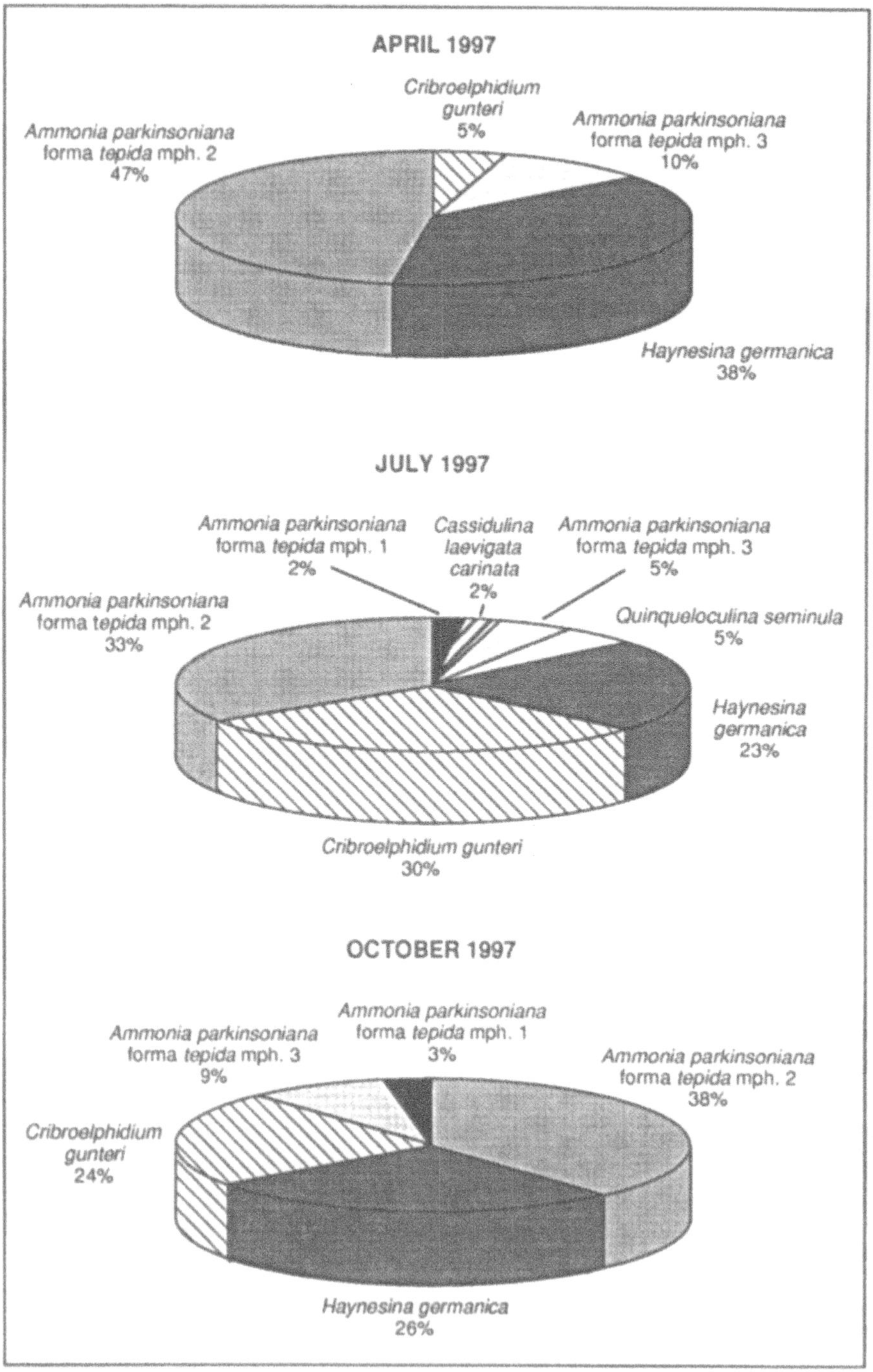

FIGURE 5. Average percentages of species, formae, and morphotypes constituting the living deformed assemblages.

TABLE 5. Different Modes of Abnormality in Test Morphology Observed in Living Assemblages from the Goro Lagoon [1: reduced chamber(s) size, 2: aberrant chamber shape, 3: distorted chamber arrangement or change in coiling, 4: abnormal additional chamber(s), 5: abnormally protruding chamber(s), 6: Siamese twins, 7: nondeveloped test]

Species/formae/morphotypes	Abnormality						
	1	2	3	4	5	6	7
Ammonia parkinsoniana forma *tepida* mph. 1	•	•	•				
Ammonia parkinsoniana forma *tepida* mph. 2	•	•	•	•	•	•	•
Ammonia parkinsoniana forma *tepida* mph. 3	•	•	•	•	•	•	•
Cassidulina laevigata carinata	•	•					
Cribroelphidium gunteri	•	•	•	•	•	•	•
Haynesina germanica	•	•	•	•		•	•
Quinqueloculina seminula	•	•	•				

Siamese twins, and (7) nondeveloped test (Table 5 and Figs. 2 and 3). The first three are observed more frequently, and some morphological abnormalities co-occurred in the same specimen. However, occasionally the abnormality was so severe that it was impossible to determine which part of the test was affected, and sometimes even impossible to determine the foraminiferal taxon.

4.2. Sediment Geochemistry

Trace element concentrations and TOC percentages measured in the spring and autumn vary from station to station (Table 6). However, the relative proportion of different trace elements is similar in all the stations, suggesting the same source(s) of pollution for the Goro Lagoon over the surveyed time interval.

As is well known, the distribution and concentration of trace elements in an aquatic environment depends on different factors (e.g., chemical composition, size of the deposited particulate matter, biological activity at the sea bottom) and is strongly influenced by some physical and chemical parameters (e.g., salinity, temperature, Eh, and pH) that determine their stability. However, no relationship was found in this study between the heavy metal concentrations and values of the physical and chemical parameters recorded by Colombo *et al.* (1991, 1994) throughout the Goro Lagoon. There was also no clear correlation between the faunal density of living assemblages and the concentration of trace elements and TOC percentages (Figs. 6–8). These findings suggest that there is no relation between the benthic foraminiferal life on bottom sediments of the Goro Lagoon and the relative development of abnormal morphologies, on the one hand, and levels of organic matter and trace elements, on the other.

TABLE 6. Concentration of Selected Trace Metals and TOC Values in Sediments at Each Selected Station in April and October 1997 in Relation to Reference Values Provided for Some Elements by the EPA

	Trace metal									
Station	V ppm	Cr ppm	Co ppm	Ni ppm	Cu ppm	Zn ppm	Hg ppm	Pb ppm	Th ppm	TOC wt%
April 1997										
1	77	199	15	104	38	108	0.2	32	7	1.21
3	106	208	18	123	56	177	0.4	50	9	0.18
4	98	190	18	118	50	157	0.35	41	12	0.07
6	43	106	8	75	25	72	0.053	22	9	0.89
9	93	190	18	115	37	104	0.53	28	16	1.22
10	97	193	17	113	48	148	0.06	31	11	1.82
15	76	205	13	107	32	111	0.17	23	9	0.85
16	97	199	16	124	52	144	0.55	39	9	1.55
October 1997										
1	72	141	8	94	39	117	0.24	31	4	2.20
3	103	202	17	116	53	170	0.36	55	10	2.00
4	100	203	18	120	50	162	0.33	45	11	2.10
6	103	207	22	123	52	159	BDL[a]	36	7	2.90
9	73	202	12	105	30	94	0.12	22	13	0.80
10	89	185	14	105	43	139	0.27	32	13	1.90
15	76	198	11	104	34	114	0.15	18	13	0.90
16	87	169	13	113	48	151	0.33	35	11	2.50

Element	Nonpolluted	Moderately polluted	Very polluted
Cr	<25	25–75	>75
Ni	<20	20–50	>50
Cu	<25	25–50	>50
Zn	<90	90–200	>200
Cd	—	—	>6
Hg	<1	—	>1
Pb	<40	40–60	>60

[a]BDL = Below detection limit.

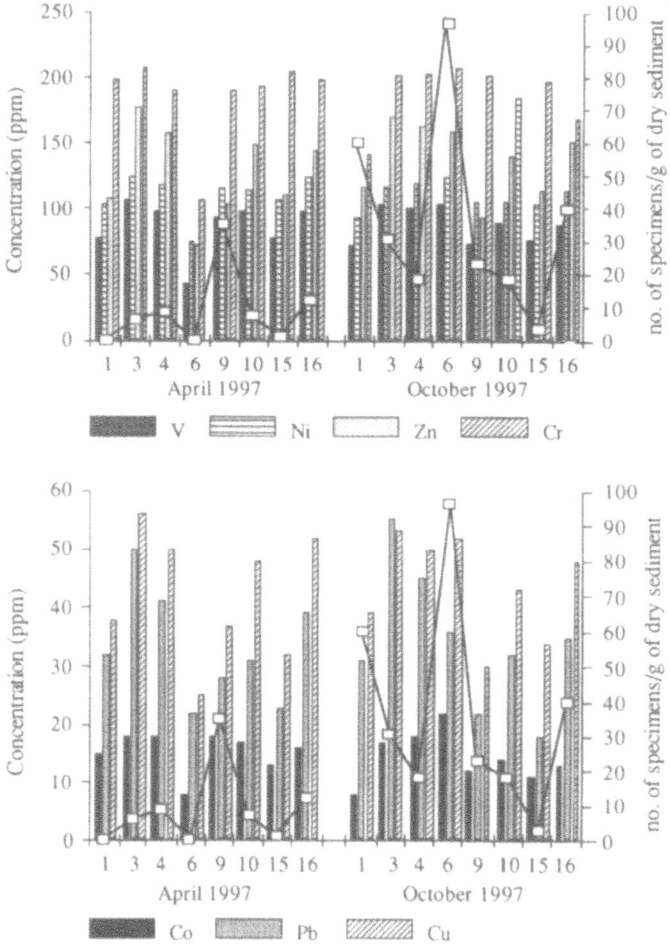

FIGURE 6. Concentration of V, Ni, Zn, Cr, Co, Pb, and Cu at selected stations in April and October 1997 plotted against faunal density of living assemblages.

All sites display marked enrichment in Cr and Ni values that are higher than the average values of unpolluted alluvial plain pelites in the neighboring area (see Fagioli *et al.*, 1991, 1994). Sediments were moderately enriched in Cu and Zn and occasionally showed higher concentrations of Pb. According to the reference values provided for some elements by the Environmental Protection Agency (EPA) (see Prater and Anderson, 1977) and also taking into consideration trace element concentrations supplied by Fagioli *et al.* (1991, 1994), it can be said that the Goro Lagoon has been a moderately to very polluted at least with Cr, Ni, Cu, Zn, and Pb for the last 10 years.

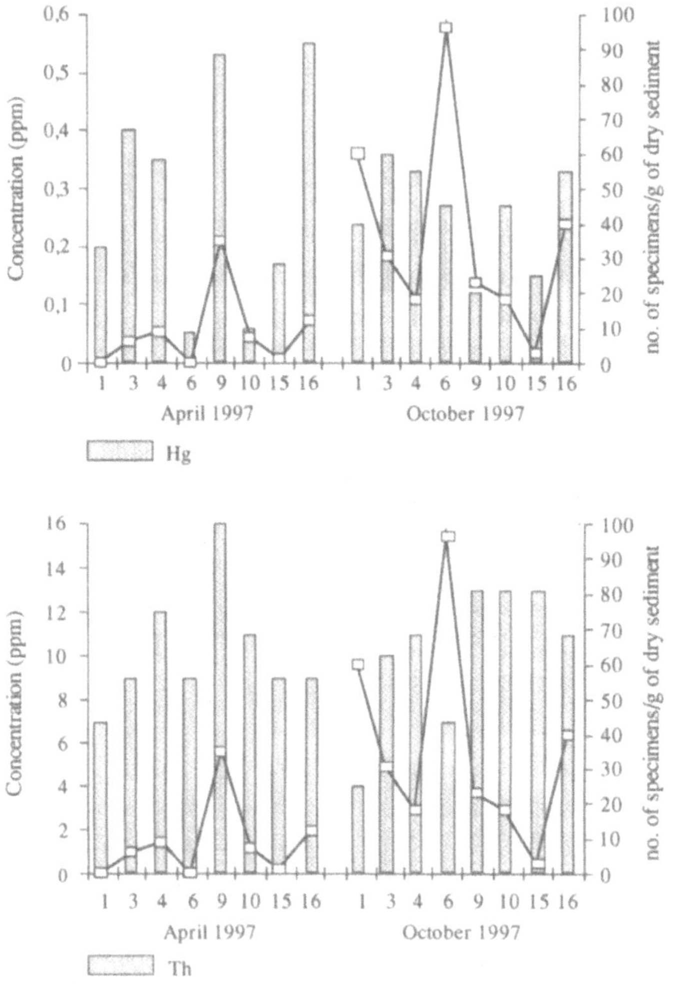

FIGURE 7. Concentration of Hg and Th at selected stations in April and October 1997 plotted against faunal density of living assemblages.

No definite correlation was noted between the percentages of living abnormal specimens and concentrations of V, Cr, Co, Ni, Cu, Zn, Hg, and Pb (Fig. 9). Although not particularly high (4–16 ppm, with an average value of 10 ppm), Th concentrations display a significant positive relationship with the percentage of living abnormal specimens (Fig. 10), but this finding is not surprising. In fact, in spite of its radioactivity, Th is not particularly dangerous in natural environments but becomes especially so by affecting skeletal components (Bowie and Plant, 1983). Unfortunately, mechanisms controlling

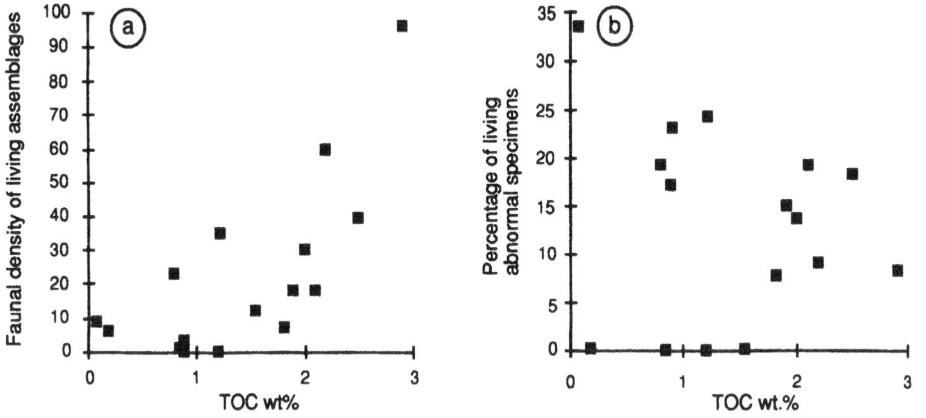

FIGURE 8. Scatter diagrams of TOC wt.% at selected stations in April and October 1997 plotted against faunal density of living assemblages (a) and percentage of living abnormal specimens (b).

its distribution and concentration in the Goro Lagoon are unknown. According to Albani *et al.*(1995), in the nearby Venice Lagoon the concentration of Th is on the average 5.6 ppm and, as other elements such as Mn, P, V, Ga, Rb, and Nb, it appears to be related mainly to inland sources, including agriculture, although some additional local input cannot be excluded. As is well known, Th is an actinide metal that is now used chiefly for nuclear and electronic applications. However, the possibility that the high concentrations recorded over the last years (up to 119 ppm during summer 1996, author's unpublished data) might somehow be related to recent nuclear explosion(s) (i.e., the 1986 Chernobyl reactor incident) both directly and indirectly by redistribution of fallout products cannot be excluded.

5. Conclusions

1. This study confirms and supports the feasibility of studying benthic foraminifera as a technique for the *in situ* continuous monitoring of trace metal pollution. The use of time-averaged surface assemblages indicates that foraminifera have responded relatively rapidly to certain environmental stresses in the Goro Lagoon.
2. A remarkable number of specimens exhibiting a variety of morphological abnormalities occurs within very low-diversity living assemblages of the Goro Lagoon.
3. The occurrence of these teratological forms, also in accord with previous studies, may be attributed to both environmental stress resulting from anthropogenic trace metal pollution and from natural effects (i.e., extreme conditions and/or rapid changes in ecological parameters).

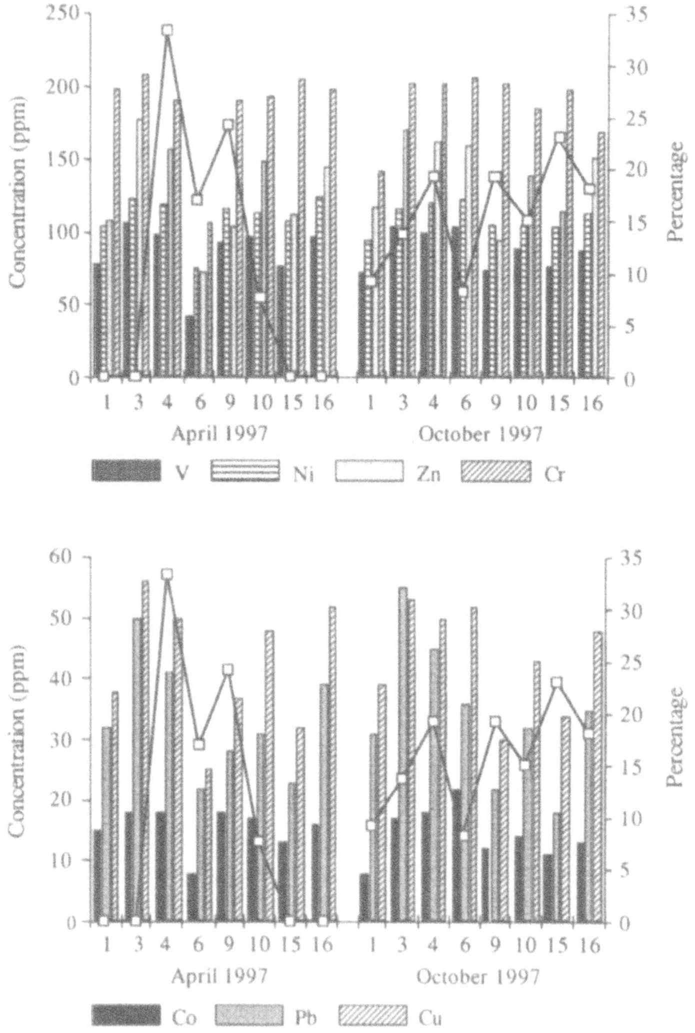

FIGURE 9. Concentration of V, Ni, Zn, Cr, Co, Pb, and Cu at selected stations in April and October 71997 plotted against percentage of abnormal specimens.

4. Higher percentages of abnormal specimens occur where the salinity of the water is lowered.
5. Increased levels of heavy metals such as Cr, Ni, Cu, Zn, and Pb do not have a marked effect upon foraminiferal distribution and morphological abnormalities. However, a clear positive correlation between the concentrations of Th and percentages of deformed specimens was found, showing that this metal may have played a part in the development of the morphological abnormalities. Further studies over a longer

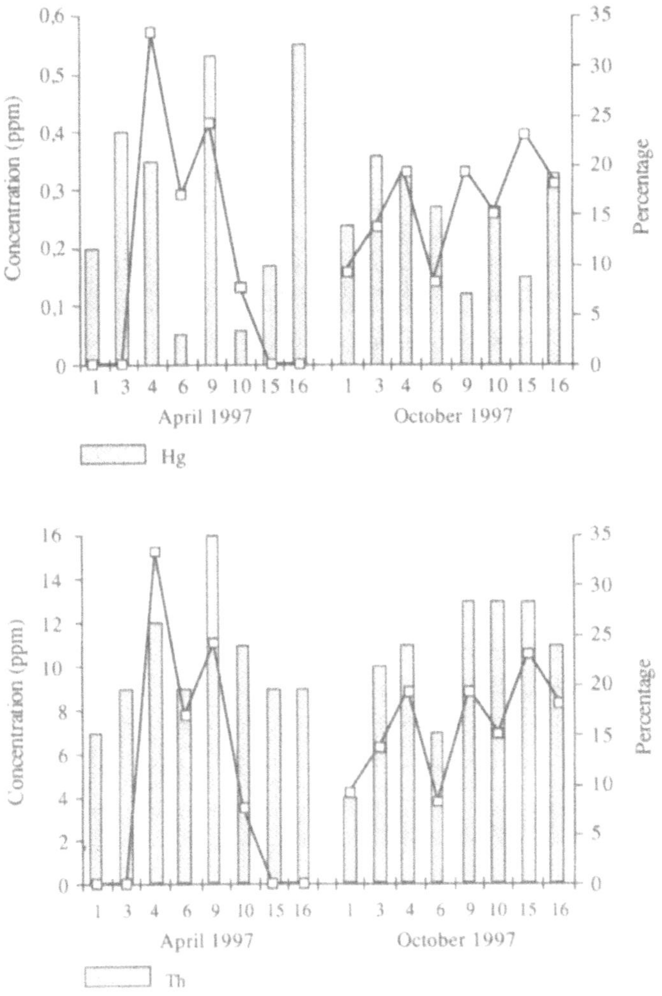

FIGURE 10. Concentration of Hg and Th at selected stations in April and October 1997 plotted against percentage of abnormal specimens.

surveyed time interval and at different localities and settings would help toward a better understanding of this relationship.

6. The morphological abnormalities observed in the Goro Lagoon over the surveyed time interval appear to be the result of a combination of factors and it is very difficult at this time to isolate any single specific cause.
7. Controlled laboratory culture experiments coupled with studies of the biochemical and crystallographic mechanisms of the development of test deformities can help to resolve the morphological reaction of the foraminiferal tests to specific degrees and types of pollution.

Appendix-Faunal List

Species, Formae, and Morphotypes Mentioned in the Text with Author Attributions

Ammonia beccarii (Linnaeus) forma *beccarii* Linnaeus = *Nautilus beccarii* Linnaeus, 1758.

Ammonia parkinsoniana (d'orbigny) forma *parkinsoniana* (d'orbigny) = *Rosalina parkinsoniana* d'orbigny, 1839

Ammonia parkinsoniana (d'Orbigny) forma *tepida* Cushman = *Rotalia beccarii* (Linnaeus) var. *tepida* Cushman, 1926.

Ammonia perlucida (Heron-Allen and Earland) = *Rotalia perlucida* Heron-Allen and Earland, 1913.

Cassidulina laevigata carinata Silvestri, 1896.

Cribroelphidium gunteri (Cole) = *Elphidium gunteri* Cole, 1931.

Cribroelphidium poeyanum (d'orbigny) forma *decipiens* (Costa) = *Polystomella decipiens* Costa, 1856.

Elphidium advenum (Cushman) = *Polystomella advena* Cushman, 1922.

Haynesina germanica (Ehrenberg) = *Nonion germanica* Ehrenberg, 1840.

Haynesina granosa (D'orbigny) forma *lidoense* Cushman = *Elphidium lidoense* Cushman, 1936.

Quinqueloculina seminula (Linnaeus) = *Serpula seminula* Linnaeus, 1758

Trochammina inflata (Montagu) = *Nautilus inflatus* Montagu, 1808

ACKNOWLEDGMENTS: I express my thanks to two anonymous reviewers for their helpful comments and constructive criticism. I am particularly grateful to the editor for his kind assistance and suggestions. Thanks are due to G. Gabbianelli and to A. Mordenti for providing all the facilities to carry out this study. I also express my gratitude to S. Galeotti for help in manuscript preparation. Assistance provided by M. Bruciatelli, L. Fantozzi, D. Gentiloni Silverj, P. Sartini, and P. Pellegrino is thankfully acknowledged for sampling and computer work. Thanks are due to P. Ferrieri for his SEM assistance. This study has been supported by the Italian MURST (ex 60% grant).

References

Albani A. D., Rickwood P. C., Favero V. M., and Barbero R. S., 1995, The geochemistry of recent sediments in the Lagoon of Venice: Environmental implications, *Atti Ist. Veneto Sci., Lett. Arti* **153**:235–321.

Almogi-Labin, A., Perelis-Grossovicz, L., and Raab, M., 1992, Living *Ammonia* from a hypersaline inland pool, Dead Sea area, Israel, *J. Foram. Res.* **22**:257–266.

Alve, E., 1991, Benthic foraminifera in sediment cores reflecting heavy metal pollution in Srfjord, Western Norway, *J. Foram. Res.* **21**:1–19.

Alve, E., 1995, Benthic foraminifera response to estuarine pollution: a review, *J. Foram. Res.* **25**:190–203.

Ashraf, M.S. 1997, Thanatocoenosic foraminiferal assemblage and impact of heavy metals Pollution on *Ammonia beccarii* (s.l.), Manzalah Lagoon, Egypt, Abstract Volume EUG 9, Strasbourg, France, 23–27 March 1997, p. 663.

Banerji, R. K., 1992, Heavy Metals and Benthic Foraminiferal Distribution along Bombay Coast, India, in: *Studies in Benthic Foraminifera* (Y. Takayanagi and T. Saito, eds.), Tokai University Press,Tokyo, pp. 151–157.

Barbanti, A., Ceccherelli, V.U., Frascari, F., Rosso, G., and Reggiani, G., 1992, Nutrient release from sediments and the role of bioturbation in the Goro Lagoon (Italy), in: *Marine Coastal Eutrophication* (R. A. Vollenweider, R. Marchetti, and R. Viviani, eds.), Elsevier, Amsterdam, pp. 475–487.

Bencivelli, S., and Castaldi, N. (a cura di), 1991, *Studio integrato sull'ecologia della Sacca di Goro in relazione ai problemi dell'inquinamento, della conservazione e dello sviluppo, 10 anno di ricerche,* Franco Angeli s.r.l., Milano, 263 pp.

Bencivelli, S., Castaldi, N., and Finessi, D. (a cura di), 1994, *Studio integrato sull'ecologia della Sacca di Goro in relazione ai problemi dell'inquinamento, della conservazione e dello sviluppo, 20 anno di ricerche,* Franco Angeli s.r.l., Milano, 303 pp.

Bernhard, J. M., 1988, Postmortem vital staining in benthic foraminifera: duration and importance in population and distributional studies, *J. Foram. Res.* **18**:143–146.

Boltovskoy, E., and Boltovskoy, A., 1968, Foraminiferos y tecamebas de la parte inferior del Rio Quequén Grande (sistemàtica, distribución, ecologia). Museo Argentino de Ciencias Naturales, *Rev. Hidrobiol.* **2**(4):127–164.

Boltovskoy, E., and Wright, R., 1976, *Recent Foraminifera,* W. Junk, The Hague, 515 pp.

Boltovskoy, E., Scott, D. B., and Medioli, F., S., 1991, Morphological variations of benthic foraminiferal tests in response to changes in ecological parameters: A review, *J. Paleontol.* **65**:175–185.

Bowie, S. H. U. and Plant, J. A., 1983, Natural radioactivity in the environments, in: *Applied Environmental Geochemistry* (I. Thornton, ed.), Academic Press, London, pp. 481–494.

Caralp, M. H., 1989, Size and morphology of the benthic foraminifer *Melonis barleeanum:* Relationship with marine organic matter, *J. Foram. Res.* **19**:235–245.

Ciabatti, M., 1967, Ricerche sull'evoluzione del Delta Padano, *Giorn. Geol.* **34**:1–26.

Closs, D., and Madeira, M. L., 1968, Seasonal variations of brackish foraminifera in the Patos Lagoon, southern Brazil, Universidade do Rio Grande do Sul, Escola de Geologia, Publicaçao especial 15, pp. 1–51.

Coccioni, R., Gabbianelli, G., Gentiloni Silverj, D., Fonti, P., Kaminski, M.A., Monechi, S., and Tateo, F., 1997, Benthic foraminiferal to heavy metal pollution in the Goro Lagoon (Italy), *1st Int. Conf. on Application of Micropaleontology in Environmental Sciences,* Tel Aviv, June 15–20, 1997, Abstracts Volume, pp. 47–48.

Colombo, G., Camerota, C., Bisceglia, R., Zaccaria, V., Gaiani, V., and Carrieri, A., 1991, Variazioni spaziali e temporali delle caratteristiche fisico-chimiche delle acque e della biomassa fitoplanctonica della Sacca di Goro, in: *Studio integrato sull'ecologia della Sacca di Goro in relazione ai problemi dell'inquinamento, della conservazione e dello sviluppo, 10 anno di ricerche* (S. Bencivelli and N. Castaldi, a cura di), Franco Angeli s.r.l., Milano, pp. 9–38.

Colombo, G., Bisceglia, R., Zaccaria, V., and Gaiani, V., 1994, Variazioni spaziali e temporali delle caratteristiche fisico-chimiche delle acque e della biomassa fitoplanctonica della Sacca di Goro nel quadriennio 1988–1991, in: *Studio integrato sull'ecologia della Sacca di Goro in relazione ai problemi dell'inquinamento, della conservazione e dello sviluppo, 20 anno di ricerche* (S. Bencivelli, N. Castaldi, and D. Finessi, a cura di), Franco Angeli s.r.l., Milano, pp. 9–82.

Dal Cin, R., and Pambianchi, P., 1991, I sedimenti della Sacca di Goro (Delta del Po), in: *Studio integrato sull'ecologia della Sacca di Goro in relazione ai problemi dell'inquinamento, della conservazione e dello sviluppo, 10 anno di ricerche* (S. Bencivelli and N. Castaldi, a cura di), Franco Angeli s.r.l., Milano, pp. 253–263.

De Stigter, H. C., Jorissen, F.J., and Van der Zwaan, G.J., 1998, Bathymetric distribution and microhabitat partitioning of live (Rose Bengal stained) benthic foraminifera along a shelf to bathyal transect in the southern Adriatic Sea, *J. Foram. Res.* **21**:40–65.

Ellison, R. L., Broome, R., and Ogilvie, R., 1986, Foraminiferal response to trace metal contamination in the Patapsco River and Baltimore Harbour, Maryland, *Mar. Pollut. Bull.* **17**(9):419–423.

Fagioli, F., Landi, S., Locatelli, C., and Vecchietti, R., 1991, Valutazione del carico di metalli nelle acque, nei sedimenti e nei principali organismi accumulatori, in: *Studio integrato sull'ecologia della Sacca di Goro in relazione ai problemi dell'inquinamento, della conservazione e dello sviluppo, 10 anno di ricerche* (S. Bencivelli and N. Castaldi, a cura di), Franco Angeli s.r.l., Milano, pp. 135–172.

Fagioli, F., Righini, F., Landi, S., and Locatelli, S., 1994, Valutazione del carico di metalli nelle acque, nei sedimenti e nei principali organismi accumulatori, in: *Studio integrato sull'ecologia della Sacca di Goro in relazione ai problemi dell'inquinamento, della conservazione e dello sviluppo, 20 anno di ricerche* (S. Bencivelli, N. Castaldi, and Finessi, D., a cura di), Franco Angeli s.r.l., Milano, pp. 177–226.

Franzini, M., Leoni, L., and Saitta, M., 1972, Revisione di una metodologia analitica per fluorescenza-X basata sulla correzione completa degli effetti di matrice, *Rend. Soc. It. Min. Petrol.* **31**:365–378.

Franzini, M., Leoni, L., and Saitta, M., 1975, A simple method to evaluate the matrix effects in X-ray analysis, *X-Ray Spectrom.* **1**:151–154.

Furssenko A., V., 1978, Vvedenie v izuchenie foraminifer (*Introduction to the Study of Foraminifera*) (V.I Gudina, ed.), Nuaka, Novosibirsk, 242 pp.

Gentiloni Silverj, D., 1997, Benthic foraminiferal response to heavy metal pollution in the Goro Lagoon (Italy), Unpublished B.Sc. Project Report, University College London, Department of Geological Sciences, 53 pp.

Gooday, A. J., 1986, Meiofaunal foraminifera from the bathyal Porcupine Seabight (northeast Atlantic) size structure, standing crop, taxonomic composition, species diversity and vertical distribution in the sediment, *Deep-Sea Res.* **33**:1345–1373.

Jorissen, F. J., 1988, Benthic foraminifera from the Adriatic Sea: Principles of phenotypic variation, *Utrecht Micropaleontal Bull.* **37**:1–176.

Jorissen F. J., de Stigter, H. C., and Widmark, J.G.V., 1995, A conceptual model explaining benthic foraminiferal microhabitats, *Mar. Micropaleontol.* **26**:3–15.

Lee J. J., and Anderson, O. R., 1991, *Biology of Foraminifera*, Academic Press, London, 368 pp.

Leoni, L., and Saitta, M., 1976, X-ray fluorescence analysis of 29 traces elements in rock and mineral standard, *Rend. Soc. It. Min. Petrol.* **32**:497–510.

Leoni, L., Menichini, M., and Saitta, M., 1982, Determination of S, Cl and F in silicate rocks by X-ray fluorescence analyses, *X-Ray Spectrom.* **11**:156–158.

Lidz, L., 1965, Sedimentary environment and foraminiferal parameters: Nantucket Bay, Massachusetts, *Limnol. Oceanogr.* **10**:392–402.

Loeblich A. R., Jr., and Tappan, H., 1988, *Foraminiferal Genera and Their Classification*, Van Nostrand Reinhold Company, New York, 970 pp.

Lutze, G. F., and Altenbach, A., 1991, Technik und Signifikanz der Lebendfärbung benthischer Foraminiferen mit Bengalrot, *Geol. Jahrb.* **128**:251–265.

Murray, J. W., 1991, *Ecology and Palaeoecology of Benthic Foraminifera*, Longman Scientific & Technical, Harlow, 397 pp.

Naidu, T. Y., Rao, D. C., and Rao, M. S., 1985, Foraminifera as pollution indicators in the Visakhapatnam Harbour Complex, east coast of India, *Bull. Geol. Min. Met. Soc. India* **52**:88–96.

Pambianchi, P., Rosso, G., and Frascari, F., 1994, Nutrienti e sostanza organica nei sedimenti della Sacca di Goro, in: *Studio integrato sull'ecologia della Sacca di Goro in relazione ai problemi dell'inquinamento, della conservazione e dello sviluppo, 20 anno di ricerche* (S. Bencivelli, N. Castaldi, and D. Finessi, a cura di), Franco Angeli s.r.l., Milano, pp. 269–289.

Poag, C. W., 1978, Paired foraminiferal ecophenotypes in Gulf Coast estuaries: ecological and paleoecological implication, *Trans. Gulf Coast Assoc. Geol. Soc.* **28**:395–421.

Prater L. B., and Anderson M. A., 1977, A 96 hour bioassay of other creek Ohio, *Water Pollut. Control Fed.* **49**:2099–2106.

Pugnetti, A., Viaroli, P., and Ferrari, I., 1992, Processes leading to dystrophy in a Po River Delta lagoon (Sacca di Goro): phytoplankton-macroalgae interactions, in: *Marine Coastal Eutrophication* (R.A. Vollenweider, R. Marchetti, and R. Viviani, eds), Elsevier, Amsterdam, pp. 445–456.

Rinaldi, A., Montanari, G., Ghetti, A., Ferrari, C.R., and Ferrari, A., 1992, Eutrophy and dystrophy in the Goro Lagoon (Italy), in: *Marine Coastal Eutrophication*, (R. A. Vollenweider, R. Marchetti, and R. Viviani, eds.), Elsevier, Amsterdam, pp. 457–470.

Schafer, C. T., Smith, J. N., and Loring, D. H., 1980, Recent Sedimentation events at the head of Saguenay Fjord, Canada, *Environ. Geol.* **3**:139–150.

Seiglie, G. A., 1964, Significación de los foraminiferos anormales de la laguna Unare, *Lagena* **1**:6.

Sharifi, A. R., Croudace, I. W., and Austin, R. L., 1991, Benthic foraminiferids as pollution indicators in Southampton Water, southern England, U.K, *J. Micropaleontol.* **10**(1):109–113.

Stubbles S., 1993, Recent benthic Foraminiferida as indicators of pollution in Restronguet Creek, Cornwall, *Proc. Ussher Soc.* **8**:200–204.

Vollenweider, R. A., Marchetti, R., and Viviani, R. (eds.), 1992, *Marine Coastal Eutrophication*, Elsevier, Amsterdam, 1043 pp.

Walker, D., Linton, A. E., and Schafer, C. T., 1974, Sudan Black B: A superior stain to Rose Bengal for distinguishing living from non-living foraminifera, *J. Foram. Res.* **4**:205–215.

Walton, W. R., 1952, Techniques for recognition of living foraminifera, *Contr. Cushman Found. Foram. Res.* **3**:56–60.

Watkins, J. G., 1961, Foraminiferal ecology around the Orange County, California, ocean sewer outfall, *Micropaleontology* **7**:199–206.

Wright, R., C., 1968, Miliolidae (foraminiferos) recientes del estuario del Rio Quequén Grande (Prov. de Buenos Aires), *Rev. Mus. Argentino Cien. Nat. Bernardino Rivadavia, Hidrobiol.* **2**:225–256.

Yanko, V., Kronfeld, J., and Flexer, A., 1994, Response of benthic foraminifera to various pollution sources: Implications for pollution monitoring, *J Foram. Res.* **24**:1–17.

Yanko, V., Ahmad, M., and Kaminski, M.A., 1998, Morphological deformities of benthic foraminiferal tests in response to pollution by heavy metals: Implications for pollution monitoring, *J. Foram. Res.* **28**:177–200.

Zampi, M., and D'Onofrio, S., 1984, I foraminiferi dello stagno di S. Gilla (Cagliari), *Atti Soc. Tosc. Sci. Nat., Mem.* **91**:237–277

Zampi, M., and D'Onofrio, S., 1987, I foraminiferi della laguna di Levante (Orbetello, Grosseto), *Atti. Soc. Tosc. Sci. Nat. Mem.* **93**:101–127

Zaninetti, L., 1982, Les foraminifères des marais salants de Salin-de-Girand (Sud de la France): milieu de vie et transport dans le salin, comparaison avec les microfaunes marines, *Géologie. Mediter.* **9**:447–470.

Chapter 4

Impact of Anthropogenic Environmental Change on Larger Foraminifera

Tarawa Atoll, Kiribati, South Pacific

MICHAEL T. EBRAHIM

1. Introduction

Tarawa Atoll, the seat of government and main population center for the Republic of Kiribati, is located in the western Pacific Ocean at latitude 1°28′N and longitude 173°00′E (Fig. 1). The 35 islands of the republic are divided into

MICHAEL T. EBRAHIM • School of Earth Sciences, Victoria University of Wellington, Wellington, New Zealand.

Environmental Micropaleontology, Volume 15 of *Topics in Geobiology*, edited by Ronald E. Martin. Kluwer Academic/Plenum Publishers, New York, 2000.

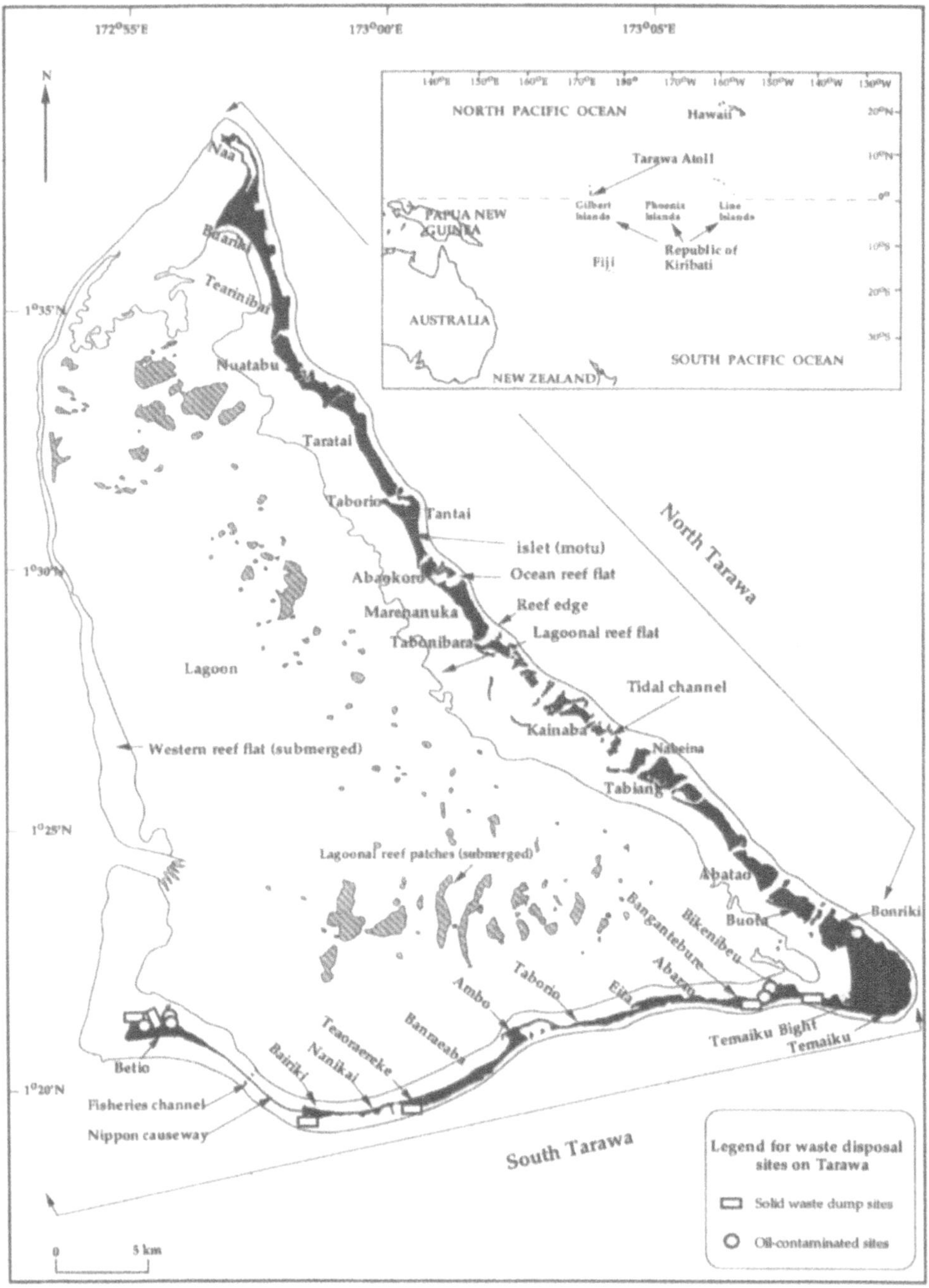

FIGURE 1. Location of Tarawa Atoll, islet (motu) names, morphological features, and sites for waste disposal.

three main groups, the Gilbert, Line, and Phoenix islands, with Tarawa Atoll located in the central Gilbert Group.

Tarawa is a low-lying atoll that consists of small islets (motu) no more than 3 m above sea level. As a result, it is extremely vulnerable to external environmental influences that lead to coastal erosion. Coastal erosion is a natural process and is affected by climatic events such as El Niño. The postulated increase in sea level within the next century (Wyrtki, 1990; Nunn, 1994) may also lead to erosion. Coastal erosion is also initiated and/or accelerated by artificial disturbances of the natural environment from human activity. For Tarawa, such disturbances include causeway construction, mining and removal of sand, and the impact of pollution.

The vulnerability of the atoll to the effects of environmental change has been of concern for many years (UNESCO, 1994), as Tarawa supports over 41% of the nation's total population of 75,900, with 27,300 people living on South Tarawa and 4000 on North Tarawa (Ministry of Environment and Natural Resources, pers. com. 1998). The small islet of Betio (the country's main port and commercial center; Fig. 1) had a population of 10,344 concentrated on 1.54 km^2 of land, and is the most densely populated part of Kiribati.

The present chapter discusses the overall sedimentary pattern of the atoll, with particular emphasis on foraminifera and their importance as a major component of sediment supply of different environments. The focus is mainly on the larger foraminiferal species (mostly 0.5–1.0 mm; typical adults of these species reach 1–2 mm in diameter) that supply approximately 35–50% of the sediment to land and beach areas. The presence and health of live larger foraminifera is of concern. Any change in the population abundance of these foraminiferal species will affect sediment supply to the islets and threaten atoll stability (see also in this volume Hallock, Ch. 5; Eagar, Ch. 6; Ishman, Ch. 16). Data strongly suggest that the difference in foraminiferal populations between North and South Tarawa may be the result of anthropogenic influences.

2. Physical Setting

Tarawa atoll is nearly triangular in shape with numerous low islets on the windward northeastern and southern sides (North and South Tarawa, respectively). The western (leeward) rim lacks motu, but there are several reef passages, which allow exchange of water between the lagoon and open ocean (Fig. 1).

Tarawa covers a total area of 490 km^2. The total land, lagoon, and intertidal areas are 30 km^2, 344 km^2, and 116 km^2, respectively. The land consists mostly of unconsolidated carbonate sand, which overlies coralline limestone and cemented carbonate conglomerate. The motu are typically less than 400 m wide (often less than 200 m) and are surrounded by extensive reef flats 800 m or more wide on both the lagoon and ocean sides. The lagoon reef flat of North Tarawa is significantly wider than average and extends about 1.5 km from the motu. Tarawa Lagoon is relatively shallow, with an average depth of 7 m and a maximum of 24 m (Richmond, 1997). Intermediate depths of 10–20 m occur in the central and western portions of the lagoon, and sand shoals, patches, and pinnacle reefs cover much of the lagoon floor.

3. Human Changes to the Environment

3.1. Causeway Construction

The islets forming Tarawa were originally separated by a natural system of channels between the open ocean and Tarawa Lagoon. Most of these passages are now closed by causeways, forming a barrier between the open ocean and the lagoon. A 28-km road now runs the full length of South Tarawa.

On North Tarawa, a number of causeways were constructed between the late 1970s and the late 1980s. In South Tarawa, causeways have been constructed across ocean–lagoon passages between Betio and Bairiki (Nippon Causeway), Bairiki and Nanikai (Bairiki Causeway), Nanikai and Teaoraereke (Andersen Causeway), Ambo and Taborio (Stewart Causeway), and at Temaiku Bight (Ananau Causeway). Nippon Causeway is the longest (3.4 km; Fig. 1) and was completed in 1987 under the sponsorship of the Japan International Co-operation Agency (1985). Part of this project included construction of an artificial channel to allow small fishing boats to pass between Tarawa Lagoon and the ocean (Fig. 1).

3.1.1. Effects of Causeway Construction

Construction of causeways, while practical for road services, appears to have had significant long-term social and ecological effects as well as affecting sediment distribution. Discussion with the inhabitants of Tarawa indicates that causeways have produced problems with land ownership regarding newly formed or disappearing land. This is compounded by an increasing population that now has easier access to remote areas and that has resulted in overfishing.

Causeway construction has also had an impact on sediment distribution. The structures are limiting or stopping tidal flow between the islets and thus preventing transport of sand between the lagoon and oceanside beaches (Gillie, 1991). Any reduction of the sand that nourishes these beaches will affect shoreline stability (Harper, 1989*a*, *b*). A hydrodynamic and water quality model suggests that the maximum flushing of the lagoon occurs in the central part of the western reef while poor circulation occurs in South Tarawa (Kimmerer and Walsh, 1981; Leva *et al.* 1993).

3.2. Water Pollution

Tarawa lagoon is significantly contaminated by human waste, particularly in the southeast where pollutants include fecal coliform bacteria. There are three major sources of pollution: (1) Bacterial wastes, which include sewer outfalls and domestic and other solid wastes [up to 75% of the estimated 6500 tons/year of the domestic waste is composed of benign organic matter (Royds Consulting Ltd. 1996]; (2) hydrocarbon contamination including hydrocarbon wastes and spills from commercial activities; (3) medical wastes, which include improper disposal of biohazardous hospital wastes such as tissues, cultures, stools, and plastics (Kaly, 1996). Other sources of pollution are related to the exposure of war debris and toxic wastes (Wilson, 1994). Locations of existing official dump sites for solid waste disposal and the main oil-contaminated sites in South Tarawa are indicated in Fig. 1. However, unregulated waste dumping is widespread throughout the intertidal area along South Tarawa. The main reason for the buildup of pollution is inadequate flushing of lagoonal water by fresh oceanic water, with particularly poor circulation occurring along South Tarawa. Water quality in the lagoon has declined since the 1977 cholera outbreak (Naidu *et al.* 1991). In South Tarawa, reopening a natural passage at the southeast corner (now the Ananau Causeway) has been suggested to reduce the public health risk from existing contamination of the southeast part of the lagoon (Royds Consulting Ltd. 1996). This recommendation was based on a hydrodynamic model which suggests that tidal flow will reestablish water circulation in the southern lagoon and will reduce pollutant concentrations by approximately 10%.

3.3. Sediment Removal

Sediments required for causeway construction, infrastructure, and road services are mainly derived from beaches, with limited production from reef flats and nearshore lagoon sediments. Sediment consumption rates have substantially increased from 17,616 m^3 in 1989 to 42,649 m^3 in 1993 (Biribo and Smith, 1994). Growing demand for sediment for construction is resulting in major coastal erosion problems; exploitation of the physical resources is outstripping sand supply (Forbes and Hosoi, 1995).

4. Methods of Study

4.1. Sample Collection

The study of the sediments of Tarawa is based on 164 surficial samples obtained by dredging and hand-sampling. Sampling sites were systematically selected to show sediment composition across all of the atoll's environments and to investigate possible anthropogenic influences. The sampling framework was designed to investigate areas where the expected differences owing to human impact would be highlighted in foraminiferal populations, and also to investigate those areas where natural tidal channels connecting the lagoon and the ocean are still open. Samples were collected along the reef flats, beaches, motu, and causeways of South Tarawa. These samples represent areas subjected to the most anthropogenic impact. Samples from North Tarawa were collected along the reef flats on ocean and lagoon sides, through tidal channels, and on beaches and motu. These samples represent relatively undisturbed environments.

Many samples were stained with Rose Bengal dissolved in alcohol to enable identification of living protoplasm in foraminiferal tests. Also, numerous areas of the back-reef flat, and ocean reef front were examined for living larger foraminifera.

4.2. Grain Size Analysis

Grain size distribution was obtained by dry-sieving each sample into eight grain size fractions: gravel [>4 mm (pebble) and >2 mm (granule)], five classes of sand [>1.0 mm (very coarse sand), >0.5 mm (coarse sand), >0.25 mm (medium sand), >0.125 mm (fine sand), and >0.063 mm (very fine sand)], and mud (<0.063 mm). This simplified sieving program provided a manageable number of fractions for compositional analyses.

4.3. Clast Composition Determination

Component analysis was undertaken on fractions coarser than 0.25 mm. For each fraction where sufficient material was available, a minimum of 300 clasts was each assigned to one of twenty-seven categories that make up the sediments. The composition of each sample was then calculated by dividing the number in each category for each size fraction by the total count of all the grains within that fraction and multiplying by the fraction mass. The results, therefore, represent 1200 or more identified clasts per sample. Statistical computations were performed within Excel spreadsheets.

5. Results

Grain size analyses show that well sorted, coarse sand predominates on the coasts of North and South Tarawa. Poorly sorted, medium sand dominates the land areas, and sediments within the lagoon are mainly poorly sorted, medium and fine sand, except near patch reefs, where sediments are coarse and poorly sorted.

The average composition of all samples studied, in descending order of abundance, is coral (41.2%), foraminifera (27.2%), mollusks (17.9%), and the calcareous green alga *Halimeda* (9.4%). The remainder includes fragments of crustaceans (1.9%), echinoids (1%), worm tubes (0.9%), bryozoans (0.2%), unidentified grains (0.2%), and beach rock (0.1%).

Sediment composition varies greatly among the environments sampled. Figure 2 illustrates the overall proportions of the main sediment components for each different environment. Coral debris increases in abundance progressively from the lagoon toward the ocean reef-flat areas. The abundance of foraminiferal tests in the sediments increases from offshore to onshore on both sides of the islets, and reaches highest concentrations in land samples. *Halimeda* and molluskan fragments are also important contributors. The proportion of *Halimeda* fragments is highest in the lagoon and decreases progressively toward the land areas. The amount of molluskan material is highest in the lagoon and decreases onshore, with the smallest amounts recorded in samples collected from tidal channels and motu areas.

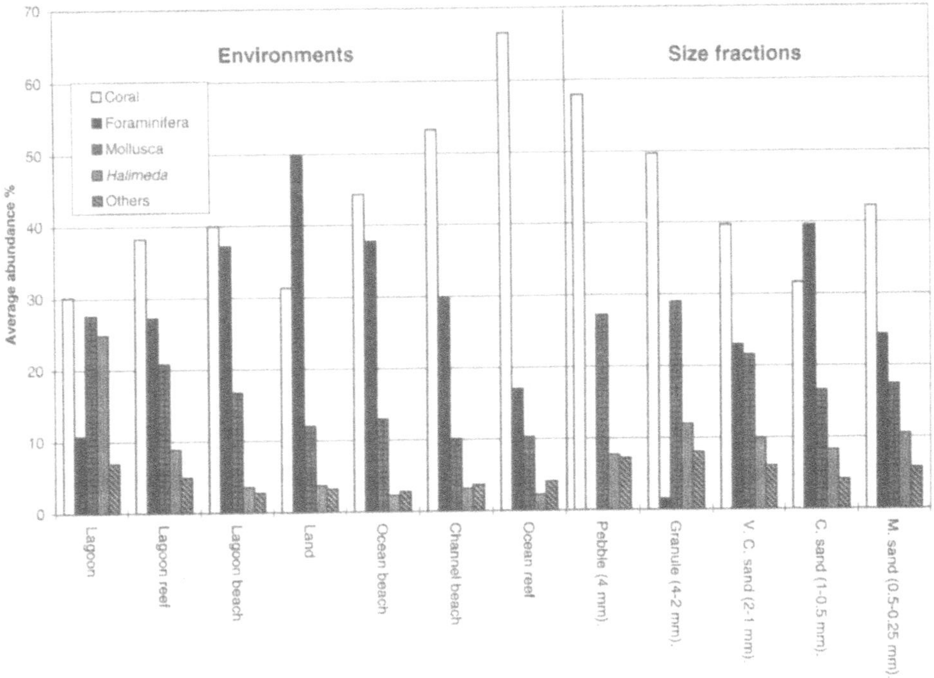

FIGURE 2. Average concentration of sediment components both in different environments and in various size fractions on Tarawa Atoll. The order in which environments are presented synthesizes east–west transects starting in the lagoon and ending on the other side of the islets on the ocean reef flat (cf. Fig. 3). The skeletal constituent category referred to as "Others" includes crustaceans, echinoids, polychaete worm tubes, unidentified grains, bryozoans and human debris. Size fractions are arranged from large (left) to small (right); V.C. sand (very coarse sand); C. Sand (coarse sand); M. sand (medium sand).

Figure 2 also shows the relationship between size fractions and average abundance of sediment contributors. The concentration of foraminifera is highest in coarse sand (40%), moderate in very coarse (23%) and medium sand (24%), and negligible in other size fractions.

5.1. Distribution of Foraminifera in Tarawa Sediments

The only published studies on foraminifera of the Gilbert Group were made for Onotoa Atoll, southeast of Tarawa, by Todd (1957, 1961). The present study focuses on foraminifera because they constitute the largest single component of sediment in motu areas (50%), and the second largest component in the overall sediment distribution of the atoll. Foraminiferal tests make up, on average, 27% of all grains identified and are abundant in sediments of reef patches, tidal channels and their beaches, ocean and lagoon beaches, and reef flats. Their concentration is lowest in the lagoon, particularly in the deeper parts (Fig. 3).

5.1.1. Species of Foraminifera

Nearly all species recorded in this study are well known in the shallow water sediments of the tropical Pacific (e.g., Cushman *et al.*, 1954; Todd and Low, 1960). Species of the suborder Rotaliina are the most abundant in quantitative terms, while small miliolids are the most diversified, but also the least abundant. In general, foraminifera are most diverse in the lagoon (approximately 30–40 species) with a few species concentrated in shallow water.

Foraminiferal tests in Tarawa sediments are predominantly sand-sized skeletal material (Fig. 2). The atoll has a small group of genera that are characteristic of tropical seas and have larger-sized tests compared to the general population of foraminifera. Typical adults of these genera reach 1–2 mm in diameter, although the largest test sizes exceed 2 mm.

Quantitatively six species, *Calcarina spengleri* (67%), *Marginopora vertebralis* plus *Sorites marginalis* (8%), *Amphistegina lessonii* plus *A. lobifera* (6%), and *Baculogypsina sphaerulata* (6%), make up most of the foraminiferal material identified (88%). The remainder includes species of the suborder Textulariina (4.8%), small rotaliines (4%), *Heterostegina depressa* (2.7%), and small miliolids (1.5%) (Fig. 4).

5.1.2. Average Abundance of Tests of Different Foraminiferal Species in Different Environments

The average abundance of different foraminiferal tests found in the different environments is shown in Fig. 4. *Calcarina spengleri* comprises, on average, more than 82% of the total of foraminiferal tests identified from land samples. Some of the latter samples have an average *C. spengleri* abundance of more than 90% (e.g., 101 (99.2%), 107 (91.2%), and 135 (91.8%);

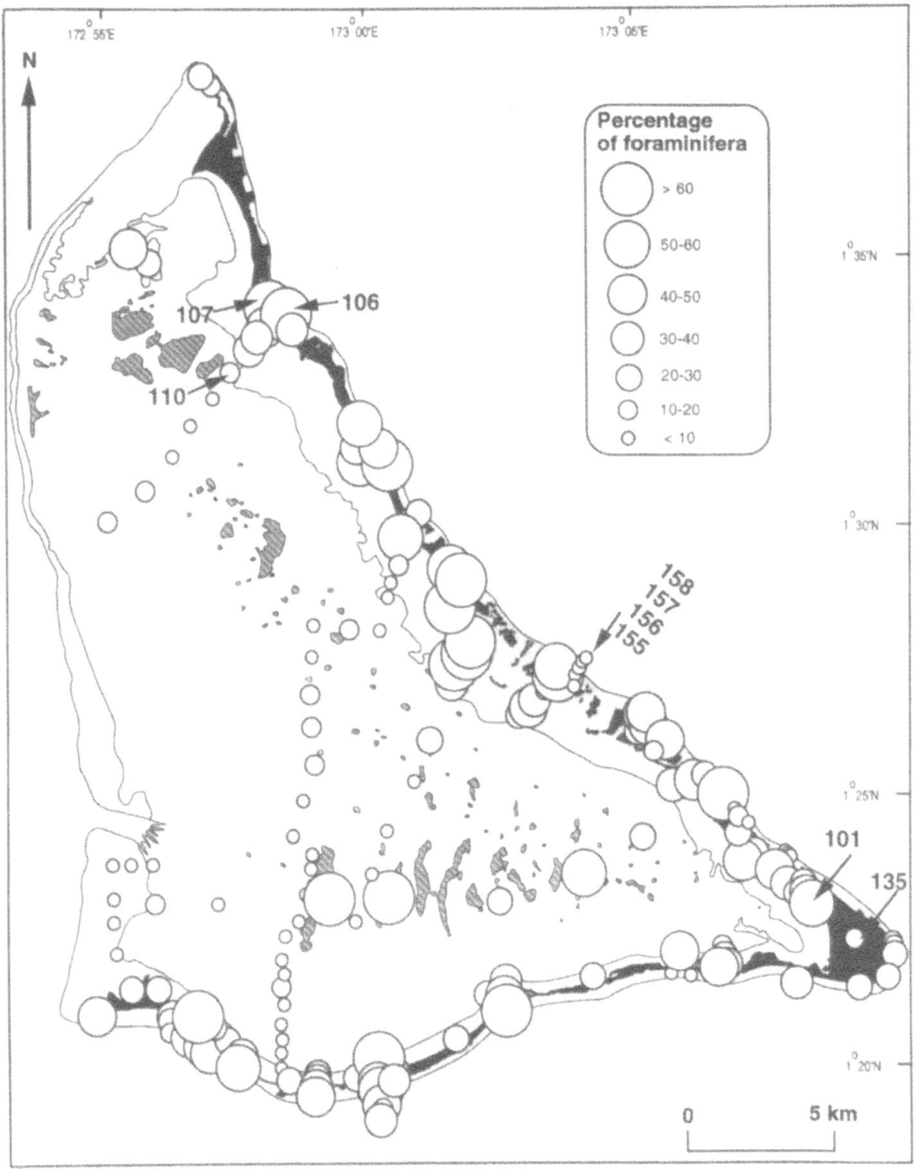

FIGURE 3. Contribution of foraminiferal tests to sediments of Tarawa Atoll. Foraminifera make the greatest contribution to sediments of land areas, the reef flats, and around reef patches. Foraminiferal concentrations are lowest in the lagoon, particularly in deeper areas. Sample sites referred to in the text are indicated.

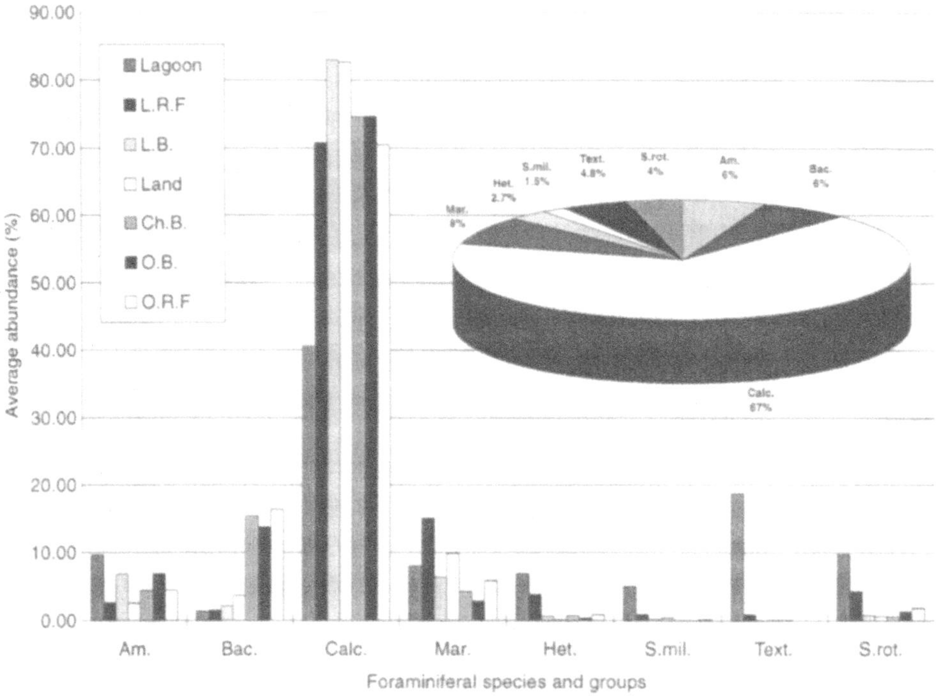

FIGURE 4. Average abundance of different foraminiferal species or groups of species in different environments. The pie graph shows the average percentage obtained for the different species/ families of foraminifera with respect to total foraminifera identified in all samples studied (Calc. = *Calcarina spengleri*; Bac. = *Baculogypsina sphaerulata*; Am. = *Amphistegina lessonii* and *Amphistegina lobifera*; S.rot. = Small rotaliines; Text. = Textulariina; S.mil. = Small miliolids; Het. = *Heterostegina depressa;* Mar. = *Marginopora vertebralis* and *Sorites marginalis*; L.R.F = lagoon reef flat; L.B. = lagoon beach; Ch.B. = channel beach; O.B. = ocean Beach, and O.R.F. = ocean reef flat).

Fig. 3]. On ocean reef flats, beaches, and tidal channels, sands are predominantly derived from tests of *C. spengleri,* but also show substantial contributions from tests of *B. sphaerulata* and *A. lessonii.* In general, intertidal samples are dominated by larger foraminifera. Smaller foraminifera are usually present but are numerically and volumetrically insignificant.

5.2. Living Foraminifera on Tarawa Atoll

No living foraminifera are found on lagoon or ocean reef flats of South Tarawa (Collen 1995; per. obs. 1998). However, the amount of these larger foraminifera in many South Tarawa sediments (Fig. 3) indicates that they have been derived from a local source and should be living nearby in abundance. Dead specimens typically appear highly abraded. The only live specimens recovered in South Tarawa were taken from the reef at the southeastern corner of the atoll in 1995. Here, living specimens of *B. sphaerulata* and *C. spengleri* contain an unusually large proportion of malformed specimens.

The two sites where living foraminifera were found in North Tarawa are on the ocean-facing facies around Nuatabu and Kianaba (Fig. 1), where samples 106 and 154–158, respectively, were collected (Figs. 1 and 3), with *B. sphaerulata* and *C. spengleri* again being the most abundant. Both sites are close to tidal channels where the lagoon has a free connection to the open ocean. Here, some of the algae important to the habitat of epiphytic larger foraminifera occur sporadically. Within the shallow lagoon areas where the water depth is less than 1.5 m some *C. spengleri* were observed living directly on reef rock rather than on algae (sample 110; Fig. 3).

The ocean beach sand of North Tarawa contains dense concentrations of *C. spengleri* and *B. sphaerulata*. These tests are well preserved, with all tests possessing all their spines. The well-preserved appearance and lack of any abnormalities or abrasions not only implies the closeness of the living populations, but also the ecologic health of the area. In contrast, tests of *B. sphaerulata* and *C. spengleri* from beach and reef-flat sand on South Tarawa show deformation of a large part of their total assemblage, and exhibit morphological deformation, abraded appearance, and a lack of spines. While test deformation may occur naturally in a small proportion of the population as a result of excess calcification or epiphytic habit, a high number of deformed tests is possibly the result of polluted conditions (Alve, 1995).

6. Discussion and Conclusion

Foraminifera make a major contribution to the sediment budget of Tarawa. They constitute 50% of motu sediments, more than 35% of beach sands, and approximately 12% of lagoon sediments. The abundance of foraminiferal tests in South Tarawa sediments implies that live larger foraminifera should also be plentiful, yet it appears that they are either absent or very scarce in the reefs and reef flats of South Tarawa.

This chapter suggests that the main reason for the absence of live foraminifera is anthropogenic pollution. Benthic foraminifera have long been demonstrated to be sensitive to coastal pollution. Bandy *et al.* (1964) and Yanko *et al.* (1994) showed an inverse relationship between contamination and live foraminiferal abundance in polluted localities, and Alve (1995) has shown that morphological deformation is typical in polluted localities.

Along South Tarawa, the natural environment has been heavily modified over the last 40 years. The concentration of human population is considered a relative measure of pollution. Population increase has been dramatic, with the growth rate climbing from 0.4% per annum to the current rate of 2% per annum. More than half of the population of Tarawa lives on three islets: Betio (10,344), Bikenibeu (4885), and Bairiki (2153), with land areas 1.54 km^2,

1.81 km^2, and 0.46 km^2, respectively (Royds Consulting Ltd, 1996). The increasing population has placed greater demands on the natural resources of the atoll, from the removal of sediments for construction to dumping of waste materials. The small size of the land (motu) area and lagoon mean that human impact is concentrated in a comparatively small area, but as water circulation to and from the ocean is limited by causeways, pollution has become a persistent problem.

Evidence of pollution's effects is widespread: most of the coral and calcareous algae around South Tarawa reef flat are dead, with only a few live patches along the extreme seaward edge of the ocean reef flat (pers. obs. 1998). Kaly (1996) found decreases in coral abundance and abnormalities in fish caught near sewer outfalls, and reported detrimental impacts on 44% of the 150 organisms and physical features examined that were exposed to sewage.

In this study, dead foraminifera were markedly more abundant than live in polluted localities, and one species in particular appeared to be very sensitive to contamination. *B. sphaerulata* accounts on average for 20–>35% of total foraminifera collected along North Tarawa's ocean-facing facies; in South Tarawa, the same facies close to pollution sources have typically less than 5% *B. sphaerulata* concentrations, and none over 10%.

However, the presence of morphological deformations in many tests is not as obviously linked to pollution, although Alve (1995) has shown that deformation may be associated with pollution. In this study, live foraminiferal populations were recovered from an ocean reef site containing a high proportion of *B. sphaerulata* and *C. spengleri* with test abnormalities. The site was near the most windward point of the atoll, where there is continual water exchange and most other reef biota (including coral and fish) are flourishing. In the absence of a pollution source, a natural mechanism such as solar irradiation or temperature sensitivity might be responsible (Williams *et al.* 1997). However, it still seems plausible that at least some of the test deformations present in South Tarawa foraminifera were associated with pollution, as the proportion of deformed specimens from North Tarawa was negligible. It is possible that the "rogue" population is simply more sensitive to a low level of some pollutant than the other reef biota.

The absence of live foraminifera in South Tarawa leads to the bigger issue of the sediment cycle. Without a source of live foraminifera, it seems evident that a major component of South Tarawa's sediment is at risk. Live foraminifera from ocean-side facies have access to the lagoon through North Tarawa's tidal channel; South Tarawa's channels are blocked by causeways. South Tarawa is also removing sediment for infrastructure development. This has implications for the stability of the islets and, hence, for the people who live on them.

Reestablishment of the sediment cycle would require reestablishment of living foraminiferal populations around South Tarawa. Lutze (1965) was the first to report that the return of favorable environmental conditions after an adverse period will lead to rapid recolonization by benthic foraminifera. Similarly, Collen (1996) has reported rapid recolonization of the ocean reef flats of Funafuti Atoll, Tuvalu, by larger foraminifera following their complete removal by tropical cyclone Bebe in 1972. This implies that if pollution can be reduced in Tarawa, foraminiferal populations will recover rapidly and the sedimentary cycle can be reestablished within a short period of time. Reopening selected channels (perhaps by building culverts under the causeways) would be one way to both decrease pollutant concentrations and restart sediment transport.

In conclusion, it is quite possible that anthropogenic changes have had a serious impact on Tarawa's natural environment. The effect on the foraminifera population and sediment budget is twofold:

1. Human-induced pollution and limited water circulation appear to be affecting live foraminiferal distribution and abundance. This is directly reducing the availability of an important sediment component.
2. The sediment cycle itself is being disrupted by anthropogenic removal of sediment and restriction of sediment transport through motu channels.

However, if some of these detrimental effects of human activity can be mitigated, the sediment cycle is likely to reestablish itself in a short space of time.

ACKNOWLEDGMENTS: I thank Profs. Peter Barrett and Euan Smith and Dr. John Collen for their comments; particular thanks to Prof. Ronald Martin of the University of Delaware for reviewing the manuscript and the valuable suggestions; Dr. George Scott at the Institute of Geological and Nuclear Sciences for the valuable discussions; Mr. Robert Smith and Dr. R. Howorth of the South Pacific Applied Geoscience Commission (SOPAC), who provided bathymetric data; Ms Naomi Biribo at the Ministry of Environment and Natural Resources, and Ms Baranika Kamaie at the Ministry of Works and Energy, Tarawa Atoll, Kiribati, for providing logistical support during the field work. The study that this chapter presents was supported by New Zealand Ministry of Foreign Affairs and Trade, SOPAC, and the Internal Grants Committee of Victoria University of Wellington.

References

Alve, E., 1995, Benthic foraminiferal responses to estuarine pollution: A review. *J. Foram. Res.* **25**:190–203.

Bandy, O. L., Ingle, J. C., and Resig, J. M., 1964, Foraminifera, Los Angeles County Outfall Area, California, *Limnol. Oceanogr.* **9**:124–137.

Biribo, N., and Smith, R., 1994, Sand and gravel usage, South Tarawa, Kiribati, 1989–1993, South Pacific Applied Geoscience Commission (SOPAC) Preliminary Report No. 75, Suva, Fiji, p. 14.

Collen, J. D., 1995, Preliminary comment on visit to Funafuti, Majuro and Tarawa atolls, June–July 1995, South Pacific Applied Geoscience Commission (SOPAC) Preliminary Report No. 81, Suva, Fiji, p. 6.

Collen, J. D., 1996, Recolonisation of reef flats by larger foraminifera, Funafuti, Tuvalu. *J. Micropalaeontol.* **15**:130.

Cushman, J. A., Todd, R., and Post R., 1954, Recent foraminifera of the Marshall Island, Bikini and nearby atolls, *Oceanogr. (Biol.). U.S. Geol. Surv. Prog. Pap.* 260-H, pp. 319–384, pls. 82–93, figs. 116–118.

Forbes, D. L., and Hosoi, Y., 1995, Coastal erosion in South Tarawa. Report of mission to determine shoreline recession rates at Tungaru Central Hospital & King George V/Elaine Bernacchi School, erosion and sedimentation along Nippon Causeway, and coastal protection strategies in Kiribati, June–July 1995, South Pacific Applied Geoscience Commission (SOPAC) Technical Report No. 225, Suva, Fiji, p. 91.

Gillie, R. D., 1991, Beach profile surveys on Betio and Bairiki and along the Nippon Causeway, South Tarawa, Kiribati, January 1991, South Pacific Applied Geoscience Commission (SOPAC) Preliminary Report No. 37, Suva, Fiji, p. 22.

Harper, J. R., 1989*a*, Follow-up survey of the Betio/Bairiki causeway, Tarawa, Republic of Kiribati, South Pacific Applied Geoscience Commission (SOPAC) Technical Report No. 86, Suva, Fiji, p. 22.

Harper, J. R., 1989*b*, Evaluation of beach profile data from Betio and Bairiki, Republic of Kiribati, South Pacific Applied Geoscience Commission (SOPAC) Technical Report No. 94, Suva, Fiji, p. 136.

Japan International Co-operation Agency (JICA), 1985, Basic design study report on Betio-Bairiki causeway—fisheries channel project in the Republic of Kiribati, JICA Report CR (2) 85–66, p. 103 & 41 appendices (in Japanese, English language version available as Report CR (2) 85–67).

Kaly, U. L., 1996, South Tarawa sanitation and public health project, Republic of Kiribati, Initial Environmental Examination (IEE) and an Environmental Impact Assessment (EIA), Report No. TA 2641–KIR, p. 126.

Kimmerer, W. J., and Walsh, T. W., 1981, Tarawa Atoll Lagoon: Circulation, nutrient fluxes, and the impact of human waste, *Micronesia* **17**:161–179.

Leva, D., Chen, C. W., and Gomez, L. E., 1993, Hydrodynamic and water quality modelling of Tarawa Lagoon, Republic of Kiribati progress report, Systech Engineering, Inc. report prepared for the Republic of Kiribati and the United States Agency for International Development, p. 39.

Lutze, G. F., 1965, Zur Foraminiferen-fauna der Ostsee, *Meyniana* **15**:75–142.

Naidu, S., Aalbersberg, W. G. L., Brodie, J. E., Fauvao, V. A., Maata, M., Naqasima, M., Whippy, P., and Morrison, R. J., 1991, Water quality studies on selected South Pacific lagoons, UNEP Regional Seas and Studies No. 136, and South Pacific Regional Environmental Programme (SPREP) Reports and Studies No. 49, p. 35.

Nunn, P. D., 1994, *Oceanic Islands*, Blackwell, Oxford, p. 413.

Richmond, B. M., 1997, Reconnaissance geology of the atoll islets of the Gilbert Islands, Kiribati: Coastal and Environmental Geoscience Studies of the Southwest Pacific Islands, in: (A. M. Sherwood, comp.) South Pacific Applied Geoscience Commission (SOPAC) Technical Bulletin No. 9, Suva, Fiji, pp. 25–50.

Royds Consulting Ltd., 1996, Sanitation and Public Health Project, Republic of Kiribati, Final Report, prepared for Asian Development Bank (ADB), No. TA 2497–KIR, p. 248.

Todd, R., 1957, Smaller foraminifera, in geology of Saipan, Mariana Islands; pt. 3, Palaeontology, *U.S. Geol. Survey Prof. Pap.* 280–H, pp. 265–320.

Todd, R., 1961, Foraminifera from Onotoa Atoll, Gilbert Islands, *U.S. Geol. Survey Prof. Pap.* 354-H, pp. 171–191.

Todd, R., and Low, D., 1960, Smaller Foraminifera from Eniwetok drill holes, *U.S. Geol. Survey Prof. Pap.* 260-X, pp. 799–861.

UNESCO, 1994, Island Agenda. An Overview of UNESCO's Work on Island Environments, Territories, and Societies, UNESCO, Paris, France. Adolarc, p. 130.

Williams, D. E., Hallock, P., Talge, H. K., Harney, J. N., and McRae, G, 1997, Responses of *Amphistegina gibbosa* populations in the Florida Keys (U.S.A.) to a multi-year stress event (1991–1996), *J. Foram. Res.* **27**:264–269.

Wilson, C., 1994, Kiribati State of the Environment Report. Report for South Pacific Regional Environment Programme (SPREP) as part of the Kiribati National Environmental Strategy (NEMS), *Apia*, p. 46.

Wyrtki, K., 1990, Sea-level rise: The facts and the future, *Pacif. Sci.* **44**:1–16.

Yanko, V., Kronfeld, J., and Flexer, A., 1994, Response of benthic foraminifera to various pollution sources: Implications for pollution monitoring, *J. Foram. Res.* **24**:1–17.

Chapter 5

Larger Foraminifera as Indicators of Coral-Reef Vitality

PAMELA HALLOCK

1. Introduction

Human activities are impacting ecosystems on a global scale; Vitousek *et al.* (1997*a*) asserted that no ecosystem on the Earth's surface is free of human influence. The land, atmosphere, and hydrosphere have all been altered to varying degrees. Up to 50% of the Earth's land area has been transformed or degraded (Vitousek *et al.*, 1997*a*). As a result of stratospheric ozone depletion, intensities of biologically damaging ultraviolet radiation (UVB) at 20°N latitude between April and August now exceed the June 1969 (summer solstice) maximum (Shick *et al.*, 1996). Carbon dioxide concentration in the atmos-

PAMELA HALLOCK • Department of Marine Science, University of South Florida, St. Petersburg, Florida 33701.

Environmental Micropaleontology, Volume 15 of *Topics in Geobiology*, edited by Ronald E. Martin. Kluwer Academic/Plenum Publishers, New York, 2000.

phere has increased by nearly 30% since the beginning of the Industrial Revolution (Schimel *et al.*, 1995), with potential influences ranging from climate changes to altered ocean chemistry. Human activities have effectively doubled the annual transfer of nitrogen from the atmospheric pool of N_2 to biologically-available fixed nitrogen (Schnoor *et al.*, 1995; Vitousek *et al.*, 1997*b*). Much of this fixed nitrogen, along with nitrous oxides from fossil-fuel burning (e.g., Prinn *et al.*, 1990), is washed by the rains into aquatic systems.

Coastal systems are among those most impacted by human activities. Approximately 60% of the Earth's humans live within 100 km of an ocean (Vitousek *et al.*, 1997*a*). Coastal waters are typically the receptacles of runoff from land, including dissolved and particulate inputs resulting from human activities. Increased sediment loads from denuded terrestrial environments can have devastating, although generally localized, impacts on coastal systems (e.g., Hatcher *et al.*, 1989). Nutrient input into coastal waters tends to increase in direct proportion to human populations in the watershed (Walsh, 1984). The effects of nutrient runoff range from expanding "dead zones" in the delta vicinity (e.g., Rabalais *et al.*, 1996; Sen Gupta *et al.*, 1996) to chlorophyll-enriched plumes that can be detected hundreds to thousands of kilometers from river mouths (e.g., Müller-Karger *et al.*, 1989; Hallock *et al.*, 1993*a*). Some chemical pollutants may have only local influence, while others may be concentrated through food chains, with possible effects on organisms thousands of kilometers from the source (e.g., references in Colborn *et al.*, 1996). Sediments, nutrients, and chemical pollutants can also be transported via the atmosphere, entering the ocean as dust particles or carried by rainwater (e.g., Fanning, 1989; Martin *et al.*, 1989).

1.1. Human Impacts on Coral Reefs

Coral reefs are not exempt from anthropogenic influence. Bryant *et al.* (1998) estimated that approximately 60% of the Earth's coral reefs are threatened by human activities that include nutrient and other chemical pollution, sedimentation, destructive fishing practices, and shipping. Moreover, most coral reefs naturally occur in clear, relatively sediment-free, nutrient-poor subtropical and tropical coastal waters (e.g., Birkeland, 1997). Because reef-building corals are predominantly long-lived colonial organisms, subtle changes in water quality can have sublethal effects ranging from reduced growth and reproduction rates of coral colonies to chemical interference with the metamorphosis of free-swimming coral larvae to attached juvenile corals (Richmond, 1993). Such factors can render reef-building corals less competitive than macroalgae, sponges, and other nonsymbiotic or noncalcifying organisms.

Another consequence of the dependence of coral reefs on clear water is the potential susceptibility of reef-dwelling organisms to sublethal effects of increasing intensities of UVB radiation that are the result of stratospheric ozone depletion (e.g., Shick *et al.*, 1996). The pervasive misconception that

UVB is rapidly absorbed by seawater is based on measurements in coastal waters that contain significant concentrations of UVB-absorbing phytoplankton and dissolved organic matter (Bricaud *et al.*, 1981). In fact, in clear oceanic and reefal waters, potentially damaging intensities of UVB can penetrate to tens of meters of water depth (Smith and Baker, 1979; Gleason and Wellington, 1993).

Potential ways that UVB can influence reef communities and individual reef organisms are many. UVB-tolerant organisms such as cyanobacteria may produce nuisance blooms (e.g., Butler *et al.*, 1995). Corals living in shallow waters produce chemical sunscreens to protect themselves from UVB damage (e.g., Shibata, 1969; Dunlap *et al.*, 1986). Bleaching, a stress response whereby the coral expels its zooxanthellae, occurs in response to elevated intensities of UVB, particularly if water temperature rises simultaneously (e.g., Glynn *et al.*, 1993; Reaka-Kudla *et al.*, 1994). Elevated water temperatures increase metabolic demands, so the rate of sunscreen production declines, leaving the corals more susceptible to UVB damage. Other sublethal effects of UVB include damage to photosynthesis, intracellular functions, and behavioral responses, as well as to general fitness (e.g., Hadar and Worrest, 1991; Shick *et al.*, 1996).

Although no conclusive evidence for UVB damage in field populations of corals has been reported (Glynn, 1996; Shick *et al.*, 1996), it is not unreasonable to suspect that the proliferation of disease outbreaks (e.g., Peters, 1997; Santavy and Peters, 1997; Richardson *et al.*, 1998) and of coral bleaching events (e.g., Williams and Bunckley-Williams, 1990; Glynn, 1996; Brown, 1997*a*) worldwide in the 1980s and 1990s have been exacerbated by sublethal stress caused by increasing UVB dose rates. Reef decline has been observed in areas, for example, of the Florida Keys (e.g., Dustan and Halas, 1987; Porter and Meier, 1992) that are protected from destructive overfishing and isolated from the effects of terrestrial sedimentation and where conventional measures of water chemistry cannot detect a reduction in water quality (Szmant and Forrester, 1996). Furthermore, adult corals can often tolerate subtle changes in water quality that preclude coral recruitment and therefore long-term survival of the reef-building community (e.g., Richmond, 1993). Such observations have generated questions of how to distinguish anthropogenically diminished coral-reef vitality from natural fluctuations in coral communities.

1.2. Reef-Dwelling Foraminifera as Indicators

I proposed (Hallock, 1996, 1998) that reef-dwelling foraminifera, especially larger taxa that host algal symbionts, hold substantial potential as indicators of reef vitality. This proposal was based on several practical and physiological arguments:

1. Larger foraminifera are widely used as paleoenvironmental indicators in studies of fossil limestones.
2. Smaller foraminifera have long proven useful in pollution studies.

3. Physiological analogies between zooxanthellate corals and foraminifera with algal symbionts result in similar environmental requirements.
4. Similar kinds of stress symptoms have been observed in foraminiferal populations and assemblages as those reported for corals and coral-reef communities.
5. The relatively short life spans of symbiont-bearing foraminifera as compared with long-lived colonial corals can potentially enable differentiation between long-term reef decline associated with declining water quality and temporary reef decline associated with natural episodic mortality events.
6. Records of environmental change, as indicated by changes in foraminiferal assemblages, are available in reef sediments.
7. A readily identifiable genus, *Amphistegina*, is nearly circumtropical in distribution and is widely abundant in healthy reef environments.
8. The amenity of some foraminifera, especially selected species of *Amphistegina*, to culture and experimentation provides the potential for developing laboratory and field bioassay protocols for assessing suitability of an environment for reef growth.
9. Foraminifera with algal symbionts, particularly *Amphistegina* spp., are relatively small and abundant in comparison to macroinvertebrates, permitting statistically significant sample sizes to be collected quickly and inexpensively, for either assemblage assessment or for experimental purposes, with minimal impact on reef resources.

The first two practical arguments will not be developed here. The first goal of this chapter is to summarize the physiological analogies and differences between larger foraminifera and corals that provide the basis for use of foraminifera as tools in reef research. The second goal is to provide examples of how reef-dwelling foraminifera have been used or potentially can be utilized in studies of reef vitality.

2. Analogies and Differences between Foraminifera and Corals

2.1. Response to Nutrification

The principal physiological analogy between reef-building corals and larger foraminifera is the dependence of both groups on algal symbionts to enhance growth and calcification (e.g., Lee and Anderson, 1991). I recognized this analogy (Hallock, 1981*a*) when I used examples from both groups to develop a model to predict the energetic benefits of algal symbiosis. I concluded that this mode of life was highly advantageous when dissolved nutrients (i.e., NH_N^+, NO_2^-, NO_3^-, PO_4^{3+}) and particulate food resources were scarce. Subsequent papers by my colleagues and myself applied this insight to both foraminiferal ecology and evolution (e.g., Hallock, 1982, 1985; Hallock *et al.*, 1991) and to coral-reef ecology and carbonate sedimentation (e.g., Hallock and Schlager, 1986; Hallock 1988*a*; Hallock *et al.*, 1993*a*). Subsequent physiological studies of corals (e.g., Falkowski *et al.*, 1993; Steven and Broadbent, 1997) and of larger foraminifera (Lee, 1998) have shown that fixed-nitrogen limitation is crucial to maintenance of the host–symbiont relationship.

We (Hallock and Schlager, 1986; Hallock, 1987; Hallock *et al.*, 1993*a*) (Birkeland (1977, 1988, 1997) independently recognized that nutrient availability in a locale or region is a major control on benthic community structure, especially in subtropical and tropical seas. We predicted that, as the nutrient supply increases, reef-building coral domination of the benthos will gradually give way to macroalgal domination as algal symbiosis becomes less advantageous. Somewhat higher nutrient flux will promote phytoplankton blooms in the water column, limiting light penetration to the benthos and promoting the dominance of the benthos by nonsymbiotic filter-feeding animals such as sponges, ascidians, and bivalves, and of detritus-feeding echinoderms and crustaceans that do not directly require sunlight for survival.

I predicted (Hallock, 1987, 1988*a*) that benthic foraminiferal assemblages should respond analogously to nutrient flux. In very low nutrient marine environments, such as those found around most Pacific atolls, larger foraminifera totally dominate sand-sized sediments in reef systems (e.g., McKee *et al.*, 1956; Hallock, 1981*b*). But as nutrient supplies increase, bioeroded coral fragments, calcareous algae, molluscan debris, and smaller herbivorous and detritivorous foraminifera become increasingly common as sediment constituents. As the environment becomes unsuitable for the survival of foraminifera with algal symbionts, their dead tests will become increasingly rare in the sediments, and the remnants become increasingly corroded (e.g., Cottey and Hallock, 1988). This takes place under conditions of "nutrification," i.e., increase in nutrient flux that results in change in community structure (e.g., Cockey *et al.*, 1996), but usually not in a measurable increase in dissolved nutrients in the water column because the planktic and benthic communities are able to incorporate and utilize all available nutrients (e.g., Laws and Redalje, 1979). True "eutrophication," which is nutrification to the degree that organic carbon buildup occurs in bottom waters and sediments, results in further change in benthic community structure: domination by opportunistic taxa that can tolerate episodic anoxia (e.g., Alve, 1995).

Nutrification to the extreme of eutrophication should seldom occur in high-energy open-shelf and reef-margin environments, where mixing processes are active. Therefore, an important application of larger foraminifera may be to differentiate between nutrification-induced decline in coral dominance in a reef environment as compared with decline in response to episodic mortality events such as temperature extremes or hurricanes, which are independent of water quality. One reason that foraminiferal assemblages should be particularly useful in such cases is the fundamental difference between long-lived colonial corals and, by comparison, the relatively short generation times of larger foraminifera of a few months to a year or two (Cockey *et al.*, 1996). Larger foraminiferal populations would be expected to decline in response to chronic decline in water quality, and to do so faster than long-lived adult corals. On the other hand, larger foraminiferal populations should rebound quickly after episodic mortality events such as temperature extremes and hurricanes. Furthermore, there is no reason to suspect that foraminifera were directly affected by the so-called white-band disease that devastated acroporid populations Caribbean wide in the 1980s (Gladfelter, 1982; Peters, 1997). This disease may have resulted from the introduction of a new pathogen into immunologically naive populations and appears to be independent of water quality (Goreau *et al.*, 1998). In fact, the dominant larger foraminifera in reef environments, *Amphistegina* spp., dwell on dead-coral rubble and phytal substrates. If water quality is suitable, decline in coral cover should result in an increase in habitat for *Amphistegina* and therefore increased abundance of their tests in the sediments.

2.2. Comparisons of Existing Problems in Corals and Foraminifera

Ongoing problems contributing to the decline of coral reefs include bleaching (e.g., Glynn, 1996; Brown, 1997*a*), disease outbreaks (Peters, 1997; Santavy and Peters, 1997; Richardson *et al.* 1998), and predator outbreaks (e.g., Birkeland and Lucas, 1990; Carpenter, 1997). If these problems are specific to reef-building corals, and not associated with declining water quality [e.g., as suggested by Szmant and Forrester (1996) for declining coral populations along the Florida reef tract], the results could still include community changes from coral or coral–algal dominance to dominance by macroalgae, with loss in reef-building capability. Epiphytic, symbiont-bearing foraminifera should benefit from increased algal cover, if other aspects of the environment are suitable. However, some larger foraminifera are, in fact, exhibiting stress responses very similar to those of reef-building corals, which indicates that some problems, at least on the Florida reef tract, are environmental rather than coral-specific.

2.2.1. Bleaching

According to Glynn (1996), bleaching in zooxanthellate corals and other reef-dwelling organisms involves the temporary or permanent loss of photosynthetic microalgae and/or algal pigments. Small-scale bleaching events are commonly associated with specific disturbance events such as temperature, light, or salinity extremes. Large-scale mass bleaching events are more difficult to explain, though present evidence implicates elevated sea-surface temperatures associated with global climatic phenomenon, especially El Niño/Southern Ocean Oscillation events. Moreover, corals also bleach in response to exposure to elevated intensities of biologically damaging UVB radiation (Glynn, 1996), and are particularly susceptible to the combination of elevated temperature and UVB (Glynn *et al.*, 1993; Reaka-Kudla *et al.*, 1994). Thus, a logical hypothesis for the onset of mass bleaching events worldwide beginning in the 1980s is the combined stress caused by global warming and ozone depletion.

Bleaching (Fig. 1a, b) was not reported in *Amphistegina* field populations prior to 1988, despite extensive collection of *Amphistegina* populations around the world in the 1970s and 1980s (e.g., Hallock Muller, 1974; Hallock and Larsen, 1979; Hallock, 1984; Hallock *et al.*, 1986*b*). In February 1988, a small sample from the Bahamas, collected during a post-bleaching coral survey, was found with several "mottled" specimens. Collections in the Florida Keys in early May 1991 produced normal specimens, but in late June yielded many specimens that appeared granular to slightly mottled, which indicated the onset of symbiont loss. In September 1991, *Amphistegina* populations throughout the Florida Keys were extensively mottled and bleached (Hallock *et al.*, 1993*b*). In subsequent years (1992–1996), onset of symbiont loss occurred each year in March, proportions of the populations exhibiting symbiont loss peaked in June–July, and populations began to recover in late summer (Talge *et al.*, 1997; Williams *et al.*, 1997). Hyperthermal stress is not a factor in symbiont-loss observed in *Amphistegina*, because sea-surface temperatures in the Florida Keys were typically lowest in March, when symbiont loss symptoms were increasing, and highest in August/September, when recovery was evident in the populations (Hallock *et al.*, 1995).

Bleaching in *Amphistegina* results from deterioration and digestion of the diatom endosymbionts, followed by autolysis of organelles and vacuolation and deterioration of the host cytoplasm (Talge and Hallock, 1993, 1995). The process is degenerative and occurs over several days to a month or more. Affected populations also exhibit anomalously high proportions of shell breakage (up to 40% of affected populations as compared to 5% in pre-bleaching-event populations) (Toler and Hallock, 1998), anomalous proportions of deformed individuals, and both predation (Hallock and Talge, 1994) and cyanobacterial infestation not previously observed (Hallock *et al.*, 1995).

Observation of mottling, breakage, and infestation in *Amphistegina* populations as far away as the West Australian shelf (Talge *et al.*, 1997) argues against local pollution as a causal mechanism. On the West Australian shelf in January 1996, the proportions of specimens exhibiting symbiont loss increased from less than 10% of populations at 23°S latitude, to more than 30% at 30°S (Hallock *et al.*, 1996). The worldwide occurrence of symbiont loss, the latitudinal trend seen in Australia, the seasonal cycle in the Florida Keys, the tendency for diminishing severity of symptoms with depth, and the types of damage (e.g., photosynthesis, shell-protein synthesis, reproductive damage, and loss of defense mechanisms) are all consistent with biologically damaging UVB radiation as a causal mechanism. Laboratory experiments support this hypothesis (Hallock *et al.*, 1995).

Thus, bleaching in *Amphistegina* is not identical to that in corals. However, if increasing intensities of UVB associated with ozone depletion are causing bleaching in these reef-dwelling foraminifera, the logical assumption

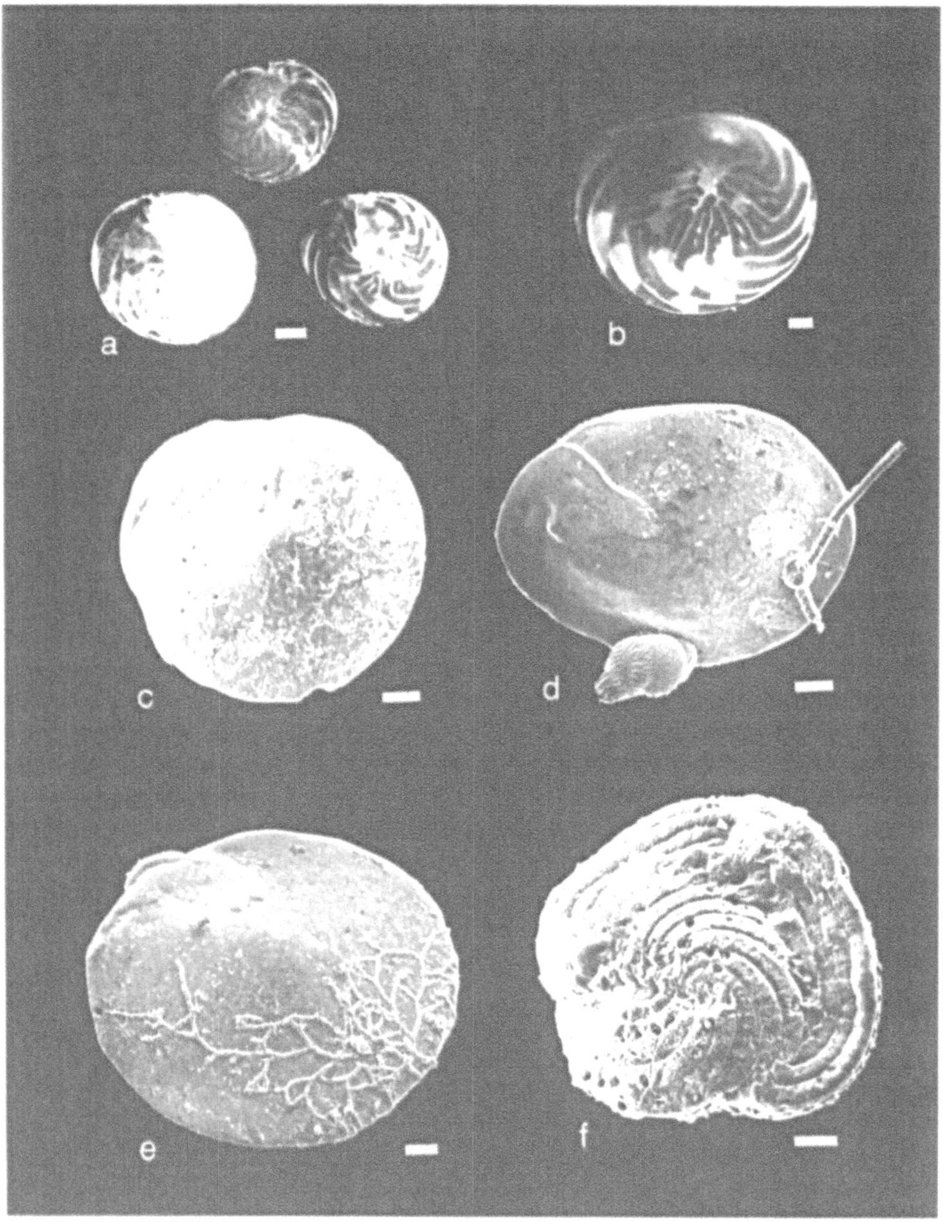

FIGURE 1. (a) Clockwise from top: normal, mottled, and bleached *Amphistegina gibbosa* from Florida Keys; (b) mottled *A. lessonii* from the West Australian Shelf; (c) scanning electron micrograph (SEM) of endolithic cyanobacterial infestation in the test of an *A. gibbosa* specimen collected live from the Florida Keys; (d) SEM of epiphytic red algae and *Floresina amphiphaga* predation on the test of an *A. gibbosa* specimen collected live from the Florida Keys; (e) SEM of an unidentified epiphytic infestation in the test of an *A. gibbosa* specimen collected live from the Florida Keys; (f) SEM of endolithic and epiphytic infestation on the test of an *Cyclorbiculina compressa* specimen collected live from the Florida Keys (all scale bars = 0.1 mm). (Reprinted from Hallock, P., *et al.*, 1995, A new disease in reef-dwelling foraminifera: Implications for coastal sedimentation. *J Foram. Res.* **25**(3):280–286; reprinted with permission of the *Journal of Foraminiferal Research.*)

is that UVB stress may be contributing to the susceptibility of corals to temperature stress (e.g., Glynn *et al.*, 1993; Reaka-Kudla *et al.*, 1994), and perhaps is also contributing to disease outbreaks.

2.2.2. Disease Outbreaks

Disease is rapidly emerging as the single greatest threat to reef-building corals (Shinn 1989; Goreau *et al.*, 1998). Caribbean *Acropora* spp. (elkhorn and staghorn corals), once the major reef-builders of western Atlantic and Caribbean reefs (e.g., Macintyre and Glynn, 1976; Shinn, 1988), have been devastated throughout the region by white-band disease (e.g., Gladfelter, 1982; Peters, 1997), which causes tissue sloughing, leaving bare skeleton which is then colonized by algae and sessile invertebrates. Unusual aggregates of Gram-negative bacteria were found in the tissues of affected corals (Peters, 1984), but the pathogen has yet to be identified with certainty. Black-band disease, discovered in the 1970s (Rützler *et al.*, 1983), is a complex association of microorganisms including photosynthetic cyanobacteria, and sulfate-reducing and sulfide-oxidizing bacteria (Richardson *et al.*, 1997). Numerous previously unrecognized diseases have appeared through the 1980s and 1990s (Santavy and Peters, 1997; Richardson *et al.*, 1998; Goreau *et al.*, 1998). Diseases have been more commonly reported from western Atlantic and Caribbean corals than from Indo-Pacific corals (Peters, 1997). Diseases often appear in corals stressed by bleaching or nutrient pollution, although the near-elimination of acroporids Caribbean-wide by white-band disease was apparently independent of water quality (Goreau *et al.*, 1998).

Amphistegina spp., and other foraminifera with algal symbionts, are also falling victim to previously unknown diseases. Several years ago we (Hallock *et al.*, 1995) described symbiont loss and associated symptoms in *Amphistegina* as a new disease. The process of symbiont digestion and cytoplasmic deterioration is degenerative and progressive (Talge and Hallock, 1995). Affected populations exhibit cyanobacterial infestations, both intracellular and endolithic (Fig. 1c), not previously observed in live foraminifera (Hallock *et al.*, 1995; Toler and Hallock, 1998). Thus, like corals, foraminifera stressed by bleaching appear unable to defend themselves from infestations by cyanobacteria and other microorganisms (Fig. 1d, e) that may normally be benign. *Heterostegina* spp. collected from the West Australian Shelf (Hallock *et al.*, 1996) similarly exhibited cyanobacterial infestation, and *Cyclorbiculina compressa* collected from the Florida Keys in the 1990s were very commonly epiphytized and infested (Fig. 1f).

2.2.3. Predator Outbreaks

The most famous coral predator is *Acanthaster plancii*, the Crown-of-Thorns starfish. Outbreaks of COT, as this starfish is often called, were first recorded in the 1950s in Japan and 1960s on Australia's Great Barrier Reef

(Brown, 1997*b*). These outbreaks appear to be related to several factors including nutrient input by terrestrial runoff that triggers plankton blooms and thereby provides greater food supplies and higher survival rates for COT larvae (Birkeland and Lucas, 1990). In the Galapagos, the sea urchin *Eucidaris thouarsii* can be sufficiently abundant to influence reef community structure (Glynn *et al*, 1979). In several Caribbean locations, the predatory snail, *Corallophila abbreviata*, has been responsible for significant coral mortality, especially where corals are stressed (Hayes, 1990). In the western Pacific, outbreaks of predatory snails, *Drupella* spp., have killed up to 95% of the corals in localized patches, though the causes of the outbreaks are not well understood (Carpenter, 1997). More recently, unusual feeding behavior by parrotfishes in Bonaire has been observed to cause significant mortality. James Cervino (pers. commun.) postulated that the parrotfish are attracted to disease-weakened coral colonies and is testing this hypothesis experimentally.

A predatory foraminifer, *Floresina amphiphaga* (Fig. 1e) preys upon *Amphistegina gibbosa* in the Florida Keys and probably elsewhere. Related species (Revets, 1990) have been reported from reef areas worldwide (Hottinger *et al.*, 1993; Hallock and Talge, 1994). In samples collected prior to 1991, this predator was rare. Following the onset of bleaching symptoms in 1991, this species appeared in unprecedented abundance, feeding on stressed *A. gibbosa* (Hallock and Talge, 1994). Unpublished laboratory experiments demonstrated that the predator preferentially selects symptomatic prey.

2.3. Changes in Reef Community Structure

Cockey *et al.* (1996) studied decadal-scale changes in foraminiferal assemblages in sediments off Key Largo, Florida, an area where studies of coral cover indicate decline in reef vitality (Dustan and Halas, 1987; Porter and Meier, 1992), and where studies of water and sediment chemistry show substantial decline in water quality nearshore but no detectable decline on the reef margin (Szmant and Forrester, 1996). Cockey *et al.*'s (1996) strategy was to compare foraminiferal assemblages in sediments collected in 1982 and 1991–1992 with a published study of sediments from the same traverses (Fig. 2) collected in 1960–1961 (Rose and Lidz, 1977). The foraminiferal assemblage changes are profound and far more dramatic than anticipated over this relatively short time span of 30 years. The foraminiferal assemblages in the sediments changed from predominantly symbiont-bearing taxa (*Amphistegina* and the Soritidae) to predominantly smaller herbivorous and detritivorous miliolid and rotalid taxa (Fig. 3). This change is consistent with community changes in the macrobenthos (Dustan and Halas, 1987; Porter and Meier, 1992) and cannot be explained by the loss of coral to coral-specific diseases, unless those diseases were accompanied by, and possibly promoted by, decline in the suitability of the environment for organisms dependent upon algal symbiosis for growth and calcification.

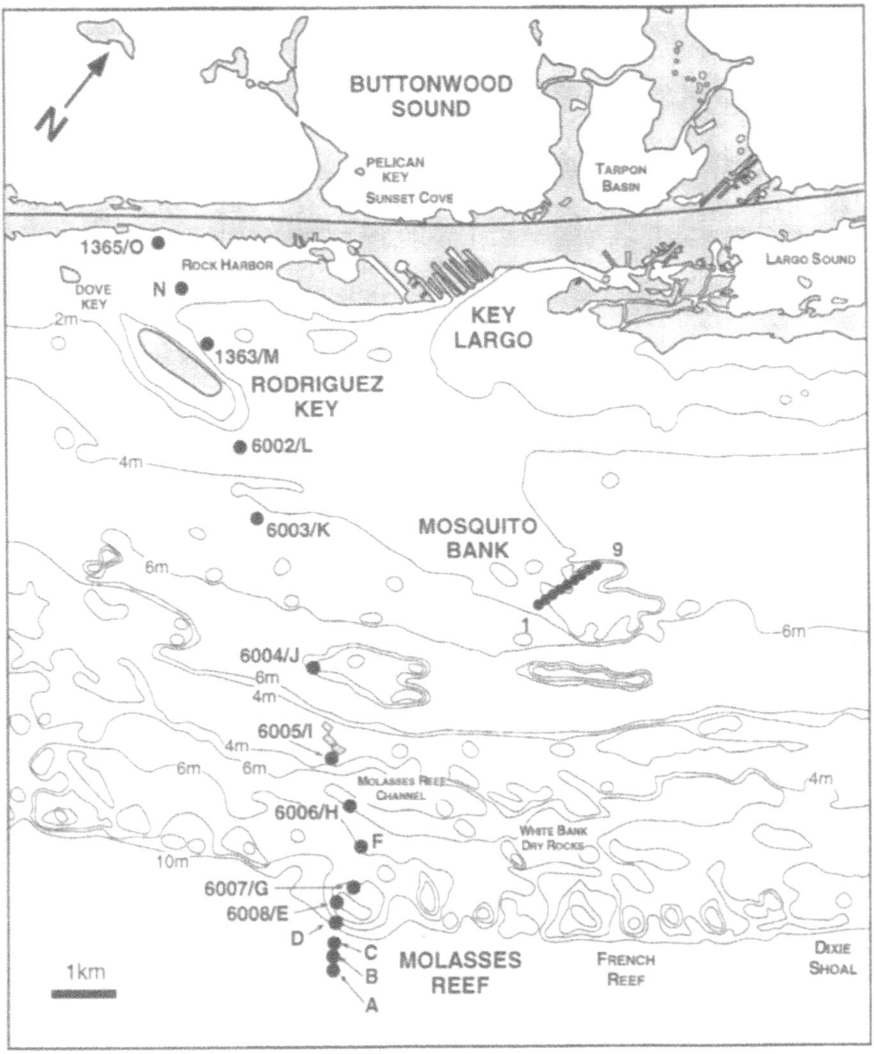

FIGURE 2. Sites on the open shelf and reef margin off Key Largo, Florida, where foraminiferal assemblages in surface sediments were sampled in 1960 by Rose and Lidz (1977) and in 1991–1992 by Cockey *et al.* (1996), with the results shown in Fig. 3. (Source: E. Cockey, P. Hallock, and B. H. Lidz, 1996, *Decadal-Scale Changes in Benthic Foraminiferal Assemblages Off Key Largo, Florida, Coral Reefs,* Vol. 15, pp. 237–248; reprinted with permission of Springer-Verlag.)

3. Potential Application of Foraminifera to Reef Studies

Deterioration of coastal environments is an issue of scientific and public concern worldwide. Two major problems in addressing this issue are lack of historical data for comparison with modern conditions and lack of bioassay organisms that can be used to test the effects of environmental stresses.

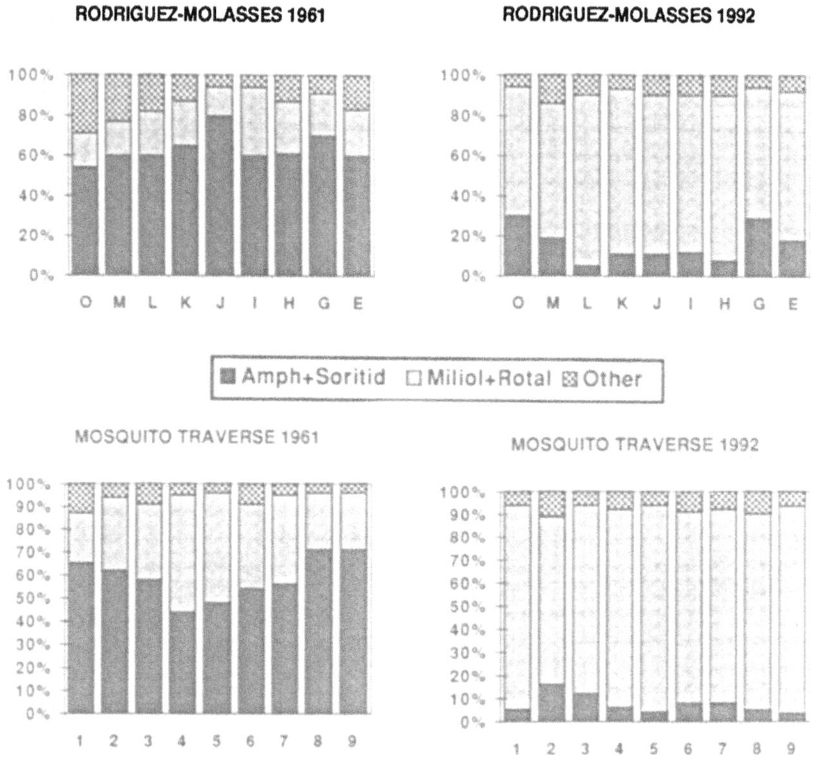

FIGURE 3. Changes in relative abundances of foraminiferal tests in sediments between 1961 and 1992. "Amphi + Soritid" refers to Amphisteginidae and Soritidae; "Miliol + Rotal" refers to Miliolidae and Rotaliidae; "Other" includes the Alveolinidae, Cornuspiridae, Elphidiidae, Glibigerinidae, and Textulariidae. (Source: E. Cockey, P. Hallock, and B. H. Lidz, 1996, *Decadal-Scale Changes in Benthic Foraminiferal Assemblages Off Key Largo, Florida, Coral Reefs*, Vol. 15, pp. 237–248; Reprinted with permission of Springer-Verlag.)

Foraminifera exhibit a combination of characteristics that together provide unique potential for utilization as "ecosystem indicators" in modern environmental research:

1. Foraminifera are abundant, diverse, and widespread in marine ecosystems, exploiting a great variety of environments, substrates, and nutritional modes.
2. Their shells are important sediment constituents, and record environmental conditions through assemblage compositions, shell morphologies, and geochemistry.
3. Living populations and surface-sediment assemblages can be used to assess the current state of a benthic ecosystem.
4. Foraminifera can be studied from sediment cores to assess decadal-, century-, and millennial-scale changes in community structure at sites of interest, providing both an historical record and a target for environmental mitigation.

5. Foraminiferal assemblages from sediment cores collected at sites of interest can be compared with assemblages from cores at reference (control) sites as a means of assessing the suitability of reference sites.
6. Some species can be readily maintained in culture, so laboratory protocols can be established to determine responses of selected taxa to physical stresses and pollutants of concern.
7. Field transplant studies can be designed to determine if individuals or assemblages from reference sites exhibit increased stress or mortality at sites of concern.
8. Foraminifera are relatively small in size (typically 0.1–10 mm) and widely abundant (often 10^4–10^6 per m^2), and therefore statistically significant sample sizes can be collected quickly and inexpensively, for either assemblage assessment or for experimental studies, with minimal environmental impact.

While other kinds of organisms may be useful in some of the same ways, only foraminifera can be used in all of these ways. For example, soft-bodied organisms cannot provide the historic perspective. Shelled macroinvertebrates such as molluscs may be significant sediment constituents, but are usually represented by shell debris rather than by shells identifiable to genus and species. Cores drilled from individual coral heads can provide detailed records of growth rates for each individual cored but cannot provide the record of community response that is available from the sediment record. Furthermore, the field costs and environmental impacts of collecting statistically significant surface samples and sediment cores for foraminiferal studies are only a fraction of the costs and impacts of studies of macroorganisms.

3.1. The Sediment Record

The single most useful approach to answering questions such as "has the benthic community changed?" and, if so, "are the changes anthropogenically induced or are they natural environmental fluctuations?" is to utilize the sediment record. There are four basic ways to use the sediment record, and foraminiferal studies can apply any one way or any combination (Fig. 4):

1. Examination of surface-sediment samples from an area, comparing foraminiferal assemblages with local environmental parameters (e.g., Angel *et al.*, 2000).
2. Comparison of surface sediments at a "site of interest" (i.e., a known or suspected site of anthropogenic impact) with selected "control" or "reference" sites thought to presently have environmental conditions similar to that of the site of interest prior to impact (e.g., Yanko *et al.*, 1998).

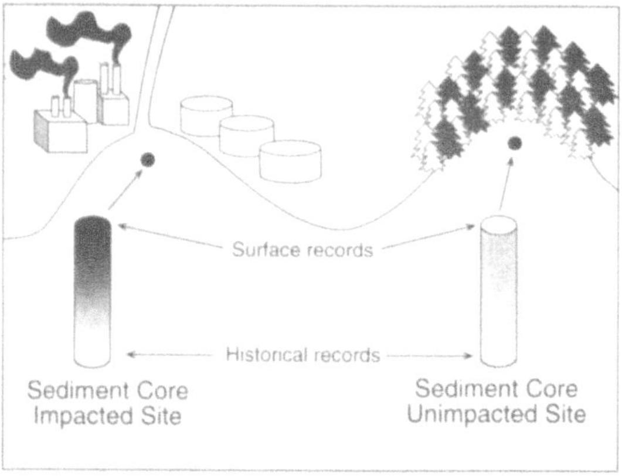

FIGURE 4. Cartoon showing how foraminiferal assemblages in sediment cores can be compared from anthropogenically impacted and relatively unimpacted (reference) sites: (a) comparison of surface samples between sites, (b) comparisons of assemblages prior to impact to assess the value of the reference site, and (c) record of natural, preimpact variability.

3. Comparison of data on surface sediments from a site or area of interest with data similarly collected on samples studied several decades previously (e.g., Cockey *et al.*, 1996).
4. Examination of sediment cores from a site of interest, determining changes through time (e.g., Scott *et al.*, 1995).

The most useful and broadly applicable approach is to collect and study short sediment cores to assess environmental changes over time. Even when reference sites are used in a study, there is always the question as to whether the reference site is truly representative of the site of interest. Comparison of upcore changes at the reference site compared with the site of interest allows assessment of the suitability of the reference site and provides information on temporal changes and natural variability at both sites. Collection and analysis of short cores is being utilized by the U.S. Geological Survey team currently studying environmental changes in Florida Bay (e.g., Brewster-Wingard *et al.*; 1996; Ishman, 1997, Ch. 16 in this volume). The major challenge of this approach is dating the cores and time-resolution. ^{14}C does not provide sufficient resolution for decadal-scale studies, and reef sediments may lack sufficient terrestrial material for accurate ^{210}Pb dating. Active bioturbation of the sediments further limits resolution, especially where sedimentation rates are very low.

3.2. Environmental Requirements and Tolerances

The habitat requirements of many foraminiferal groups are sufficiently well known as to be generally useful in interpreting environmental changes.

Murray (1973) showed that relative proportions of the three most common benthic groups, the agglutinate Textulariida, the porcelaneous Miliolida, and the perforate Rotaliida, when plotted on a ternary diagram, provide clues as to the environments in which these organisms live. The agglutinated forms, particularly those that produce organic or ferrigenous adhesives, generally dominate environments where seawater is undersaturated with respect to $CaCO_3$. The Miliolida, on the other hand, secrete imperforate shells of high-magnesium calcite, which are most easily produced where $CaCO_3$ saturation is high, especially in warm or hypersaline environments. Foraminifera with more complex rotalid shells have been able to adapt to a great range of environments, so order-level data is less useful. However, readily identifiable families, genera, and morphotypes are often characteristic of certain environments (Table 1).

For example, the larger foraminifera, which have algal endosymbionts, typically dominate relatively pristine reef and carbonate shelf environments in subtropical and tropical areas. They are poor competitors where nutrification occurs (e.g., Hirshfield *et al.*, 1968; Cockey *et al.*, 1996). There are three modern families of larger rotalid foraminifera, which are easily recognized by their sizes and shapes. The genus *Amphistegina* (Family Amphisteginidae) (Figs. 1 and 5*a*) is the most consistently abundant throughout subtropical and tropical regions, with the exception of the Eastern Tropical Pacific. The

TABLE 1. Informal Groups of Particular Use in Environmental Analyses

Taxon or group	Environment significance	Sample reference
Larger foraminifera	Dependence on algal symbiosis, characteristic of reefs/carbonate shelves	Hallock, 1998
Larger Rotalids	See above, depths 0–100+ m	
Amphistegina	Probable UVB damage includes symbiont loss and calcification abnormalities	Hallock *et al.*, 1995
Larger Miliolids	See above; depths 1–30 m	Hallock and Peebles, 1993
Soritids	Abundant on phytal substrates	
Peneroplids	Particularly tolerant of hypersalinity	Murray, 1973
Opportunistic taxa	Rapidly respond to habitat disturbance often dominate close to outfalls	Alve, 1995
Ammonia	Low/mid latitude opportunists	
Buccella	Temperate opportunists	
Eggerella/Eggerelloides	Temperate opportunists	
Elphidium	Temperate opportunists	
Trochammina	Temperate opportunists	
Attached epiphytes	Phytal substrates (e.g., macroalgae, seagrasses, mangrove roots)	Alve, 1995

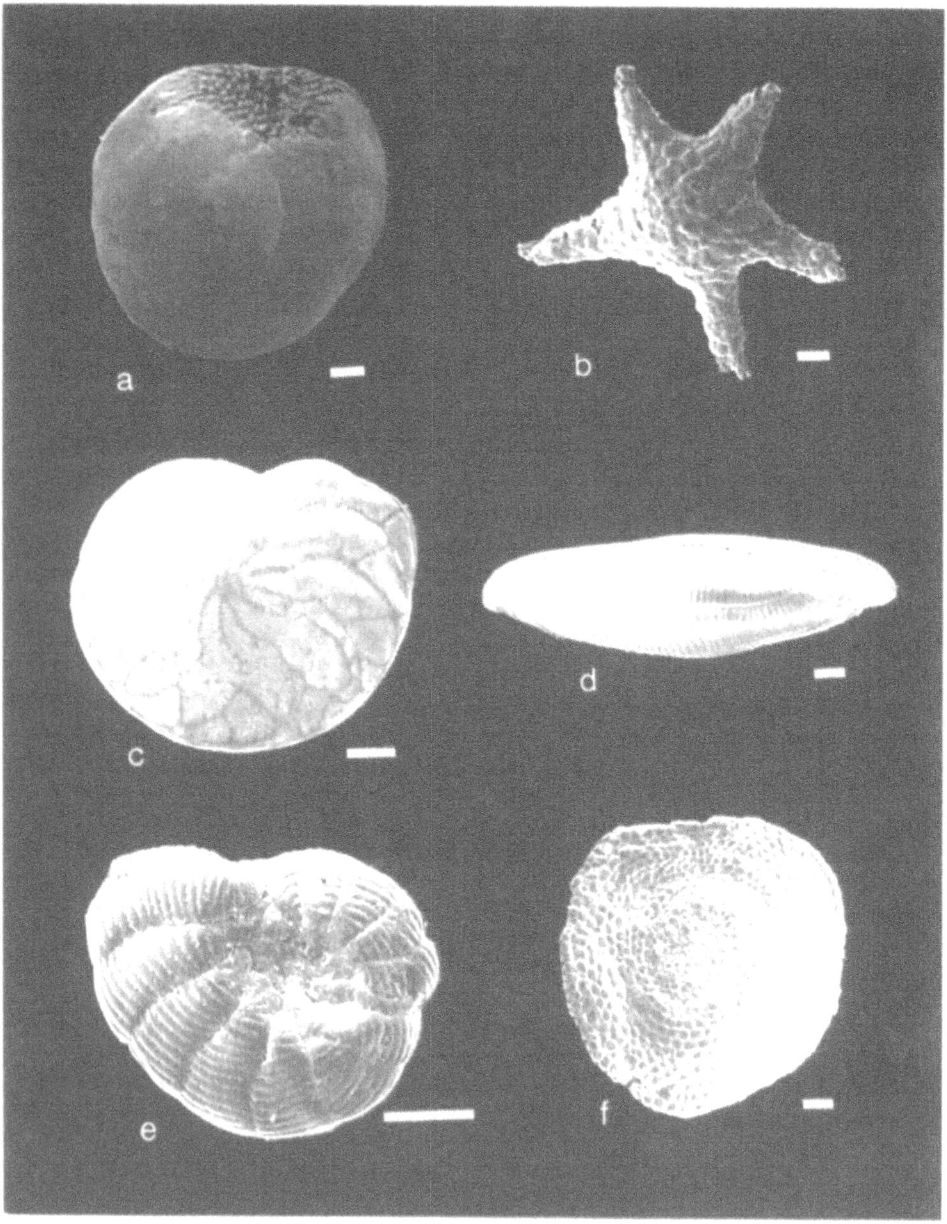

FIGURE 5. SEM's of representative larger foraminifera (all scale bars = 0.1 mm): (a) Amphisteginidae—*A. lobifera*; (b) Calcarinidae—*Baculogypsina sphaerulata*; (c) Nummulitidae—*Heterostegina antillarum*; (d) Alveolinidae—*Alveolinella quoyii*; (e) Peneroplidae—*Peneroplis elegans*; (f) Soritidae—*Sorites orbiculus*.

"star-sand" Calcarinidae (Fig. 5*b*) are mostly restricted to the East Indian–West Pacific area, while Nummulitidae (Fig. 5*c*) are seldom abundant at safe working depths for Scuba divers (<30 m) (Hallock, 1984; Hohenegger, 1994). The three larger miliolid families are readily distinguishable from other taxa, though not necessarily from each other. Like the Calcarinidae, the Alveolinidae (Fig. 5*d*) are of minimal concern outside the East Indian–Western Pacific region. The Peneroplidae (Fig. 5*e*) and Soritidae (Fig. 5*f*) have representatives that are tolerant of hypersalinity, and several are only abundant on phytal substrates and therefore are particularly useful in areas where changes in seagrass cover is of concern.

At the other end of the environmental spectrum are opportunistic genera of smaller foraminifera such as *Ammonia*, *Buccella*, *Eggerella*, and *Elphidium* (Table 1), many of which can tolerate a variety of stresses including nutrient and heavy-metal pollution (Alve 1995). In between are the majority of smaller miliolid and rotalid taxa, whose abundances will tend to increase with increasing food supplies until organic loading creates oxygen stress (e.g., Cockey *et al.*, 1996). Then the opportunistic taxa take over. At the pollution source, there may be a barren zone where even opportunistic taxa cannot survive. Thus, changes in relative abundance of foraminiferal tests in the sediments is an important parameter (Alve 1995).

Alve (1995), Yanko *et al.* (1994, 1998, 1999), and others also have shown that test abnormalities are a common stress response in foraminifera that have been documented in populations exposed to hypersalinity and a variety of chemical pollutants. We (Hallock *et al.*, 1995; Toler and Hallock, 1998) have shown that test abnormalities are also common in stressed *Amphistegina*.

Although foraminiferal studies are sometimes included in routine environmental research, they could certainly be more widely applied. One challenge is the complexity of foraminiferal taxonomy, which can deter managers from including foraminiferal studies in environmental protocols. Besides requiring foraminiferal specialists, detailed taxonomic work may not be cost effective. Development of protocols that rely primarily on indicator species and morphogroups (e.g., Hallock, 1996, 1998) is one way to address taxonomic limitations.

3.3. *Amphistegina* Densities

One strategy to increase the application of foraminiferal studies to reef research is to utilize a single, relatively identifiable taxon or morphogroup. *Amphistegina* spp. are among the most common reef-dwelling organisms worldwide. Two species, *A. lobifera* (Fig. 5a) and *A. lessonii* (Fig. 1b), are abundant on reefs and associated hard substrate environments throughout the Indo–Pacific except for the eastern tropical Pacific (Hallock 1988b, c). *A. lobifera* lives most abundantly at depths less than 10 m; *A. lessonii* is most common at depths from 5–40 m (Hallock, 1984; Hohenegger, 1994). Three other deeper-dwelling species occur but are not important for this discussion.

In Hawaii, dead shells of these two species make up nearly a quarter of the nearshore sediment (Hallock Muller, 1976); on Kapingimarangi Atoll, their contribution is closer to 90% (McKee *et al.*, 1956). In the western Atlantic and Caribbean, *A. gibbosa* (Fig. 1a, c) is the ecological vicariate of *A. lessonii* (Hallock *et al.*, 1986*b*).

Amphistegina individuals commonly live on closely cropped coraline and filamentous algae on reef substrate. They also live on some macrophytes, particularly if there is other epiphytic growth. They are generally abundant on reef rubble, specifically the nodules common on and at the base of reef and live-bottom substrates. Population densities are low in the most exposed, high-energy, reef-margin environments (Hallock, 1984), or where runoff or high bioerosion rates flood the substrate with muddy sediments, where fleshy algae and *Halimeda* dominate the substratum, or where excess organic matter accumulates in the sediments (Hallock, 1988*a*).

Amphistegina individuals host diatom endosymbionts in an interdependent relationship very similar to that found between corals and their zooxanthellae (Lee and Anderson, 1991). The golden-brown to olive-green color of the diatom symbionts, combined with the relatively large size of foraminifera (1–3 mm adult diameter) make living *Amphistegina* very easy to identify and to distinguish from dead tests. These protists have been extensively studied both in the field and in the laboratory since 1970 (e.g., Hallock Muller, 1974; Lee *et al.*, 1980; Röttger *et al.*, 1980; Hallock *et al.*, 1986*b*, 1995; Williams *et al.*, 1997). Habitat observations made while collecting *Amphistegina*, combined with laboratory observations of the sensitivity of these protists to algal overgrowth, were the basis for the series of papers on why algal symbiosis and mixotrophic nutritional modes characteristic of coral reefs appear to be adaptations to low-nutrient environments and why communities shift to predominance of autotrophic and heterotrophic modes as nutrient supplies increase (Hallock, 1981*a*, 1987, 1988*a*; Hallock and Schlager, 1986; Hallock *et al.*, 1993*a*).

Living *Amphistegina* populations are easily sampled by collecting reef rubble (Hallock Muller, 1974; Hallock, 1981*b*; Williams *et al.*, 1997). Population abundances are compared by counting the number of *Amphistegina* collected on a piece of rubble and estimating the area of the bottom covered by that piece (Table 2). In healthy reef conditions, *Amphistegina* are typically abundant; often up to several hundred living individuals can be found on rubble covering 100 cm^2 of bottom (i.e., densities of 10^2–10^3 per 100 cm^2, e.g., Hallock *et al.*, 1986*a*). On the other hand, in the 1970s in Hawaii, off Honolulu (Fig. 6), where disposal of millions of gallons of sewage daily promoted macroalgal growth, *Amphistegina* densities were lower by about an order of magnitude (10^1/100 cm^2), and were not found living in nutrient-stressed south or central Kaneohe Bay (Hallock, 1984).

Research in Caribbean and western Atlantic reefs has reinforced and refined observations made on Indo–Pacific reefs. With only *A. gibbosa*, population densities in back-reef areas seldom compare with *A. lobifera* densities in the Indo–Pacific, and overall densities tend to be somewhat

TABLE 2. Application of Live *Amphistegina* Densities as Reef-Vitality Indicators

Density (d)	Score	Interpretation
$d > 100$ *Amphistegina*/100 cm^2	4	Environment excellent for *Amphistegina* and probably for reef growth
$d = 50–100$ *Amphistegina*/100 cm^2	3	Environment good for *Amphistegina* and probably for reef growth
$d = 10–50$ *Amphistegina*/100 cm^2	2	Environment marginal for *Amphistegina* and for long-term reef growth
$0 < d < 10$ *Amphistegina*/100 cm^2	1	Environment marginal for *Amphistegina* and questionable for reef growth
Amphistegina not found	0	Environment unsuitable for *Amphistegina* and very questionable for reef growth

lower. Yet the same trends are evident. For example, in the 1980s, on reefs off La Parguera, Puerto Rico, abundances of live *Amphistegina* appeared to be related to reef vitality (Hallock, 1996). Though the inner reefs had substantial coral cover, sponges and macroalgae appeared to be taking over the substratum, and the corals were heavily bioeroded (Hallock, 1988*a*). Biota were draped with muddy mucus. Rubble, where it could be found, was overgrown by sponges. Living *Amphistegina* were rare. Midshelf reefs had fewer sponges and more *A. gibbosa*. And on the forereefs of the outer reef arc, coralline-algal

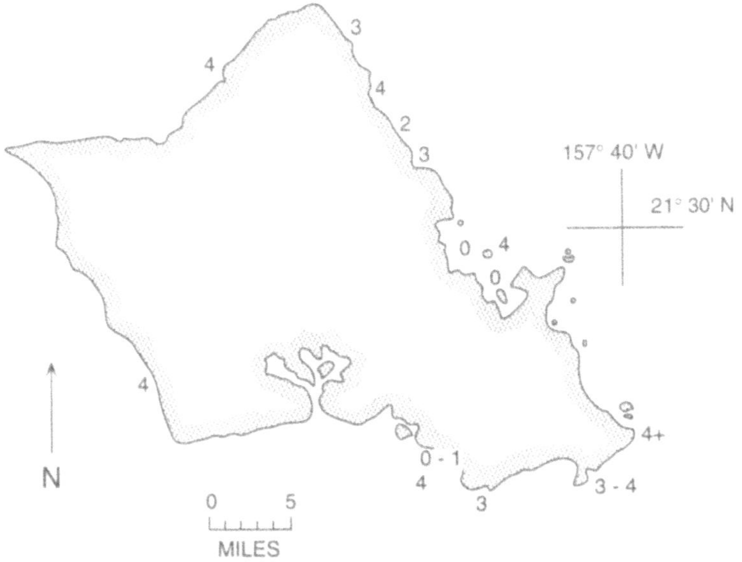

FIGURE 6. Density scores (see Table 2) of live *Amphistegina* spp. on reef rubble collected from depths of 2 to 10 m around Oahu, Hawaii, between 1970 and 1976.

rubble and resident *A. gibbosa* occurred abundantly within the healthy coral community.

Thus, densities of living *Amphistegina* spp. on reef rubble (Table 2) have the potential to be used as a simple, low-cost indicator of reef vitality (Hallock, 1996). Because their life span is a few months, these protists respond more directly to environmental degradation than do longer-lived hermatypic corals. For example, Cockey *et al.*, (1996) found that proportions of *A. gibbosa* and other larger foraminiferal tests in sediments off the upper Florida Keys were already in decline in the early 1980s. Yet, because of the limitations of chemical measurements in reef environments (e.g., Laws and Redalje, 1979), reef researchers continue to argue whether upper Keys coral reefs have declined owing to declining water quality or only because of coral-specific diseases and mortality events (e.g., Szmant and Forrester, 1996).

3.4. Potential Use of Foraminifera in Bioassay Studies

The foraminifera have tremendous potential for use as bioassay organisms, particularly for coral-reef ecosystems. As discussed in several symposia on reef monitoring (e.g., D'Elia *et al.*, 1991; Crosby *et al.*, 1996), environmental changes can have subtle, sublethal effects on coral-reef communities. Because *Amphistegina* spp. are amenable to culture and easily transported in the field, these foraminifera can readily be used to test stress factors affecting coral-reef systems (Table 3; see also in this volume Geslin *et al.*, Ch. 9; Bresler and Yanko-Hombach, Ch. 10).

A variety of factors are known to cause stress in *Amphistegina*, including algal overgrowth, temperature shock, salinity stress, elevated UVB radiation, and heavy metals (Hallock Muller, 1974; Hallock, 1979, 1981*c*; Hallock *et al.* 1986*b*, 1995; and unpublished observations). Responses can be evaluated in short-term (e.g., mortality, autolysis, stress-protein production), intermediate (e.g., growth rates, symbiont loss), and long-term experiments (e.g., reproduc-

TABLE 3. Stressors Known or Predicted to Influence Coral-Reef Ecosystems

Temperature—heat/cold shock, prolonged hyperthermia/hypothermia
Salinity—hypo/hypersalinity shock, prolonged hypo/hypersalinity
Temperature/salinity—combined effects, especially warm or cold hypersaline plumes
Nutrients (e.g., ammonia, nitrates, and phosphates)
Effluent dilutions (i.e., sewage, stormwater, etc.)
Sediments
Nuisance algal blooms
Hypoxia
Pesticides
Heavy metals
Hydrocarbons and dispersants
Biologically damaging ultraviolet radiation

tive responses, calcification anomalies). Their micro- to meiofaunal sizes (<0.1 mm to >1 cm diameter) allow individual responses to be monitored without requiring large and costly infrastructures; i.e., a statistically significant test group of 10 to 100 individuals can be maintained in petri dishes or tiny mesh enclosures.

3.4.1. Laboratory Studies

Stress in *Amphistegina* can be assessed using (a) visual, (b) cytological, and (c) biochemical protocols that address mechanistic cause–effect hypotheses.

Visual protocols include assessing vitality/mortality using pseudopodial activity, symbiont loss by color changes, and growth by weekly diameter measurements (Hallock Muller, 1974; Hallock, 1979, 1981*c*; Hallock *et al.* 1986*b*, 1995). Observations of loss of symbiont color in field populations of *Amphistegina* spp. provide a basis for visually ranking experimental responses in these protists (Hallock *et al.*, 1995). "Normal," healthy specimens are a rich diatom golden-brown to olive green. In the early stages of stress, specimens are often "granular" in appearance or "slightly mottled" with a few small white spots. As symbiont-loss progresses (Fig. 1a), white spots increase in number and size, from "mottled" (evident spots but >50% golden) to "very mottled" (<50% golden-brown), to "bleached," i.e., where the specimen is mostly white with some granular darkening near the aperture. Granular to mottled specimens cytologically exhibit progressive deterioration and digestion of the diatom endosymbionts, followed by autolysis of organelles, particularly Golgi, and vacuolation and deterioration of the host cytoplasm (Talge and Hallock, 1993, 1995). The test of a "very mottled" or "bleached" specimen is mostly filled with water, as most of the cytoplasm has broken down.

Cytological protocols include quantification of organelles, symbionts, and large vacuoles, and locations of acid phosphatase activity. Both light (LM) and transmission electron microscopy (TEM) provide useful information (e.g., Koestler *et al.*, 1985; Hallock *et al.*, 1993*b*; Talge and Hallock, 1995). LM permits low-resolution examination of serial sections of whole specimens, with biochemical identification and location of particular cell components.

Specimens are prepared for TEM analysis using modifications of established preparation techniques (e.g., Talge and Hallock, 1995). Mottled *A. gibbosa* typically exhibit lysosomes surrounding remaining endosymbionts, which are usually in various stages of degradation (Talge and Hallock, 1995). The presence, within symbiont vacuoles, of the common hydrolytic (digestive) enzyme acid phosphatase (e.g., Alberts *et al.*, 1993), indicates digestion of the symbionts by the host (Lee and Anderson, 1991). Similarly, the presence of free acid phosphatase in the cytoplasm confirms autolysis of the host cell. The presence and locations of lysosomal digestive activity in control and stressed foraminifera can be determined biochemically by localization of acid phosphatase activity using β-glycerophosphate. Methods developed by Bowser *et al.* (1992) have been adapted for use on *Amphistegina* specimens. Stereology can be used on light or transmission electron micrographs to statistically quantify and describe cellular organization and structure (e.g., Bozzola and Russell, 1992; Fagerberg, 1988).

Biochemical protocols are currently in developmental stages for use with reef-dwelling foraminifera. Techniques that show promise for environmental research include epifluorescence microscopy, spectrofluorimetry (Bernhard *et al.*, 1995), and ATP assay (e.g., Bernhard, 1993), which were developed to assess viability in field-collected foraminifera. Fortunately, distinguishing live *Amphistegina* visually, based on pseudopodial response, is generally very easy.

Quantifying sublethal stress in the short-term experiments is more challenging, a task for which stress-protein analysis techniques hold promise. Organisms protect themselves from cellular damage by a stress-protein response (Sanders, 1993), in which cells produce increased levels of stress proteins (e.g., heat-shock proteins or chaperonins) while repressing production of other proteins. Stress proteins are found in small quantities in some unstressed cells, but they are synthesized in greater quantities in cells under stress. Two potential approaches are: (a) to measure intracellular levels of stress proteins (Sanders, 1993), and (b) to measure expression of heat-shock genes by mRNA analysis (e.g., Pichard and Paul, 1991, 1993). Although stress proteins have not been studied to any extent in foraminifera, Stress-90, Stress-70, chaperonin, and low molecular-weight stress proteins have been identified in most taxa examined, ranging from bacteria to humans, including protozoa (Sanders, 1993). S. K. Toler (unpublished) has identified production of Stress-70 by *Archaias angulatus* after exposure to a 35°C water temperature for 1 h. She quantified the amount of stress proteins produced following the stress event using immunoblotting, adapting procedures described by Bradley and Ward (1989) and Cochrane *et al.* (1991). Because stress proteins are highly conserved and cross-reactive, necessary antibodies are commercially available.

There are many more potential strategies for biochemical protocols. Yanko and co-workers are developing fluorescein tracer techniques (e.g., Bresler and Yanko, 1995) and are also examining magnesium/calcium ratios, after noting that damaged tests tend to be enriched in magnesium (e.g., Yanko *et al.*, 1998). Much more basic research is needed in this area to fully exploit the potential for the use of foraminifera in environmental studies.

3.4.2. Field Bioassay Studies

A variety of questions face reef managers, or will in the near future, for which field bioassay experiments could provide answers, or at least insights. For example, at a site of documented decline in coral populations, are water quality and other environmental characteristics still suitable for reef development? That is, if coral larvae or juvenile corals were transplanted on that site, would the water quality support their survival and growth? This will become an increasingly important question in the next decade as techniques for collecting coral larvae and for propagating corals in the laboratory make such mitigation projects more feasible. Reef-dwelling foraminifera with algal symbionts, particularly *Amphistegina* spp., have several characteristics that indicate their potential for use in field bioassay experiments:

1. Their environmental requirements are similar to those of corals, as discussed previously.
2. They are relatively small compared with most reef invertebrates, yet are large enough to be seen and easily manipulated at low magnification on a stereomicroscope.
3. Statistically significant sample numbers can be placed in a relatively small containment area or device, thereby limiting the impact of the experiment on the environment.
4. They can be collected from loose reef rubble so the environmental impact of sampling is minimal. In addition, reef rubble, with associated foraminifera, can be transplanted to sites of concern easily, inexpensively, and with minimal environmental impact.
5. Assessment protocols developed for laboratory experiments are directly translatable to field experiments. Furthermore, specimens exposed to stresses in the field, and their controls, can be subsequently monitored in laboratory culture to assess longer-term responses.
6. Their nearly circumtropical distribution and abundance provide the potential for their use almost anywhere reefs occur. Therefore, experiments using visual protocols can be designed and carried out very inexpensively in developing countries with limited financial resources but substantial human resources.

Bioassay experiments with *Amphistegina* can be applied in many of the same situations that corals or coral larvae can be used. However, experiments using live coral require much larger containers for far fewer specimens. Collection of statistically significant numbers of individual corals or coral plugs may have unacceptable environmental impact in regions where corals are already at risk. Further, while coral larvae are similar in size to *Amphistegina*, larvae are only available immediately after spawning, which for many species is once per year. *Amphistegina* are typically available throughout the year. Thus, increased emphasis on basic physiological research and molecular studies will enhance the usefulness of *Amphistegina* and other larger foraminifera in both field and laboratory bioassay applications.

4. Larger Foraminifera and Global Change

As noted in the introduction to this chapter, human activities are altering environments globally. Anthropogenic changes include problems that are local in nature but are globally pervasive, such as coastal sedimentation resulting from deforestation, and changes that are truly global, including rising atmospheric CO_2 concentrations and stratospheric ozone depletion. Larger foraminiferal populations circumtropically are being affected by these changes, as are the coral reefs that many larger foraminifera live on or around.

In most cases, the environmental changes that are detrimental to reef-building corals are detrimental to larger foraminifera: sediments and nutrients reduce water transparency, smother slower-growing benthic organisms, and favor fast-growing, noncalcifying taxa. Although little is known about the effects of chemical pollutants on reef-dwelling organisms, what is known is not encouraging (e.g., Richmond, 1993; Peters, 1997). Recent research on symbiont loss and associated symptoms in *Amphistegina* spp. indicate that these foraminifera may be more directly impacted by increasing intensities of biologically damaging UVB even than corals (Hallock *et al.*, 1995). On the other hand, larger foraminifera appear to be less sensitive than corals to elevated sea-surface temperatures that have developed under extreme El Niño/Southern Ocean Oscillation events in recent years, which may be harbingers of accelerated global warming.

Perhaps the most intriguing questions regarding larger foraminifera and global change relates to differences in shell mineralogies between corals and foraminifera. Doubling to tripling of current CO_2 levels, to atmospheric concentrations comparable to those during the Early Eocene (e.g., Berner, 1994), are predicted for the 21st century (e.g., Watson *et al.* 1990). Higher atmospheric CO_2 will lower the pH and carbonate saturation states of sea-surface waters. These changes in ocean chemistry are predictably more serious for aragonitic corals, calcareous green algae, and high-Mg calcite miliolid larger foraminifera than for low-Mg calcite larger rotalid foraminifera (e.g., Hallock, 1997).

5. Summary

Coral reefs are at risk worldwide from consequences of impacts including coastal sedimentation and nutrification, climate destabilization associated with global warming, and ozone depletion. Reef-dwelling foraminifera, especially larger taxa that host algal symbionts, hold substantial potential as indicators of reef vitality for a variety of practical and scientific reasons:

1. Larger foraminifera are widely used as paleoenvironmental indicators in studies of fossil limestones.
2. Smaller foraminifera have long proven useful in pollution studies.
3. Different taxa have evolved to exploit the great variety of environments, substrates, and nutritional modes in marine systems.
4. Foraminiferal shells morphologically and geochemically record environmental conditions.
5. Physiological analogies between zooxanthellate corals and foraminifera with algal symbionts result in similar environmental requirements.

6. Similar kinds of stress symptoms have been observed in foraminiferal populations and assemblages as those reported for corals and coral-reef communities.
7. The relatively short life spans of symbiont-bearing foraminifera as compared with colonial coral can potentially enable differentiation between long-term reef decline associated with declining water quality and temporary reef decline associated with natural episodic mortality events.
8. Records of environmental change, as indicated by changes in foraminiferal assemblages, are available in reef sediments.
9. A readily identifiable genus, *Amphistegina*, is nearly circumtropical in distribution, is widely abundant in reef environments, and has great potential as an indicator species.
10. Foraminifera with algal symbionts, particularly *Amphistegina* spp., are relatively small and abundant, permitting statistically significant sample sizes to be collected relatively quickly and inexpensively with minimal impact on reef resources.
11. The amenity of some foraminifera, especially selected species of *Amphistegina*, to culture and experimentation provides the potential for developing laboratory and field bioassay protocols for assessing environmental suitability for reef growth.

These characteristics justify routine application of foraminiferal assemblages in surface sediments and sediment cores to address a variety of environmental questions in reef and coastal environments. These characteristics also argue for intensification of physiological and molecular research efforts to identify responses that will facilitate use of *Amphistegina* and other larger foraminifera as bioassay organisms in reef research.

ACKNOWLEDGMENTS: Aspects of this research have been or are being funded by NOAA's National Undersea Research Center at the University of North Carolina at Wilmington's Florida Keys Program (Contracts No. 9120, 9221, 9204.4, 9322, 9515, 9609, and 9703-66); by U.S. National Science Foundation Grants OCE-9203278 and OCE-9214614; by the U.S. Environmental Protection Agency NCERQA Grant No. GAD #R825869; and by the 1994 W. Storrs Cole Memorial Research Award from the Geological Society of America.

References

Alberts, A., Bray, D., Lewis, J., Raff, M., Roberts, K., and Watson, J. D., 1993, *Molecular Biology of the Cell*, Garland, New York.

Alve, E., 1995, Benthic foraminiferal responses to estuarine pollution: a review, *J. Foram. Res.* **25**:190–203.

Angel, D. L., Verghese, S., Lee, J. J., Zuber, D., and Symons, A., 2000, Impact of a net cage fish farm on the benthic Foraminifera in the northern Gulf of Eilat (Aqaba, Red Sea), *J. Foram. Res.* **30**:54–65.

Berner, R. A., 1994, Geocard II: A revised model for atmospheric CO_2 over Phanerozoic time, *Am. J. Sci.* **294**:56–91.
Bernhard, J., 1993, Experimental and field evidence of Antarctic foraminiferal tolerance of anoxia and hydrogen sulfide, *Mar. Micropaleontol.* **20**:203–213.
Bernhard, J. M., Newkirk, S.G., and Bowser, S.S., 1995, Towards a non-terminal viability assay for foraminiferan protists, *J. Euk. Microbiol.* **42**:357–367.
Birkeland, C., 1977, The importance of the rate of biomass accumulation in early successional stages of benthic communities to the survival of coral recruits, *Proc. 3rd Int. Coral Reef Symp., Miami* **1**:16–21.
Birkeland, C., 1988, Geographic comparisons of coral-reef community processes, *Proc. 6th Int. Coral Reef Symp., Townsville, Australia,* **1**:211–220.
Birkeland, C., 1997, Geographic differences in ecological processes on coral reefs, in: *Life and Death of Coral Reefs* (C. Birkeland, ed.), Chapman and Hall, New York, Ch. 12, pp. 273–287.
Birkeland, C., and Lucas, J. S., 1990, *Acanthaster plancii: Major Management Problem of Coral Reefs*, CRC Press, Boca Raton.
Bowser, S. S., Alexander, S. P., Stockton, W. L., and DeLaca, T., 1992, Extracellular matrix augments mechanical properties of pseudopodia in the carnivorous foraminiferferan *Astrammina rara*: Role in prey capture, *J. Protozool.* **39**:724–732.
Bozzola, J. J., and Russell, L. D., 1992, *Electron Microscopy*, Jones and Bartlett, Boston.
Bradley, B. P., and Ward, J. B., 1989, Detection of a major stress protein using a peptide antibody, *Mar. Environ. Res.* **28**:471–475.
Bresler, V., and Yanko, V., 1995, Chemical ecology: a new approach to the study of living benthic epiphytic foraminifera, *J. Foram. Res.* **25**:267–279.
Brewster-Wingard, G. L., Ishman, S. E., Edwards, L. E., and Willard, D. A., 1996, Preliminary report on the distribution of modern fauna and flora at selected sites in north-central and northeastern Florida Bay, U.S. Geological Survey Open-File Report 96–732.
Bricaud, A., Morel, A., and Prieur, L., 1981, Absorption by dissolved organic matter of the sea (yellow substance) in the UV and visible domains, *Limnol. Oceanogr.* **26**:43–53.
Brown, B. E., 1997*a*, Coral bleaching: causes and consequences, *Proc. 8th Int. Coral Reef Symp., Panama* **1**:65–73.
Brown, B. E., 1997*b*, Disturbances to reefs in recent times, In: *Life and Death of Coral Reefs* (C. Birkeland, ed.), Chapman and Hall, New York, Ch. 15, pp. 354–379.
Bryant, D., Burke, L., McManus, J., and Spalding, M., 1998, *Reefs at Risk: A Map-Based Indicator of Potential Threats to the World's Coral Reefs*, World Resources Institute, Washington, D.C.
Butler, M. J. IV, Hunt, J. H., Herrnkind, W. F., Childress, M. J., Bertelsen, R., Sharp, W., Matthews, T., Field, J. M., and Marshall, H. G., 1995, Cascading disturbances in Florida Bay, U.S.A: Cyanobacterial blooms, sponge mortality, and implications for juvenile spiny lobsters *Panuirus argus, Mar. Ecol. Progr. Ser.* **129**:119–125.
Carpenter, R. C., 1997, Invertebrate predators and grazers, in: *Life and Death of Coral Reefs* (C. Birkeland, ed.), Chapman and Hall, New York, Ch. 9, pp. 198–248.
Cochrane, B.J., Irby, R.B. and Snell, T.W., 1991, Effects of copper and tributyltin on stress protein abundance in the rotifer *Brachionus plicatilis, Comp. Biochem. Physiol.* **98C**:385–390.
Cockey, E. M., Hallock, P., and Lidz, B.H., 1996, Decadal-scale changes in benthic foraminiferal assemblages off Key Largo, Florida, *Coral Reefs* **15**:237–248.
Colborn, T., Dumanoski, D., and Myers, J. P., 1996, *Our Stolen Future,* Dutton, New York.
Cottey, T. L., and Hallock, P., 1988, Test surface degradation in *Archaias angulatus, J. Foram. Res.* **18**:187–202.
Crosby, M. P., Gibson, G. R., and Potts, K. W. (eds.), 1996, A Coral Reef Symposium on Practical, Reliable, Low-Cost Monitoring Methods for Assessing the Biota and Habitat Conditions of Coral Reefs, January 26–27, 1995. Office of Ocean and Coastal Resource Management, National Oceanic and Atmospheric Administration, Silver Spring, MD.
D'Elia, C. F., Buddemeier, R. W., and Smith, S. V., 1991, Workshop on coral bleaching, coral reef systems and global change. Rept. Proc., Maryland Sea Grant College Publ. No. UM-SG-TS-91–03, College Park, MD, pp. 1–49.
Dunlap, W. C., Chalker, B. E., and Oliver, J. K., 1986, Bathymetric adaptations of reef-building

corals at Davies Reef, Great Barrier Reef, Australia. III. UVB absorbing compounds, *J. Exp. Mar. Biol. Ecol.* **104**:239–248.

Dustan, P., and Halas, J. C., 1987, Changes in the reef-coral community of Carysfort Reef, Key Largo, Florida:1974–1982, *Coral Reefs,* **6**:91–106.

Fagerberg, W., 1988, Appendix II. Stereology, in: *Introduction to Biological Electron Microscopy: Theory and Techniques* (C. J. Dawes, ed.), Ladd Research Industries, Inc., Burlington, Vermont, pp. 265–281.

Falkowski, P. G., Dubinsky, Z., Muscatine, L., and McCloskey, L., 1993, Population control in symbiotic corals, *BioScience* **43**:606–611.

Fanning, K. A., 1989, Influence of atmospheric pollution on nutrient limitation in the ocean, *Nature* **339**:460–463.

Gladfelter, W. G., 1982, Whiteband disease in *Acropora palmata*: Implications for the structure and growth of shallow reefs, *Bull. Mar. Sci.* **32**:639–643.

Gleason, D., and Wellington, G., 1993, Ultraviolet radiation and coral bleaching, *Nature* **365**:836–838.

Glynn, P., 1996, Coral reef bleaching: facts, hypotheses and implications, *Global Change Biol.* **2**:495–510.

Glynn, P. W., Wellington, G. M., and Birkeland, C., 1979, Coral growth in the Galapagos: Limitation by sea urchins, *Science* **203**:47–49.

Glynn, P. W., Imai, R., Sakai, K., Nakano, Y., and Yamazato, K., 1993, Experimental responses of Okinawan (Ryukyu Islands, Japan) reef corals to high sea temperature and UV radiation, *Proc. 7th Int. Coral Reef Symp.*, Guam **1**:27–37.

Goreau, T. J., Cervino, J., Goreau, M., Smith, G., Hayes, R., Richardson, L., Williams, E., Nagelkerken, I., Porter, J., Porter, K., Garzon-Ferreira, J., Quirolo, C., McField, M., Gil, D., Bruckner, A., Santavy, D., Peters, E., Littler, M., Littler, D., Patterson, K., and Campbell, J., 1998, Rapid spread of diseases in Caribbean coral reefs, *Rev. Biol. Trop.* **46**:157–171, suppl. 5.

Hadar, D-P., and Worrest, R. C., 1991, Effects of enhanced solar ultraviolet radiation on aquatic ecosystems, *Photochem. Photobiol.* **53**:717–725.

Hallock, P., 1979, Trends in test shape with depth in large, symbiont-bearing foraminifera, *J. Foram. Res.* **9**:61–69.

Hallock, P., 1981*a*, Algal symbiosis: A mathematical analysis, *Mar. Biol.* **62**:249–255.

Hallock, P., 1981*b*, Production of carbonate sediments by selected foraminifera on two Pacific coral reefs, *J. Sed. Petrol* **51**:467–474.

Hallock, P., 1981*c*, Light dependence in *Amphistegina, J. Foram. Res.* **11**:40–46.

Hallock, P., 1982, Evolution and extinction in larger Foraminifera, *3rd N. Am. Paleontol. Conv., Proc.* **1**:221–225.

Hallock, P., 1984, Distribution of selected species of living algal symbiont-bearing foraminifera on two Pacific coral reefs, *J. Foram. Res.* **14**:250–261.

Hallock, P., 1985, Why are larger Foraminifera large? *Paleobiology* **11**:195–208.

Hallock, P., 1987, Fluctuations in the trophic resource continuum: A factor in global diversity cycles? *Paleoceanography* **2**:457–471.

Hallock, P., 1988*a*, The role of nutrient availability in bioerosion: Consequences to carbonate buildups, *Palaeogeogr. Palaeoclim. Palaeoecol.* **63**:275–291.

Hallock, P., 1988*b*, Interoceanic differences in foraminifera with symbiotic algae:A result of nutrient supplies?, *Proc. 6th Int. Coral Reef Symp., Townsville, Australia* **3**:251–255.

Hallock, P., 1988*c*, Diversification in algal symbiont-bearing foraminifera: A response to oligotrophy? *Rev. Paleobiol.*, Spec. Publ. No. 2, *Benthos '86*, pp. 789–97.

Hallock, P., 1996, *Amphistegina* (Foraminiferida) densities as a practical, reliable, low-cost indicator of coral reef vitality, in: *A Coral Reef Symposium on Practical, Reliable, Low-Cost Monitoring Methods for Assessing the Biota and Habitat Conditions of Coral Reefs,* January 26–27, 1995 (M. P. Crosby, G. R. Gibson, and K. W. Potts, eds.), Office of Ocean and Coastal Resource Management, National Oceanic and Atmospheric Administration, Silver Spring, MD, pp. 37–43.

Hallock, P., 1997, Reefs and reef limestones in Earth history, in: *Life and Death of Coral Reefs* (C. Birkeland, ed.), Chapman and Hall, New York, pp. 13–42.

Hallock, P., 1998, Foraminifera as ecosystem indicators: Phase 1—A marine benthic perturbation index; Phase 2—Bioassay protocols, *Proc. 1996/97 STAR Grants Ecological Assessment/ Ecosystem Indicators Program Review, February 3—5, 1998, Las Vegas, Nevada.* U.S. Environmental Protection Agency, pp. 9–10.

Hallock, P., and Larsen, A. R., 1979, Coiling direction in *Amphistegina, Mar. Micropaleontol.* **4**:33–44.

Hallock, P. and Peebles, M. W., 1993, Foraminifera with chlorophyte endosymbionts:Habitats of six species in the Florida Keys, *Mar. Micropaleontol.* **20**:277–92.

Hallock, P., and Schlager, W., 1986, Nutrient excess and the demise of coral reefs and carbonate platforms, *Palaios* **1**:389–398.

Hallock, P., and Talge, H. K., 1994, A predatory foraminifer, *Floresina amphiphaga*, n. sp., from the Florida Keys, *J. Foram. Res.* **24**:210–213.

Hallock, P., Cottey, T. L., Forward, L. B., and Halas, J. C., 1986*a*, Population dynamics and sediment production by *Archaias angulatus* (Foraminiferida) in Largo Sound, Florida, *J. Foram. Res.* **16**:1–18.

Hallock, P., Forward, L. B., and Hansen, H. J., 1986*b*, Environmental influence of test shape in *Amphistegina, J. Foram. Res.* **16**, 224–31.

Hallock, P., Premoli Silva, I., and Boersma, A., 1991, Similarities between planktonic and larger foraminiferal evolutionary trends through Paleogene paleoceanographic changes, *Palaeogeogr. Palaeoclim. Palaeoecol.* **83**:49–64.

Hallock, P., Müller Karger, F. E., and Halas, J. C., 1993*a*, Coral reef decline—Anthropogenic nutrients and the degradation of western Atlantic and Caribbean coral reefs, *Res. Explor.* **9**:358–378.

Hallock, P., Talge, H. K., Smith, K. and Cockey, E. M., 1993*b*, Bleaching in a reef-dwelling foraminifera *Amphistegina gibbosa, Proc. 7th Int. Coral Reef Symp., Guam* **1**, 44–49.

Hallock, P., Talge, H. K., Cockey, E. M., and Muller, R. G., 1995, A new disease in reef-dwelling foraminifera: implications for coastal sedimentation, *J. Foram. Res.* **25**, 280–1286.

Hallock, P., Bone, Y., Collins, L. B., and James, N. P., 1996, Larger Foraminifera of the west Australian shelf: Modern maladies and Pleistocene paleodepths, [Abstract], Geological Society of America Annual Meeting, Denver, CO, October 1996, *GAAPBC* **28**:129.

Hallock Muller, P., 1974, Sediment production and population biology of the benthic foraminifer *Amphistegina madagascariensis, Limnol. Oceanogr.* **19**:802–809.

Hallock Muller, P., 1976, Sediment production by shallow-water benthic foraminifera at selected sites around Ohau, Hawaii, *Marit. Sed., Spec. Publ.* **1**:263–265.

Hatcher, B. G., Johannes, R. E., and Robertson, A. I., 1989, Review of research relevant to the conservation of shallow tropical marine systems, *Oceanogr. Mar. Biol. Ann. Rev.* **27**:337–414.

Hayes, J. A., 1990, Prey preference in a Caribbean corallivore, *Coralliophila abbreviata* (Lamarck) (Gastropoda, Coralliophilidae), *Bull Mar. Sci.* **47**:557–560.

Hirshfield, H. I., Charmatz, R., and Henson, L., 1968, Foraminifera in samples taken from Eniwetok Atoll in 1956, *J. Protozool.* **15**:497–502.

Hohenegger, J., 1994, Distribution of living larger foraminifera NW of Sesoko-Jima, Okinawa, Japan, *P.S.Z.N.I. Mar. Ecol.* **15**:291–334.

Hottinger, L., Halicz, E., and Reiss, Z., 1993, *Recent Foraminiferida from the Gulf of Aqaba, Red Sea.* Slovenska Akademija Znanosti in Umetnosti, Ljubljana.

Ishman, S. E., 1997, Ecosystem history of south Florida: Biscayne bay sediment core descriptions, U.S. Geological Survey Open-File Report 97–437.

Koestler, R. J., Lee, J. J., Reidy, J., Sheryll, R. P., and Xenophontos, X., 1985, Cytological investigation of digestion and reestablishment of symbiosis in the larger benthic foraminifer *Amphistegina lessonii, Endocyt. Cell Res.* **2**:21–54.

Laws, E. A. and Redalje, D. G., 1979, Effect of sewage enrichment on the phytoplankton populations of a tropical estuary, *Pacif. Sci.* **33**:129–144.

Lee, J. J., 1998, Living sands:Larger Foraminifera and their endosymbiotic algae, *Symbiosis* **25**:71–100.

Lee, J. J., and Anderson, O. R., 1991, Symbiosis in Foraminifera, in: *Biology of Foraminifera* (J. J. Lee and O. R. Anderson, eds.), Academic Press, New York, pp. 157–220.

Lee, J. J., McEnery, M. E., and Garrison, J. R., 1980, Experimental studies of larger foraminifera and their symbionts from the Gulf of Elat on the Red Sea, *J. Foram. Res.* **10**:31–47.

Macintyre, I. G., and Glynn, P. W., 1976, Evolution of a modern Caribbean fringing reef, Galtea Point, Panama, *Am. Assoc. Petrol. Geol., Bull.* **60**:1054–1072.

Martin, J.-M., Elbaz-Poulichet, F., Guieu, C., Loye-Pilot, M.-D., and Han, G., 1989, River versus atmospheric input of material to the Mediterranean, *Nature* **321**:427–428.

McKee, E. D., Chronic, J., and Leopold, E. B., 1956, Sedimentary belts in the lagoon of Kapingimarangi Atoll, *Am. Assoc. Petrol. Geol. Bull.* **43**:501–562.

Müller-Karger, F. E., McClain, C. R., Fisher, T. R., Esaias, W. E., and Varela, R., 1989, Pigment distributions in the Caribbean Sea: Observations from space, *Prog. Oceanogr.* **23**:23–64.

Murray, J. W., 1973, *Distribution and Ecology of Living Benthic Foraminiferids*, Crane, Russak and Co., New York.

Peters, E. C., 1984, A survey of cellular reactions to environmental stess and disease in Caribbean scleractinian corals, *Helgol. Meeresun.* **37**:113–137.

Peters, E. C., 1997, Diseases of coral-reef organisms, in: *Life and Death of Coral Reefs* (C. Birkeland, ed.), Chapman and Hall, New York, pp. 114–139.

Pichard, S. L., and Paul, J. H., 1991, Detection of gene expression in genetically engineered microorganisms and natural phytoplankton populations in marine environments by mRNA analysis, *Appl. Environ. Microbiol.* **57**:1721–1727.

Pichard, S. L., and Paul, J. H., 1993, Gene expression per gene dose, a specific measure of gene expression in aquatic microorganisms, *Appl. Environ. Microbiol.* **59**:451–457.

Porter, J. W., and Meier, O. W., 1992, Quantification of loss and change in Floridian reef coral populations, *Am. Zool.* **32**:625–640.

Prinn, R., Cunnold, D., Rasmussen, R., Simmonds, P., Alyca, F., Crawford, A., Fraser, P., and Rosen, R., 1990, Atmospheric emissions and trends of nitrous oxide deduced from 10 years of ALE-GAGE data, *J. Geophys. Res.* **95**:18,369–385.

Rabalais, N. N., Turner, R. E., Justic, D., Dortch, Q., Wiseman, W. J., and Sen Gupta, B. K., 1996, Nutrient changes in the Mississippi River and system responses on the adjacent continental shelf, *Estuaries* **19**, 386–407.

Reaka-Kudla, M. L., O'Connell, D. S., Regan, J. D., and Wicklund, R. I., 1994, Effects of temperature and UVB on different components of coral reef communities from the Bahamas, *Proc. Coll. Global Aspects of Coral Reefs* (R. N. Ginsburg, Compiler), Rosenstiel School of Marine and Atmospheric Science, University of Miami, Miami, FL, pp. 126–131.

Revets, S., 1990, The genus *Floresina*, gen. nov., *J. Foram. Res.* **20**:157–161.

Richardson, L. L., Kuta, K. G., Schnell, S., and Carlton, P. G., 1997, Ecology of the black band disease microbial consortium, *Proc. 8th Int. Coral Reef Symp., Panama* **1**:597–600.

Richardson, L. L., Goldberg, W. M., Kuta, K. G., Aronson, R. B., Smith, G. W., Ritchie, K. B., Halas, J. C., Feingold, J. S., Miller, S., 1998, Florida's mystery coral-killer identified, *Nature* **392**:587–588.

Richmond, R. H., 1993, Coral reefs:present problems and future concerns resulting from anthropogenic disturbance, *Am. Zool.* **33**:524–536.

Rose, P. R., and Lidz, B. H., 1977, Diagnostic foraminiferal assemblages of shallow-water modern environments:south Florida and the Bahamas, *Sedimentia VI*, pp. 1–56.

Röttger, R., Irwan, A., Schmaljohann, R., and Franzisket, L., 1980, Growth of the symbiont-bearing foraminifera *Amphistegina lessonii* d'Orbigny and *Heterostegina depressa* d'Orbigny (Protozoa), *Endocytobiol. Endosymb. Cell Biol.* **1**:125–132.

Rützler, K., Santavy, D. L., and Antonius, A., 1983, The black band disease of Atlantic reef corals: III. Distribution, ecology, and development, *P.S.Z.N.I. Mar. Ecol.* **4**:329–358.

Sanders, B. M., 1993, Stress proteins in aquatic organisms: An environmental perspective, *Crit. Rev. Toxicol.* **23**:49–75.

Santavy, D. L., and Peters, E. C., 1997, Microbial pests:coral disease in the western Atlantic, *Proc. 8th Int. Coral Reef Symp., Panama* **1**:607–612.

Schnoor, J. L., Galloway, J. N., and Schlesinger, W. H., 1995, Nitrogen fixation: Anthropogenic enhancement–environment response, *Global Biogeochem. Cycles* **9**:235.

Scott, D. B., Schafer, C. T., Honig, C., and Younger, D. C., 1995, Temporal variations of benthic foraminiferal assemblages under or near aquaculture operations: Documentation of impact history, *J. Foram. Res.* **25**:224–235.

Sen Gupta, B. K., Turner, R. E. and Rabalais, N. N., 1996, Seasonal oxygen depletion in continental-shelf waters of Louisiana: Historical record of benthic foraminifers, *Geology* **24**, 227–230.

Shibata, K., 1969, Pigment and a UV-absorbing substance in corals and a blue green alga living in the Great Barrier Reef, *Plant Cell Physiol. Tokyo* **10**:325–335.

Shick, J. M., Lesser, M. P., and Jokiel, P. L., 1996, Effects of ultraviolet radiation on corals and other coral reef organisms, *Global Change Biol.* **2**:527–545.

Shimel, D., Enting, I. G., Heimann, M., Wigley, T. M. L., Raynaud, D., Alves, D., and Siegenthaler, U., 1995, CO_2 and the carbon cycle, in: *Climate Change 1994* (L. G. Houton, L. G. Meira Filho, J. Bruce, H. Lee, B. A. Callander, E. Haites, N. Harris, and K. Maskell, eds.), Cambridge University Press, Cambridge, pp. 35–71.

Shinn, E. A., 1988, The geology of the Florida Keys, *Oceanus* **31**:47–53.

Shinn, E. A., 1989, What is really killing the corals, *Sea Frontiers* March-April, pp. 72–81.

Smith, R. C., and Baker, K. S., 1979, Penetration of UVB and biologically-effective dose-rates in natural waters, *Photochem. Photobiol.* **29**:311–323.

Steven, A. P. L., and Broadbent, A. D., 1997, Growth and metabolic responses of *Acropora palifera* to long-term nutrient enrichment, *Proc. 8th Int. Coral Reef Symp., Panama* **1**:867–872.

Szmant, A. M., and Forrester, A., 1996, Water column and sediment nitrogen and phosphorus distribution patterns in the Florida Keys, USA, *Coral Reefs* **15**:21–41.

Talge, H. K., and Hallock, P., 1993, Observation of symbiont loss in benthic foraminifera, *Scanning* **15**(III):86–88.

Talge, H. K., and Hallock, P., 1995, Cytological examination of symbiont loss in benthic foraminifera, *Amphistegina gibbosa, Mar. Micropaleontol.* **26**:107–113.

Talge, H. K., Williams, D. E., Hallock, P., and Harney, J. N., 1997, Symbiont loss in reef foraminifera: consequences for affected populations, *Proc. 8th Int. Coral Reef Symp., Panama* **1**:589–594.

Toler, S. K. and Hallock, P., 1998, Shell damage in *Amphistegina* populations: Potential indicator of environmental stress, *Mar. Micropaleontol.* **34**:107–115.

Vitousek, P. M., Mooney, H. A., Lubshenco, J., and Melillo, J. M., 1997*a*, Human domination of Earth's ecosystems, *Science* **277**:494–499.

Vitousek, P. M., Aber, J. D., Howarth, R. W., Likens, G. E., Matson, P. A., Schindler, D. W., Schlesinger, W. H., and Tilman, G. D., 1997*b*, Human alteration of the global nitrogen cycle:Sources and consequences, *Ecol. Appl.* **7**:737–750.

Walsh, J. J., 1984, The role of ocean biota in accelerated ecological cycles: A temporal view, *BioScience* **34**:499–507.

Watson, R. T., Rodhe, H., Oeschger, H., and Siegenthaler, U., 1990, Greenhouse gases and aerosols, in: *Climate Change: The IPCC Scientific Assessment* (J. T. Houghton, G. J. Jenkins, and J. J. Ephraums, eds.), Cambridge University Press, Cambridge, pp. 1–40.

Williams, D. E., Hallock, P., Talge, H. K., Harney, J. N., and McRae, G., 1997, Responses of *Amphistegina gibbosa* populations in the Florida Keys (U.S.A.) to a multi-year stress event (1991–1996), *J. Foram. Res.* **27**:264–269.

Williams, E. H. Jr., and Bunckley-Williams, L., 1990, The world-wide coral reef bleaching cycle and related sources of coral mortality, *Atoll Res. Bull.* **335**:1–71.

Yanko, V., Kronfeld, J., and Flexer, A., 1994, Response of benthic foraminifera to various pollution sources:Implications for pollution monitoring, *J. Foram. Res.* **24**:1–17.

Yanko, V., Ahmad, M., and Kaminski, M., 1998, Morphological deformities of benthic foraminiferal tests in response to pollution by heavy metals: Implications for pollution monitoring, *J. Foram. Res.* **28**:177–200.

Yanko, V., Arnold, A. J., and Parker, W. C., 1999, The effects of marine pollution on benthic foraminifera, in: *Modern Foraminifera,* B. Sen Gupta, ed.), Kluwer Academic, Dordrecht, Ch. 13, pp. 217–235.

Chapter 6

Ostracoda in Detection of Sewage Discharge on a Pacific Atoll

STEPHEN H. EAGAR

1. Introduction

Ostracoda have long been known to be useful as indicators of variations in the environment, such as salinity, pH, Eh, and temperature, and more recent studies have focused on pollution. Initially, these studies concentrated on freshwater species and, indeed, most of the subsequent work has also been done using freshwater species. Ostracodes were used to show the influence of a sewage discharge into a stream in Israel (Rosenfeld and Ortel, 1983). Later, other studies worked on the effects of pesticides on freshwater species in ricefields (Lim and Wong, 1986). Pioneering work was done in France by Bodergat (1978) on the effects of cerium on the environment through studying

STEPHEN H. EAGAR • Research School of Earth Sciences, Victoria University of Wellington, Wellington, New Zealand.

Environmental Micropaleontology, Volume 15 of *Topics in Geobiology*, edited by Ronald E. Martin. Kluwer Academic/Plenum Publishers, New York, 2000.

the valves of marine water species and also on the effects of industrial and urban water from a large city in Japan on the marine ostracodes (Bodergat and Ikeya, 1988). Another study by Bodergat *et al.* (1997) was done on natural pollution or eutrophication. Following changes observed in the faunas at Tarawa Atoll, Republic of Kiribati, (Eagar, 1998), where the number of species had recently declined, a study was undertaken in New Zealand on the extreme situation, of a direct sewer discharge from a large city into the marine environment (Eagar, 1999; see also in this volume Ebrahim, Ch. 4; Hallock, Ch. 5; Rosenfeld *et al.*, Ch. 7; Schornhikov, Ch. 8; Ishman, Ch. 16).

2. Setting

Tarawa Atoll is the seat of government for the Republic of Kiribati (Fig. 1). The atoll is triangular in shape with two sides emergent (maximum height above sea level 2.0 m) in a series of islets joined by causeways. The population is approximately 30,000 with 27,000 people living on the southern limb of the triangle.

The Tarawa Lagoon is 30 km from north to south and 25 km from east to west. The maximum depth is 25 m and the mean is 6 m with tidal flats up to 2 km wide. The sediments (generalized) are coral in the west, *Halimeda* (green calcareous alga) in the east, with a decrease in grain size eastward. The bottom is muddy. The tidal flats on the oceanic south side range from 160 m at Bikenibeu to 800 m at Betio. The depth of the ocean is up to 4000 m. Tarawa is in a region of equatorial upwelling, where relatively cool deep nutrient-rich water rises to the surface as warmer waters move poleward under the

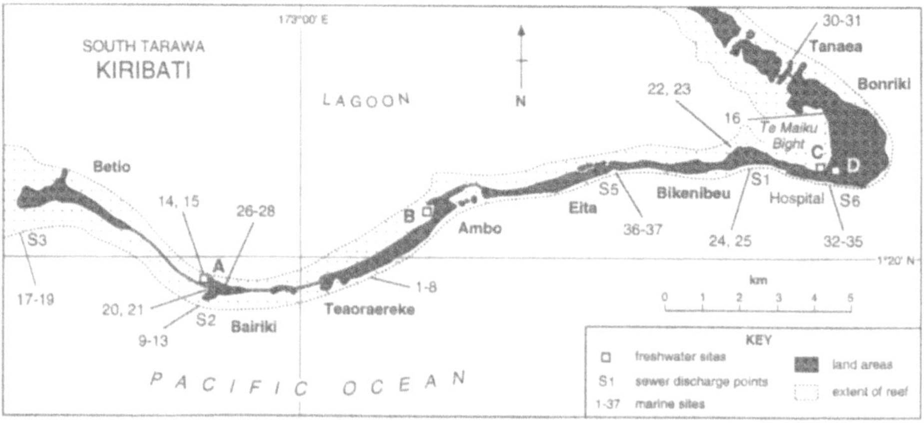

FIGURE 1. Map of South Tarawa. showing sample locations. A–D are the nonmarine sites and 1–37 are the marine localities. S1–S6 are the sewer discharge points. Dark shaded areas are the islets and the stippled areas show the reef exposed at low tide.

TABLE 1. Facilities Used as Toilets in South Tarawa (from Biosystems, 1995)

	Number	Percent
Households	3297	100
Toilet block/homeflush/communal	2239	68
Water latrine	392	12
Lagoon beach	998	28
Ocean beach	1059	32
Other	298	9

influence of the trade winds and the Earth's rotation. The currents in the top 300 m flow west but are variable. The equatorial current moves east below the surface and south of Tarawa. The tidal range is about 1.5 m. The sea temperature on the reef is 30°C and the salinity 36‰.

Pollution occurs on a large scale, but there was no aquatic environment found anywhere that was devoid of live animals. The pollution is almost entirely anthropogenic, with piles of rubbish on the beaches in dumps, other rubbish strewn around the foreshore, a large percentage of the population using the intertidal zone as a toilet, and some pit latrines that infiltrate the freshwater lens (Table 1).

A sewage reticulation scheme was installed in 1983/1984 with saltwater flushing and three outlets onto the oceanic reef. These discharge with the aid of pumps capable of pumping 12 liters/sec through 200-mm-diameter pipes on the edge of the reef flat, which ranges from 160–800 m in width. There are two other private schemes, one at the Tungaru Hospital, Bikenibeu (200-mm diameter) and one at the Moroni High School, Eita (100-mm diameter). No actual discharge values for any pipes were available. The flow is erratic as the pipes are subject to frequent blockages.

Studies have been done on the water quality (Naidu *et al.*, 1991) and the results show high levels of fecal coliform, *Salmonella*, and *Cholera* bacteria in the lagoon and well waters. How these bacteria affect the ostracodes, whether it be by directly infecting them or by affecting their food, is not known. The same report records some heavy metals present in the lagoon, including mercury, copper, chromium, cadmium, and lead.

3. Methods

Following the work on a sewer outfall in Wellington, New Zealand (Eagar, 1999), the sewer outfalls on Tarawa were the subject of investigation. However, the situation at Tarawa is different as the pipes are smaller, are distributed over several outlets, and are flushed with seawater. Using the localities given in Naidu *et al.* (1991), because analyses were already available, the sewage discharge sites were revisited in 1998. One site, S4 at Betio village, no longer exists, but there are additional ones at Tungaru Hospital and the Moroni High School (Fig. 1).

Initially, the sampling was conducted at the Atoll Research Programme, University of the South Pacific at Teaoraereke, as it was considered to be sufficiently remote from any discharge points to preclude any anthropic influence. The purpose was to establish a control area to ascertain the baseline ostracode fauna present. This was supported by samples from the open coast at Tanaea (Nos. 30, 31). Following examination of this site, the sewage discharge points (S1–S6) were examined. When the material was looked at through a microscope, it was clear that the sewer pipes were not the only areas of contamination, so the lagoon reef and the reef adjacent to the causeway linking Bairiki and Betio were also tested by sampling in a series of traverses.

The best and simplest way to collect ostracodes to check for pollution is to gather the same type and volume (about 150 ml) of seaweed. Failing that, the sediments in the affected area are collected and used. The material is then washed into a coarse sieve over a fine one under running water, concentrated on one edge of the sieve, transferred into a petri dish, and examined under a stereoscopic microscope. The ostracodes are picked living and transferred into ethanol for storage. On Tarawa, these same methods were used and the reef platform was noted to have a zoning of the seaweeds. In any unpolluted area, closest to the high tide, there is an algal mat. Seaward from this zone is the brown funnelweed (*Padina gymnospora*) followed by the grapeweed (*Caulerpa racemosa*) and sometimes a fibrous weed (*Halymenia* sp) at the edge of the reef. There is very little seaweed on the oceanic coasts at Betio and Bairiki as there are few pools on the edge of the reef, which is covered in coral rubble.

4. Ostracodes

Twenty species of ostracodes (Table 2) were found living in several environments. As Hartman and Kühl (1978) noted two decades ago none of the ostracode faunas have any abnormalities, or even variations in the ornamentation as described by Eagar (in press).

4.1. Freshwater Ostracoda

Freshwater interstitial species and heavy metal pollution have been studied by Plenet *et al.* (1992). Ponds on Tarawa are uncommon, are expressions of the freshwater lenses on the land surface, and are heavily contaminated with bacteria, but it is not known whether they contain any heavy metals. (The lagoon marine waters do contain some metal contamination, probably from rubbish dumps). Only two out of the four ponds sampled have ostracodes; the

FIGURE 2. Open ocean fauna: (1) *Cytherelloidea fijiensis* (Brady), left valve; (2) *Morkhovenia cuneola* (Brady), right valve. (3) *Quadracythere* (*Tenedocythere*) *deltoides* (Brady); (4) *Bosasella* sp. left valve; (5) *Microxestoleberis gracilis* (Brady), right valve; (6) *Neonesidea* sp, right valve; and (7) *Paranesidea* sp, right valve (scale bar: 0.1 mm).

FIGURE 3. Lagoon and mixed fauna: (1) *Loxoconcha huahineensis* Hartmann left valve; (2) *Touroconcha marcida* (Brady), left valve; (3) *Tongacythere* sp, right valve; (4) *Callistocythere crenata* (Brady), left valve; (5) *Loxoconcha heronislandensis* Hartmann, left valve; (6) *Tanella ochracea* (Brady), right valve; (7) *Kotoracythere inconspicua* (Brady), left valve; and (8) *Paradoxostoma* sp, (broken), right valve (scale bar: 0.1 mm).

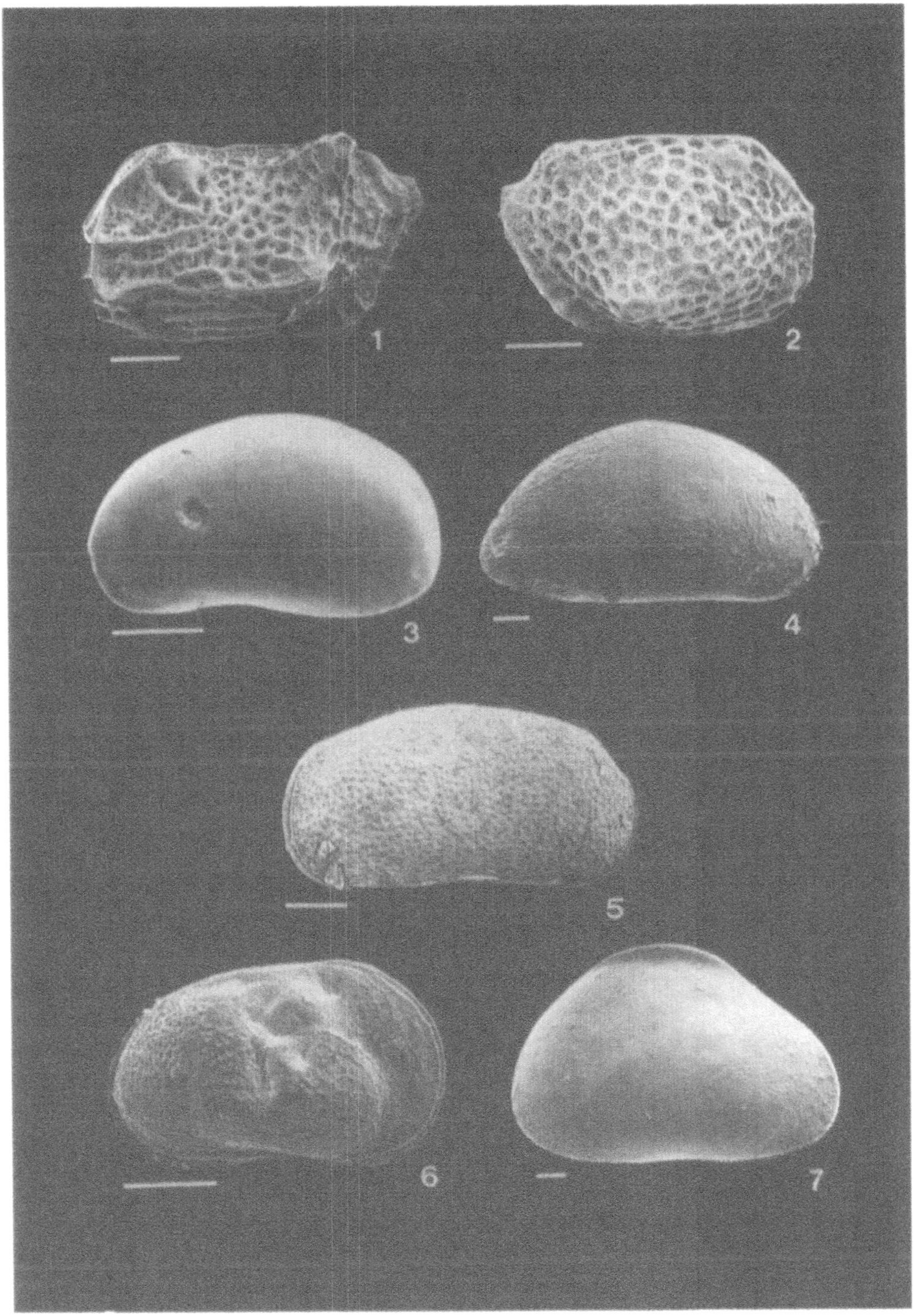

FIGURE 4. Polluted/freshwater material: (1) *Touroconcha marcida* (Brady), left valve; (2) *Loxoconcha huahineensis* Hartmann, right valve; (3) *Xestoleberis* sp, right valve; (4) *Phlyctenophora* sp, right valve; (5) *Hansacypris consobrina* (Brady), left valve; (6) *Limnocythere* cf. *notodonta* Vavra, right valve; and (7) *Cyprinotus cingalensis* Brady, right valve (scale bar: 0.1 mm).

TABLE 2. Distribution of Marine Ostracoda[a]

	Ocean reef																					Lagoon reef				
Species	17	18	19	12	11	13	1	3	4	5	6	7	35	36	37	24	25	32	33	34	30	14	26	27	23	15
Bosasella		1	1																							
Tenedocythere	3	4	6	4	4	7	8			14	5	1	4	3	5	5	5	4	5	3	51	5	14			
Callistocythere		1					1										2	1			1		4			
Xestoleberis	1	7	1	2						1			1	1	2	3	1	3		1	6					1
Microxestoleberis		1					1																			
Touroconcha	4	2	9	6		2	12	6		14	3	8	5	2	3	3	5	7	2	6	8	7	84	1		
L. huahineenses			4										2		1					3	9	5	13	6	9	3
L. heronislandensis								1		1	1								1			12				
Phlyctenophora			7																		1				1	21
Paranesidea					1	2	1			5	1				2		2	1		2	1					
Neonesidea			1					2												1						
Tongacythere						2						1														
Cytherelloidea						1															1					
Paradoxostoma				1						1												1				
Morkhovenia																						3				
Tanella																						1				
Kotorocythere										1	1										2	2	4			
Hansacypris																						5				
Myodocopids		X			X	X	X	X	X	X	X	X			X						X					

[a]Numbers of live specimens complete with appendages and all instars are arranged from west to east. The myodocopids are undifferentiated and their presence is indicated by an X. For locality details see the appendix. Sewer outfall samples are: 17, 11, 36, 24, 33.

one at Bairiki (sample A) has *Cyprinotus cingalensis* in great profusion, and the one at Ambo (sample B) has *Limnocythere* cf. *notodonta*, but this species is considerably less abundant.

4.2. Marine Ostracoda

The fauna comprises monospecific genera with the exception of *Loxoconcha*. As I indicated in an earlier work (Eagar, 1998), there are species that are found only in the lagoon, others only on the oceanic reef, and some that are ubiquitous. The names I used then (Eagar, 1998) have been retained for uniformity. Four species are found living inshore in the lagoon samples: *Loxoconcha huahineensis*, *Tanella ochracea*, *Phlyctenophora* sp., *Morkhovenia* sp. On the oceanic shore there are: *Quadracythere* (*Tenedocythere*) *deltoides*, *Bosasella* sp, *Cytherelloidea fijiensis* (Brady), *Neonesidea* sp., *Paranesidea* sp., *Microxestoleberis* sp. and *Tongacythere* sp. *Callistocythere crenata*, *Xestoleberis* sp., *Paradoxostoma* sp., *Touroconcha marcida* (Brady), and *Kotorocythere inconspicua* (Brady) are common to both environments.

Several myodocopid species were found (in 1998) and these are indicative of clean, open marine conditions. They will be the subject of a separate paper and will not be dealt with any further here.

5. Discussion

Changes in the structure of the ostracode fauna were first noted (Eagar, 1998) from samples taken in 1982 and 1983 prior to the construction of the Nippon causeway linking Betio and Bairiki in 1986–1988 and samples collected in 1995. No live ostracodes were collected in any material from 1982 to 1995 and fewer species have been recovered with each successive collecting trip. It was considered (Eagar, 1998) that the effects of the population on the lagoon waters, together with the linking of the causeways, were causing the decline of the ostracodes, as compared with the faunas in the northern and less populated part of the lagoon. This is still the case, but the effects are thought to be less severe than first suggested. The anthropogenic impact is confined to the shoreline immediately adjacent to the populated islets on South Tarawa.

A comprehensive study was done by Biosystems (Biosystems Analysis, Inc., 1995) as part of an aid scheme from 1992 to 1994 and a management plan was developed for the lagoon. Some biota, water quality, and circulation in the lagoon were looked at as part of that study. As the focus was on the water quality for the purposes of the present chapter, the facilities used by the population are summarized in Table 1. It is important to realize that the human population is projected to increase from its current level to 35,000–46,0000 in the next 15 years.

It was hoped that the increased use of flush toilets would increase the water quality of both the freshwater lenses and the inshore marine environments. A monitoring program was set up to check the quality of the freshwater lenses, but never implemented (Naidu, 1991), so no continuous data are available. The sources of contamination for the freshwater (and ultimately the marine environment) are feces (from humans, dogs, pigs, and poultry), which are found in wells and ponds. The pollutants are represented as fecal coliform and streptococcal levels in the sea of up to five times the U.S. Environmental Protection Agency acceptable level for bathing and recreation. The pollution levels are up to 29,000 fecal colliform/ 100 ml in the sea and up to 37,000 in wells (Naidu 1991). Since shellfish are filter feeders they are useful indicators of the presence of human sewage and thus of such human pathogenic organisms as *Vibro cholera*, *Shigella* spp., *Salmonella* spp., and Hepatitis A virus (Naidu *et al.*, 1991). Ostracodes are scavengers, but no direct relationship is known at this stage between the numbers of ostracodes and the microscopic (pathogenic) organisms. However, it is more likely that the increase in nitrate and phosphate levels do have a direct influence, as the algae, which form an important part of the diet and habitat of the benthic ostracodes, are depleted in the areas of contamination.

5.1. Lagoon Environment

Because of the size of the lagoon, which covers 359 km^2 and has an average depth of 7 m, anthropogenic contamination is confined to the nearshore areas with an increasing dilution effect offshore. This agrees with the observations made on samples from the lagoon in front of the Otintai Hotel at Bikenibeu (Nos. 22, 23) and Bairiki (Nos. 26–28).

Some species of ostracodes are common to the lagoon and oceanic coasts, but the reduced numbers can be still be an indication for the detection of contamination. Sites 14 and 26 have several species normally found on the oceanic reef. These are localities that are flushed with oceanic water coming through the channel cut through the Betio–Bairiki causeway and the anticlockwise inflow of oceanic water from the west. Site 14 is in an area of active accretion and site 26 is sufficiently remote for the pollutants not to affect them. Similarly, the samples from west to east in the lagoon series (Table 2) show a decline in both numbers and species diversity.

I consider that the presence of *Hansacypris consobrina* (Brady) indicates seepage of freshwater into the marine environment and that its presence is evidence of a natural effect. It has been found at Taborio, southeast Betio, and Te Maiku on the open coast (Eagar, 1998). In the study for this chapter it was found in sample 14, which is adjacent to a perched water table (freshwater sample A) by the Bairiki–Betio causeway.

5.2. Open Coast Environment

Only a few species are restricted to the open coastline, where the size of the faunas (six to ten species) and the numbers of specimens recovered are much greater than in the lagoon, with the exceptions of sites 14 and 26, as discussed above. *Loxoconcha huahineensis* tends to be confined to the lagoon and *L. heronislandensis* to the ocean side.

The distribution of the ostracodes around the sewer discharges behave in a way similar to those in New Zealand (Eagar, 1999), but with far less impact, as the sanitary system is flushed with seawater. In both countries there are fewer species and lower numbers around the pipes. The recovery zone is closer to the pipes at Tarawa. The diversity of species around the discharge points is reduced to three species at Tarawa. *Tenedocythere* sp. and *Touroconcha marcida* are tolerant of the sewage without an appreciable reduction in numbers.

5.3. Freshwater Environment

Only two ponds (sites A, B) yielded ostracodes and in both cases a single, but different species was found. *Cyprinotus cingalensis* Brady was found at site A (adjacent to a domestic rubbish dump) and *Limnocythere* cf. *notodonta* (Vavra) at site B. The genus *Limnocythere* is cosmopolitan in distribution and is recorded as tolerating a range of salinities. If the ponds were unpolluted, one would expect to find more species.

6. Summary

1. The total number of ostracode species found on Tarawa is small, not because of any anthropogenic reason, but because the fauna arrived a relatively short time ago, having migrated, probably from the west, where there are more species in common with Tarawa. Atolls, as opposed to emergent high volcanic or rocky islands, tend to have about 20% of the number of species compared to Pelau (Whatley and Watson, 1988) or New Caledonia (Hoibien, pers. com.).
2. The sewer outlets do not have an appreciable effect on the reef because of the saltwater flushing system. The elimination of freshwater from the system, the small volume of discharge, and the broad distribution of the outlets do not concentrate the material and makes minimal impact on the reef. There is a slight reduction in all biota on the reef edge adjacent to the sewer pipes.
3. There are polluted areas such as the area adjacent to a village between the Nippon causeway (which links Bairiki with Betio) and the oceanic

reef, and the area up to 50 m seaward from the high tide mark in the lagoon between Bairiki and Te Maiku Bight. No live ostracodes were collected from either of these areas.

4. The circulation of the lagoon water is limited in the eastern part of the lagoon toward Te Maiku Bight, where there is also little seaweed. There is also a buildup of nitrates along parts of the lagoon further northward and it is considered to be causing the increased growth of the seagrass seen on aerial photographs over the past 50 years. The nitrates are likely to be the reason for the low species diversity and numbers.
5. The freshwater ponds are sometimes clean enough to support ostracodes. However, there are restricted numbers of species and distribution of specimens. Some waters are devoid of ostracodes. Further studies need to be made on the freshwater species.

The environment at Tarawa is not beyond redemption and would recover relatively quickly from its present state if remedial measures were taken. Long-term monitoring of the effects of a sewage outfall over 30 years (Stott *et al.*, 1996) has shown such a recovery. It is important that baseline data be recorded and monitored with a regular sampling program so the long-term effects become known in Tarawa. These programs are simple and effective because no large expenditure is necessary to gain the knowledge.

7. Appendix—List of Samples/Localities

Freshwater material

A. Adjacent to Mary's Motel and the beginning of the Nippon causeway, lagoon side: small pond.
B. Ambo, opposite Eita and Viana stores, adjacent to the road on the lagoon side: pond with babai growing on margin.
C. Road junction of Te Maiku and Bonriki airport: fish pond, adjacent to the east side of road. *No fauna.*
D. Mangrove swamp in corner of road Te Maiku Bight, same locality as sample C, *No fauna.*

Marine material

1. Atoll Research Programme, Teaoraereke: ocean reef flat, pool toward high tide.
2. Atoll Research Programme, Teaoraereke: ocean reef flat, pool on reef edge. *No fauna.*

3. Atoll Research Programme, Teaoraereke: ocean fringing reef.
4. Atoll Research Programme, Teaoraereke: ocean reef flat.
5. Atoll Research Programme, Teaoraereke: ocean reef pool.
6. Atoll Research Programme, Teaoraereke: fish trap toward ocean fringing reef.
7. Atoll Research Programme, Teaoraereke: fish trap toward ocean fringing reef.
8. Atoll Research Programme, Teaoraereke: fish trap toward ocean fringing reef. *No fauna.*
9. Bairiki, behind the President's Maneaba: between beach rock on ocean reef platform. *No fauna.*
10. Bairiki, behind the President's Maneaba: between beach rock on ocean reef platform. *No fauna.*
11. Bairiki sewer outfall (S2): ocean reef at pipe.
12. Bairiki sewer outfall (S2): ocean reef 50 m to east of pipe.
13. Bairiki sewer outfall (S2): ocean reef 50 m to west of pipe.
14. Bairiki, west Mary's Motel: lagoon reef platform, bottom sediments.
15. Bonriki, lagoon reef flat near borrow pits for airport runway.
16. Atoll Research Programme, ocean reef platform, Teaoraereke: light trapping. *No fauna.*
17. Betio sewer (S3): ocean reef edge near pipe.
18. Betio sewer (S3): ocean reef edge 50 m from pipe
19. Betio sewer (S3): pool on ocean reef platform.
20. Bairiki, south side of causeway: reef platform. *No fauna.*
21. Bairiki, south side of causeway: reef platform. *No fauna.*
22. Bikenibeu, in front of Otintai Hotel: lagoon reef, funnelweed. *No fauna.*
23. Bikenibeu, in front of Otintai Hotel: lagoon reef, green weed.
24. Bikenibeu, sewer (S1) ocean reef edge, near pipe.
25. Bikenibeu, sewer (S1) ocean reef edge, 50m from pipe.
26. Bairiki, opposite Australian High Commissioner's residence: lagoon reef platform: low-tide green weed 200 m from shore.
27. Bairiki, opposite Australian High Commissioner's residence: lagoon reef platform, funnelweed 150 m from shore.
28. Bairiki, opposite Australian High Commissioner's residence: lagoon reef platform, anaerobic sediment under coral rubble 50 m from shore. *No fauna.*
29. Betio/Bairiki causeway: channel for fishing boats, light trapped. *No fauna.*
30. Tanaea: ocean reef edge.
31. Tanaea: ocean reef platform. *No fauna.*
32. Tungaru hospital sewer outfall: 50 m east of pipe.
33. Tungaru hospital sewer outfall: near pipe.
34. Tungaru hospital sewer outfall: 50 m west of pipe
35. Moroni High School, Eita: 50 m east of sewer pipe.
36. Moroni High School, Eita: near pipe.
37. Moroni High School, Eita: 50 m west of pipe.

ACKNOWLEDGMENTS: The author wishes to sincerely thank the Ministry of Foreign Affairs and Trade, New Zealand, and South Pacific Applied Geological Commission (SOPAC), Fiji, for financing the fieldwork for this work; Naomi Biribo-Atauea, Ministry of Natural Resources Development, Tarawa, Kiribati; Temakei Tebano, University of the South Pacific Atoll Research Programme, Tarawa, Kiribati for the use of the facilities; Dr Koen Martens, Royal Belgian Institute of Natural Sciences, Brussels for information on the freshwater species.

References

Biosystems Analysis, Inc., 1995, Management plan for Tarawa Lagoon, Republic of Kiribati: Volume 2 (each paper is individually paginated).

Bodergat, A-M., 1978, Un marqueur instantine de la pollution en cérium dans la zone de Cortiou (Rade sud de Marseille, Bouches-du-Rhone) *Aurila speyeri* Brady, 1868 (Ostracoda, Podocopida, *Doc. Lab. Géol. Fac. Sci., Lyon H.S* **4**:71–85.

Bodergat, A-M., and Ikeya, N. 1988, Distribution of recent Ostracoda in Ise and Mikawa Bays, Pacific Coast of Central Japan, in: *Evolutionary Biology of Ostracoda, Proc. 9th Int. Symp. on Ostracoda*, (Hanai, T., Ikeya, N. and Ishizaki, K. eds.), Elsevier, Amsterdam, pp. 413–428.

Bodergat, A-M., Rio, M., and Ikeya, N, 1997, Tide versus eutrophication: Impact on ostracodes populations structure of Mikawa Bay (Japan), *Rev. Micropaléontologie* **40**:3–13.

Eagar, S. H., 1998, Recent Ostracoda from Tarawa, Kiribati, Pacific Ocean, in: What about Ostracoda? Actes du 3ᵉ Congrès Européen des Ostracodologistes, 1996. *Bull. Centre Rech. Elf Explor. Prod., Mémoire* **20**:55–67.

Eagar, S. H., 1999, Distribution of Ostracoda around a coastal sewer outfall: A case study from Wellington, New Zealand, *J. Royal Soc. New Zealand* **29**:257–264.

Eagar, S. H., in press, Intraspecific variation in the shell ornament of benthic Ostracoda (Crustacea) from Kiribati, *Oceanologica Acta.*

Hartmann, G., and Kühl, C., 1978, Zur Variabilität der Oberflächenornamente der Schlenlebender Ostracoden-Populationen, *Mitteil. Hamburg. Mus. Inst.* **75**:221–223.

Lim, R. P., and Wong, M. C., 1986, The effects of pesticides on the population dynamics and production of Stenocypris major Baird (Ostracoda) in ricefields, *Arch. Hydrobiol.* **106**:421–427.

Milau, B., Dekens, N., and Wouters, K., 1997, Evaluation de l'utilisation des ostracodes comme bio-indicateurs potentiels de pollution application aux eaux de la Slack (Boulonanais, France), *Ecologie* **28**:3–12.

Naidu, S., Aalbersberg, W. G. L., Brodie, J. E., Fuavao, V. A., Maata, M., Nagasima, M., Whippy, P., and Morrison, R. J. 1991. Water quality studies on selected South Pacific lagoons, UNEP Regional Seas Reports and Studies No. 136, UNEP, 1991; and SPREP Reports and Studies No. 49, South Pacific Regional Environment Programme, 1991, pp. 1–99.

Plenet, S., Marmonier, P., Gibert, J., Stanford, J.A., Bodergat, A-M., and Schmidt, C. M., 1992, Groundwater hazard evaluation: A perspective for the use of interstitial and benthic invertebrates as sentinels of aquifer metallic contamination, in: *Proc. 1st Int. Conf. Ground Water Ecology* (J. A. Stanford and J. G. Simons, eds.), Florida, pp. 319–329.

Rosenfeld, A., and Ortal, R, 1983, Ostracodes as indicators of water pollution in Nahal Harod, northern Israel, in: *Applications of Ostracoda* (R. F. Maddocks, ed.) University of Houston Geoscience, Houston, pp. 229–237.

Ruiz Muñoz, F., González-Regalado, M. L., Flores, J. B., and Morales, J. A. 1997, The response of ostracod assemblages to recent pollution and sedimentary processes in the Huelva Estuary, SW Spain, *Sci. Total Environ.* **207**:91–103.

Stott, L. D., Hayden, T.P., and Griffith, J., 1996, Benthic foraminifera at the Los Angeles County Whites Point outfall revisited, *J. Foram. Res.* **26**:357–368.

Whatley, R., and Watson, K., 1988, A preliminary account of the distribution of Ostracoda in recent reef and reef associated environments in the Pelau Seribu or Thousand Island Group, Java Sea, in: *Evolutionary Biology of Ostracoda Proc. 9th Intern. Symp. Ostracoda* (T. Hanai, N. Ikeya, and K. Ishizaki, eds.), Elsevier, Amsterdam, pp. 399–411.

Chapter 7

Ostracodes as Indicators of River Pollution in Northern Israel

AMNON ROSENFELD, REUVEN ORTAL, and AVRAHAM HONIGSTEIN

1. Introduction

Organic pollution produces changes in the aquatic environment. Effluent reduces the dissolved-oxygen concentration as a result of decomposition of organic material and increases the levels of ammonia and phosphate, and both the biochemical oxygen demand (BOD) and chemical oxygen demand (COD). These affect the stream fauna so that different communities often show a zonation that characterizes distinct sections downstream of the discharge (Kolkwitz, 1950; Hynes, 1960; Liebmann, 1962; Hawkes, 1962; Chandler, 1970). This biotic zonation led to the traditional classification of water quality: a heavily polluted (polysaprobic) zone, a moderate (mesosaprobic) polluted zone (divided into two subzones, alpha and beta), and a slightly (oligosaprobic) polluted zone, also known as the recovery zone, which indicates an advanced stage of self-purification of the river.

AMNON ROSENFELD • Geological Survey of Israel, Jerusalem 95501, Israel. REUVEN ORTAL • Nature Reserves Authority, Jerusalem 94467, Israel. • AVRAHAM HONIGSTEIN • Ministry for National Infrastructures, Oil and Gas Section, Jerusalem 94387, Israel.

Environmental Micropaleontology, Volume 15 of *Topics in Geobiology*, edited by Ronald E. Martin. Kluwer Academic/Plenum Publishers, New York, 2000.

Nonmarine ostracodes are sensitive to changing physicochemical parameters. The ability of these taxa to respond rather quickly to pollution makes them good indicators of specific pollution levels, thereby showing the real state of the stream. Faunal data may even detect slight environmental changes that cannot be observed by physicochemical analysis only.

Few studies have dealt with freshwater ostracodes in response to organic pollution in riverine environments. We (Rosenfeld and Ortal, 1983) carried out a study in northern Israel, in which we found an inverse correlation between levels of organic pollution and ostracode population densities. We also observed variable ostracode assemblages that indicate different polluted zones. Bromley and Por (1975) investigated the polluted water of a stream in central Israel and found a clear zonation of the fauna downstream of the pollution source, with the ostracodes *Herpetocypris chevreuxi* Sars and *Eucpris clavata* Baird flourishing in the recovery zone. Mezquita *et al.* (1996) also found that ostracodes were absent from river samples of lowest-quality water (organic pollution) in the eastern Iberian Peninsula. Different ostracode assemblages occurred along a gradient from high pollution toward the recovery zone. In a river in northern France, Milhau *et al.* (1997) found that freshwater ostracodes are good bioindicators of water quality.

Similarly, few publications deal exclusively with the effect of pollution on marine ostracodes. Kaesler *et al.*, (1979) recognized the rapid recolonization of ostracodes after a crude oil spill in the Strait of Magellan. Bodergat (1978) found cerium within the valves of the ostracode *Aurila speyeri* (Brady 1868), as a result of industrial pollution in the Gulf of Marseille, and Ruiz Munoz *et al.* (1994) recognized the negative influence of heavy metal industrial pollution on an estuarine ostracode fauna in southwest Spain (see also in this volume, Eager, Ch. 6; and Schornikov, Ch. 8).

2. Materials and Methods

The Harod Valley (Fig. 1) stretches from the Yizre'el Valley to the Jordan Valley and is named after its major stream. The climate of the region is subtropical and semiarid, with an average annual rainfall of 400 mm. Nahal Harod (nahal is stream or river in Hebrew) originates in the upper Yizre'el Valley, 500 m above sea level, and descends to 230 m below sea level, where it flows into the lower Jordan River. Nahal Harod drains the eastern part of the Yizre'el Valley, the Bet-Shean Valley, the Harod Valley, and the surrounding mountains. The length of the river is 35 km and the drainage basin is about 180 km^2 in area. The upper section of the stream is dry throughout most of the year. Nahal Harod cuts through Quaternary alluvium, travertine, and basalts. Domestic sewage and industrial wastes are discharged into Nahal Harod at several points (Fig. 1, indicated by arrows). The effluent of the town Bet Shean

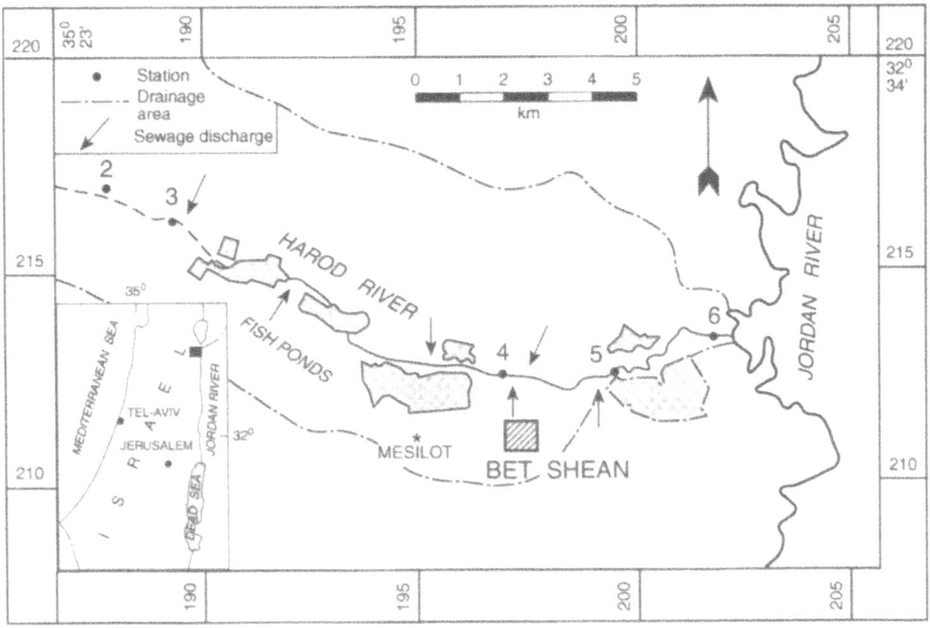

FIGURE 1. Location of Nahal Harod, northern Israel. Studied area marked on insert.

is discharged regularly (affecting mainly station 5), whereas its emission at the other stations is intermittent.

Nahal Harod is a fully anthropogenic system. The influx of most of its water is from recycling in fish ponds, irrigation systems, and sewage flushing, while rain floods are very rare and short-lived in this region. The river carries water throughout the year, but its highstand is during the summer months owing to intensive irrigation. The natural discharge of the base flow of Nahal Harod before its regulation was about 80 Mcm/year (Mcm = million cubic meters). Recently (end of the 1990s), after piping of all the springs, the base flow has decreased to only about 10–12 Mcm/year (including drainage, fishponds, and sewage water; data from Bet Shean Valley Water Authority).

The water and sediment samples were routinely collected by the Pollution Control Unit (Nature Reserves Authority, Israel). Four stations (stations 3,4,5, and 6; Fig. 1) along Nahal Harod were monitored and sampled for ostracodes over a period of sixteen years (1978–1994). The repeatability of the results is shown in many locations by random sampling during a long observation period.

The water quality at station 2 (Harod Bridge) was measured, too, but no ostracode samples were taken there. The other stations are located near sources of pollution. Station 3 (Heftzi-Bah Bridge; altitude of −80 m) has a muddy substrate with some gravel with 5- to 10-cm water depth and is about 1 m wide. Station 4 (Bet-Shean Bridge; altitude −130 m) has a hard substrate of mud and gravel, a water depth of 15–30 cm, and is about 1.70 m wide. Station 5 (Roman Bridge; altitude −210 m) has a muddy gravel substrate and a water depth of 15–30 cm and is about 3.00 m wide. Station 6 (Patrol Road Bridge; altitude −250 m) has a variable sandy to muddy substrate with gravel, a water depth of 25–80 cm, and is about 3.50 m wide. An unpolluted reference station was sampled and monitored at Mesilot (Fig. 1, Tables 1 and 2), where some clean water from a spring flows into Nahal Harod.

The ostracode samples were collected using a plankton handnet of 212-μm mesh by swiping three times through the water thereby covering approximately 0.25 m^2 bottom area for quantitative analysis. The live ostracodes were treated with formalin and picked in the laboratory. Juvenile and adult specimens with soft parts were counted together. The ostracode material is deposited in the Inland Water Ecological Laboratory, Zoology Department, The Hebrew University, Jerusalem, designated under IES numbers.

The laboratory and field physicochemical determinations are given in Table 1. These include temperature (°C.), water velocity (in cm/sec), pH, BOD (in mg O_2/liter; laboratory data), nitrate, phosphate, ammonia and chloride (in mg/liter), DO% (percentage saturation of dissolved oxygen; field measurements), and ABS (detergent content in mg/liter).

The water analyses were carried out in strict accordance with the Standard Methods of American Public Health Association (A.P.H.A., 1971). The BOD determination constitutes an empirical test, in which standardized laboratory procedures are used to determine the relative oxygen requirements of wastewater effluents and polluted water over an incubation period of 5 days (A.P.H.A., 1971). The BOD parameter defines the level of pollution in the sample and, in fact, defines the dysoxic–anoxic condition of the river.

Saprography (saprobic scales from different regions) is a matter of geography. Measurements of pollution in Europe (Lange-Bertalot, 1979) show much lower BOD values than the numbers used in India (Rama Rao *et al.*, 1978). Therefore, in this work we adopted the pollution system used in Israel by the Pollution Control Unit in the Nature Reserves Authority (Ortal, 1977). The BOD values of the proposed pollution levels are as follows: >50 mg O_2/liter for very heavy pollution (polysaprobic); 15–50 mg O_2/liter for heavy pollution (polysaprobic-alpha mesosaprobic); 8–15 mg O_2/liter for moderate pollution (beta mesosaprobic); and 5–8 mg O_2/liter for slight pollution (oligosaprobic). In almost clean waters the BOD values are 0–5 mg O_2/liter.

The distribution of the ostracodes (Table 2), their population densities (PD, in specimens per sample), and diversity (number of species) in Nahal Harod during the investigation with respect to pollution levels (BOD) are shown in Fig. 2 (on specific dates) and Fig. 3 (average values for the time intervals).

3. Results and Discussion

3.1. Physicochemical Factors

The temperature of the water at all stations during the years 1978–1994 ranged from 24° to 35°C during the summer and from 15° to 20°C during the winter. The pH was slightly alkaline. The velocity of the stream was slow to moderate at station 3 and moderate to strong at stations 4, 5, and 6. The average chlorinity was 1000 to 1200 mg Cl/liter (salinity of 1.6–1.9‰), which makes Nahal Harod oligohaline. According to the measured physicochemical factors (Table 1, Figs. 2 and 3), the conditions at stations 3, 4, and 6 were variable during the years 1978–1994. Influx of sewage to stations 3 and 4 was irregular, whereas station 5 regularly received the most polluted discharge from the town of Bet Shean. Thus, the BOD values at station 5 were consistently higher and the dissolved oxygen concentration lower throughout the investigated period, thereby defining the station as a heavy polluted one. Different BOD concentrations were measured at stations 3 and 4 at different seasons, ranging from 0 mg O_2/liter in March 1982 to 30 mg O_2/liter BOD in June 1981. Station 6, which was relatively remote from sites of sewage discharge, showed variable BOD values, but normally the water quality was better as a result of the influx of clean spring water and of natural self-purification processes. However, this station is often disturbed by road construction, and the channel of the river has been temporarily blocked. The BOD of the cleanest water at the Mesilot station was low (1.5 mg O_2/liter).

The high concentrations of ammonia and phosphate at station 5 show clearly the effect of effluent discharge. At stations 3 and 4 these factors were variable (Table 1). The measured physicochemical factors did not always reflect the real state of the river as a result both of limitations inherent in the measuring methods and transient changes in environmental factors owing to momentary sewage releases. Therefore, the fauna is an even better indicator of pollution levels. Saprobiological levels can be defined on the basis of biotic assemblages and supplemented by physicochemical parameters.

According to the BOD measurements, station 5 could be defined as alpha-mesosaprobic to polysaprobic, and stations 3, 4, and 6 fall within the oligosaprobic to mesosaprobic (alpha and beta) categories, depending on effluent discharges. Since the pollution originates from a number of sites along the river, self-purification occurs only in the last section of the river at station 6.

The greatest degree of pollution was measured at all stations during the summer. During the winter it was somewhat lower, even at station 5, probably because of rainfall dilution. Although the BOD values (as well as other factors) varied throughout the year, in general, the pollution level remains more or less constant at each station during the year. According to the BOD levels station 5 was the most heavily polluted one throughout the year, whereas station 6 was the least polluted, and stations 3 and 4 were moderately polluted. The increase in organic pollution levels during the last three decades is well-illustrated in Fig. 4; the 1990s showed the highest levels at all stations.

TABLE 1. Physicochemical Parameters of Stations in Nahal Harod during the Years 1978–1994

Station	Date	Temperature C°	pH	BOD mgO_2/liter	ABS mg/liter	Cl mg/liter	NO_3 mg/liter	DO %	NH_4 mg/liter	PO_4 mg/liter
3	12.21.78	17.3	7.1	11	0	1874	1.4	44	0.1	7
3	5.14.79	24.9	7.6	8	0.1	5144	4.5	53	0.4	—
3	7.10.79	26.8	7.5	6	0.21	742	1.3	69	0.7	0.8
3	5.22.80	25.1	7.8	19	0.28	377	—	8	3.96	6.78
3	6.2.81	34.5	7.4	30	0.16	4165	—	—	—	—
3	3.30.82	20.4	8.1	0	0	1160	8.5	85	0.06	0.16
3	10.10.94	28.5	8.5	14	0.05	726	—	7.3	3	0.72
4	12.21.78	15.4	7.5	0	0.4	1296	3.2	100	0	0.37
4	5.14.79	25.2	7.6	—	0.44	1732	3.4	128	3.2	—
4	7.10.79	27	7.6	15	0.21	1775	1.7	82	1.35	0
4	11.21.79	19.6	8.1	6	0.4	2747	1.3	92	1.15	0.15
4	5.22.80	24	8	—	0.28	1657	71	90	0	0.96
4	6.2.81	25.7	8	27	0.1	1985	—	—	—	—
4	3.30.82	17.9	8.4	0	0	446	4.2	101	0.02	0.07
4	7.6.94	27.7	7.8	28	0.11	1429	—	4.9	9	1.8
4	10.10.94	24.9	8.5	24	0	1384	—	8.5	0.8	0.5

5	12.21.78	16.1	7.6	18	0.1	1225	1.9	76	0	0.52
5	5.14.79	25.3	7.6	40	0.14	1331	0.4	54	7.68	—
5	7.10.79	27.8	7.6	66	0.5	1274	0.6	57	12.35	4.8
5	11.21.79	19.8	8	31	0.1	1580	1.1	70	3.71	0.17
5	5.22.80	24.5	7.9	32	0.66	1306	20	68	1	1.36
5	6.2.81	24.9	8	26	0.2	1512	—	—	—	—
5	3.30.82	17.1	8.3	39	0.46	351	21.5	70	14	2.64
5	7.6.94	27.6	8.4	55	0.27	434	—	2.9	12.3	1.6
5	10.10.94	26	8.4	44	0.08	1066	—	3.4	5.3	3.9
6	12.21.78	18.8	7.2	3	0.1	1328	2.3	56	0	0.1
6	7.10.79	25.6	7.6	0	0.02	1381	7.2	92	1	0.45
6	11.21.79	21	7.8	3	0	1739	—	84	0.61	0.4
6	5.22.80	24.9	8	0	0.12	1405	69	82	2.18	0.99
6	6.2.81	24.6	8	17	0.1	1330	—	—	—	—
6	3.30.82	22.9	7.7	13	0	386	2.5	35	0.83	0.55
6	7.6.94	24.8	7.7	22	0.09	1543	—	3.5	13.4	1.9
6	10.10.94	26.3	8.1	11	0.02	1570	—	5.4	7	1.8
*	8.8.94	27.8	7.4	1.5	0	1325	—	8.7	0.1	0.82

—Not determined.
*Reference station at Mesilot.

TABLE 2. Distribution of Ostracode Species in Nahal Harod, Their Population Density (PD), Diversity (Div), and Pollution Levels (BOD) during the Years 1978–1994

					Ostracodes					
Station	Date	BOD mgO_2/liter	PD	Div	*H. Incongruens*	*H. salina*	*I. bradyi*	*C. neglecta*	*C. aculeata*	*C. torosa*
3	12.21.78	11	21	3	6	12	3			
3	5.14.79	8	300	2	50	250				
3	7.10.79	6	689	2	96	593				
3	5.22.80	19	11	3	3	5	3			
3	6.2.81	30	18	2	8	10				
3	3.30.82	0	280	1	280					
3	10.10.94	14	3	1	3					
4	12.21.78	0	132	1	132					
4	5.14.79	—	41	1	41					
4	7.10.79	15	1	1	1					
4	11.21.79	6	98	1	98					
4	5.22.80	—	2	1		2				
4	6.2.81	27	6	1	6					
4	3.30.82	0	37	5	14	2	17	2	2	
4	7.6.94	28	3	1	3					
4	10.10.94	24	14	2	11	3				
5	12.21.78	18	30	1	30					
5	5.14.79	40	23	2	22	1				
5	7.10.79	66	3	1	3					
5	11.21.79	31	2	1	2					
5	5.22.80	32	0	0						
5	6.2.81	26	17	2	11	6				
5	3.30.82	39	1	1	1					
5	7.6.94	55	3	1	3					
5	10.10.94	44	8	2	6	2				
6	12.21.78	3	0	0						
6	7.10.79	0	6	1	6					
6	11.21.79	3	167	1	167					
6	5.22.80	0	1	1	1					
6	6.2.81	17	1	1	1					
6	3.30.82	13	0	0						
6	7.6.94	22	6	1	6					
6	10.10.94	11	6	1	6					
*	8.8.94	1.5	11	2		4				7

—Not determined.
*Clean water station at Mesilot.

3.2. Ostracode Fauna

The biotic system responds directly to pollution as defined by BOD and, therefore, the ostracodes can serve as a complementary indicator for the definition of water quality. Change in the structure of the communities over time is one of the criteria for estimating changes in pollution levels. Methods

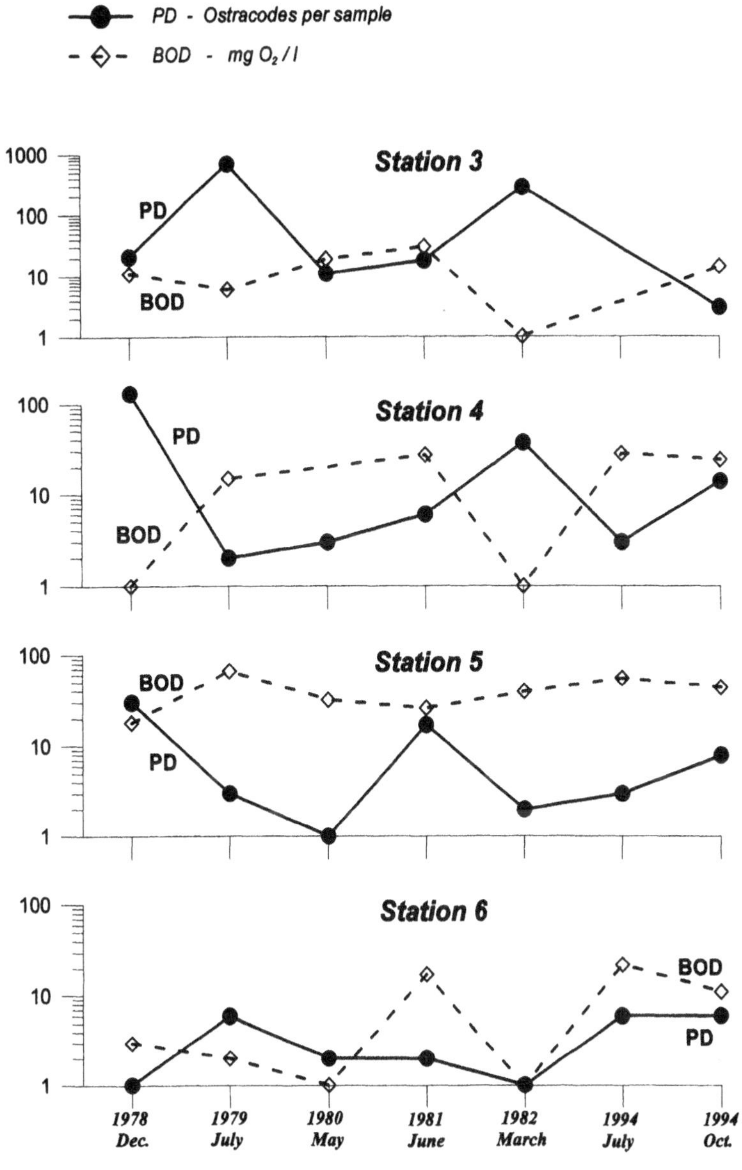

FIGURE 2. Ostracode population densities (PD = specimens/sample) and biochemical oxygen demand (BOD = mg O_2/liter) at four stations in Nahal Harod during the years 1978–1994.

for interpretation of biological data are based upon the response of biocenoses to pollution stress (by the disappearance of sensitive species and the survival of tolerant ones). Biological taxa may be used to assess the quality of water; moreover it can indicate the presence of micropollutants that cannot be detected directly through physicochemical analysis (Ghetti, 1980).

Six ostracode species were found in Nahal Harod (Table 2). *Heterocypris incongruens* Gauthier 1938, and *Heterocypris salina* (Brady 1868) were the

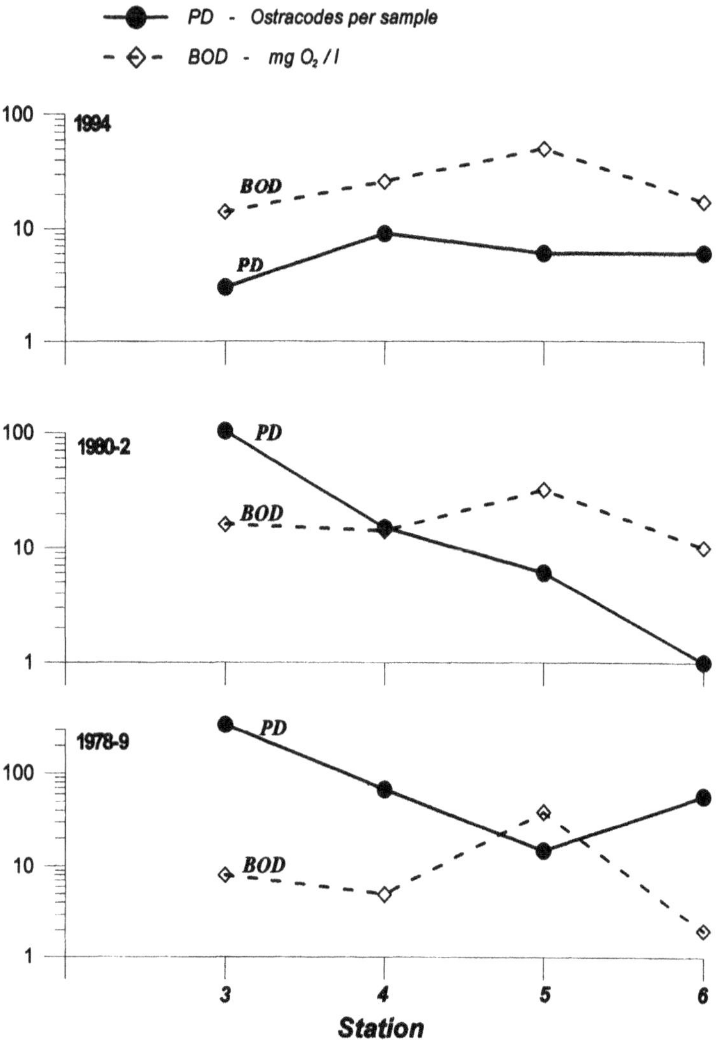

FIGURE 3. Ostracode population densities (PD) and biochemical oxygen demand (BOD) of stations in the Nahal Harod according to time intervals (for data see Table 2).

dominant ostracode species, whereas *Ilyocypris bradyi* (Sars 1890), *Candona neglecta* Sars 1887, *Cypridopsis aculeata* (Lilljeborg 1853), and *Cyprideis torosa* (Jones 1857) were rare. [For taxonomy and synonymy see Klie (1938), Bronstein (1947), Wagner (1957), and Purper and Wuerdig-Maciel (1974).]

Quantitative population analysis of the ostracodes (density and diversity) can be used to determine pollution levels. Ostracodes were more commonly found in low to moderately polluted waters and were rare or absent in the heavily polluted zones (Table 2). Different ostracode species can also indicate pollution levels since each ostracode species exhibits its own tolerance limits.

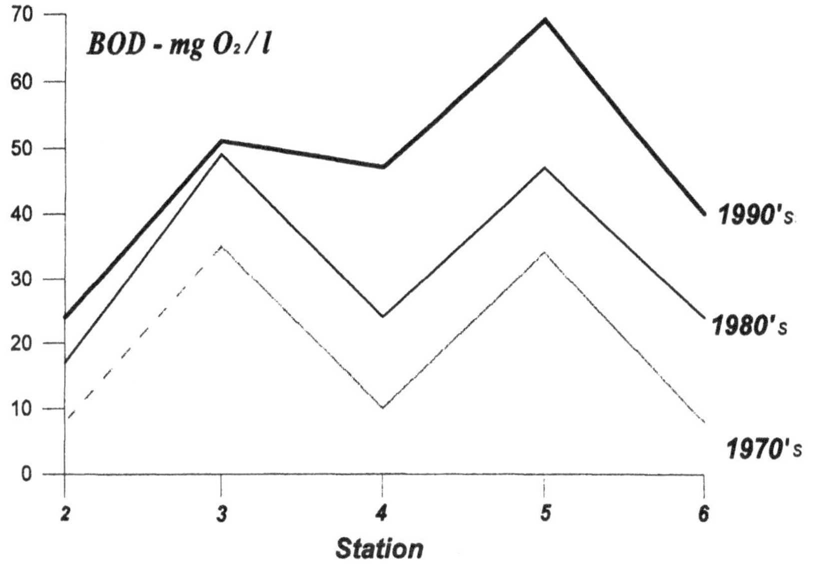

FIGURE 4. Biochemical oxygen demand (BOD) of stations in Nahal Harod according to decades (1970s, 1980s and the 1990s). (Physicochemical factors were measured three times a year; for data see Table 2).

We recorded the presence of *H. incongruens* and *H. salina* in polluted rivers (Rosenfeld and Ortal, 1983). The BOD tolerance range of *H. incongruens* is from 0 to 66 mg O_2/liter, whereas the tolerance range of *H. salina* is from 0 to 26 mg O_2/liter BOD. On the other hand, *I. bradyi, C. neglecta, C. aculeata,* and *C. torosa* seem to be clean-water dwellers: all of them are common in freshwater environments (Klie, 1938) and were found at station 4 in March 1982, and at the Mesilot station with BOD of 0–1.5 mg O_2/liter. *C. torosa* was also observed in the Jordan River. Similar to our results, Margalef (1953, 1965) and Mezquita *et al.* (1999) found in rivers in Spain that *H. incongruens* is very tolerant of low oxygen and high organic matter content. The ostracode species *H. salina* was recorded by Mezquita *et al.* (1999) as a species that can tolerate certain amounts of organic pollution, while *I. bradyi* was found in river sites with good water quality, thus agreeing with the present results.

The population density of the ostracodes in Nahal Harod can clearly be related to the level of organic enrichment (Table 2). High population densities (>100 specimens per sample) were recorded at stations where the BOD was low (0–8 mg O_2/liter, almost effluent-free; Station 3 in May and July 1979; March 1982; station 4 in December 1978, and November 1979). High population densities of ostracodes correspond to a low to moderate pollution level. The lowest population densities were recorded at Station 5, where high BOD values indicated heavy to very heavy pollution. It seems that the deficiency of ostracodes in station 6 (Figs. 2 and 3) is due to mechanical road construction disturbances (input of sandy material and low water transparency), rather than to physicochemical factors.

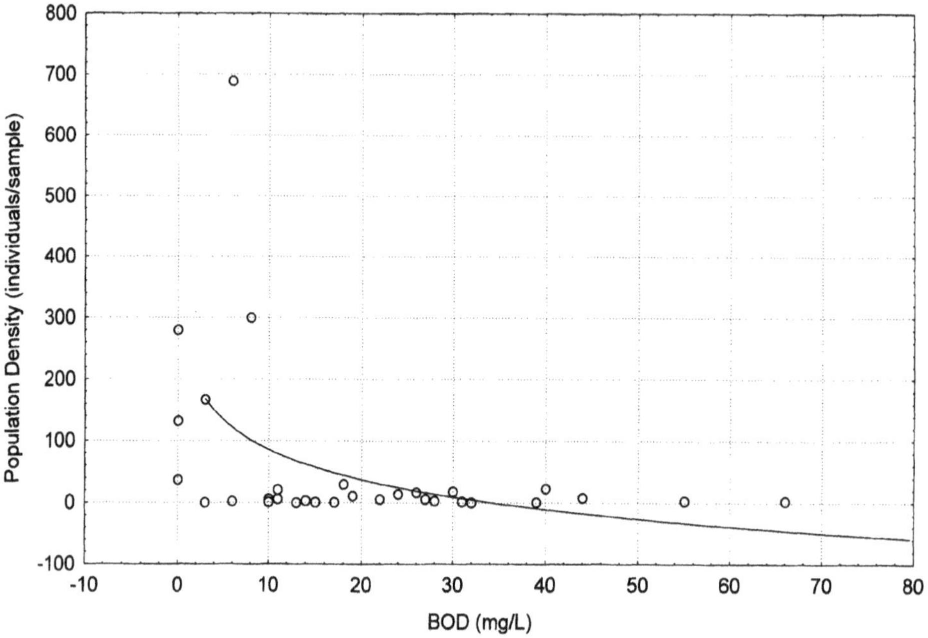

FIGURE 5. Curvilinear regression of pollution parameter (BOD) vs. ostracode population density (PD) in Nahal Harod.

The ostracode population density and the BOD values are correlated in a curvilinear regression graph (Fig. 5). For the data analysis (Software: "Statistica," StatSoft, version 5.0, 1995), a linear regression with log transformation of the independent variable BOD was used, omitting missing values and data from station 6 owing to sampling problems; see above). The log transformation was chosen in favor of a simple linear regression after a curved-fitting analysis. The logarithmic response was most likely due to the fact that freshwater ostracodes are found to accumulate in large numbers under favorable environmental conditions. The best fit was found with log base 10 (Statistica—nonlinear estimation module). The intercept equation (Y) value is $Y = 468.898 + 305.676 \times \log\ 10(X) + \text{eps}$. The regression parameter (r) is $r = -0.5583$ and $r^2 = 0.3117$ ($P < 0.013$). This analysis enables us to use the ostracode population density data to describe organic pollution levels in streams of 0–40 mg/liter BOD.

The ability of nonmarine ostracodes to respond quickly to environmental perturbations is attributed to their sensitivity to environmental changes and to their relatively short life cycle (a few generations per year) in comparison with many other taxa (Rosenfeld, 1979).

Juveniles as well as adults of *H. incongruens* and *H. salina* were found together at stations 3, 4, and 6, whereas no adults were found at station 5. *H. incongruens* is the dominant species in heavily polluted water (Stations 4 and 5), whereas in moderately polluted water (Station 3), *H. salina* is dominant. *H.*

incongruens was also found in small numbers at station 6. It is possible that the relatively smaller number of *H. incongruens* compared to *H. salina* (1:5 ratio) at station 3 reflects biotic competition.

Preliminary examinations of the ostracode populations from other polluted freshwater rivers in the central coastal plain of Israel show that the dominant ostracode species were also *H. incongruens* and *H. salina*. This suggests that both may be used as indicator species of organic pollution.

4. Conclusions

Ostracodes are found to be sensitive to different levels of pollution. *Heterocypris incongruens* and *H. salina* may be used as indicators for determining slight to moderate pollution levels. These ostracodes are scarce or absent in heavily polluted zones, where conditions appear to be lethal. An inverse relationship was found between both population density and diversity of ostracodes and pollution levels. Peak BOD values correspond to low population densities, whereas low BOD values correspond to high population densities. The significant probability value in a curvilinear regression analysis reinforces this observation. The ostracode *H. incongruens* appears to have a broad tolerance to organic pollution, ranging from oligosaprobic to polysaprobic conditions. *H. salina* has a smaller tolerance range, from oligosaprobic to mesosaprobic, whereas the low pollution tolerance of *Candona neglecta, Cypridopsis aculeata, Ilyocypris bradyi* and *Cyprideis torosa* restricts them to clean or oligosaprobic waters. In a stream where effluent is irregularly discharged, momentary measured physicochemical factors can reflect a false picture of water quality, whereas the ostracode fauna clearly shows the real state of the river.

Further research with a higher sampling rate throughout the years is needed to determine definitively the influence of various types of pollution on ostracode abundance and distribution.

ACKNOWLEDGMENTS: We thank the Water Commission Authority (Ministry of Agriculture) for financing water analyses. Thanks are also due to the Nature Reserves Authority (Israel) for allowing us to use the biological samples and the physicochemical data. The authors are grateful to the reviewers of the manuscript for their constructive remarks and comments. Thanks are due to Ms. S. Greenberg (Nature Reserves Authority) and Dr. Z. Minster (Geological Survey of Israel, Jerusalem) for computerizing the data, and to Dr. L. Whittaker (Nature and Natural Parks Protection Authority, Jerusalem) for help in the statistical analysis. Dr. H. J. Bromley and Dr. G. N. Herbst (Zoology Department The Hebrew University, Jerusalem) and Mr. H. Glassman (Water Pollution Control Unit, Nature and Natural Parks Protection Authority, Northern District, Israel) kindly assisted in the sampling.

References

A.P.A.H., 1971, American Public Health Association: Standard Methods (13th ed.), Washington, D.C., 879 pp.

Bodergat, A. M., 1978, Un marquer instane de la pollution en cérium dans la zone de Cortiou (Rade sud de Marseille, Bouches-Rhône) *Aurila speyeri* Brady 1868 (Ostracoda, Podocopida), *Doc. Lab. Géol. Fac. Sci. Lyon, H. S.* **4**:71–85.

Bromley, H.J., and Por, F.D., 1975, The metazoan fauna of a sewage-carrying wadi, Nahal Soreq (Judean Hills, Israel), *Freshwater Biol.* **5**:121–133.

Bronstein, Z. S., 1947, Faune de l'URSS—*Ostracodes des eaux douces*, Edition de l'Academie des Sciences de l'URSS, Moscow, Leningrad, 339 pp [in Russian].

Chandler, J. R., 1970, A biological approach to water quality management, *Water Pollut. Cont.* **69**:415–421.

Ghetti, P. F., 1980, Biological indicators of the quality of running water, *Bull. Zool.* **47**:381–390.

Hawkes, H. A., 1962, Biological aspects of river pollution, in *River Pollution: 2. Causes and Effects* (R. Klein, ed.) Butterworths, London, pp. 311–437.

Hynes, H. B. N., 1960, *The Biology of Polluted Waters*, Liverpool University Press, Liverpool.

Kaesler, R. L., Smith, S., and Whatley, R. C., 1979, Ostracoda and petroleum pollution in the Strait of Magellan, in *Taxonomy, Biostratigraphy and Distribution of Ostracodes* (Krstic, N., ed.), *Proc. 7th Int. Symp. Ostracoda*, Serbian Geological Society, Belgrade, pp. 237–242.

Klie, W., 1938. Ostracoda, in: *Die Tierwelt Deutschlands und der angrenzenden Meeresteile, Vol. 34: Krebstiere Order Crustacea*, Verlag Gustav Fischer, Jena, 230 pp.

Kolkwitz, R., 1950, *Oekologie der Saprobien. Ueber die Beziehungen der Wasserorganismen zur Umwelt: Schriftenreihe Ver. Wasser- Boden- und Lufthygiene*, Vol. 4, pp. 1–64.

Lange-Bertalot, H., 1979, Pollution tolerance of diatoms as criterion for water quality estimation, *Nova Hedweg.* **64**:285–304.

Liebmann, H., 1962, *Handbuch der Frischwasser- und Abwasserbiology*, R. Oldenbourg, Munich, 588 pp.

Margalef, R., 1953, Los crustaceos de las aguas continentales ibericas, Inst. Forestal Investigaciones y Experiencias, Madrid.

Margalef, R., 1965, Los organismos indicadores en la limnologia, Inst. Forestal Investigaciones y Experiencias, Madrid.

Mezquita, F., Sanz-Brau, A., and Miracle, M.R., 1996, New data on freshwater ostracod assemblages (Crustacea, Ostracoda) from Cuenca (central Spain), *Bull. Soc. Nat. Luxembourg* **97**:239–247.

Mezquita, F., Hernandez, R., and Rueda, J., 1999, Ecology and distribution of ostracods in polluted Mediteranean river: *Palaeogeog. Palaeoclim. Palaeoecol.* **148**:87–103.

Milhau, B., Dekens, N., and Wouters, K., 1997, Evaluation de l'utilisation des Ostracodes comme bio-indicateurs potentiels de pollution. Application aux eaux de la Slack (Boulonnais, France), *Ecologie* **28**:3–12.

Ortal, R., 1977, Biological key for pollution mapping of the streams of Israel: unpubl. report data, Nature Reserves Authority, Israel, Jerusalem, 3 pp.

Purper, I., and Wuerdig-Maciel, N. L., 1974. Occurrence of *Heterocypris incongruens* Ramdohr, 1808—Ostracoda—in Rio Grande do Sul, Brazil: Discussion on the allied genera, *Cyprinotus, Hemicypris, Homocypris and Eucypris*, Vol. 3, Pesquisas, Porto Alegre, pp. 69–91.

Rama Rao, S. V., Singh, V. P., and Mall, L. P., 1978, Pollution studies of the River Khan (Indore), India. 1: Biological assessment of pollution, *Water Res.* **12**:555–559.

Ruiz Munoz, M.,Gonzales-Regalado, L., and Borrego, J., 1994, Ostracodos y contaminacion: el estuario el rio Odiel (Huelva, SO Espania), *Col. Paleontol.* **46**:175–189.

Rosenfeld, A., 1979, Seasonal distribution of recent ostracodes from Kiel Bay, Western Baltic Sea, *Meyniana* **31**:59–82.

Rosenfeld, A. and Ortal, R., 1983, Ostracodes as indicators of water pollution in Nahal Harod, northern Israel, in: *Applications of Ostracoda* (R. F. Maddocks, ed.), Department of Geosciences, University of Houston, pp. 229–237.

Wagner, C. W., 1957, *Sur les Ostracodes du Quaternaire Récent des Bays-Bas et leur utilisation dans l'étude géologique des dépôts Holocènes*, Mouton & Co., Paris, 259 pp.

Chapter 8

Ostracoda as Indicators of Conditions and Dynamics of Water Ecosystems

EUGENIJ I. SCHORNIKOV

1. Introduction

Ostracoda are a diverse (>40,000 species) class of crustaceans. They are known from the Cambrian, and recent ostracodes dwell in all possible aquatic biotopes, from oceanic hadal to humid land biotopes and subterranean waters, in which they form specific complexes of species.

Ostracods have been widely used in petroleum exploration because they are excellent indicators in bio- and ecostratigraphy, paleoecology, paleobathymetry, paleoceanography, paleogeography, and paleoclimatology (De Deckker *et al.*, 1988; Whatley and Maybury, 1990). Freshwater and oligohaline ostracodes have been used less frequently as indicators of the quality of fresh (Dubowsky, 1927; Makrushin, 1974; Sladeček, 1978; Ghetti, 1980; Rosenfeld and Ortal, 1983) and estuarine waters (Pascual, 1991; Ruiz Munos *et al.*, 1994) and environmental disturbance (Danielopol *et al.*, 1990; Greiger, 1993).

Despite their lack of use in environmental studies, ostracodes are excellent bioindicators because they react to relatively small-scale environmental perturbances that might be survived by other taxa (see in this volume Eager, Ch. 6; Rosenfeld *et al.*, Ch. 7). As pollution increases, eurytopic ostracode

EUGENIJ I. SCHORNIKOV • Institute of Marine Biology, Far Eastern Branch, Russian Academy of Sciences, Vladivostok 690041, Russia.

Environmental Micropaleontology, Volume 15 of *Topics in Geobiology*, edited by Ronald E. Martin. Kluwer Academic/Plenum Publishers, New York, 2000.

species tend to become dominant. Finally, the opportunistic species also die out. Ostracodes do not withstand high pollution and do not live in polysaprobic and hypersaprobic zones, but such conditions are obvious without the aid of meiofauna. Ostracodes can also be used to compare the results of anthropogenic stress with preanthropogenic (historical) conditions. Results of such studies can then be used to make recommendations for environmental restoration. In the last several years, we have undertaken investigations of the use of the recent marine coastal, freshwater, and subterranean ostracodes as indicators of environmental conditions (Schornikov, 1990, 1991, 1995, 1996). The data discussed herein are mainly concerned with studies of marine coastal and freshwater ostracodes.

2. Notes on Methodology

In collecting ostracodes, we use a small (36-cm-width) dredge of personal construction with a nylon sack (meshes of 0.15 × 0.15 mm) into which we insert a sieve (meshes of 1.5 × 1.5 mm). This simple apparatus allows us to obtain extremely rich surface samples in different biotopes, samples that are much richer than bottom grabs.

It is best to fix samples with 75% alcohol containing Rose Bengal dye. Ostracodes that were alive during collection are dyed pink and are easily distinguished from shells containing dead chitinous remnants of the soft body, which may be preserved for a very long time after death.

In heavily polluted regions characterized by intense sediment deposition, such as marine ports, it is very useful to obtain sediment cores at the same time that the surface samples are collected. Cores may be used to distinguish the dynamics of transformation of ostracode complexes. In quantitative analyses of heterogeneous substrates such as sediment, rock, and algae, we have used the living percentage of individuals and shells of each species. This method allows adequate comparison of faunas independent of sample volume and collection method. Samples collected usually contain several tens of thousands of ostracodes represented by separate valves and whole shells. Samples are therefore split into manageable sizes (Galtsova, 1971).

3. Distribution of Ostracoda in Marine Ports

Investigations of the marine fauna of the port of Vladivostok were carried out in April–May 1992. We studied 46 samples collected by small dredge at the stations shown in Fig. 1. Also, cores 8 cm in diameter were taken to 40 and 60 cm subbottom depth from stations 6 and 8 and to 16 m subbottom depth from the Golden Horn Bay.

The remnants of 70 ostracode species were found, mostly as valves and empty shells. Only three species, *Euphilomedes nipponica* Hiruta 1976,

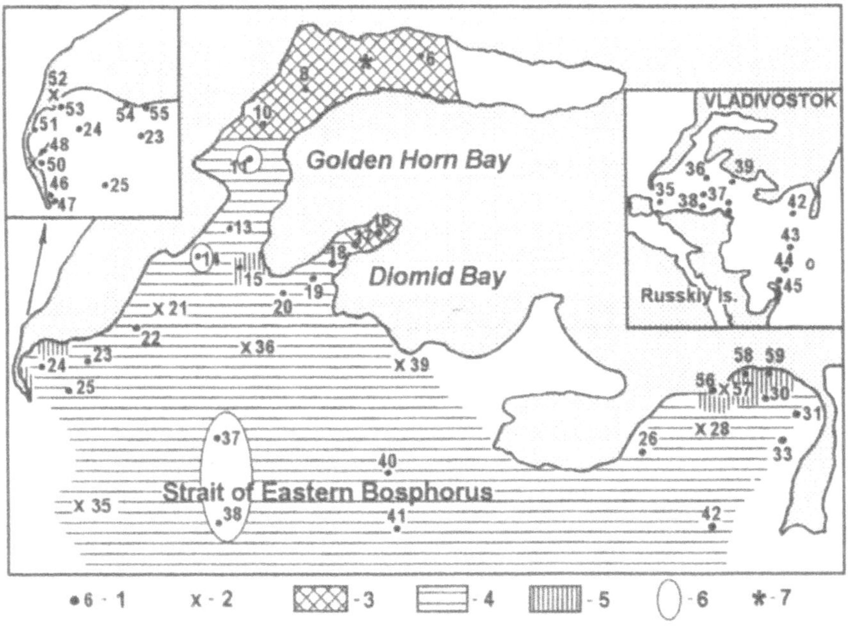

FIGURE 1. Distribution of ostracode complexes in the Vladivostok port area: (1) station numbers; (2) stations without ostracodes; (3) zone of heaviest pollution-I; (4) zone of heavy pollution-II; (5) zone where living ostracodes were found-III; (6) zones where the remnants of ostracode cold-water complex were found; (7) position of well. See text for further discussion.

Sclerochilus verecundus Schornikov 1981, and *Loxocauda* sp., were found alive. *E. nipponica* was found in the surf at station 55 (1 m depth, 33 specimens) and 15 (20 m depth, 1 specimen), and also in the bay at stations 56–58 and 30 (1–10 m, 3–59 specimens). *E. nipponica* lives above bottom during most of its life, but rises to the sea surface during spawning. Possibly because of its high dispersal ability, it is the most ubiquitous species in the region studied. As this species is not closely connected with sediments, poisons in the sediment do not play a decisive role in its life cycle.

All other species found belonged to the order Podocopida, which does not possess swimming organs and crawls slowly on or in the substratum. As the Podocopida are closely connected with the substratum, they are more sensitive to anthropogenic stress. On the other hand, *Sclerochilus verecundus* and *Loxocauda* sp. were found alive at Stations 46 (9 and 1 specimens, respectively) and 50 (7 specimens) at depths up to 1 m. The first species dwells in fouling assemblages of rocks and other hard substrata, and the second one inhabits vegetation, mainly *Zostera*.

Ostracodes were not found at some stations (Fig. 1), a phenomenon that could be the result of a number of different reasons. For example, in unpolluted areas algae-dwelling ostracodes are numerous on algae, but in heavily polluted areas, they avoid algae covered by oil film. At station 39, however, near the shore of a cape, where the traces of large-scale construction are apparent, the bottom is composed of solid clay and is barren of both meio- and macrofauna. Evidently, this is anthropogenic sediment eroded from construction sites. At station 52, where everyday rubbish accumulates and dead *Zostera* is evident, ostracodes are absent because of the abundance of putrid organics, although their absence may also be the result of organic matter decay and dissolution.

According to the complexes of Ostracoda and their remnants, we have divided the region of the Vladivostok port into three zones (Fig. 1; Table 1):

Zone I. The zone of heaviest pollution, which is situated in the inner parts of Golden Horn and Diomid Bays. Here, where silts are impregnated by oil products and saturated by coal particles and everyday rubbish, only single ostracode valves or shells are found. In the Golden Horn Bay anthropogenic deposits make up 1.5–2 m. No ostracodes are found in cores up to 60 cm deep, but they appear in the interval of 0.84–0.87 m (Table 1). It is likely that many lived here after the founding of Vladivostok in 1863, and possibly up to the 1930s, but that they disappeared because of oil pollution in connection with the transition oil-burning vessels in the mid-20th century.

Zone II. The zone of heavy pollution continues into Zone II. Here, silts are polluted primarily by coal particles, and there are no living ostracodes. However, sediments also contain the shells of relatively eurytropic warm-water ostracods, which inhabit nearby shallow Ussuriisky and Amursky Bays: e.g., *Cytheromorpha acupunctata* (Brady 1880), *Spinileberis quadriaculeata* (Brady 1880), *Pontocythere subjaponica* (Hanai 1959), *Bicornucythere bisanensis* (Okubo 1975), *Aurila disparata* Okubo 1980, and *Robustaurila ishizakii* (Okubo 1980), and *Xestoleberis hanaii* Ishizaki 1968.

Zone III. Zone III, in which ostracodes are normally found alive. In spite of heavy oil pollution (observed visually) ostracodes survive here, evidently owing to strong wave mixing and water aeration.

At three sites, Stations 11 (11 m depth), 14 (11 m), and 37–38 (23 m), the valves and shells of another ostracode complex were found: *Palmenella* cf.

TABLE 1. Ratios of Ostracode Fauna[a]

Species	Well core	Station	Quantity
Howeina camptocytheroidea Hanai 1957	8s 7v = 23v = 45%		
Bicornucythere bisanensis (Okubo 1975)	3s 5v = 11v = 21%	6	1s
Hemicythere? emarginata (Sars 1866)	5v = 10%		
Hemicythere orientalis Schornikov 1973	2s = 4v = 8%	8	2s
Yezocythere hayashii Hanai et Ikeya 1991	4v = 8%	16	1s
Hemicythere gorokuensis Ishizaki 1966	1s = 2v = 4%		
Howeina sp.	1s = 2v = 4%		
Semicytherura sp.	0	10	1v

[a]Quantity: s, shells; v, valves and % from well core (interval of 0.84–0.87 m) and surface samples (stations) in Zone I.

limicola (Norman 1865), *Heterocytherideis* cf. *sorbiana* (Jones 1857), *Sarsicytheridea* cf. *bradii* (Norman 1865), *Acantocythereis mutsuensis* Ishizaki 1971, *Hemicythere? emarginata* (Sars 1866), *Baffinicythere* cf. *ishizakii* Irizuki 1996, *Robertsonites* sp. 2 Tabuki 1986, *Elofsonia* cf. *concinna* (Jones 1856), *Hemicytherura* aff. *clatrata* (Sars 1866), and *Cytheropteron* cf. *arcuatum* (Brady, Crosskey, and Robertson 1874), sensu Cronin and Ikeya, 1987. However, in the surface deposits of Amursky Bay, we did not find a single valve of these ostracodes. Here, cold-water species are found that are widely distributed to the north; in cold waters many of them may penetrate into polluted areas. But in the Peter the Great Bay they are extremely sensitive to the oxygen concentration and water clarity, and spread now only in the open part of the bay at a depth more than 15 m (usually more than 40 m). Most probably, cold-water ostracodes penetrated into the strait of Eastern Bosphorus during the last cool episode in South Primorie ct 2200 BP years. With the subsequent temperature rise, they did not die out, and continued to live owing to water clarity and favorable oxygen regime until anthropogenic pollution appeared. In a core these ostracodes were found in the interval of 4.70–0.84 m; therefore, they appear to have lived here until very recently.

Similar patterns of ostracode distribution have been observed in Sevastopol Bay and Krabovaya Inlet (Shikotan Island, South Kuril Islands), where the port of Krabozavosk is located. Also, Faradzov (1966) notes that in the Caspian Sea, near the oil trades (from Baku Bay up to Alyaty Cape), ostracodes were completely absent during autumn 1964.

4. Lacustrine and Subterranean Ostracodes

In 1989, the Khanka Lake basin situated in the South Primorie on the boundary between China and The Soviet Union were extremely polluted by fertilizers, herbicides, and other poisons used in rice cultivation. In 1990, we studied the ostracode fauna of the Russian part of Khanka Lake and adjacent water bodies to monitor the anthropogenic impact on aquatic ecosystems. The ostracode fauna in the open part of the Khanka Lake was shown to be in a critical condition. Only 5 specifically lacustrine ostracode species were found here. In total, 73 ostracode species were found in the region studied, and 53 of them are new to science. The ostracode complexes, which are characteristic for different biotopes, were noted, and the tendencies of their destruction were revealed according to the anthropogenic impact. According to these data the scale of their indicator abilities was elaborated.

In 1994–1995 we studied the ostracode fauna of the water bodies of Ussuri Nature Reserve in connection with plans for coal strip-mining and the prospect of related heavy pollution. Among 26 ostracode species, 6 representatives of the genera *Vestalenula, Cryptocandona, Mixtocandona, Nannocandona,* and *Cavernocypris* lived in the streams under the sediment layer in subterranean waters. These genera react to minor acid rain, dust, and smoke atmospheric events, refuse dumps situated in the upper parts of the stream basins, etc. These particular genera also indicate suitable habitat for the breeding of salmon.

Similar studies were conducted near the salmon hatchery off the Ryazanovka River (flowing into Peter the Great Bay, the Sea of Japan) and the Azabache Lake on the Kamchatka Pennisula. In the first case the sprawning grounds of *Oncorhynchus masou* (Brevoort) and in the second case those of *Oncorhynchus nerka* (Wallbaum) were studied. The subterranean ostracode species clearly mark the bottom parts of water bodies, that are suitable for the breeding of salmon.

ACKNOWLEDGMENTS: This study was partially supported by Korea Science and Engineering Foundation Grant 975-0400-002-2. I thank D. A. Sokolenko for assistance during the execution of this work, and Dr. A. V. Moshchenko for critical review of versions of the manuscript.

References

Danielopol, D. L., Carbonel, P., and Colin, J. P. (eds.), 1990, *Cytherissa* (Ostracoda)—the *Drosophila* of paleolimnology, *Bull. Inst. Géol. Bassin d'Aquitaine* **47/48**:1–310.

De Deckker, P., Colin, J. P., and Peypouquet, J. P. (eds.), 1988, *Ostracoda in the Earth Sciences*, Elsevier, Amsterdam.

Dubowsky, N. W., 1927, Materials for cognition of fauna and biology of Ostracoda of the Sewerny Donetz River basin, *Trud. Khark. Tovar. Doslid. Pryr.* **5**(2):107–120, Kharkov [in Russian].

Faradzov, G. R., 1966, Seasonal variations of ostracod fauna of the western coast of Middle Caspian Sea, *Izv. Akad. Nauk Azerbaydz. SSR, Ser. Biol. Nauk* **3**:62–68, Baku [in Russian].

Galtsova, V. V., 1971, Quantitative analysis of meiobenthos, *Hydrobiol. J.* **7**(2):132–136, Kiev [in Russian].

Geiger, W., 1993, *Cytherissa lacustris* (Ostracoda, Crustacea): Its use in detecting and reconstructing environmental changes at the sediment-water interface, *Verh. Int. Verein. Limnol.* **25**:1102–1107.

Ghetti, P. F., 1980, Biological indicators of the quality of running water, *Bool. Zool.* **47**:381–390.

Makrushin, A. V., 1974, *Biological Analysis of the Quality of Water*, Zoologicheskiy Institut Russian Akademii Nauk-Vsesoyuzznoe Gidrobiologicheskoe Obschestvo, Lenningrad [in Russian].

Pascual, A., 1991, Ostracodes in Biscayan estuaries (Spain): their relationship with the environment, *Cah. Micropaleontol. N.S.* **6**(2):5–21.

Rosenfeld, A., and Ortal, R., 1983, Ostracodes as indicators of water pollution in Nahal Harod, northern Israel, in: *Applications of Ostracoda, Proc. 8th Int. Ostracode Symp. Houston* (R. F. Maddocks, ed.), Department of Geosciences, University of Houston, pp. 229–237.

Ruiz Munos, F., Gonzalez-Regalado, M. L., and Borrego, J., 1994, Ostracodos y contaminacion: el estuario del rio Odiel (Huelva, SO Espana), *Col. Paleontol., Madrid.* **46**:175–189.

Schornikov, E. I., 1990, Ostracods-bioindicators of condition of water ecosystems, in: *Ecologicheskie problemy ohrany jivoi prirody: Tezysy Dokladov Vsesoyuzoi Conferencii* (V. A. Krassilov, ed.), Part 3, Moscow, pp. 235–236 [in Russian].

Schornikov, E. I., 1991, Ostracods of the Khanka Lake Basin, *Prog. Abstr. 11th Int. Symp. on Ostracoda*, Warrnambool, Victoria, Australia, p. 78.

Schornikov, E. I., 1995, Subclass Ostracoda, in: Vshivkova, T. S., *Hydrobiological Investigations in Ussuri Nature Reserve.* Part I. Freshwater Fauna (Biodiversity and Biogeographical Analysis) Dal'nauka, Vladivostok [in Russian], pp. 18–19.

Schornikov, E. I., 1996, Ostracodes as Indicators of Water Ecosystems Dynamics, *Bull. Far East. Branch Rus. Acad. Sci.* **5**:36–42, Vladivostok [in Russian].

Sladeček, V., 1978, Indicator value of freshwater Ostracoda, *Acta hydrohim. hydrobiol.* **6**(6):561–565.

Whatley, R. C., and Maybury, C. (eds.), 1990, *Ostracoda and Global Events*, Chapman and Hall, London.

III

Physiological Responses of Foraminifera to Pollution

Chapter 9

Environmental Variation and Foraminiferal Test Abnormalities

EMMANUELLE GESLIN, VERONIQUE STOUFF, JEAN-PIERRE DEBENAY, and MAURICE LESOURD

EMMANUELLE GESLIN, VERONIQUE STOUFF, and JEAN-PIERRE DEBENAY • Laboratoire de Géologie, Equipe EA 2644, Faculté des Sciences, Université d'Angers, 49045 Angers Cedex, and Laboratoire d'Etudes des Bio-Indicateurs Marins (LÉBIM), Ker Châlon, 85350 Ile d'Yeu, France. MAURICE LESOURD • CNRS and SCME, Faculté de Médecine, Rue de Haute Reculée, 49 045 Angers Cedex, France.

Environmental Micropaleontology, Volume 15 of *Topics in Geobiology*, edited by Ronald E. Martin. Kluwer Academic/Plenum Publishers, New York, 2000.

1. Introduction

Foraminifers are increasingly used as bioindicators of environments. Their community structure provides information on the general characteristics of the environment, especially in highly changing paralic environments (e.g., Hayward and Hollis, 1994), and some species are sensitive to specific environmental parameters. Test morphology may also be related to environmental characteristics and is sometimes used as a bioindicator. The size and the density of pores, e.g., have been considered as indicators of dissolved oxygen concentration (Sen Gupta and Machain-Castillo, 1993).

Abnormal tests have been reported from both fossil (e.g., Amicis, 1895; Cann and De Deckker, 1981; Caralp, 1989) and on recent specimens (e.g., Arnal, 1955; Vénec-Peyré, 1981; Alve, 1991; Yanko *et al.*, 1994, 1998). They are, however, often related to environmental stress which may be due, for instance, to hyposalinity or hypersalinity. (Table 1 and Fig. 1). Deformations may also develop in areas where strong hydrodynamics may cause damage to the tests (Table 1 and Fig. 1). In the recent literature, morphological abnormalities are considered to result mainly from various anthropogenic activities (Table 1 and Fig. 1) (see the review in Yanko *et al.*, 1999). Interestingly, most of the papers relating abnormalities to pollution have been published only recently (since 1980), whereas most of the papers published before 1980 related abnormalities to natural environmental stress (Fig. 2). Nevertheless, even some of the earlier authors as well as more recent papers have suggested that they could be used as bioindicators of pollution (Seiglie, 1975; Vénec-Peyré, 1981; Alve, 1991; Sharifi *et al.*, 1991; Yanko *et al.*, 1994; Yanko, 1997; see also in this volume Hallock, Ch. 5; and Bresler and Yanko-Hombach, Ch. 10).

The aim of this chapter is to discuss the possible causes and origins of abnormal test formation and the possible use of foraminiferal test abnormalities as bioindicators of pollution. It is based on qualitative and quantitative studies of abnormal tests in various natural environments (nonpolluted and polluted), in culture (normal conditions and hypersaline conditions), and on observations of the wall texture of abnormal tests.

2. What Should Be Considered as "Abnormal" in Environmental Studies?

It is sometimes complicated to determine exactly what constitutes an abnormality because of the subjective determination of the limit between normal and abnormal forms (e.g., Fig. 3: *12* to *15*). Attempts have been made to develop an objective mathematical basis to test abnormalities using image analysis such as Fourier analysis (Bonetti *et al.*, 1996). However, the complexity of the three-dimensional malformations makes the utilization of this technique hazardous.

TABLE 1. Survey of Published Papers Reporting Abnormal Tests of Recent Foraminifers[a]

Stress origin	Reported causes of abnormality	References
Anthropogenic origin	Pollution of heavy metals	Banerji, 1989, 1992; Alve, 1991; Sharifi *et al.*, 1991; Yanko *et al.*, 1994, 1998; Coccioni *et al.*, 1997; Kravchuk *et al.*, 1997
	Chemical pollutants	Seiglie, 1971; Setty and Nigam, 1984; Jorissen, 1988; Jayaraju and Reddi, 1996
	Pollution by hydrocarbons	Whitcomb, 1978; Vénec-Peyré, 1981
	Pollution by supply of organic matter such as domestic sewage	Watkins, 1961; Setty and Nigam, 1983 Yanko *et al.*, 1994, 1998
	Thermal pollution	Seiglie, 1975
	General pollution or unspecified	Lidz, 1965; Seiglie, 1968; Boltovskoy and Boltovskoy, 1968; Stubbles, 1996; Bonetti *et al.*, 1997
Natural ecologic origin	Hypersalinity	Seiglie, 1964; Sellier de Civrieux, 1968; Ayala-Castañares and Segura, 1968; Resig, 1974; Brasier, 1975; Cann and De Deckker, 1981; Zaninetti, 1982, 1984; Almogi-Labin *et al.*, 1992
	Hyposalinity	Hofker, 1971
	Variation of salinity	Arnal, 1955; Sellier de Civrieux, 1968; Tufescu, 1968; Closs and Madeira, 1968; Brasier, 1975
	Input of freshwater with changes in °C, S‰, pH, nutrients...	Arnal, 1955; Eichler *et al.*, 1996; Sousa *et al.*, 1997
	Unfavorable trophic conditions	Heron-Allen and Earland, 1910; Boltvoskoy, 1954, 1956
	Low level of oxygen	Lutze, 1964; Pujos-Lamy, 1973
	Input of natural trace elements	Boltovskoy, 1956
	General[b]	Seiglie, 1964; Lidz, 1965; Seiglie, 1968; Le Campion, 1968; Madeira-Falcetta, 1974; Pujos, 1976; Scott and Medioli 1980*a,b*; Hallock *et al.*, 1995
Mechanical origin	Hydrodynamics and regeneration phenomena	Boltovskoy, 1957; Vilela and Koutsoukos, 1992; Vilela, 1994; Geslin *et al.*, 1998*b*

[a]Reports have been grouped by cause. Experimental studies are not included. Studies with particular approaches such as cytological study (e.g., Yanko, 1995) or crystallographic study (Geslin *et al.*, 1998*a*) have also been omitted.

[b]In these cases, authors did not state ecological parameters which may induce abnormalities. Causes are reported such as "estuarine or brackish water environments," "changing of water conditions," "abnormal ecological conditions."

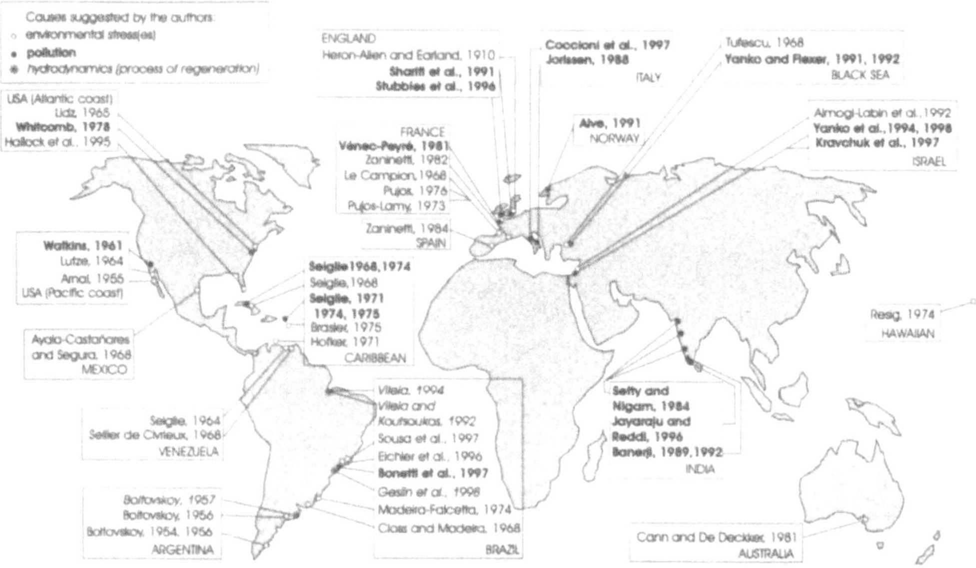

FIGURE 1. Location of studies which have reported abnormal tests in natural environments.

Numerous classifications of morphological test abnormalities have also been proposed (Zaninetti, 1982; Caralp, 1989; Alve, 1991; Almogi-Labin *et al.*, 1992; Yanko *et al.*, 1994, 1998; Geslin *et al.*, 1998*a*). However, these classifications based on morphological descriptions (e.g., Fig. 4) regardless of the origin and cause of abnormalities have led to a multitude of abnormality categories. Moreover, Geslin *et al.* (1998*a*), who classified abnormal tests of *Ammonia* into nine types, pointed out that several types may occur on the same test and that some abnormalities are so severe that it is impossible to determine which part of the test is affected. Sometimes it is even impossible to determine the foraminiferal genus.

A strong morphological variability (ecophenotypes) is also known in several species of *Ammonia* (e.g., Schnitker, 1974; Jorissen, 1988) and of *Elphidium* (e.g., Feyling-Hanssen, 1972; Goubert, 1997), both of which frequently occur in coastal polluted environments. This variability generally corresponds to the presence or absence of a morphological character (e.g., umbilical plug, pustulae) (Fig. 3: *3* and *4*) but can also affect the whole test, which may be small, poorly ornamented, or thin-walled (see reviews in Boltovskoy and Wright, 1976, and in Boltovskoy *et al.*, 1991). Such characteristics affecting a large part of the population and resulting from permanent environmental conditions may give information on the environment but will not be considered as morphological abnormalities.

Epiphytic species such as *Rosalina globularis*, *Lobatula lobatula* (=*Cibicides lobatulus*), and *Planorbulina mediterranensis* (Fig. 3: *1* and *2*) show a high variability of test morphology resulting from the irregularity of the substrate

(Boltovskoy and Wright, 1976). As a result, abnormal specimens of these species cannot be used to indicate environmental stress.

Very small deformation may be observed in tests where a chamber may be reduced or overdeveloped. As the construction of a chamber generally needs only about 1 hr, a reduced or overdeveloped chamber may correspond to momentary changes in very local environmental conditions. Thus, these small abnormalities may be used to detect short-term environmental impacts but should not be considered as characterizing strong widespread environmental stress.

Even obvious abnormal shapes, sizes, or dispositions of one or several chambers are not always induced by environmental stress. For example, globular chambers may be formed by some planktonic (Bé, *et al.*, 1983) or benthic species (Hottinger *et al.*, 1993). Severe malformations may also follow the processes of reproduction, such as in *Ammonia tepida* in laboratory culture in Angers (France) (Stouff, 1998); specimens may survive after asexual reproduction (schizogony) and construct new chambers with abnormal sizes, shapes, and dispositions (Fig. 3: *5* to *7*). Thus, a good knowledge of the biology of the studied species is also necessary to detect abnormalities that have no direct relation to the environment. Severe abnormalities may also be brought about by mechanical trauma (Bé and Spero, 1981); (Fig. 3: *8* to *11*), but are generally easy to distinguish by scanning electronic microscope (SEM) investigation because they are often characterized by the presence of scars, irregular contours of crushed or repaired chambers, or by the construction of new

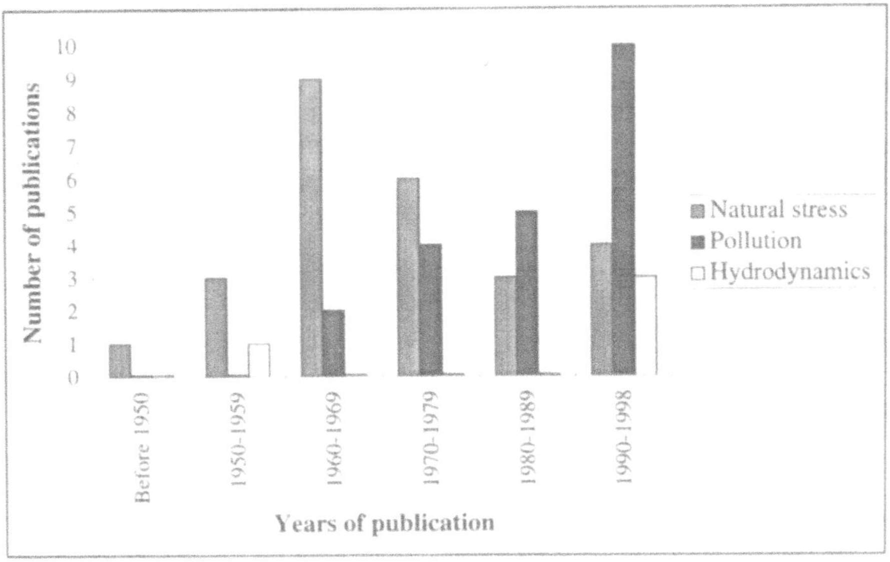

FIGURE 2. Number of publications mentioning abnormal tests in relation to natural stress, pollution or hydrodynamic factors.

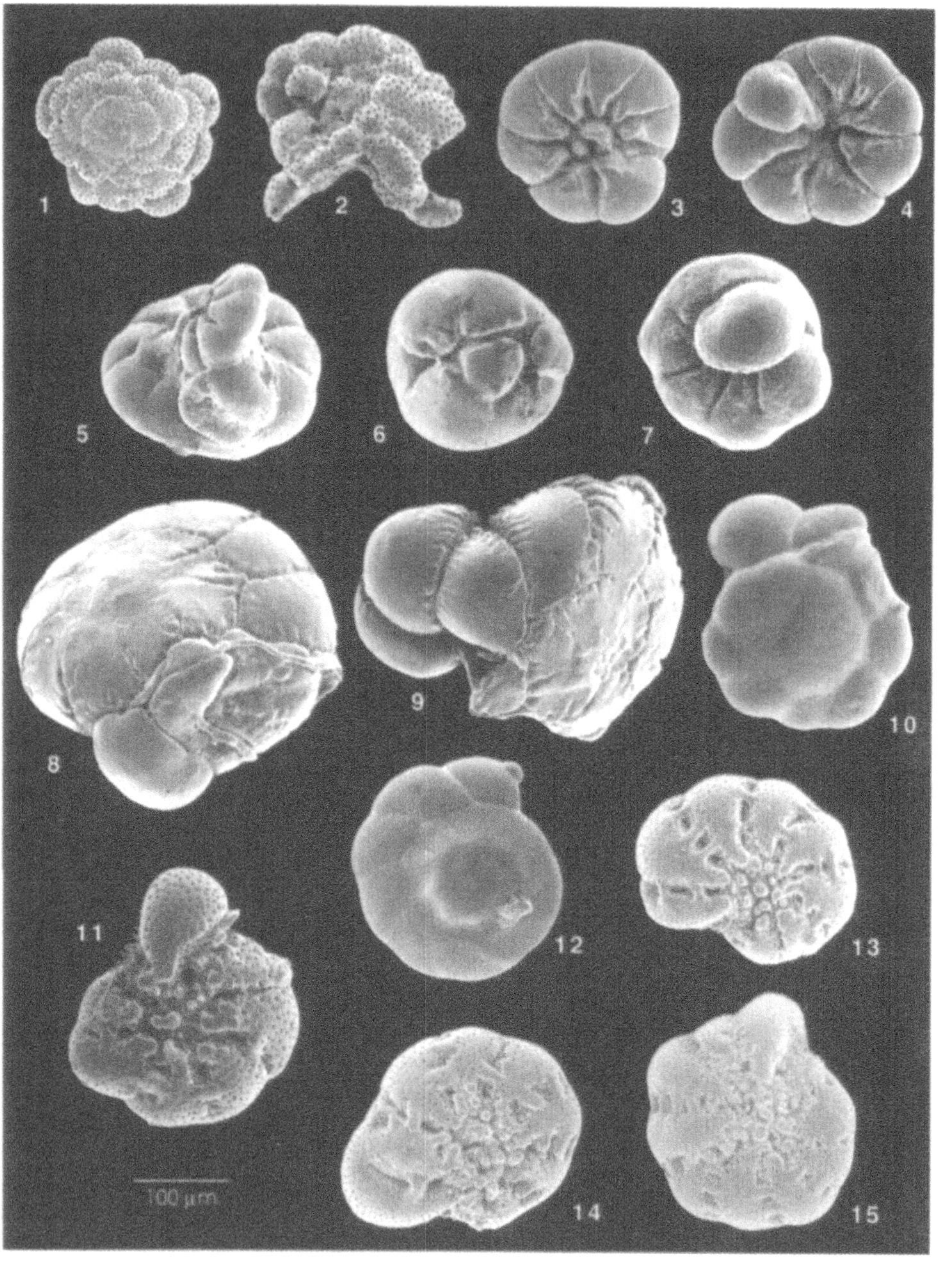

FIGURE 3. ***1,2*** *Planorbulina mediterranensis*, an epiphytic species which may exhibit a high variability in test morphology resulting from the irregularity of the substrate. ***3,4*** *Ammonia tepida*: (*3*) Specimen with an umbilical plug; (*4*) Specimen without umbilical plug. ***5–7*** Abnormal morphologies of *A. tepida* induced by asexual reproduction process (schizogony): (*5*) Schizont with juvenile fixed on the umbilical side of the parental test; (*6*) Schizont with abnormal shape of chambers; (*7*) Schizont with additional globular chambers on the umbilical side. ***8–11*** Abnormal tests resulting from physical injury showing scars of regeneration: (*8–9*) Abnormal

chambers in a coiling plane different from the original ones (see Boltovskoy, 1957, 1965; the review in Boltovskoy and Wright, 1976) These deformations result either from strong hydrodynamics or from the action of predators (Vilela and Koutsoukos, 1992; Geslin *et al.*, 1998*b*; Aktürk, 1976). Such tests also have to be systematically detected and ignored during environmental studies using test abnormalities as indicators of stress.

This review shows that test abnormalities may result from various causes and that a careful examination is necessary to distinguish abnormalities used for detecting environmental stress. SEM investigations and good illustrations are necessary in such studies.

3. Material and Methods

3.1. Sample Collection in Natural Environments

Samples of superficial sediment were collected in different paralic environments (Tables 2 and 3). Sampling was done by foot in intertidal zones or by using a surface grab from a boat in deeper areas. After collection, samples (50 cm^3 of sediment) were washed through a 50-μm sieve; dried (40°C), and floated in carbon tetrachloride (CCl_4) to separate foraminifers from detrital sediment. This procedure is easy, fast, and inexpensive.

Samples of superficial sediment containing living *Ammonia tepida* were collected in the Bay of Bourgneuf (west Atlantic coast of France). Specimens of *A. beccarii* and *Elphidium crispum* were collected regularly on algae in the rocky intertidal zone of the Ker Châlon beach located outside the laboratory of marine bioindicators (LEBIM) in Yeu Island (France).

3.2. Culture Procedures

Living *Ammonia tepida* were maintained in a petri dish with the original surface sediment. Half of the volume of water was renewed about once a month with microfiltered seawater.

To study the influence of salinity, adult schizonts of *A. tepida* (characterized by a large proloculus, the lobed shape of the last three chambers and a test size of 300 μm or more in diameter) were regularly selected from the petri

---

tests of *Ammonia beccarii*; (*10*) Abnormal test of *A. tepida*; (*11*) Abnormal test of *Elphidium* aff. *gunteri*. ***12–15*** Small abnormalities on individuals that are at the limit between normal and abnormal tests: (*12*) Specimen of *Ammonia tepida* with a small additional chamber; (***13–15***) Specimens of *E.* aff. *gunteri*; (*13*) The last chamber has a reduced size; (*14*) The last chamber is overdeveloped; (*15*) The penultimate chamber has a reduced size. (All photographs are on the same scale, scale bar = 100 μm.)

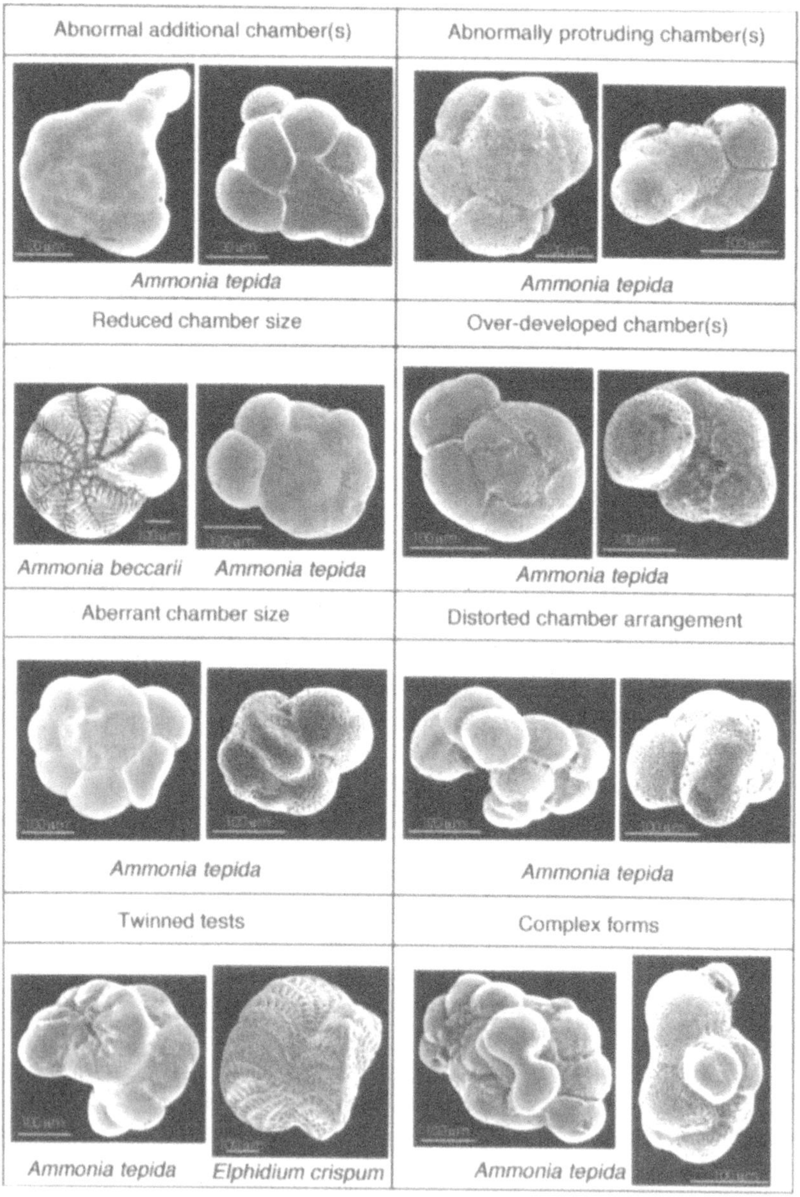

FIGURE 4. Different categories of morphological abnormalities described in the literature (e.g., Alve, 1991; Yanko *et al.*, 1994; Geslin *et al.*, 1998*a*).

TABLE 2. Main Topics of Study and Number of Abnormal Hyaline Tests Examined in Each Paralic Environment

Origin of samples	Main topic	Number of tests used for morphological study under stereomicroscope	Among these: Number of tests photographed using SEM	Among these: Number of crushed tests used for textural study using SEM
Rio Guaraú, Reserve of Juréia (São Paulo, Brazil)	Quantitative (%) + Qualitative (morphology)	350 (all abnormal hyaline tests from 6 stations)	350	0
Rio Una, Reserve of Juréia (São Paulo, Brazil)	Quantitative (%)	380 (all abnormal hyaline tests from 8 stations)	0	0
Itacolomi Lagoon, Reserve of Juréia (São Paulo, Brazil)	Quantitative (%)	65 (all abnormal hyaline tests from 10 stations)	0	0
Bertioga channel (São Paulo, Brazil)	Quantitative (%)	120 (all abnormal hyaline tests from 12 stations)	10	0
Araruama Lagoon (Rio de Janeiro, Brazil)	Qualitative (morphology under extreme saline conditions)	150	50	0
Golfe du Morbihan (French Atlantic coast)	Textural (crystallization)	51	28	18
Baie de Bourgneuf (French Atlantic coast)	Textural (crystallization)	4	0	4
La Barre de Monts (French Atlantic coast)	Textural (crystallization)	18	3	12
Lagune de Gourine (Mediterranean lagoon)	Textural (crystallization)	28	16	12
Éítang du Prévost (Mediterranean lagoon)	Textural (crystallization)	3	0	3

TABLE 3. Environmental Characteristics of the Brazilian Paralic Environments Where Abnormal Hyaline Tests Were Collected

Origin of samples	Environment	Salinity	Anthropogenic activities
Rio Guaraú Juréia-Itatins Ecological Reserve São Paulo, Brazil Duleba *et al.*, in press	Estuary in a tropical forest	30‰ near the mouth decreasing upward to 5‰	Only native inhabitants; very little anthropogenic activity
Rio Una Juréia-Itatins Ecological Reserve São Paulo, Brazil Duleba *et al.*, in press	Estuary in a tropical forest	16–30‰	Only native inhabitants; very little anthropogenic activity
Itacolomi Lagoon Juréia-Itatins Ecological Reserve São Paulo, Brazil Duleba *et al.*, in press	Small lagoon in a tropical forest isolated from the sea by a sand spit bordered by mangroves	Very low at the moment of sampling	Only native inhabitants; very little antropogenic activity
Bertioga Channel São Paulo, Brazil Eichler *et al.*, 1998	Channel (25 km) with two sea connections bordered with mangroves	24–38‰	Tourist activities; near polluted areas (Santos harbor, Cubatão industrial area)
Araruama Lagoon Rio de Janeiro, Brazil Unpublished data	Hypersaline lagoon connecting to sea by a straight channel	~70‰	Tourist and industrial activities

dish. They were transfered to small glass crystallizing jars containing water microfiltered through a membrane filter of 0.2 μm, where the reproduction processes were observed. Part of these cultures were maintained in normal saline seawater (37%), while the other part was maintained in hypersaline water (50%). The same experimental procedure was carried out with adult specimens of *Elphidium crispum* maintained in normal saline water (37%) and in hypersaline conditions (42%). As the salinity tolerance of this marine species was unknown, salinity was limited at 42%. Salinity was monitored regularly and corrected when necessary by adding distilled water. One or two drops of a diatom culture (*Amphiprora* sp.) was added every week. All cultures were maintained at 20°C in an incubator subjected to vertical lighting (2 × 20 W) on a 12-hr light/dark cycle.

In order to observe the hypersaline effect on adult foraminifers, ten normal *A. tepida* with more than 12 chambers from a normal salinity culture (37%) were transfered into a hypersaline culture (50%).

3.3. Quantitative Field Studies

In four Brazilian environments, all hyaline abnormal tests were counted to determine the percentage (relative abundance) of abnormal tests (Table 4). These percentages were calculated only for stations with more than 100 hyaline foraminifers in order to obtain statistically reliable counts.

TABLE 4. Percentages of Abnormal Hyaline Tests

Environment	Percentage of abnormal hyaline tests[a]
Rio Guaraú Ecological Reserve (Brazil) Geslin *et al.*, 1998*b*	2–10% (n = 367)
Rio Una Ecological Reserve (Brazil) Duleba *et al.*, in press	3–29% (n = 383)
Itacolomi Lagoon Ecological Reserve (Brazil) Duleba *et al.*, in press	1–4% (n = 64)
Bertioga Channel (Brazil) Eichler *et al.*, 1998	0–4% (n = 123)

[a] n = number of abnormal hyaline tests.

3.4. Morphological and Textural Studies

Abnormal tests were picked from treated samples and put on stubs prepared with carbon conductive adhesive tape for SEM observation. Specimens obtained in cultures were picked with an automatic micropipette, rinsed in distilled water, washed with a fine brush, and dried in a microslide before being placed on a stub. All SEM observations and photographs were conducted in the SCME (Service Commun de Microscopie Electronique) of the University of Angers with a SEM Jeol 6301F.

Textural studies of abnormal test wall were carried out in order to study the test biomineralization process in relation to the environment. For SEM investigations, the abnormal parts of the test were crushed with a scalpel directly on the stub prepared with the carbon conductive adhesive tape. This experimental procedure avoids the dispersion of test fragments.

4. Results

4.1. Percentage of Abnormal Hyaline Tests

In nonpolluted Brazilian environments, percentages of abnormal tests may be as high as 29% in the Rio Una (Table 4), while in the most polluted of these environments (Bertioga), there are very few (0 to 4%).

The number of abnormal tests (roughly 500) observed in culture dishes with original sediment in normal saline conditions (37%) is about 1%. Under hypersaline conditions (50%), about half of the juveniles exhibited malformations or an unusual chamber size in the early stages of their development (Stouff, 1998). Similar results have been obtained on a few cultures of *Elphidium crispum* under normal and hypersaline (42%) conditions.

4.2. Abnormal Test Morphology

The majority of malformations observed in natural environments characterized by fluctuations of salinity (Table 3) correspond to: (1) a protuberance on the spiral side near the proloculus (Fig. 5: *1* and *2*); (2) an abnormal size or shape of one to several chambers (Fig. 5: *3* to *7*); (3) a growth in two whorls resulting in double tests (Fig. 5: *8* and *9*); and (4) complex forms where the perturbation cannot be determined precisely (Fig. 5: *10* to *12*). Deformed tests exhibiting regeneration scars will be discussed later.

Malformed specimens in hypersaline natural environments (Araruama lagoon—Rio de Janeiro, Brazil) exhibit either : (1) a protuberance on the spiral side near the proloculus (Fig. 6: 1); (2) an abnormal size or shape (Fig. 6: *2* to *4*) or arrangement of one to several chambers; (3) double or triple tests (Fig. 6: *5* and *6*); or (4) complex forms probably resulting from fusions of embryos (Fig. 6: *7* and *8*).

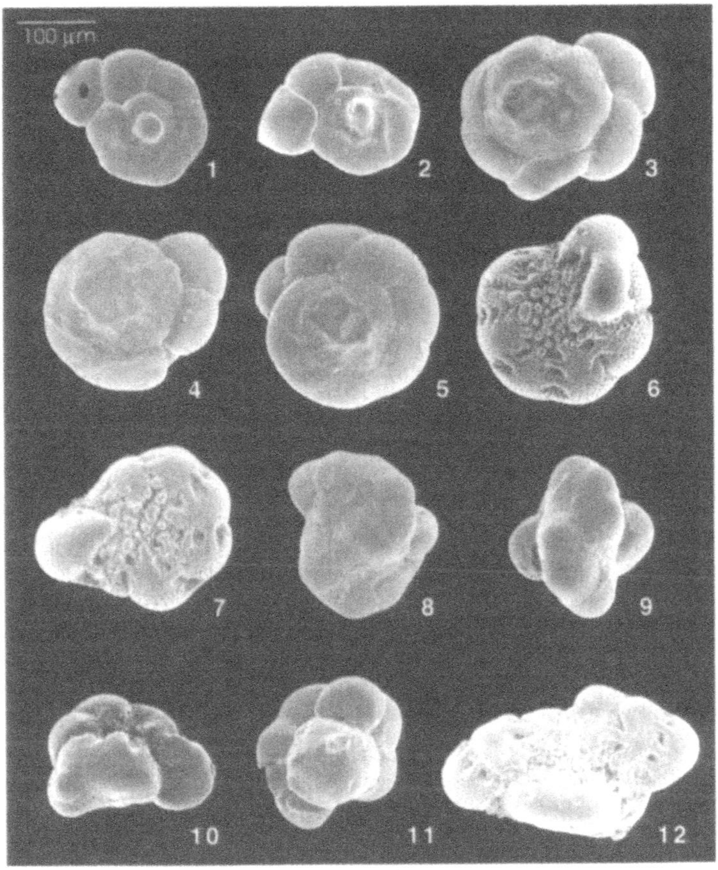

FIGURE 5. Abnormal tests collected in natural environments characterized by fluctuations of salinity (Rio Guaraú, Brazil; Golfe du Morbihan, France). (*1,2*) *Ammonia tepida* exhibiting a protuberance on the spiral side near the proloculus. (*3*) *A. tepida* with an aberrant chamber shape. (*4,5*) *A. tepida* with an aberrant chamber size. (*6*) *Elphidium* aff. *gunteri* with an aberrant chamber shape. (*7*) *E.* aff. *gunteri* with an abnormal size or shape of one chamber. (*8,9*) *A. tepida* with two whorls resulting in double tests. (*10,11*) Complex forms of *A. tepida*. (*12*) Complex form of *E.* aff. *gunteri*. (All photographs are on the same scale, scale bar = 100 μm.)

Abnormalities observed under normal saline culture conditions (37%) include the presence of a protuberance on the spiral side near the proloculus (Fig. 7: 1), an abnormal arrangement of the first chambers (Fig. 7: *2*) or double tests with normally shaped chambers (Fig. 7: *3* to *6*). Under hypersaline culture conditions, abnormalities are more frequent which means about 1% of the tests are abnormal in normal saline conditions and about 50% in hypersaline conditions.

Based on about 100 abnormal juveniles from hypersaline cultures, malformations may be grouped into four categories (Stouff, 1998):

1. Perturbations affecting size or shape of the first chambers (Fig. 8: *1* to *3*).

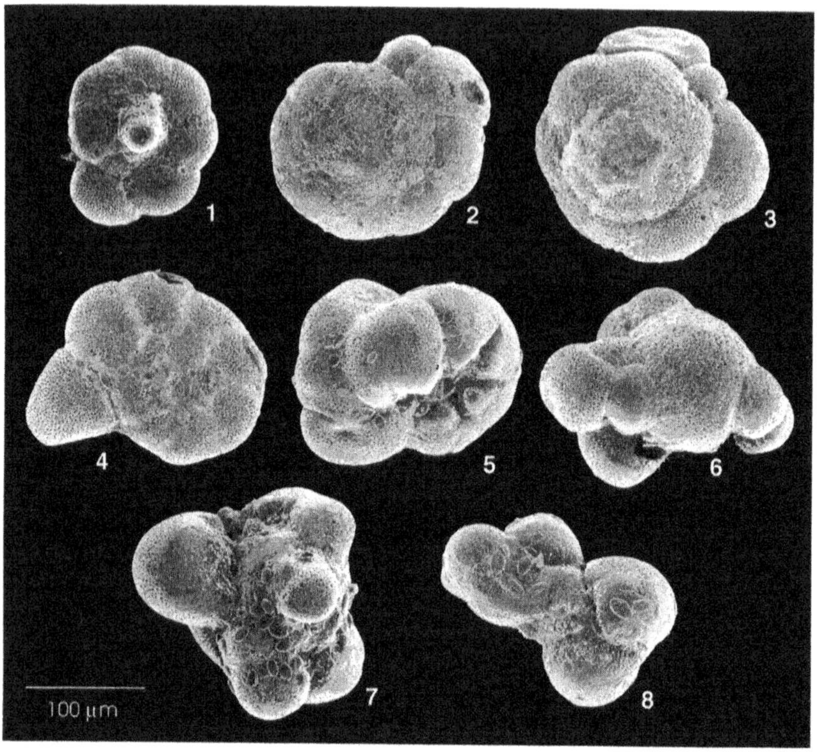

FIGURE 6. Abnormal tests of *Ammonia tepida* collected in a hypersaline environment (Araruama Lagoon, Brazil). (*1*) Protuberance on the spiral side near the proloculus. (*2*) Reduced size of the last chamber. (*3*) Reduced size of the penultimate chamber and abnormal shape of the last chamber. (*4*) Abnormal shape of the last chamber. (*5*) Double test. (*6*) Triple test. (*7,8*) Complex forms. (All photographs are on the same scale, scale bar = 100 μm.)

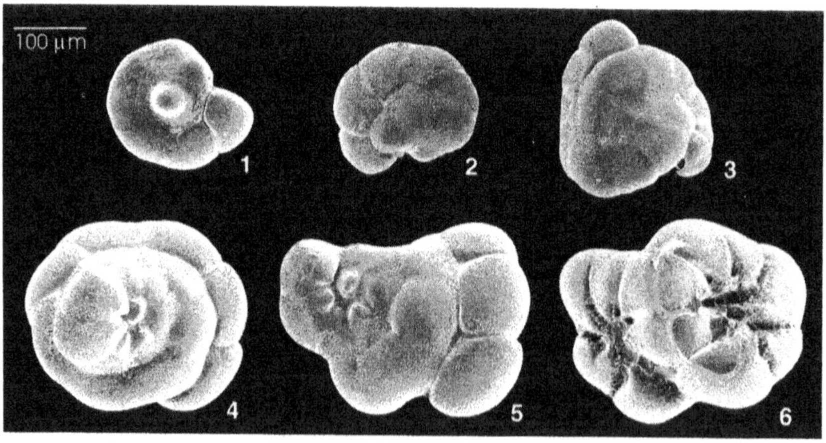

FIGURE 7. Abnormal tests of *Ammonia tepida* from cultures maintained under normal saline conditions (37‰). (*1*) Juvenile specimen exhibiting a protuberance on the spiral side near the proloculus. (*2*) Juvenile specimen with an abnormal arrangement of the first chambers. (*3–6*) Double tests. (All photographs are on the same scale, scale bar = 100 μm.)

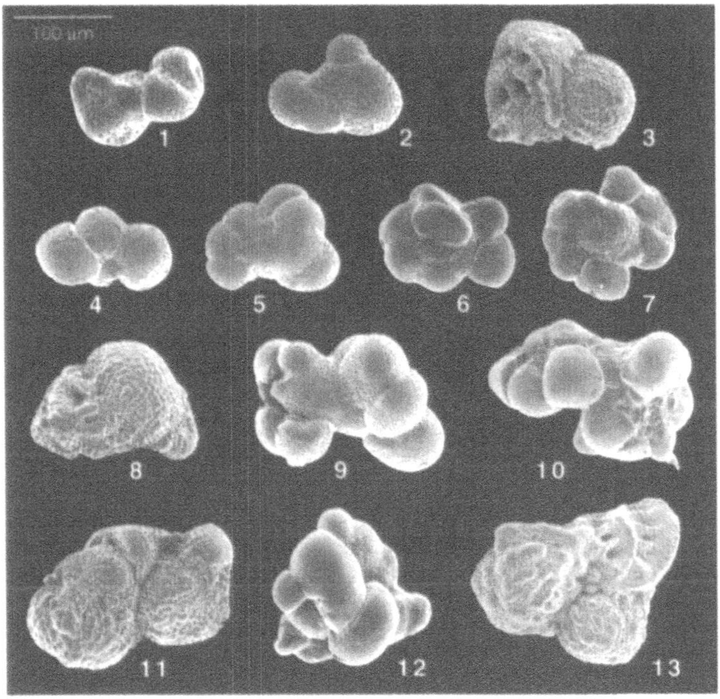

FIGURE 8. Abnormal juveniles picked from in hypersaline cultures. (*1*) Juvenile of *A. tepida* with aberrant chamber shape. (*2*) Juvenile of *A. tepida* with aberrant chamber size and shape. (*3*) Juvenile of *Elphidium crispum* with aberrant chamber shape. (*4–6*) Juvenile of *A. tepida* exhibiting a modification of the coiling plane of the first chambers. (*7*) Juvenile of *A. tepida* with two whorls resulting from two second chambers. (*8*) Juvenile of *E. crispum* with two second chambers which may result in two whorls. (*9,10*) Double tests of *A. tepida* probably resulting from fusion of embryos. (*11*) Fusion of two embryos of *E. crispum*; two proloculi are clearly visible. (*12*) Complex form of *A. tepida* resulting from fusion of embryos. (*13*) Abnormal test of *E. crispum* resulting from fusion of embryos. (All photographs are on the same scale, scale bar = 100 μm.)

2. Modifications of the coiling plane of the first chambers (Fig. 8: *4* to *6*).

3. Development of two different whorls: the origin of the two whorls appears clearly as the consequence of the formation of two second chambers or of two third chambers (Fig. 8: *7* and *8*).

4. The fusion of embryos or juveniles and the development of complex abnormal forms (Fig. 8: *9* to *13*): these were examined in cultures after asexual reproduction (schizogony) and correspond to early fusions of embryos or juveniles. Most of these abnormalities in juvenile tests correspond to those observed in a hypersaline natural environment (the Araruama lagoon—Brazil). Morever, two out of ten normal specimens picked from a normal salinity culture (37%) and transferred into a hypersaline culture (50%) formed test abnormalities. One of ten constructed chambers larger than the ones formed previously in normal saline conditions (Fig. 9: *1a* and *1b*), and another one developed many chambers with a disturbed arrangement (Fig. 9: *2a* and *2b*).

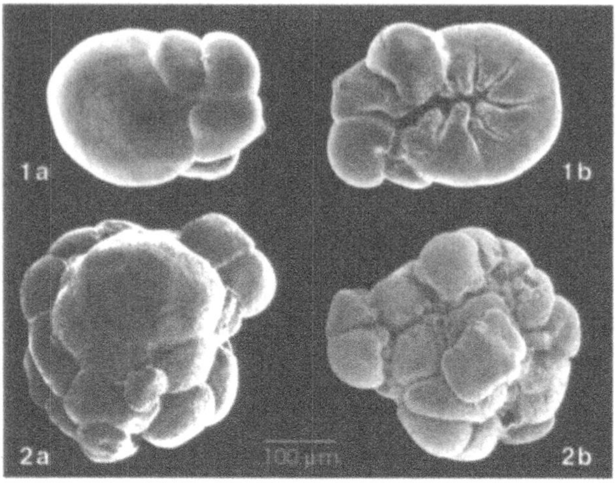

FIGURE 9. Adult tests of *Ammonia tepida* placed in hypersaline cultures only at the adult stage. (*1*) Abnormal size and arrangement of the last four chambers, *1a*-Spiral side, *1b*-Umbilical side of the same specimen; (*2*) Complex form, *2a*-Spiral side, *2b*-Umbilical side of the same specimen. (All photographs are on the same scale, scale bar = 100 μm.)

4.3. Textural Study

The wall texture of normal tests of *Ammonia* was described by Debenay *et al.* (1996). The various calcitic elements that compose the wall develop through various stages of assembly of crystallites (a few nanometers in diameter) into arrangements of increasing complexity (needle to crystal) oriented normally to the wall surface (Fig. 10: *1*). The last stage of arrangement is the formation of large cleaved crystals, mainly around the pores (Fig. 10: *2*).

When a new chamber is constructed, the crystallites are deposited on both the inner and outer sides of an organic template (primary organic sheet), leading to the formation of an outer lamella and a thinner inner lamella (Fig. 11). The outer lamella covers all of the outer surface of the preexisting test. It is composed of crystallites grouped in arrangements deposited in continuity with those of the previous lamellae (Fig. 10: *3*). Each outer lamella is separated from the previous one by a thin organic layer termed the "interlamellar organic sheet" (Fig. 11). The surface of the test is smooth and regularly perforated with pores (Fig. 10: *4*). The common boundaries between crystals can sometimes be observed on the surface, showing "jigsaw-puzzle-like sutures."

SEM observations of abnormal crushed tests distinguished two types of abnormal texture (Geslin *et al.*, 1998*a*):

1. A disorganized texture was observed in the wall of an "overdeveloped chamber" of an abnormal *Ammonia*, in contrast to normal tests and to the normal chambers of this particular abnormal test: the cleavages are not

regularly oblique to the pores, but have various orientations showing that the test is an aggregate of irregularly arranged crystals (Fig. 10: *5*).

2. Empty cavities were observed in the wall of some abnormal tests of *Ammonia*. Two cavities were observed between the inner and the outer lamellae in the same part of the wall (Fig. 10: *6* and *7*). In one of them, a small amount of amorphous material probably corresponds to organic matter (Fig. 10: *7*). At some distance of these aberrant textures, the wall exhibits a normal crystalline organization as described above. In another abnormal wall, numerous cavities have been observed in different areas located between two successive outer lamellae.

5. Discussion

5.1. Significance of Percentages of Abnormal Tests

In the literature, the presence of abnormal tests in polluted areas has sometimes been reported without a percentage (e.g., Watkins, 1961; Seiglie, 1964; Setty and Nigam, 1984; Jorissen, 1988), whereas other authors have specified the percentages of abnormal tests (e.g., Alve, 1991 [1–3% with a maximum of 7%]; Coccioni *et al.*, 1997 [$\leqslant$10%]; Seiglie, 1975 [5%]; Lidz, 1965 [30%]; Yanko *et al.*, 1994 [2–3%]; Yanko *et al.*, 1998 [3.5%]). These numbers show great variability and it is necessary to determine the limit above which the percentage is abnormal. In cultures in the Angers laboratory (France), under normal conditions, 1% of abnormalities is usually observed. Thus a low percentage of abnormalities cannot be considered as an indicator of abnormal environmental conditions, all the more so as the limits defining normal and abnormal may be difficult to determine. Morever, some abnormalities that may result from intrinsic processes, such as reproduction without direct relation to environmental conditions, have often been considered to result from environmental stress. For example, Seiglie (1975) described a great number of abnormal tests that he ascribed to organic pollution. However, the abnormal tests he described look like the abnormal specimens induced by reproductive processes (schizogony) in the Angers laboratory.

The detailed study of each abnormal specimen selected among the samples of the nonpolluted Rio Guaraú (Brazil) reveals that half of the abnormal tests show obvious regeneration scars (Fig. 3: *11*), which probably result from strong tidal currents and water discharge (Geslin *et al.*, 1998*b*). These observations suggest that mechanical causes may induce a great abundance of abnormal tests and that they can only be differentiated from those caused by pollution via SEM investigation.

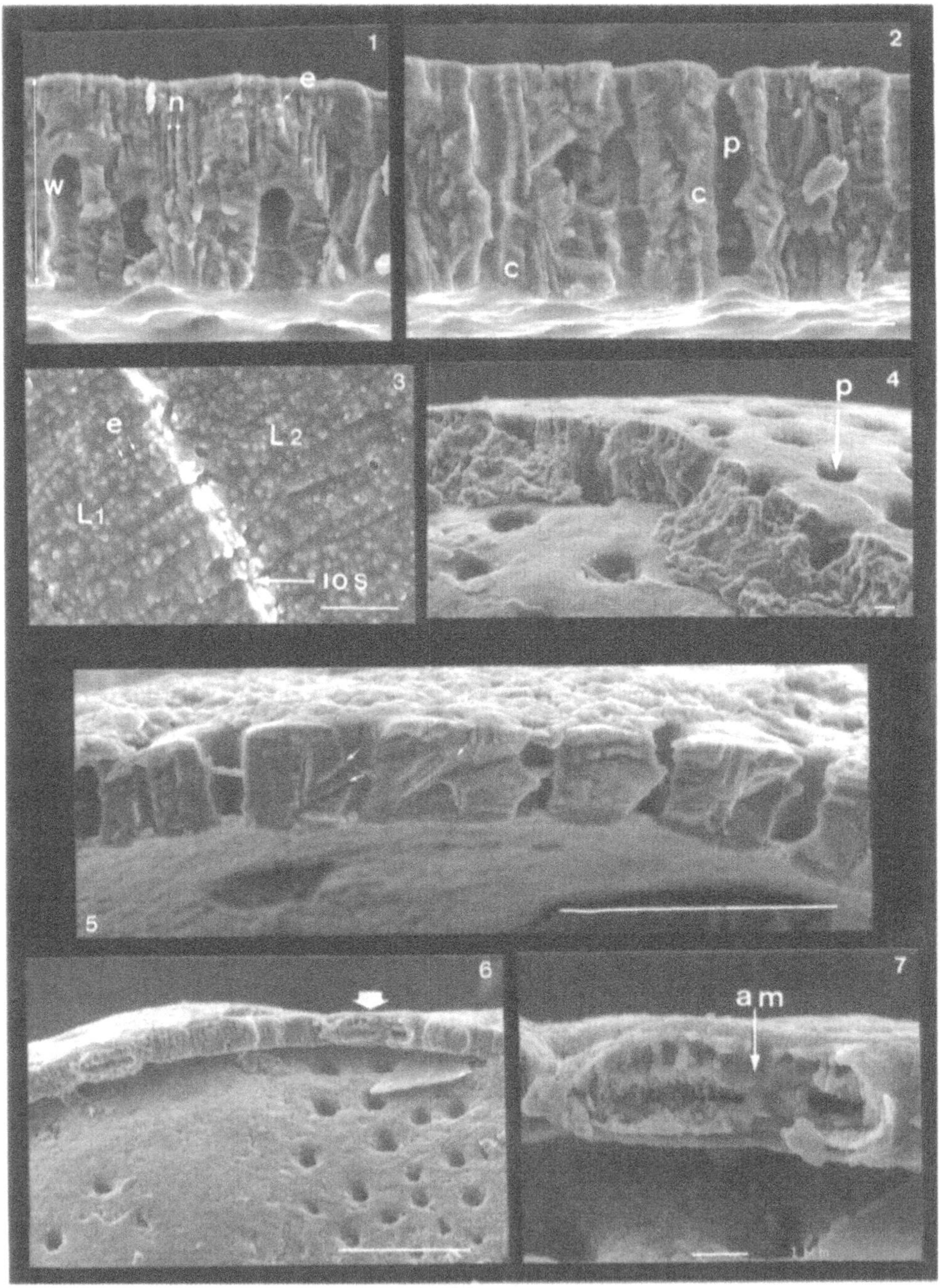

FIGURE 10. Tests of *Ammonia*: (**1–4**) Section and surface of the wall of normal tests (*1,2*) Details of wall texture. (*1*) The wall (w) is made up of primary calcitic elements (e) termed crystallites; needles (n) oriented normally to the wall surface, consist of smoothed rows of crystallites (scale bar = 1 μm). (*2*) Needles can gather into large cleaved crystals (c) around the pores (p) (scale bar = 1 μm). (*3*) Detail of the wall composition: the crystallites (e) grouped in arrangements are

5.2. Comparison of Results from Cultures and Natural Environments

In cultures and in natural environments, the same kinds of morphological abnormalities were generally observed (e.g., Fig. 5: *1*; Fig. 6: *1*, and Fig. 7: *1*). However, adult tests with an abnormal size or shape of one or more chambers, which are often reported in natural environments (e.g., Almogi-Labin *et al.*, 1992; Alve, 1991; Yanko *et al.*, 1998), are very rare in cultures and may result from variability of the natural environment.

5.3. Influence of Pollution

Several authors have related the presence of abnormal specimens to pollution (Seiglie, 1975; Vénec-Peyré, 1981; Alve, 1991; Sharifi *et al.*, 1991; Yanko *et al.*, 1994, 1998). However, the proportion of abnormal tests is sometimes very low and cannot be considered as significantly higher than in most natural nonpolluted environments. On the other hand, on the southwest Louisiana shelf in the vicinity of oil rigs, Locklin and Maddocks (1982) noted that "the small number of abnormal individuals found in these samples does not seem excessive, and it appears that most of these deformities can be explained by mechanical causes. There are no sufficient abnormalities to indicate environmental stress caused by pollution." Indeed, the percentage of abnormal tests is higher in nonpolluted than in polluted areas in Brazilian paralic environments. Thus, the percentage of abnormal tests cannot necessarily be considered as a direct criterion of pollution levels.

5.4. Influence of Salinity

The influence of salinity on test malformation is debatable. Malmgrem (1984) did not report a strong influence of salinity on test morphology, even under strong hypersaline conditions (92%), and Scott *et al.* (1976) reported that the morphology of *Ammonia beccarii* seems not to be affected by variations in salinity in lagoons in southern California (15 to 68%). However, abnormal tests are often reported in the literature under hypersaline conditions [e.g., Seiglie, 1964; Ayala-Castañares and Segura, 1968; Cann and De

---

deposited in continuity with those of the previous lamellae. The two lamellae (L1 and L2) are separated by an organic layer called the interlamellar organic sheet (IOS) (scale bar = 10 μm). (*4*) The surface of the test is smooth and regularly perforated with pores (p) (scale bar = 1 μm). (*5–7*) Wall texture of abnormal tests. (*5*) Different calcitic elements are oriented irregularly (arrows) (scale bar = 10 μm). (*6*) The wall shows two empty cavities (arrows) (scale bar = 10 μm). (*7*) Detail of *6*, a small amount of amorphous material (am) is observed in the cavity and probably corresponds to organic matter (scale bar = 1 μm).

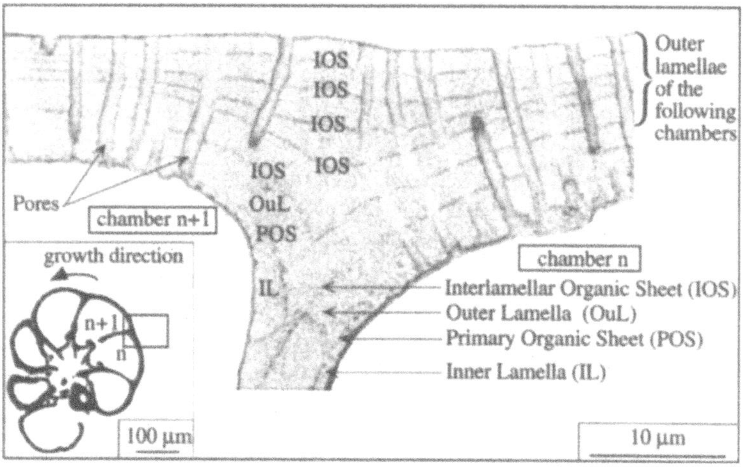

FIGURE 11. Schematic section of a normal test of *Ammonia* and detail of an intersection between two chambers.

Deckker, 1981 (Pl. 3, Figs. 27 and 32); Zaninetti, 1982; Almogi-Labin *et al.*, 1992 (Pl. 1, Figs. 13 and 14)]. In hypersaline cultures, we found roughly 50% of early juveniles that underwent early ontogenic perturbations leading to malformations of later juvenile and adult tests, as opposed to 1% in normal saline conditions (Stouff, 1998). These observations in both natural environments and cultures strongly suggest that hypersalinity may induce malformations during ontogenic development. Thus, foraminiferal morphology as a bioindicator of pollution must be applied cautiously in hypersaline conditions, since it is necessary to differentiate between the impact of natural stress and the impact of pollution. For example, it is difficult to assert that the abnormalities of *Adelosina cliarensis* reported by Yanko *et al.* (1998) in hypersaline (about 40%) waters polluted by Cd (40 ppm) result from pollution rather than hypersalinity.

5.5. A Possible Explanation from the Textural Study

The textural study shows how the mechanism of biomineralization may be perturbed. Two types of abnormal texture result from different mechanisms: the first type from a mineralogical perturbation (trapping of impurities in the crystalline framework), and the second from an organic perturbation (abnormal production of jelly). Textural studies should perhaps supplement the interpretation of abnormalities in relation to pollution impacts.

The abnormal texture of "crystalline disorganization" can result from the introduction of alien elements into the crystalline framework. This hypothesis is in agreement with the observations of Sharifi *et al.* (1991), who showed that abnormal tests contained a high proportion of metals such as Cu and Zn. In regions polluted by heavy metals, Yanko and Kronfeld (1993) found that the abnormal tests exhibited an increase in the Mg/Ca ratio and S content. These authors suggested that the trace metal poisoning leads to a breakdown in the organism's biochemical ability to selectively incorporate Ca against the high Mg/Ca gradient in seawater, which in turn causes an increase in the Mg concentration in the tests.

The empty cavities can be compared with: (1) the cavities found in the wall of the *Protelphidium* aberrant test by Vénec-Peyré (1984) after an oil spill; and (2) the interlamellar cavities seen in oyster shells contaminated by tributyltin (TBT) pollution (Alzieu, 1991). In this case, the abnormality consists of waferlike chambering of the shell with the formation of an interlamellar "jelly" that latter disappears, leaving an empty cavity. The presence of cavities in the test of *Ammonia* may result from the same kind of phenomenon. It can be described in three successive phases: (1) the production of "jelly" in organic layers; (2) the calcification on both the inner and outer sides of the "jelly" during the construction of the last chamber or over the "jelly" during the deposition of the outer lamella that covers the preexisting shell; and (3) the disappearance of the "jelly," thereby leaving an empty cavity.

6. Conclusion

This study demonstrates that abnormalities of foraminiferal tests may result from various processes, including reaction to environmental stress such as salinity, and not just pollution. It is therefore necessary to make a reliable distinction between natural and anthropogenic impacts before using abnormal tests in environmental monitoring.

References

Aktürk, S. E., 1976, Traumatic variation in the *Globorotalia menardii* d'Orbigny group in late Quaternary sediments from the Caribbean, *J. Foram. Res.* **6**(3):186–192.

Almogi-Labin, A., Perelis-Grossovicz, L., and Raab, M., 1992, Living *Ammonia* from a hypersaline inland pool, Dead Sea area, Israel, *J. Foram. Res.* **22**(3):257–266.

Alve, E., 1991, Benthic foraminifera in sediment cores reflecting heavy metal pollution in Sørfjord, Western Norway, *J. Foram. Res.* **21**(1):1–19.

Alzieu, C., 1991, Environment problems caused by TBT in France: Assessment, regulations, prospects, *Mar. Environ. Res.* **32**:7–17.

Amicis, G., 1895, Sopra alcune forme nuove di Foraminiferi del Pliocene Inferiore, *Soc. Toscana Sci. Nat., Atti, Mem.* **14**:18–30.

Arnal, R. E., 1955, Significance of abnormal Foraminifera, *Geol. Soc. Am. Bull.* **66**:1641.

Ayala-Castañares, A., and Segura, L. R., 1968, Ecologia y distribucion de los foraminiferos recientes de la Laguna Madre, Tamaulipas, Mexico, Universidad Nacional Autonoma de Mexico, *Boll. Inst. Geol.* **87**:1–90.

Banerji, R. K., 1989, Foraminifera and discrimination of polluted environment along the Bombay Coast, Kalia Prabha, *Micropaleontology of the Shelf Sequences of India, Indian Col. Micropaleontol. Stratigr.* University of Delhi, Vol. 12, pp. 98–117.

Banerji, R. K., 1992, Heavy metals and benthic foraminiferal distribution along Bombay coast, India, *Benthos'90, Sendai, Studies in Benthic Foraminifera,* Tokai University Press, Tokyo, pp. 151–158.

Bé, A. W. H., and Spero, H. J., 1981, Shell regeneration and biological recovery of planktonic foraminifera after physical injury induced in laboratory culture, *Micropaleontology* **27**:305–316.

Bé, A. W. H., Anderson, O. R., Faber, W. W., JR., and Caron, D. A., 1983, Sequence of morphological and cytoplasmic changes during gametogenesis in the planktonic foraminifer *Globigerinoides sacculifer* (Brady), *Micropaleontology* **29**:310–325.

Boltovskoy, E., 1954, Foraminíferos de la Bahia San Blas, *Rev. Inst. Nac. Invest. Cíen. Nat., Geol.* **3**:247–300.

Boltovskoy, E., 1956, Applications of chemical ecology in the study of the foraminifera, *Micropaleontology* **2**(4):321–325.

Boltovskoy, E., 1957, Las anormalidades en los caparazones de foraminíferos y el "Indice de regeneramiento," *Ameghiniana* **1**:80–84.

Boltovskoy, E., 1965, *Los foraminíferos Recientes*, Editorial Universitaria de Buenos Aires, Argentina, p. 510.

Boltovskoy, E., and Boltovskoy, A. , 1968, Foraminíferos y Tecamebas de la parte inferior del río Quequén Grande (sistemática, distriducíon, ecología), *Rev. Mus. Argentino Cíen. Nat. "Bernardino riva davia," Hidrobiol.* **2**(4):127–164.

Boltovskoy, E., and Wright, R., 1976, *Recent Foraminifera*, W. Junk, The Hague, p. 515.

Boltovskoy, E., Scott, D. B., and Medioli, F. S., 1991, Morphological variations of benthic foraminiferal tests in response to changes in ecological parameters: A review, *J. Paleontol.* **65**(2):175–185.

Bonetti, C., França, C. A., and Bonetti Filho, J., 1996, Fourier analysis applied to the study of deformation in foraminifers tests, *Conf. Environmetrics in Brazil*, July 22–26, 1996, São Paulo, Brazil, Abst.

Bonetti, C., Debenay, J.-P., and Eichler, B. B., 1997, Deformation in benthic foraminifers tests from a polluted Brazilian coastal system, *1st Int. Conf. Application of Micropaleontology in Environmental Sciences*, Tel Aviv, Israel, June 15– 20, 1997, Abst. p. 41–42.

Brasier, M. D., 1975, Morphology and habitat of living benthonic foraminiferids from Caribbean carbonate environment, *Rev. Española Micropaleontol.* **7**(3):567–569.

Cann, J. H., and De Deckker, P., 1981, Fossil Quaternary and living foraminifera from athalassic (non-marine) saline lakes, southern Australia, *J. Paleontol.* **55**(3):660–670.

Caralp, M. H., 1989, Size and morphology of the benthic foraminifer *Melonis barleeanum*: Relationships with marine organic matter, *J. Foram. Res.* **19**(3):235–245.

Closs, D., and Madeira, M. L., 1968, Seasonal variations of brackish foraminifera in the Patos Lagoon, southern Brazil, *Univ. Rio Grande do Sul, Escola Geol., Publ. especial* **15**:1–51.

Coccioni, R., Gabbianelli, G., Gentiloni Silverj, D., Fonti, P., Kaminski, M. A., Monechi, S., and Tateo, F., 1997, Benthic foraminiferal response to heavy metal pollution in the Goro Lagoon (Italy), *1st Int. Conf. Application of Micropaleontology in Environmental Sciences*, Tel Aviv, Israel, June 15–20, 1997, Abst., p. 47–48.

Debenay, J.-P., Guillou, J.-J., and Lesourd, M., 1996, Colloidal calcite in foraminiferal tests: Crystallization and texture of the test, *J. Foram. Res.* **26**(4):277–288.

Duleba, W., Sousa, S. H. M., Debenay, J.-P., Eichler, P. B., and Eichler, B. B., in press, Caracterização ambiental dos estuários a lagunas da Estação Ecológica Juréia–Itatins, com base no estudo das associações de foraminíferos e tecamebas, *in: Ambiente, flora e fauna da Estação Ecológica Juréia-Itatins* (O. A. V. Marques, W. Duleba, and F. P. Por, eds.), São Paulo.

Eichler-Coelho, P. B., Duleba, W., Eichler, B. B., and Coelho-Júnior, C., 1996, Determinação do impacto ecológico do Valo Grande (Iguape, SP) a partir das associações de foraminíferos e tecamebas, *Rev. Brasileira Biol.* **57**(3):463–477.

Eichler, P. B., Geslin, E., Eichler, B. B., and Debenay, J.-P., 1998, Resultados preliminares de foraminíferos (Protozoa) da região estuarina de Bertioga, São Paulo, *XXII Congr. Brasileiro Zool., Recife,* February 98, Abst., p. 1.

Feyling-Hanssen, R. W., 1972, The foraminifer *Elphidium excavatum* (Terquem) and its variant forms, *Micropaleontology* **18**:337–354.

Geslin, E., Debenay, J.-P., and Lesourd, M., 1998*a*, Abnormal wall textures and test deformation in *Ammonia* (hyaline foraminifer), *J. Foram. Res.* **28**(2):148–156.

Geslin, E., Debenay, J.-P., Lesourd, M., Duleba, W., and Eichler, B. B., 1998*b*, A quantitative and qualitative study of hyaline deformed tests in a non polluted estuary (Juréia, Brazil), International Symposium on Foraminifera, Foram '98, Mexico, July 3–13, 1998, *Rev. Soc. Mexicana Paleontolog. Abst.*, **9**:38.

Goubert, E., 1997, Les *Elphidium excavatum* (TERQUEM), foraminifères benthiques, vivant en baie de Vilaine (Bretagne, France) d'octobre 1992 à septembre 1996 : morphologie, dynamique de population et relations avec l'environnement. Thèse de Doctorat, Université, de Nantes, France. Tome 1: Textes 186 pp. et annexes ; Tome 2: 70 Fig., 30 Pl.

Hallock, P., Talge, H. K., Cockey, E. M., and Muller, R. G., 1995, A new disease in reef-dwelling foraminifera: Implications for coastal sedimentation, *J. Foram. Res.* **25**(3):280–286.

Hayward, B. W., and Hollis, C. J., 1994, Brackish foraminifera in New Zealand: A taxonomic and ecologic review, *Micropaleontology* **40**:185–222.

Heron-Allen, E., and Earland, A., 1910, On the recent and fossil foraminifera of the shore-sands of Selsey Bill, Sussex: Part VI. A contribution towards the aetiology of *Massilina secans* (d'Orbigny sp.), *J. Roy. Microsc. Soc. London* **1910**:693–695.

Hofker, J., 1971, Foraminifera of the Piscadera Bay, Curaçao: Studies of Fauna Curaçao and other Caribbean Islands, Vol. 35: pp. 1–63.

Hottinger, L., Halicz, E. and Reiss, Z., 1993, Recent Foraminifera from the Gulf of Aqaba, Red Sea, Slovensha Akademija Znanosti in Umetnosti, Ljubljana, 179 pp., 230 Pl.

Jayaraju, N., and Reddi, K. R., 1996, Impact of pollution on coastal zone monitoring with benthic foraminifera of Tuticorin, south east coast of India, *Indian J. Mar. Sci.* **25**:376–378.

Jorissen, F. J., 1988, Benthic foraminifera from the Adriatic sea: Principles of phenotypic variation, *Utrecht Micropaleontol. Bull.* **37**, 176 pp.

Kravchuk, O. P., Motnenko, I., Penciner, J., Kronfeld, J., and Yanko, V., 1997, The geoecological environment of the northern Israeli shelf and its affect on foraminifera, *1st Int. Conf. Application of Micropaleontology in Environmental Sciences*, Tel Aviv, Israel, June 15—20, 1997, Abst., pp. 74–75.

Le Campion, J., 1968, Foraminifères des principaux biotopes du Bassin d'Arcachon et du proche Océan, *Bull. Cen. Etud. Recher. Sci., Biarritz* **7**(2):207–391.

Lidz, L., 1965, Sedimentary environment and foraminiferal parameters : Nantucket Bay, Massachussetts, *Limnol. Oceanogr.* **10**(3):392–402.

Locklin, J. A., and Maddocks, R. F., 1982, Recent foraminifera around petroleum production platforms on the southwest Louisiana shelf, *Trans. Gulf Coast Assoc. Geol. Soc.* **32**:377–397.

Lutze, G. F., 1964, Statistical investigations on the variability of *Bolivina argentea* Cushman, *Contr. Cushman Found. Foram. Res.* **15**(3):105–116.

Madeira-Falcetta, M. L., 1974, Ecological distribution of the thecamoebal and foraminiferal associations in the mixohaline environments of the southern Brazilian littoral, *Anaïs Acad. Brasileira Ciên.* **46**(3/4):667–687.

Malmgrem, B. A., 1984, Analysis of the environmental influence on the morphology of *Ammonia beccarii* (Linné) in southern European salinas, *Geobios* **17**(6):737–746.

Pujos, M., 1976, Ecologie des foraminifères benthiques et des thécamoebiens de la Gironde et du plateau continental Sud-Gascogne. Application à la connnaissance du quaternaire terminal de la région Ouest-Gironde: Thèse de Doctorat, Tomes 1 et 2, 269 pp.

Pujos-Lamy, A., 1973, *Bolivina subaenariensis* Cushman, indicateur d'un milieu confiné dans le Golfe de Cap-Breton, *Comp., Rend. Acad. Sci., Paris* **277**(D):2655–2658.

Resig, J. M., 1974, Recent foraminifera from a landlocked Hawaiian lake, *J. Foram. Res.* **4**(2):69–76.

Schnitker, D., 1974, Ecotypic variation in *Ammonia beccarii* (Linné), *J. Foram. Res.* **4**(4):217–223.

Scott, D. B., and Nedioli, F. S., 1980*a*, Quantitative studies of marsh foraminiferal distribution in Nova Scotia: Implications for sea level studies, *Contr. Cushman Found. Foram. Res., Spec. Publ.* **17**, 58 pp.

Scott, D. B., and Nedioli, F. S., 1980*b*, Living vs. total foraminiferal population: Their relative usefulness in Paleoecology, *J. Paleontol.* **54**:814–831.

Scott, D. B., Mudire, P. J., and Bradshaw, J. S., 1976, Benthonic foraminifera of three southern Californian lagoons: ecology and récent stratigraphy, *J. Foram. Res.* **6**(1):59–75.

Seiglie, G. A., 1964, Significacíon de los foraminíferos anormales de la Laguna Unare, *Lagena* **1**:6.

Seiglie, G. A., 1968, Foraminiferal assemblages as indicators of high organic carbon content in sediments and of polluted waters, *Am. Assoc. Petrol. Geol. Bull.* **52**(11):2231–2241.

Seiglie, G. A., 1971, A preliminary note of the relationships between foraminifers and pollution into Puerto Rican Bays, *Caribbean J. Sci.* **11**:93–98.

Seiglie, G. A., 1974, Foraminifera of Mayagüez and añasco bays and its surroundings, *Caribbean J. Sci.* **14**:1–68.

Seiglie, G. A., 1975, Foraminifers of Guayanilla Bay and their use as environmental indicators, *Rev. Española Micropaleontol.* **7**(3):453–487.

Sellier De Civrieux, J. M., 1968, Un metodo de observaciones y representaciones graficas del metabolismo (foraminíferos) con aplication al estudio de la ecologia en biotopos de salinidad altamente variable, *Bol. Inst. Oceanogr., Univ. Oriente* **7**(2):99–109.

Sen Gupta, B. K., and Machain-Castillo, M. L., 1993, Benthic foraminifera in oxygen poor habitats, *Mar. Micropaleontol.* **20**:183–201.

Setty, M. G. A. P., and Nigam, R., 1984, Benthic foraminifera as pollution indices in the marine environment of west coast of India, *Riv. Italiana Paleontol. Stratigr.* **89**(3):421–436.

Sharifi, A. R., Croudace, L. W., and Austin, R. L., 1991, Benthic foraminiferids as pollution indicators in Southampton Water, southern England, United Kingdom, *J. Micropaleontol.* **10**(1):109–113.

Sousa, S. H. M., Duleba, W., Kfouri, P., Eichler, B. B., and Furtado, V. V., 1997, Response of foraminiferal assemblages to environmental changes in the São Sebastião Channel, Northern Coast of São Paulo State, Brazil, 1997, *1st Int. Conf. Application of Micropaleontology in Environmental Sciences*, Tel Aviv, Israel, June 15–20, 1997, Abst., p. 109–110.

Stouff, V., 1998, Interêt des élevages de foraminifères en laboratoire : Etudes biologiques et ultrastructurales. Thèse de Doctorat, Université, d'Angers et EPHE, France, 405 pp.

Stubbles, S. J., Hart, M. B., Williams, C. L., and Green, J. C., 1996, Response of foraminifera to presence of heavy metal contamination and acidic mine drainage, *"Mineral, Metals and the Environment II," Conf. Institution of Mining and Metallurgy*, Prague, September 3–6, 1996, pp. 217–235.

Tufescu, M., 1968, *Ammonia tepida* (Cushman) (ord. Foraminifera). Some features of its variability in the Black Sea Basin, *Rev. Roumaine Biol. Zool.* **13**:169–177.

Vénec-Peyré, M.-T., 1981, Les Foraminifères et la pollution : étude de la microfaune de la cale de Dourduff (embouchure de la rivière de Morlaix), *Cah. Biol. Mar.* **22**:25–33.

Vénec-Peyré, M.-T., 1984, Les foraminifères et le milieu: Etude de trois écosystèmes: *Benthos'83, 2nd Int. Symp., Benthic Foraminifera*, pp. 573–581.

Vilela, C. G., 1994, Contribuição ao Conhecimento da ecologia de foraminíferos bentônicos na foz do rio Amazonas, *Anaïs Acad. Brasileira Ciên.* **66**(1):37–44.

Vilela, C. G., and Koutsoukos, E. A. M., 1992, Miliolina (Foraminiferida) em sedimentos recentes da foz do Amazonas: uma resposta comportamental a um ambiente dinâmico: *37th Congr. Brasileiro de Geologia*—SBG/SP, São Paulo, SP, Brasil, 1992, Abst., pp. 92–93.

Watkins, J. G., 1961, Foraminiferal ecology around the Orange County, California, ocean sewer outfall, *Micropaleontology* **7**(2):199–206.

Whitcomb, N. J., 1978, Effects of oil pollution upon selected species of benthic foraminiferids from the lower York River, Virginia, *Geol. Soc. Am., Progr. Abst.* **10**:515.

Yanko, V. (ed.), 1995, Benthic foraminifera as indicators of heavy metals pollution—a new kind of biological monitoring for the Mediterranean Sea: European Commission, Program Avicenne, Annual Report, #AVI CT92–0007, 270 pp.

Yanko, V., 1997, Benthic foraminifera as bioindicators of stress environment: Anthropogenic problems—Foraminiferal solution, *1st Int. Conf. Application of Micropaleontology in Environmental Sciences*, Tel Aviv, Israel, June 15– 20, 1997, Abst., pp. 117–119.

Yanko, V., and Flexer, A., 1991, Benthic foraminiferal assemblages as indicators of pollution (one example of the northwestern shelf of the Black Sea), *3rd Ann. Symp. Mediterranean Margin of Israel, Institute of Oceanography and Limnology*, Haïfa, Israel, 5 pp.

Yanko, V., and Flexer, A., 1992, Microfauna as possible indicators of hydrocarbon seepages: A new method for oil-gas trap reconnaissance, *Israel Geol. Soc.*, Annual Meeting, Ashkelon, Israel, pp. 169–170.

Yanko, V., and Kronfeld, J., 1993, Low magnesian calcitic tests of benthic foraminifera chemically mirror morphological deformations, *4th Int. Conf. Paleoceanography* Sept 21–25, 1992, Abstr.

Yanko, V., Kronfeld, J., and Flexer, A., 1994, Response of benthic foraminifera to various pollution sources: Implications for pollution monitoring, *J. Foram. Res.* **24**(1):1–17.

Yanko, V., Ahmad, M., and Kaminski, M., 1998, Morphological deformities of benthic foraminiferal tests in response to pollution by heavy metals: Implications for pollution monitoring, *J. Foram. Res.* **28**(3):177–200.

Yanko, V., Arnold, A., and Parker, W., 1999, The effect of marine pollution on benthic foraminifera, in: *Modern Foraminifera* (B. Sen Gupta, ed.), Kluwer Academic, Dordrecht, pp. 217–235.

Zaninetti, L., 1982, Les Foraminifères des marais salants de Salin–de–Giraud (Sud de la France): milieu de vie et transport dans le salin, comparaison avec les microfaunes marines, *Géol. méditerr.* **9**(4):447–470.

Zaninetti, L., 1984, Les Foraminifères du salin de Bras del Port (Santa Pola, Espagne), avec remarques sur la distribution des Ostracodes, *Univ. Barcelona, Rev. Invest. Geols.* **38/39**:123–138.

Chapter 10

Chemical Ecology of Foraminifera

Parameters of Health, Environmental Pathology, and Assessment of Environmental Quality

V. M. BRESLER and V. V. YANKO-HOMBACH

1. Introduction

Foraminifera play an important role in the global biogeochemical cycles of inorganic and organic compounds, which makes them one of the most significant of marine and marginal marine taxa (Lipps, 1983; Anderson, 1988; Lee and Anderson, 1991). The hardtests of foraminifera are preserved long after they die and can be studied in the fossil record. Therefore, foraminifera are often perceived more as a subject of geology and paleontology than of cell biology and zoology, and both paleontologists and zoologists have devoted more attention to the study of foraminiferal shells than the living specimens;

V. M. BRESLER and V. V. YANKO-HOMBACH • Institute for Nature Conservation Research, George S. Wise Faculty of Life Sciences, Tel Aviv University, Ramat Aviv 69978, Israel. Present address of V. V. Y-H: Avalon Institute for Applied Science, Winnipeg, Manitoba R3P 2G9, Canada.

Environmental Micropaleontology, Volume 15 of *Topics in Geobiology*, edited by Ronald E. Martin. Kluwer Academic/Plenum Publishers, New York, 2000.

ironically, however, molecular mechanisms of shell formation are unknown. Systematics and paleontology of foraminifera are studied in more detail than their biology, ecology, and ecotoxicology. The cytophysiology, biochemistry, molecular biology, and chemical ecology of foraminifera are particularly poorly understood.

To combine the interests of paleontologists and biologists and create an integral foraminiferology, a new scientific and methodological approach to examine and understand both the biology and the ecology of this taxon is needed. In this chapter, we propose such scientific and methodological approaches to study and understand the molecular biology, ecology, and ecotoxicology of living foraminifera. One can analyze subcellular organization, specific functions and metabolism of these unicellular organisms to understand their integrated molecular machinery mediated by different cell activities; i.e., the integrated molecular biology of the cells (Alberts, 1998). However, one can also study their interrelations with abiotic and biotic environmental factors in order to understand the fate of these organisms in a given environment, especially chemical interactions within and between species (Harborne, 1993).

2. Chemical Ecology of Foraminifera

2.1. Natural Xenobiotics in Foraminiferal Habitats

Only after World War II, did scientists realize that numerous man-made biologically active or toxic chemical compounds, particularly industrial by-products, pesticides, herbicides and drugs, reside permanently in food, air, and water (Parke, 1971). Such chemicals were named "xenobiotics" or foreign compounds, although the term is not limited to man-made compounds.

Chemical ecology studies various interactions within and between species mediated by natural xenobiotics, as well as the origin, chemical structure, and biological functions of these xenobiotics (Barbier, 1976; Rice, 1985; Harborne, 1993). The primary ecological function of natural xenobiotics is chemical protection against parasites, predators, and competitors, which became more complicated during coevolution (Whittaker and Feeny, 1971; Barbier, 1976; Albone, 1984; Rice, 1985; Harborne, 1993). Natural xenobiotics are usually divided into two groups: those that take part in the interaction between species (allochemicals) and those that take part in the chemical interaction within species (pheromones). Allochemicals may attract or repel target species; inhibit their metabolism, feeding, digestion, movement, or reproduction; change their behavior or induce morphological alterations (Table 1). These numerous attractants, repellents, deterrents, and inhibitors are crucial for selection of food plants, feeding and behavior of herbivorous animals. Pheromones regulate social behavior, sexual behavior and reproduction, and migration or forewarning about danger within a species. Thus, the formation

TABLE 1. Ecological Functions of Natural Xenobiotics

Type	Functions
1. Xenobiotics that mediate interaction between species	
Allomones (beneficial for their producers)	
Repellent	Drives away predators, parasites, or competitors
Deterrent	Inhibits feeding of predators
Arrestant	Inhibits or limits movement of predators or parasites
Attractant	Attracts pollinators, symbionts, or food animals
Suppressor	Inhibits metabolic processes in predators, parasites, or competitors
Toxins	Kills or injures predators, parasites, or competitors
Kairomones (beneficial for recipients)	
Attractant	Attracts recipients to food
Alarm signal	Warns recipients against danger
Stimulator and inductors	Induces development of protective structures
2. Xenobiotics that mediate interaction within species	
Pheromones	
Sex pheromone	Regulates sexual recognition and behavior
Social pheromone	Regulates behavior within groups or colonies and colonial life
Alarm pheromone	Warns individuals of the same species against danger
Marker pheromone	Marks territory for individuals of the same species
Regulators of population density	
Final metabolite	Inhibits development and growth of individuals of the same species

of food chains and food webs and other ecological interactions, which integrate various species into communities and ecosystems, are mediated by natural xenobiotics.

Because all living organisms from blue-green algae and bacteria to mammals produce natural xenobiotics, living organisms have evolved various antixenobiotic defense mechanisms (Whittaker and Feeny, 1971; Barbier, 1976; Bresler and Nikoforov, 1981; Albone, 1984; Bresler *et al.*, 1985, 1990; Brattsten, 1979, 1986; Lindroth, 1988; Bresler, 1989; Stoddart, 1990; Bresler and Fishelson, 1994; Bresler and Yanko, 1995*a*; Fishelson *et al.*, 1996). These mechanisms allow living organisms to adapt themselves to natural xenobiotics presented in their food and microenvironment and to use them as communicative chemical signals. Moreover, the same mechanisms can protect organisms against anthropogenic xenobiotics if they are chemically similar to natural ones. Therefore, a study of structure, affinity, and power of these defense mechanisms is crucial to understanding the key theoretical and practical problems of modern ecology and ecotoxicology (Bresler and Fishelson, 1994; Bresler and Yanko, 1995*a*; Fishelson *et al.*, 1996; Bresler *et al.*, 1998).

TABLE 2. Natural Protective Xenobiotics Produced by Some Benthic Marine Organisms

Compound	Source	Reference
Simple volatile brominated or iodinated hydrocarbons	Red, brown, and green algae	Gschwend *et al.*, 1985
Dibromophenol, dibromobenzylalkohols, and mono- to tetrabromopyrroles	Bacteria; red, brown, and green algae; Polychaetes; Hemichordates	Higa *et al.*, 1980; Faulkner, 1984*a*; King, 1986; Woodin *et al.*, 1987
Polyphenols, particularly phlorotannins	Red, brown, and green algae	Ragan and Glombitza, 1986; Van Alstyne, 1988; Harborne, 1993
Terpenoids and sesquiterpinoids	Red, brown, and green algae	Faulkner, 1984*a,b*; Harborne, 1993
Alkaloids and brominated alkaloids, particularly isoxazoline alkaloids and pyridino-acrudine alkaloids	Red, brown, and green algae; sponges; Mollusks	Faulkner, 1984*a,b*; Teeyapont *et al.*, 1993; Proksch, 1994; Avila, 1995; Fontana *et al.*, 1998; Kashman and Rudi, 1998
Brominated and nonbrominated heterocyclic compounds	Sponges; soft corals; Tunicates	Kashman and Rudi, 1998
Linear and cyclic peptides	Sponges; Tunicates	Gschwend *et al.*, 1985

Our knowledge about natural xenobiotics in different foraminiferal habitats is, however, fragmentary. A review of the literature (Table 2) demonstrates, though, that foraminifera, especially benthic foraminifera, inhabit a very complicated chemical microenvironment and are influenced by numerous natural xenobiotics. For example, brown, green, and red macroalgae and phytoplankton produce and release volatile brominated and iodinated xenobiotics. Their total production amounts to 10^{10} g/year of organobromides and more than 10^{12} g/year of organoiodides (Gschwend *et al.*, 1985). Brown and red algae also produce phenolic compounds and polyphenols as well as alkaloids and sesquiterpinoids. Marine bacteria and soft-bottom benthic infauna (e.g., Polychaeta and Hemichordata) also synthesize brominated xenobiotics. In the marine bottom environment sponges, tunicates and soft corals constitute one of the most important sources of bioactive natural xenobiotics (Table 2). Thus, marine sediment and adjacent water contain a wide variety of natural biologically active xenobiotics from macro- and microalgae, bacteria, sponges, and other invertebrates. Furthermore, the decomposition of dead seaweeds, algae, and invertebrates generates monosaccharides, small alkyl carboxylic acids, simple aromatic acids, humic acids, amino, nucleic, uronic, and muramic acids, as well as lipids and fatty acids, peptides, proteins, particularly exoenzymes, nonessential inorganic ions and hydrogen peroxide (Palenik *et al.*, 1987; Wotton, 1988; Perdue and Gjessing, 1990). Benthic foraminifera might be well adapted to these complicated chemical conditions and may effectively use some of these chemicals to protect themselves against others. For that reason, we began to study adaptive mechanisms in benthic epiphytic foraminifera, especially antixenobiotic defense mechanisms (Bresler and Yanko, 1995*a*).

Chemical composition of benthic microenvironments may vary according their community structure. For example, under the fluorescence microscope the dissected thalli of the seaweeds *Sargassum*, *Cystoseira*, *Fucus*, and *Jania* show the presence of a mucous mat covering the surface of these thalli. It contains particles of sediment and organic detritus, coralline red algae, green microalgae, bacteria, and diatoms (Bresler and Yanko, 1995*a*). The species-rich community on such a mucous mat includes ciliates, nematodes, small polychaetes and crustaceans, larvae of crustaceans and mollusks, as well as numerous epiphytic attached foraminifera such as *Pararotalia spinigera* (Le Calvez), *Rosalina macropora* (Hofker), and *Adelosina cliarensis* (Heron-Allen and Earland). The reticulopodial net of these foraminifera penetrates the mucous mat on the thalli surface. The reticulopods of benthic allogromiid foraminifera can even invade artificial substrates such as gelatin/agar gels (Bowser, 1985); thus, the reticulopodial net is in close contact with bacteria, microalgae, diatoms, and detritus particles. Numerous small oxygen bubbles produced by seaweed and microalgae are frequently located in the mucous mats, in close contact with attached foraminifera (Bresler and Yanko, 1995*a*). Such an isolated and viscous benthic epiphytic microenvironment could accumulate a higher concentration of the above-mentioned natural xenobiotics than adjacent seawater. Other special microenvironments, such as the deep-sea bottom or planktonic environments, may contain different amounts and sets of natural xenobiotics. Thus, it is necessary to study the size and chemical characteristics of different marine microenvironments.

2.2. General Antixenobiotic Defense Mechanisms in Foraminifera

When scientists began to study the transformation and fate of xenobiotics in living organisms, they discovered enzymatic systems that might make xenobiotics less toxic and more easily eliminated by transport systems. At that time pharmacologists and toxicologists considered the metabolism of xenobiotics proprietary information. At present, however, numerous biophysical, biochemical, and morphophysiological antixenobiotic defense mechanisms have been detected and described in vertebrates, especially laboratory mammals (Parke, 1971; Rapoport, 1976; Bradbury, 1979; Bresler and Nikiforov, 1981; Brattsten, 1986; Lindroth, 1988; Baba *et al.*, 1988; Bresler, 1989; Sarmento *et al.*, 1991; Stoddart, 1990). These mechanisms were also detected and studied in some invertebrate species, particularly insects and mollusks (Bresler *et al.*, 1985, 1990, 1998, in press; Bresler and Fishelson, 1994). Investigations of antixenobiotic defense mechanisms in protozoa are not as numerous and are rather fragmentary (Khan *et al.*, 1972; Murphy *et al.*, 1982; Yawetz and Agosin, 1979, 1980, 1981; Piccinni, 1989; Burton, 1991; Piccinni *et al.*, 1992). Bresler and Yanko (1995*a,b*) first investigated the possible presence of several antixenobiotic mechanisms in the marine benthic epiphytic foraminifera *Pararotalia spinigera* and *Rosalina macropora* and described the likely presence of numerous antixenobiotic defense mechanisms:

1. External diffusion barriers that prevent or limit the penetration of water-soluble xenobiotics into the cytoplasm.
2. Polysubstrate carrier-mediated membrane transport systems that eliminate xenobiotics from the foraminiferal cytoplasm.
3. Intracellular compartments that actively accumulate, store, or eliminate cationic xenobiotics.
4. Intracellular xenobiotic-binding proteins or peptides.
5. Secreted (extracellular) xenobiotic-binding proteins or peptides.
6. Enzymes that metabolize penetrated xenobiotics and decrease their toxicity.
7. Enzymes that protect foraminifera against an excess of oxygen and peroxides.
8. Enzymes that protect foraminifera against bromine and iodine ions that have penetrated their plasma membranes from seawater to cytoplasm.

The localization of these defense mechanisms in foraminiferal cells is shown schematically in Fig. 1. The following is a detailed analysis of their structure and comparative physiology, biochemistry, and ecological roles.

2.2.1. External Barriers

To examine the permeability of single cells and cell layers, fluorescent markers of different sizes and properties are widely used (Kotyk and Janacek, 1977; Bresler *et al.*, 1979; Haugland, 1996). A versatile probe, fluorescein (FLU, disodium salt), is a widely used permeability marker because at pH 7.0–9.0 in water it forms very hydrophilic negatively charged ions, which do not pass across the lipid bilayer of model membranes or intact plasma membranes (Bresler *et al.*, 1979; Bresler and Fishelson, 1994; Haugland, 1996; Schreiber *et al.*, 1997). When pieces of seaweed thalli with attached living foraminifera, *P. spinigera* and *R. macropora*, were incubated from 1 to 24 hr in seawater containing 100 μmole FLU, no FLU penetration into the foraminiferal cytoplasm was observed (Bresler and Yanko, 1995*a,b*). The mucous mat surrounding the foraminiferal cytoplasm does not interact with FLU. Freshly detached (i.e., mechanically damaged) foraminifera, which lost some of their pseudopodia, demonstrated a marked increase of permeability to FLU, but if they survived in clean seawater for 24 hr and restored their plasma membrane integrity, their permeability to FLU also ceased (Bresler and Yanko, 1995*a*). Thus, the foraminiferal plasma membrane is an impermeable barrier for hydrophilic anionic xenobiotics.

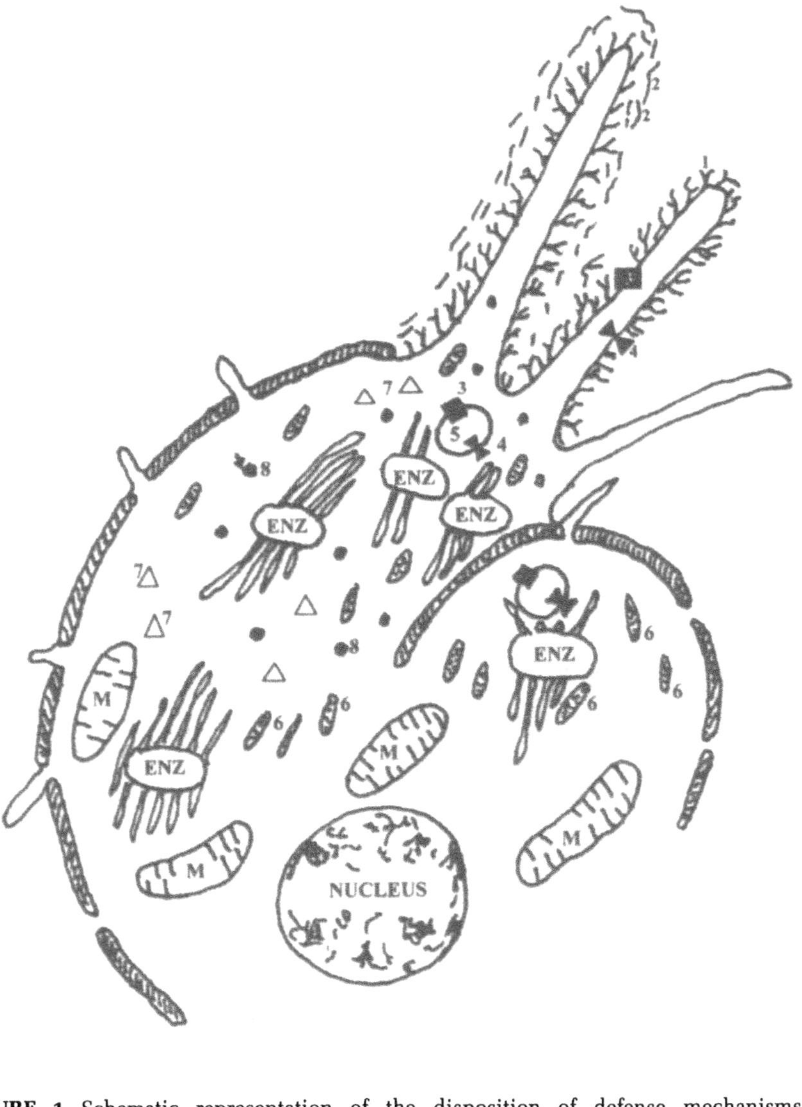

FIGURE 1. Schematic representation of the disposition of defense mechanisms in the foraminiferal cytoplasm. (1) glycocalyx, (2) mucous coat, (3) MXR (multidrug or multixenobiotic resistance) transporters, (4) SATOA (system of active transport of organic anions), (5) exocytic vesicle, (6) lysosome, (7) xenobiotic-binding proteins. (8) peroxisome, (ENZ) enzymes of detoxification and conjugation, (M) mitochondria.

It should be noted that the plasma membrane bilayer is not only impermeable to xenobiotics but also to all other water-soluble chemicals, including essential inorganic ions and metabolites. Therefore, to transport dissolved essential compounds into the cytoplasm, all cells use special mechanisms: molecular machines such as ionic pumps, carriers, and channels, or complicated subcellular structures such as pinocytic and phagocytic vesicles (Prosser, 1973; Kotyk and Janacek, 1977; Bresler and Nikiforov, 1981; Alberts *et al.*, 1983). Phagocytosis of small (bacteria) and large (algae, diatoms) prey has been studied both in some benthic and planktonic foraminifera (Andersen and Bé, 1976; Bowser and McGee-Russell, 1982; Bowser *et al.*, 1985; Travis and Bowser, 1991; Anderson and Lee, 1991). Carrier-mediated transport of solutes across plasma membranes has been less studied. It was shown that the Antarctic benthic foraminifer *Notodendrodes antarctikos* accumulates ^{14}C-labeled amino acids and glucose from seawater (DeLaca *et al.*, 1981). Experiments with the attached living Mediterranean foraminifera *P. spinigera* and *R. macropora* demonstrate that glucose added to seawater produces typical reversible alterations in the redox state of intramitochondrial NAD (Bresler and Yanko, 1995*a*). All these data indicate that foraminiferal plasma membranes contain some carriers for amino acids and sugars (glucose). However, additional studies as to their nature, specificity, and kinetic characteristics, are required. Activity and selectivity of inorganic ion transport can lead to an explanation of the presence and level of trace metals found in foraminiferal cytoplasm and shells. These mechanisms are presumably analogous to other eukaryotic uni- and multicellular organisms. They might include Na^+, K^+-ATPase, transporters or channels for essential metals, particularly calcium, magnesium, iron, copper, and zinc, and probably include secreted extracellular metal-chelating compounds. Iron in seawater is either associated with organic chelators or is present as aggregates of the almost totally insoluble iron (III) hydroxide. Aquatic prokaryotes (bacteria and cyanobacteria) and some unicellular eukaryotes (dinoflagellates) produce extracellular iron (III)-specific chelating compounds (siderophores) that enable them to solubilize and therefore acquire the iron present in the aggregates (Trick *et al.*, 1983).

The plasma membranes of all cells are permeable to hydrophobic (lipid-soluble) compounds, which penetrate into the cytoplasm via simple diffusion across the lipid bilayer (Kotyk and Janacek, 1977; Alberts *et al.*, 1983). Therefore, the examination of the plasma membrane permeability of living attached foraminifera to lipid-soluble xenobiotics was studied recently by using corresponding fluorescent markers (Acridine Orange [AO], Neutral Red [NR], Rhodamine B [RhB], and benzo[a]pyrene [BP]) and epifluorescence microscopy (Fig. 2). The data obtained show that these markers easily penetrate into the foraminiferal cytoplasm via simple diffusion through the lipid bilayer, and more lipophilic BP enters faster then less lipophilic RhB and NR. Neutral molecules of BP do not interact with the mucous mat, but the mat binds cationic RhB, NR, and especially AO. When binding the AO, it changes its color of fluorescence from green to red, which is typical for an AO interaction with mucopolysaccharides (glycosaminoglycans). This vital staining with AO shows that the external surface of the cytoplasmic body is covered by a thin homogeneous layer of mucopolysaccharide that gradually turns into a fibrous common mucous mat. The transfer of AO-treated living attached and detached foraminifera in clean seawater showed that the red fluorescence of bound AO was well preserved for the duration of the observations (48 hr).

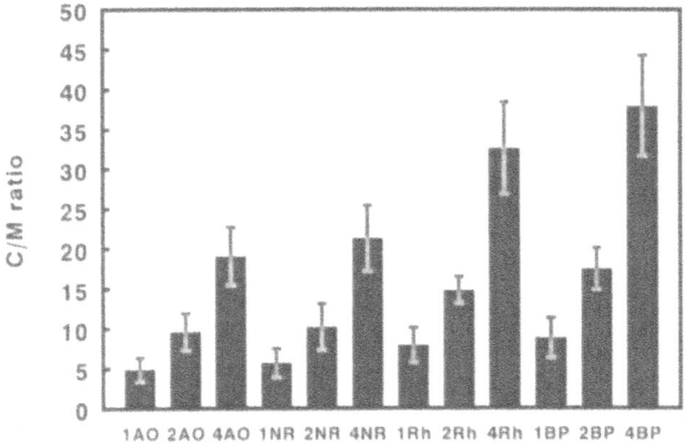

FIGURE 2. Accumulation of amphiphilic and lipophilic marker xenobiotics in cytoplasm of the youngest chamber of the benthic foraminifer *Rosalina macropora*. (1AO–4AO) incubation with 1 μM acridine orange for 1–4 hr;(1NR–4NR) incubation with 0. 25 μM neutral red for 1–4 hr. (1Rh–4Rh) incubation with 1 μM rhodamine B for 1–4 hr. (1BP–4BP) incubation with 0. 1 μM benzo(a)pyrene for 1–4 hr. Accumulation is expressed as a *C/M* ratio: The ratio of marker concentration in the cytoplasm to its concentration in the incubation medium (seawater). Each bar shows the mean $\pm$95% confidence limit for 20 animals.

These mucopolysaccharide structures represent the glycocalyx of plasma membranes and secreted mucopolysaccharides that surround the reticulopodial net of epiphytic foraminifera (Bresler and Yanko, 1995*a*). It is known that many, probably all, foraminiferal species synthesize and secrete mucopolysaccharides (glycosaminoglycans), which presumably play a significant role in their ecology, particularly in their nutritional behavior (Langer, 1992; Langer and Gehring, 1993). Our data showed that secreted mucopolysaccharides also participate in antixenobiotic defense; i.e., firmly bind cationic xenobiotics, which are then eliminated together with the mucus.

2.2.2. Transport Systems for Xenobiotic Elimination

Because the plasma membrane is permeable to hydrophobic xenobiotics, the cells might protect intracellular metabolism and structures against the compounds listed above. It was shown that cells and multicellular organisms have carrier-mediated membrane pumps, which recognize and eliminate various xenobiotics (Bresler *et al.*, 1975, 1979, 1989, 1990, 1998, in press; Di Virgilio *et al.*, 1988, 1990; Higgins, 1992; Kurelec, 1992 , 1995, 1997; Gottesman and Pastan, 1993; Deguchi *et al.*, 1997). These ATP-dependent pumps include at least two groups of carriers with different substrate and inhibitor specificity: so-called multidrug (or multixenobiotic) resistance transporters (MDRtr/MXRtr) and a system of active transport of organic anions (SATOA). MDRtr/MXRtr were first discovered in the plasma membranes of cancer cells that were resistant to a variety of anticancer drugs (Yanovich and Taub, 1983;

Sirotnak *et al.*, 1986; Gottesman and Pastan, 1993). Similar carriers, P-glycoproteins from the ATP-binding cassette (ABC) transporters superfamily, were later detected in membranes of different uni- and multicellular organisms from bacteria and protozoa to humans (Frommel and Balber, 1987; Foote *et al.*, 1989; Neyfakh *et al.*, 1991; Higgins, 1992; Kurelec, 1992, 1995, 1997; Gottesman and Pastan, 1993; Miller *et al.*, 1997; Jakob *et al.*, 1998; Bresler *et al.*, 1998, in press). They have a wide substrate specificity and eliminate a large variety of hydrophobic and amphiphilic xenobiotics, hydrophobic and amphiphilic organic cations, and hydrophobic cyclic and linear peptides. All these substrates are inhibited by verapamil and also competitively inhibit each other (Higgins, 1992; Kurelec, 1992 , 1995, 1997; Gottesman and Pastan, 1993; Tiberghien and Looz, 1996; Jakob *et al.*, 1998; Bresler *et al.*, 1998, in press). Several MXRtr substrates have inherent fluorescence, particularly daunomycin, doxorubicin, Rhodamine B, Rhodamine 123, acridine orange, ethidium bromide, and calcein-acetoxymethylester (calcein AM), which is frequently used to make MXRtr-mediated export visible and to study it microfluorometrically (Yanovich and Taub, 1983; Neyfakh *et al.*, 1991; Tiberghien and Looz, 1996; Galgani *et al.*, 1996; Smital and Kurelec, 1997; Miller *et al.*, 1997; Bresler *et al.*, 1998, in press; Jakob *et al.*, 1998).

SATOA carriers, the activities of which were detected in single cells and several specialized tissues of different Metazoa (Bresler *et al.*, 1975, 1979, 1990, 1998, in press; Bresler and Nikiforov, 1981; Di Virgilio *et al.*, 1988, 1990; Bresler and Fishelson, 1994; Paulsma *et al.*, 1996; Poot *et al.*, 1996; Deguchi *et al.*, 1997), eliminate a large variety of hydrophilic organic anionic xenobiotics with different chemical structure as well as hydrophilic anionic products of xenobiotic detoxification and conjugation from cells and tissue fluids. Transport of all these substrates is competitively inhibited by probenecid (typical inhibitor of SATOA) and the substrates also can competitively inhibit each other (Bresler and Nikiforov, 1981; Poot *et al.*, 1996; Keppler *et al.*, 1997; Schreiber *et al.*, 1997; Sweet *et al.*, 1997). Hydrophilic anionic organic dyes, particularly fluorescein, are widely used marker substrates in making SATOA visual and examining its kinetics (Bresler *et al.*, 1975, 1979, 1990; Bresler and Nikiforov, 1981; Bresler and Fishelson, 1994; Villalobos *et al.*, 1996; Schreiber *et al.*, 1997).

We Bresler *et al.*, 1975, 1979, 1985, 1990, 1998, in press; Bresler and Nikiforov, 1981; Bresler, 1989; Bresler and Fishelson, 1994) and Schreiber *et al.* (1997) postulated that SATOA plays an important ecological role in the adaptation of animals to natural xenobiotics in their food, as well as in their resistance to man-made xenobiotics. Kurelec (1992, 1995, 1997) first directed scientists' attention to the ecological role of MXRtr, especially in polluted aquatic environments. However, study of MXRtr and SATOA activity in foraminifera, especially their interaction and ecological role, has begun just recently (Bresler and Yanko, 1995*a*).

It has been shown that fluorescein liberated by esterases from fluorescein diacetate in the cytoplasm of benthic foraminifera was eliminated from the cell (Bernhard *et al.*, 1995; Bresler and Yanko, 1995*a*). After the fluorescein was liberated, it produced a diffuse green fluorescence of the cytoplasm, but for the first 30 min of the runout this diffuse fluorescence decreased, and small brightly fluorescent vacuoles and slightly fluorescent short threads appeared. After 2–4 hr, these vacuoles were transferred to the cytoplasm's periphery, and their number decreased. Thus, carrier(s) of this system are found in the plasma membrane and its derivatives (exo- or endocytic vacuoles). Bresler and Yanko (1995*a*) showed that runout of fluorescein had first-order reaction kinetics; it was inhibited by probenecid, but was insensitive to verapamil (Fig. 3). The fluorescein runout was also inhibited by low temperature or by 2,4-dinitrophenol (DNP), which is an uncoupler of respiration and oxidative ATP formation.

The literature also contains evidence that other benthic foraminifera show SATOA activity in their plasma membranes. Cultured *Allogromia* sp. NF, *A. laticollaris*, and *Elphidium* sp. that were loaded with fluorescein or its derivatives eliminate these substrates of SATOA in a similar manner when incubated in clean water (Bernhard *et al.*, 1995).

Thus, typical kinetics, inhibition by probenecid (a known specific inhibitor of SATOA), and ATP-depleting agents (low temperature, DNP) are evidence that several benthic foraminifera contain this carrier-mediated export pump phenotypically similar to those found in single cultured cells and several specialized tissues of Metazoa (Bresler *et al.*, 1975, 1979, 1990, 1998, in press; Bresler and Nikiforov, 1981; Di Virgilio *et al.*, 1988, 1990; Bresler and Fishelson, 1994; Paulsma *et al.*, 1996; Poot *et al.*, 1996; Deguchi *et al.*, 1997).

As noted above, an amphiphilic cationic marker xenobiotic RhB penetrates into the cytoplasm of living benthic foraminifera *P. spinigera* and *R.*

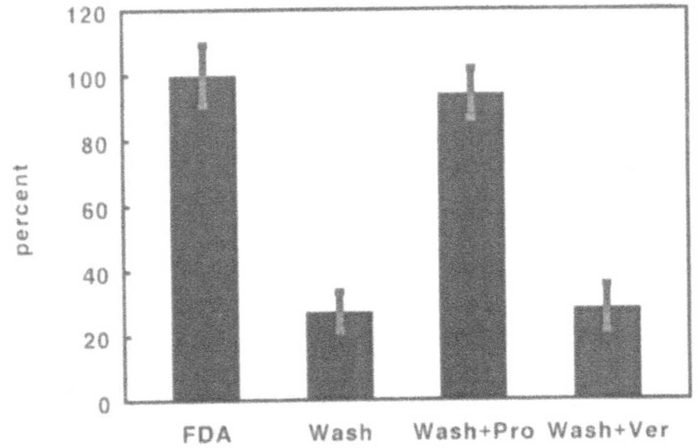

FIGURE 3. Elimination of fluorescein (FLU) by SATOA (system of active transport of organic anions) from the cytoplasm of the youngest chamber of the benthic foraminifer, *Rosalina macropora*: (FDA) Fluorescence of liberated FLU after a 15 min incubation with 1 μM fluorescein diacetate (i.e., loading with fluorescein). (Wash) incubation of loaded foraminifera in clean seawater for 1 hr. (Wash + Pro) incubation of loaded foraminifera in seawater with 50 μM of the SATOA inhibitor probenecid for 1 hr. (Wash + Ver) incubation of loaded foraminifera in seawater with 10 μM of the MXRtr inhibitor verapamil for 1 hr.

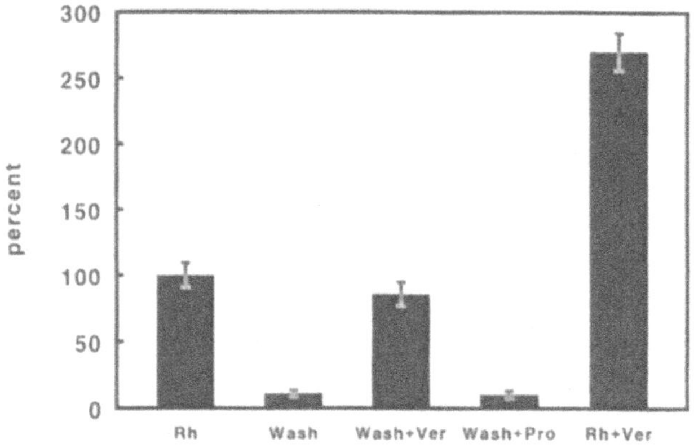

FIGURE 4*a*. Penetration and elimination of Rhodamine B (Rh) from the cytoplasm of the youngest chamber of the benthic foraminifer *Rosalina macropora*: (Rh) incubation with 10 μM Rh for 1 hr. (Wash) incubation of Rh-loaded foraminifera in clean seawater for 1 hr (Wash + Ver) incubation of loaded foraminifera in seawater with 10 μM of the MXRtr inhibitor verapamil for 1 hr. (Wash + Pro) incubation of loaded foraminifera in seawater with 50 μm of the SATOA inhibitor probenecid for 1 hr. (Rh + Ver) incubation of loaded foraminifera in seawater with 1 μM of Rh and 10 μM of the MXRtr inhibitor verapamil for 1 hr.

macropora during incubation. When such RhB-laden foraminifera are transferred into clean seawater, RhB is expelled from the cytoplasm (Fig. 4). The runout is inhibited by verapamil (a typical inhibitor of MXRtr) but is not inhibited by probenecid (a well-known inhibitor of SATOA). Incubation in water containing both RhB and verapamil exhibited a marked increase of the

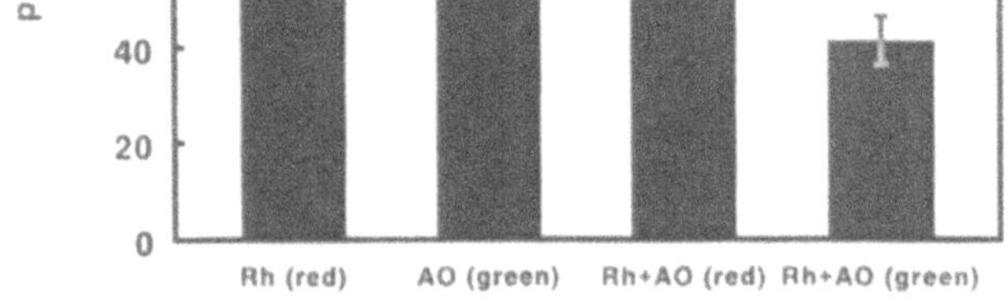

FIGURE 4*b*. Competition between Rhodamine B (Rh) and Acridine Orange (AO) for the runout from the cytoplasm of the youngest chamber of the benthic foraminifer *Rosalina macropora*: (Rh red) red fluorescence of Rh, incubation with 1 μM of Rh for 1 hr. (AO green) green fluorescence of AO, incubation with 1 μM of AO for 1 hr. (Rh + AO red) red fluorescence of Rh, incubation with 1 μM of Rh and 1 μM of AO for 1 hr. (Rh + AO green) green fluorescence of AO, incubation with 1 μM of Rh and 1 μM of AO for 1 hr.

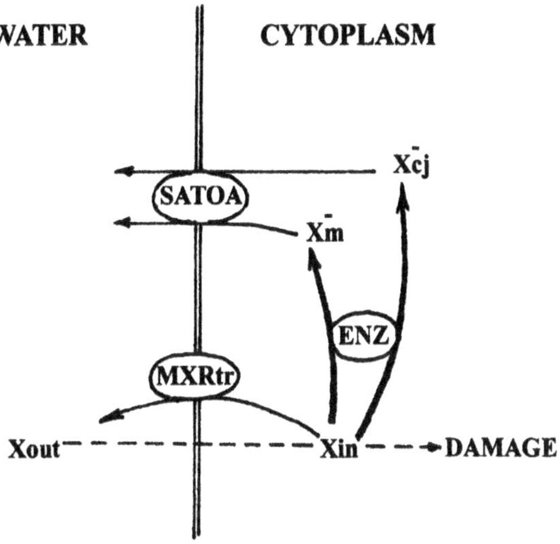

FIGURE 5. Schematic representation of cooperation between MXRtr and SATOA. (X_{out}) Hydrophobic xenobiotic in water. (X_{in}) The same xenobiotic penetrating into cytoplasm. (ENZ) enzymes of xenobiotic detoxification and conjugation. X_m^-) Anionic product of xenobiotic detoxification. (X_{cj}^-) Anionic product of xenobiotic conjugation.

RhB accumulation in the foraminiferal cytoplasm as compared to animals incubated only with RhB. Simultaneous loading of foraminiferal cytoplasm by RhB and AO inhibited the runout rate of both these markers (Fig. 4). The data obtained show the typical activity of an export pump for hydrophobic xenobiotics (i.e., MXTtr) in the plasma membrane of benthic foraminifera.

Thus, attached benthic epiphytic foraminifera display two active main export pumps, MXRtr and SATOA, that eliminate different xenobiotics congruently (Fig. 5). It is well known that the activity of these pumps is one of the main mechanisms of resistance to xenobiotics in different organisms (Foote *et al.*, 1989; Neyfakh *et al.*, 1991; Kurelec, 1992, 1995, 1997; Gottesman and Pastan, 1993). Therefore, these pumps presumably play a crucial role in the adaptation of benthic foraminifera to the complicated chemical conditions in a benthic epiphytic microenvironment described above. It should be interesting to compare the activity of these export pumps in foraminiferal populations and species living in different environmental conditions such as planktonic or deep-sea bottom communities or polluted versus nonpolluted communities. Recently, it was shown that populations of the bivalve mollusk *Donax trunculus* from clean and polluted sites along the Israeli Mediterranean shore have different phenotypic expressions of export pumps. Specimens from the clean sites exhibit the lowest mean MXRtr and SATOA activity, but the highest variability. Specimens from polluted sites show an increased mean activity of these pumps and a marked shift of the frequency polygon of MXRtr and SATOA activities to higher values. The way these alterations are expressed correlates with the pollution history of the collection sites (Bresler *et al.*, 1998, in press). This example demonstrates a key role of export pumps for survival and phenotypic selection in polluted environments. We believe that a similar selection may exist in foraminiferal populations and communities.

2.2.3. Storage of Xenobiotics in Intracellular Compartments

It is well known that acidic cell compartments, particularly lysosomes and related structures, accumulate amphiphilic cationic xenobiotics with high pK_b such as acridine orange (AO) or neutral red (NR); weak basic amines such as chloroquine; a number of tertiary amines and some peptides and heavy metals, the last probably as a complex with peptides or proteins (MacIntyre and Cutler, 1988; Stenseth and Thyberg, 1989; Bresler and Fishelson, 1994; Moriyama *et al.*, 1994; Bird and Lloyd, 1995; Millot *et al.*, 1997). Thus, several of these lysosomotropic compounds have been used for visualization of lysosomes and related organelles in living cells (Haugland, 1996). The accumulation of such compounds is an active (ATP-dependent) process that is mediated by a pH gradient between the lysosomes and the cytoplasm. Any cell injury decreases the intralysosomal accumulation of these compounds, particularly AO, NR, and chloroquine, which is widely used to assess lysosomal and cell viability and the toxic effects of pollutants (Bresler and Fishelson, 1994; Bresler and Yanko, 1995*a*,*b*; Haugland, 1996). It is commonly accepted that accumulation is determined only by dissociation of weak bases at low pH. However, lysosomes of multidrug-resistant cancer cells accumulate more AO than lysosomes of sensitive cells, and this difference is eliminated by verapamil and ammonium chloride (Millot *et al.*, 1997).

Using the marker xenobiotics AO or NR and vital microfluorometry or microspectrophotometry, it was shown that the epiphytic benthic foraminifera, *P. spinigera* and *R. macropora* actively accumulate these compounds into lysosomes and related organelles (Bresler and Yanko, 1995*a*,*b*). As an example, the ratio of intracellular NR to its concentration in an incubation medium (*C/M* ratio) was 438 ± 27 in intact attached specimens of *R. macropora*, and 263 ± 17 in "detached" specimens. Incubation with cadmium cations (0.5 μmol) decreased the *C/M* ratio to 228 ± 19 in attached specimens, and to 86 ± 14 in "detached" specimens. Quantitative NR tests allowed the assessment of the toxic action of different concentrations of heavy metals in water, with or without dissolved seaweed-derived organic material (Bresler and Yanko, 1995*b*). After surviving in clean seawater for 3–5 days, lysosomes loaded with AO or NR are transported to the reticulopodia and partially eliminated from the cytoplasm.

Thus, the intralysosomal accumulation of weak basic compounds is an important antixenobiotic defense mechanism that protects other cell structures and is involved in the elimination of xenobiotics; however, the mechanisms of elimination demand further special examination because the organelle and vesicle transport system is closely connected to cytoskeletal functions

(Travis *et al.*, 1983; Travis and Bowser, 1991), and plasma membrane deformations are mediated by actively bending microtubules (Travis and Bowser, 1990).

2.2.4. Intracellular Xenobiotic-Binding Proteins and Peptides

Both special and common xenobiotic-binding proteins have been described previously. The nature and functions of special proteins such as cyclophilin, a specific cytosolic binding protein for natural xenobiotic cyclosporin A, FK506-binding protein (FK596 or tacrolimus is a commercial name for *Streptomyces*-derived xenobiotic), or dioxin-binding proteins (Handschumacher *et al.*, 1984; Pancer *et al.*, 1993; Brown *et al.*, 1996) are unknown. They may be receptors or enzymes participating in xenobiotic metabolism. For example, ligandin, an organic anion-binding protein of the renal proximal tubules, proved to be a known enzyme, glutathione *S*-transferase (Bresler and Nikiforov, 1981); and a specific cytoplasmic receptor protein that recognizes aryl hydrocarbons and dioxins (AH-receptor) is a common mechanism for cytochrome P448 monooxygenase substrate induction in marine and terrestrial animals (Nebert and Jansen, 1979; Hahn *et al.*, 1992).

Another common group of cytoplasmic xenobiotic-binding proteins is represented by metal-binding proteins and peptides (Hamer, 1986; Kagi and Schaffer, 1988; Piccinni, 1989; Piccinni *et al.*, 1992; Bresler and Yanko, 1995*a*,*b*; Roesijardi, 1996). Metallothioneins (MTs) are inducible metal-binding proteins that control the homeostasis of essential metals, particularly copper and zinc, and play an important role in the detoxification and elimination of heavy metal ions such as cadmium and mercury. MTs can also act as scavengers of free radicals and reactive oxygen metabolites.

Chelatins, another type of metal-binding protein, also regulate the homeostasis of copper and detoxify heavy metals. Chelatins contain aromatic amino acid residues (Piccinni, 1989; Piccinni *et al.*, 1992; Polek *et al.*, 1993; Bresler and Yanko, 1995*a*,*b*; Roesijardi, 1996). MTs and chelatins have both been detected in different species, from protozoa to humans (Hamer, 1986; Kagi and Schaffer, 1988; Weber *et al.*, 1988; Piccinni, 1989; Piccinni *et al.*, 1992; Polek *et al.*, 1993; Bresler and Yanko, 1995*a*,*b*; Roesijardi, 1996).

Piccinni *et al.* (1992) investigated Cd-binding proteins in different ciliate species. Cultures were treated with different doses of Cd according to tolerance, and Cd-binding proteins were determined in the soluble fraction isolated from these cultures. Both typical Mts and chelatins were detected in the fraction. In the soluble fraction from the highly tolerant ciliate *Oxytricha granulifera*, a unique glycoprotein with high affinity for Cd was also isolated. This glycoprotein contains aromatic amino acids and low amounts of cysteine; i.e., it was different from MTs.

Microfluorometrical examination of the inherent ultraviolet fluorescence of proteins containing tryptophan, and the titration of SH-groups of MT-like proteins with fluorescein mercury acetate shows that the cytoplasm of benthic epiphytic foraminifera is characterized by an activity of both tryptophan-containing chelatins and MT-like proteins, which bind the cations of Cu^{2+}, Cd^{2+} or Hg^{2+} (Bresler and Yanko, 1995*b*). The metal-binding properties of chelatins and MT-like proteins are very similar to those in other protozoa (Weber *et al.*, 1988; Piccinni, 1989; Piccinni *et al.*, 1992). However, the further fate of heavy-metal MTs and heavy-metal chelatin complexes in the foraminiferal cytoplasm is practically unknown and must be further studied.

2.2.5. Extracellular Xenobiotic-Binding Proteins and Peptides

Most known and most general extracellular xenobiotic-binding proteins are secreted mucopolysaccharides. It was shown that the concentration of cadmium, copper, and mercury ions in seawater that produced a 50% effect within 24 h (24-h EC50) in the benthic foraminifer *P. spinigera* was twice that for attached specimens with a normal mucopolysaccharide coat than for detached specimens with a reduced secreted coat (Bresler and Yanko, 1995*b*). Preliminary experiments with fluorescent metal ions of thallium show directly that the mucopolysaccharide mat around attached foraminifera binds more of these toxic metal ions than the reduced mat of detached foraminifera. Thus, the secreted mucopolysaccharides can protect benthic foraminifera from organic and inorganic xenobiotics. An important role of mucopolysaccharides in binding and defense against inorganic and organic cationic xenobiotics is also described in aquatic bacteria, various invertebrate and vertebrate species, particularly *Zooglea ramigera*, worms, mollusks, and fish (Segner *et al.*, 1988; Arillo and Melodia, 1990; Triebskorn *et al.*, 1991; Bresler and Fishelson, 1994).

2.2.6. Detoxifying Enzymes

Detoxifying enzymes may be considered as a common mechanism of antixenobiotic defense and a main step in the metabolism of foreign compounds (Parke, 1971). Detoxifying enzymes, particularly those of the P450 gene superfamily, include several enzymes that catalyze polysubstrate monooxygenase reactions of a wide variety of xenobiotic and endogenous substrates. Polysubstrate monooxygenase (PSMO) activity is very old in evolutionary terms (Guengerich, 1990; Hahn *et al.*, 1992), and acts as a tertiary defense against hydrophobic organic xenobiotics. The PSMO system usually converts these xenobiotics into their equivalent alcohols by direct oxidation of a C–H bond, which produces more hydrophilic metabolites that can be conjugated and excreted by SATOA-like pumps (Nebert and Jansen, 1979; Hollebone *et al.*, 1995; Montellano, 1995). However, some man-made xenobiotics and the PSMO system interact less specifically, and the products can form stable covalent bonds to proteins, nucleic acids, and lipids, or the oxidation process is readily short-circuited to yield reactive lipoperoxides that damage some vital enzyme activities and physiological functions. Thus, the PSMO system cannot only protect cells against hydrophobic xenobiotics, but can also transform them into more dangerous toxic, mutagenic, and carcinogenic compounds (Nebert and Jansen, 1979; Hollebone *et al.*, 1995; Montellano, 1995).

The literature contains fragmentary data about the activity of detoxifying enzymes in protozoa but unfortunately foraminifera have never been examined (Khan *et al.*, 1972; Yawetz and Agosin, 1979, 1980, 1981; Murphy *et al.*, 1982; Berk and Roberts, 1998; Gilron and Lynn, 1998; Pratt *et al.*, 1998). Khan *et al.* (1972) examined the metabolism of the insecticide aldrin in aquatic ecosystems and showed that "mixed protozoa" transform aldrin to dieldrin. The protozoan *Tetrahymena thermophila* metabolized the fungicide pentachloronitrobenzene and showed an activity of thiol-5-methyltransferase and glutathione-*S*-transferase (Murphy *et al.*, 1982). The pathogenic protozoan *Trypanosoma cruzi* exhibited the cytochrome P450 system, whose activity was inducible by phenobarbital (Agosin *et al.*, 1976). It should be noted that both phenobarbital and the carcinogenic polycyclic compound methylcholantrene are known as typical inducers of the cytochrome P450 system in mammalian liver (Nebert and Jansen 1979; Guengerich, 1990; Hollebone *et al.*, 1995). It was shown that the microsomal fraction from *T. cruzi* exposed epoxide hydrase activity that is enhanced by about threefold by phenobarbital (Yawetz and Agosin, 1979). Activity of glutathione-*S*-transferase was also detected in *T. cruzi* and this key conjugating enzyme was isolated and purified (Yawetz and Agosin, 1980, 1981). Yawetz and Agosin (1981) concluded that epoxide hydrase transformed drugs to their epoxides, which were conjugated with glutathione by glutathione-*S*-transferase, and these metabolic reactions produced tolerance of *T. cruzi* to antiparasitic drugs.

Recently, we studied the activity of cytochrome P450 monooxygenase (ethoxyresorufin *O*-deethylase, EROD) in the cytoplasm of "detached" living benthic epiphytic foraminifera *P. spinigera* and *R. macropora* using a fluorogenic substrate of this enzyme, ethoxyresorifin, and microfluorimetry. Foraminifera were preincubated in seawater containing the polycyclic aromatic hydrocarbon (PAH), methylcholantrene. Figure 6 shows that both species demonstrate marked EROD activity and this activity increased after preincubation with methylcholantrene. Thus, benthic epiphytic foraminifera have inducible EROD activity in their cytoplasm. Our preliminary data also demonstrate that foraminifera have aryl hydrocarbon hydroxylase activity (not shown).

Other enzymes that potentially participate in the metabolism of xenobiotics, particularly enzymes of conjugation interacting with glutathione, glucuronic acid, or sulfate, require further investigation.

2.2.7. Enzymes Protecting Against an Excess of Oxygen and Peroxides

Peroxidases are ancient, general defense enzymes that protect cells against excess oxygen and excess peroxides, turning peroxide and oxygen into oxi-

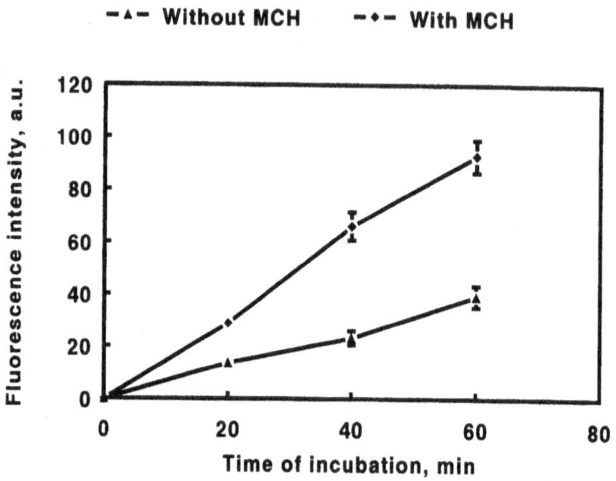

FIGURE 6. Dynamics of EROD activity in the cytoplasm of the youngest chamber of the detached benthic foraminifer *Rosalina macropora* preincubated in seawater with or without 0.1 μM of methylcholantrene.

dized compounds (Dixon and Webb, 1979). Peroxidase activity was first detected in different uni- and multicellular eukaryotic organisms (Pearce, 1968; Dixon and Webb, 1979). Peroxidase activity in the "detached" epiphytic foraminifera *P. spinigera* and *R. macropora*, was examined cytochemically (Bresler and Yanko, 1995*a*). The product of this reaction was detected in the cytoplasm of both species as numerous small blue granules, so-called peroxisomes. The reaction was inhibited by sodium azide, a known inhibitor of peroxidase (Pearce, 1968; Dixon and Webb, 1979). *P. spinigera* demonstrated a more intensive peroxidase reaction than *R. macrospora*. Peroxidase activity and peroxisomes were also detected in planktonic foraminifera (Anderson and Tuntivate-Choy, 1984; Anderson, 1988). Thus, peroxidases exist in various foraminiferal species and their activity can probably protect these species against excesses of oxygen and peroxides in the microenvironment. As noted above, a periodical excess of oxygen and peroxides is typical for epiphytic benthic microenvironments.

2.2.8. Enzymes Protecting against Bromine and Iodine

As discussed above, marine bacteria, algae, and invertebrates produce a large variety of iodinated and brominated organic xenobiotics. These organisms also demonstrate a high activity of haloperoxidases, a group of ancient and widely distributed cytoplasmic enzymes (Theiler *et al.*, 1978; Ahern *et al.*, 1980; Fenical, 1982; Van Pee and Lingens, 1985). Haloperoxidases oxidize halogens into powerful oxidants and free hypohalous acids with chemical formulas HOBr or HOI, which easily react with thiols, thioethers, amines, and other reactive molecules, transforming them into brominated or iodinated xenobiotics. These xenobiotics are used as a chemical defense against parasites and predators. Even human cells use this antiparasitic chemical weapon (Weiss *et al.*, 1986).

A marked bromoperoxidase activity was detected in the cytoplasm of living "detached" epiphytic foraminifera of the type *P. spinigera* and *R. macropora* (Bresler and Yanko, 1995*a*). The haloperoxidase activity suggests that these foraminifera can synthesize brominated xenobiotics, probably as a weapon against pests and predators. However, the primary function of haloperoxidases may be the protection of marine organisms against biologically active bromine and iodine in seawater, which easily penetrate across plasma membranes.

2.2.9. Interaction and Congruence of the Defense Mechanisms for Complete Protection

The analysis of the interactions and functional congruence of all general antixenobiotic defense mechanisms is important in assessing their ecological and ecotoxicological role (Bresler and Nikiforov, 1981; Bresler *et al.*, 1990, 1998, in press; Bresler and Fishelson, 1994; Bresler and Yanko, 1995*a*). Only the integrated activity of all general antixenobiotic mechanisms determines the sensitivity of uni- or multicellular organisms to environmental xenobiotics and their adaptive abilities. Thus, the cooperation and congruence might be a result of natural selection and a long coevolution. The cooperation and congruence of the antixenobiotic defense mechanisms is especially important for unicellular organisms such as foraminifera, because all the work and metabolic costs fall on one single cell.

MXR transporter and the lysosomes cooperatively eliminate the penetrated lipophilic or amphiphilic cationic xenobiotics, and detoxifying and conjugating enzymes transform them into more hydrophilic anionic metabolites, which are then eliminated by a second export pump, SATOA (Fig. 7). Cooperative work among these mechanisms decreases the cytoplasmic concentration of permeable xenobiotics and their metabolites to a safe level. If the permeation is high enough to prevent cooperative work from securing a safe level, xenobiotics can interact with cytoplasmic receptors, such as the aromatic hydrocarbon (Ah) receptor (Nebert and Jansen, 1979; Hahn *et al.*, 1992; Montellano, 1995), and induce a synthesis (induction) of detoxifying/conjugating enzymes (Fig. 7). It is well known that cells with optimal effectivity of the cytochromes P450, monooxygenases, Ah receptor, and conjugating enzymes are tolerant of polycyclic aromatic hydrocarbons (Nebert and Jansen, 1979; Hahn *et al.*, 1992). Therefore, it is necessary to further characterize the integrated activity of different antixenobiotic defense mechanisms in different foraminiferal specimens and species.

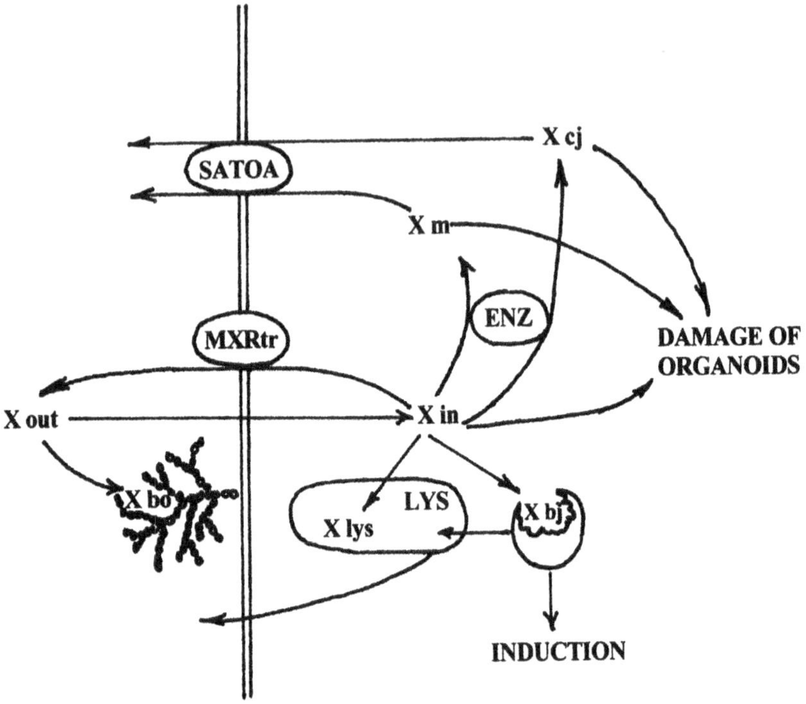

FIGURE 7. Congruent work of antixenobiotic defense mechanisms. (X_{out}) Hydrophobic xenobiotic in the water. (X_{bo}) The same xenobiotic bound by mucopolysaccharide coat. (X_{in}) The same xenobiotic penetrating into cytoplasm. (ENZ) enzymes of xenobiotic detoxification and conjugation. (X_m^-) Anionic product of xenobiotic detoxification. (X_{cj}^-) Anionic product of xenobiotic conjugation. (LYS) lysosome. (X_{lys}) Xenobiotic stored into lysosomes. (BP) xenobiotic-binding proteins. (X_{bi}) Xenobiotic bound by xenobiotic-binding proteins.

3. Antixenobiotic Defense Mechanisms and Environmental Health of Foraminifera

Any population of a species not only contains normal healthy specimens, but also some proportion of abnormal, sick, and even dying specimens, particularly specimens with inherent developmental abnormalities, genetic disorders, infectious or parasitic diseases, and diseases produced by chemical or physical environmental factors (Cotran *et al.*, 1989). The character and frequency of such pathology in any population characterizes the epidemiological situation; i.e., the interactions of the population and the biotic and abiotic environmental factors. Some authors find pathological and epidemiological analyses to be more important for ecology and ecotoxicology of aquatic animals (Meyers and Hendricks, 1984; Hinton *et al.*, 1992; Wester and Vos, 1994). This approach has been followed for many decades in classical human and mammalian toxicology, where pathology plays a central role (Cotran *et al.*,

1989). However, in the field of ecotoxicology of aquatic animals, especially invertebrates, environmental pathology and epidemiology are clearly underdeveloped as compared to human and mammalian toxicology. In the field of foraminiferal ecotoxicology, pathology of the foraminiferal shell (test) has been better studied than the pathology of cytoplasm, its organoids and nucleus (Boltovskoy *et al.*, 1991; Stubbles, 1993; Yanko *et al.*, 1994, 1998; Geslin *et al.*, 1998; see also in this volume Geslin *et al.*, Ch. 9; and Hallock, Ch. 5). Foraminiferal tests have been used to determine the former population structure and abundance of abnormal shells, and even their chemical composition, all of which allows assessment of the quality of the paleoenvironment (Boyle, 1981, 1986; Lea and Boyle, 1989, 1991; Murray, 1991; Stubbles, 1993; Langer and Gehring, 1993). However, test formation is controlled by cytoplasmic and nuclear mechanisms; therefore, shell pathology is secondary and is generated by alterations in the foraminiferal cytoplasm and nucleus.

3.1. Examination and Assessment of Foraminiferal Health

The assessment of animal health, pathology, and epidemiology in different macro- and microenvironments is the only direct means of detecting real ecological risks and alterations in the environmental quality (Meyers and Hendricks, 1984; Cotran *et al.*, 1989; Hinton *et al.*,1992; Wester and Vos, 1994). To determine the health of humans or animals, various physiological, biochemical, and especially morphological methods are used. The main principle of theoretical and practical (diagnostic) pathology asserts that any disease of an organism is a disease of its cells and can thus be detected by a microscopic examination of those cells (Cotran *et al.*, 1989). Therefore, foraminiferal pathology can be recognized via microscopic examination of cellular structures.

3.1.1. Methods for Studying Foraminiferal Health

The first problem when examining sediment samples is to distinguish live and dead specimens. Sampling and fixation produce a loss and a retraction of the reticulopodia. Conventional staining of sediment samples with Rose Bengal or Sudan Black B is widely used to distinguish live and dead foraminifera (Walker *et al.*, 1974). Proteins of the fixed cytoplasm are stained by Rose Bengal and lipids are stained by Sudan Black, whereas empty shells of dead foraminifera do not interact with these dyes. However, the shells of dead benthic specimens can absorb mucus and other organic materials that are stained by these dyes. Therefore, special vital fluorescent probes were proposed for precise distinguishing of living (viable), damaged, and dead foraminifera (Bernhard *et al.*, 1995, 1997; Bresler and Yanko, 1995*a*,*b*; Bernhard and Bowser, 1996). Viable cells are impermeable to anionic marker xenobiotics, eliminate lipophilic and amphiphilic substrates of the MXR transporter from their cytoplasm, and accumulate cationic xenobiotics in their lysosomes. Cytoplasmic enzymes, particularly esterases, rapidly hydrolyze the corresponding substrates in living cells. These specific functions or biochemical processes can be visualized in living foraminiferal specimens by using special fluorescent markers, probes, and fluorogenic enzyme substrates. This makes it possible to detect and quantitatively assess the inhibition of these processes; i.e., cell damage or death (Bresler and Yanko, 1995*a*,*b*).

Some new fluorescent probes are aldehyde fixable. For example, during the incubation of fresh samples of marine sediment with cell-permeate CellTracker™ Green CMFDA (Molecular Probes, Eugene, Oregon), cytoplasmic esterases are liberated from this fluorogenic substrate fluorescent dye, which is covalently bonded by proteins and well preserved after formaldehyde fixation. Fixed probes may even be embedded, sectioned, and examined for fluorescently labeled organisms (Bernhard and Bowser, 1996; Bernhard *et al.*, 1997). In such preparations the brightly fluorescent cytoplasm of living benthic foraminifera and other benthic organisms stands out against a black background.

Viable cells also have a typical morphology of cytoplasmic structures and organelles, including the nucleus. Conventional and epifluorescent microscopic examinations also allow detection of various morphological alterations in damaged cells, and normal, damaged, and dead foraminifera.

The health of any organism can be characterized by a set of parameters that describe its main metabolic processes, functions, and structures, particularly the activity of various antixenobiotic defense mechanisms; cell respiration, the metabolic state, and electric potential of mitochondria; total DNA, RNA, proteins, and lipid content; the state of the plasma membrane and membrane export pumps; the state of the lysosomes and related structures; enzyme activity, cell cycle, and the functional state of nuclear chromatin; the frequency of DNA damage and chromosome breaks; and the functional morphology of the cell. The determination of these parameters is done with specific fluorescent probes, markers, and fluorogenic substrates as well as light and electron microscopy. This approach was employed to unmask early reversible alterations produced in foraminiferal cytoplasm by a hypertonic seawater stimulus (Rupp *et al.*, 1986; Travis and Bowser, 1991), colchicine or cytochalasin D (Travis and Bowser, 1986, 1991), or by 2,4-dinitrophenol, copper, cadmium, and mercury ions (Bresler and Yanko, 1995*b*). This approach is especially effective for relatively large unicellular organisms such as foraminifera.

3.1.2. Pathology of the Cell

Pathological alterations in all living organisms are mainly caused by various external (environmental) chemical or physical factors, and only to a small extent by endogenous factors, mainly genetic disorders (Cotran *et al.*, 1989). Environmental chemicals may act nonspecifically on numerous cell structures and functions (like an elephant in a china shop) or only on specific cellular organoids and functions (like an arrow hitting a specific target). As an example, such natural xenobiotics as lectins, conotoxins, or tetrodotoxin act on specific plasma membrane structures and functions; rotenone acts on the mitochondrial respiratory chain; and colchicine, *Vinca* alkaloids, phalloidin, and cytochalasins all act on cytoskeletal structures and functions. The specific targets of such natural xenobiotics are so precise that these xenobiotics are widely used in cytology and molecular biology to identify and examine corresponding structures and function (Alberts *et al.*, 1983; Haugland, 1996; Alberts, 1998).

The plasma membrane, mitochondria, cytoskeleton, and lysosomes play critical roles in early responses to environmental actions, reversible and irreversible cell injury, and cell death (Cotran *et al.*, 1989). First, environmental chemicals contact the plasma membrane and its components, particularly the receptors, transporters, or lipids. Therefore, the early responses to these actions are mediated by peculiarities of the plasma membrane. Any damage to plasma membrane integrity, composition, and functions induces a runout of important cell constituents, the invasion of undesirable environmental compounds, and a disturbance of cellular homeostasis, all of which damage the cell organoids, particularly the mitochondria, lysosomes, and the cytoskeleton (Fig. 8).

Mitochondria produce energy (ATP) for all cellular needs and regulate the carbohydrate, lipid, and amino acid metabolism. Therefore, any injury generates an alteration of all energy-dependent cell structures and metabolic reactions and leads to acidification of the cytoplasm and the formation of peroxides. Damage and increased permeability of the lysosomal membrane can activate numerous lysosomal enzymes and acid hydrolyses, which cleave corresponding cell constituents, particularly RNA and DNA, thereby causing the cell to die (Fig. 8).

The structure, molecular composition, and functions of the foraminiferal cytoskeleton have been studied by many methods, particularly vital light and electron microscopy, fluorescent immunocytochemistry, and the fluorescent actin-specific probe rhodamine-phalloidin (Travis and Bowser, 1986, 1991; Bowser *et al.*, 1988). It was shown that the function of the foraminiferal cytoskeleton includes at a minimum: (1) filopod extension/retraction and bending; (2) network withdrawal; (3) intracellular transport of organelles and surface transport; and (4) reticulopodial tension and whole-cell locomotion (Travis *et al.*, 1983; Travis and Bowser, 1991; Welnhofer and Travis, 1996).

Reorganization of the cytoskeleton is one of the earliest cellular responses to a variety of environmental signals. Multiple signal transduction pathways converge to induce reorganization in order to mediate motility, shape change, and attachment to substrates (Alberts *et al.*, 1983; Bretscher, 1991; Hoyt *et al.*, 1997; Ma *et al.*, 1998). For example, it has been shown by video microscopy that 10–15 sec after introducing a hypertonic seawater stimulus, the inward flow of cytoplasm during foraminiferal reticulopods occurred only in association with cytoplasmic fibrils (Rupp *et al.*, 1986; Travis and Bowser, 1991).

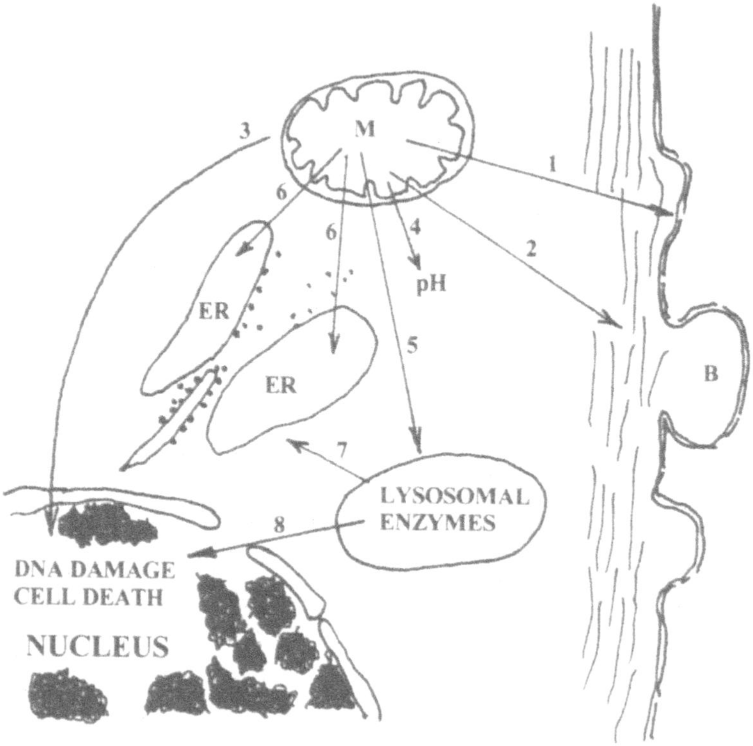

FIGURE 8. Role of plasma membrane, mitochondria, lysosomes, and cytoskeleton in reversible and irreversible cell pathology. (M) mitochondria. (ER) Endoplasmic reticulum. Damaged and swollen mitochondria induce: (1) oxidative damage of plasma membrane, (2) reorganization of cytoskeleton, retraction of reticulopodia and blebs (B) formation, (3) reorganization of nuclear envelope and alteration in chromatin structure, (4) decrease of pH, i.e., acidification of cytoplasm, (5) oxidative damage of lysosomal membrane, (6) changes of endoplasmic reticulum, (7) liberation and activation of lysosomal RNAses, and (8) liberation and activation of lysosomal DNAses.

Reticulopodal motility and structure in *Allogromia* sp. can be altered by several natural xenobiotics, particularly colchicine binding reversibly to tubulin molecules, which prevents its polymerization, and cytochalasin D, which binds reversibly to actin monomers (Travis and Bowser, 1986, 1991). The response to colchicine is initially marked by the accumulation of cytoplasm at branch points in the network; i.e., varicosity formation. Reticulopodial movement, intracellular organelle transport, and cell surface motility cease within 5 min. After the colchicine is removed, the normal pseudopodial network is restored. During incubation with cytochalasin D, the reticulopodial cytoplasm coalesces to form irregularly shaped bodies. Upon removal of the xenobiotic, cytoplasm within these bodies gradually streams into the remaining network and normal activity is restored (Travis and Bowser, 1986, 1991).

As noted above, the benthic epiphytic microenvironment contains numerous natural xenobiotics. It is possible that some of them, as well as man-made xenobiotics, can also interact with components of the cytoskeleton and alter its structure and function. Further examination of the responses of the foraminiferal cytoskeleton to these environmental xenobiotics is necessary in order to assess their role in foraminiferal pathology.

Functional, cytochemical, and morphological alterations of the plasma membrane, mitochondria, and lysosomes were detected in the benthic epiphytic foraminifera *P. spinigera* and *R. macropora* that were impacted by mechanical damage (detachment); a known permeable uncoupler, 2,4-dinitrophenol (DNP); or metal ions of copper, cadmium, and mercury (Bresler and Yanko, 1995*a*,*b*). Primary responses to DNP were detected in mitochondria: after 15 min, they changed the redox state of NADH, decreased membrane potential (Rhodamine 123 accumulation), and exhibited a marked swelling. Upon removal of DNP (washing), the redox state, potential, and morphology of the mitochondria were restored after 60 min. Cadmium produced similar but irreversible early mitochondrial response. Mechanically damaged (detached) foraminifera showed early plasma membrane responses: an enhanced permeability to the anionic marker fluorescein, a decreased pinocytosis and phagocytosis, and even the formation of blebs (small blisters) in some specimens. Similar but later alterations of the plasma membrane were detected in a foraminifer incubated for 24–72 hr in seawater containing mercury and AO or NR as a marker. Prolonged contact (24 hr) with copper, cadmium, or mercury brought about a marked decrease of both AO and NR lysosomal accumulation, a marked enlargement of some lysosomes, and the formation of vacuoles containing pyknotic mitochondria (i.e., autophagosomes). Some foraminifera also demonstrated a marked swelling and vacuolization of the cytoplasm and alterations in their nuclear structure, particularly karyopyknosis and karyorrhexis. All detected alterations are typical signs of cell injury, degeneration, and necrobiosis (Cotran *et al.*, 1989). Thus, complex cytophysiological, cytochemical, and morphological examinations may be a reliable way to detect reversible and irreversible pathological alterations in benthic foraminifera and objectively assess their health. Further examination of foraminiferal health using this methodology under experimental and field conditions can create common criteria for environmental health assessment and early warning monitoring.

3.1.3. Pathology of the Tests

The molecular machinery of test synthesis and the processing and transport of shell matrix proteins are unknown. A generally accepted theory postulates that shell wall formation and mineralization include the synthesis of: (a) specific proteins(s) that form templates for crystal nucleation, and (b) polysaccharides, which provide the crystal with form and shape (Towe and Cifelli, 1967; Hemleben *et al.*, 1986; Ter Kuile, 1991; Stathoplos and Tuross, 1994; Debenay *et al.*, 1996). Thus, calcareous foraminiferal test formation might include at least three different processes: (1) synthesis, processing, and transport of proteins that form the shell matrix; (2) their interaction with calcium and the formation of shell wall structure and texture; and (3) the control of number, form, and disposition of shell chambers. Cytochemistry, electron microscopy, and chemical dissection (extraction) show that the shells of planktonic and benthic foraminifera contain several mineral-associated proteins (Weiner and Erez, 1984; Robbins and Brew, 1990; Robbins and Healy-Wiliams, 1991; Stathoplos and Tuross, 1994). A minimum of 10–15 discrete proteinlike products from shell extracts of the planktonic foraminifer *Globorotalia menardii* were observed using gel electrophoresis.

The interactions of matrix proteins with calcium have not been studied sufficiently. Marine invertebrates, particularly foraminifera, use mainly calcium in the formation of their skeletal structures, although seawater contains only about 10 mM calcium, but 58 mM magnesium (Prosser, 1973). Therefore, they might use special transport pumps such as Ca^{2+}-ATPase in the endoplasmic reticulum to accumulate calcium in special compartments. Such compartments, similar to endoplasmic reticulum-like vesicles that form calcite crystals, were detected in imperforate larger foraminifera (Ter Kuile, 1991). The selectivity of Ca^{2+} transport and matrix protein interactions indirectly demonstrates the Mg/Ca ratio values in the shell wall and the substitution of Ca^{2+} by some other metal ions. The imperforate porcellaneous families are entirely high-magnesium calcite. The imperforate hyaline foraminifera can be aragonite, high-magnesium calcite, calcite, or a mixture of magnesium calcite and calcite. Thus, presumably the Mg/Ca ratio is controlled by species-specific (i.e., genetic) factors. These factors can include either the selectivity of ion transport to the shell matrix or the ionic selectivity of shell proteins. Any alteration of the shell Mg/Ca ratio within a given species is determined by the binding capacities of matrix proteins.

Many researchers have described the accumulation of trace metals, particularly cadmium, barium, and manganese, in foraminiferal tests (Delaney *et al.*, 1985; Boyle, 1981; Lea and Boyle, 1989, 1991; Langer and Gehring, 1993). Electron paramagnetic resonance (EPR) spectroscopy, in conjunction with thermal treatment, shows that these trace metals are associated with the organic shell material (Langer and Gehring, 1993). This indicates that trace metals penetrate into calcium-rich structures and interact with shell matrix proteins, but the transport mechanisms are as yet unknown. The accumulation of trace metals may be a result of molecular and ionic mimicry that deceives the transport mechanisms and test proteins. It is known that trace metals such as cadmium, mercury, and lead rarely exist in biological fluids and seawater as free cations but combine with numerous ligands. For example, cadmium forms the metal–anion complexes $[Cd(OH)(HCO_3)_2]^-$ and $[Cd(OH)(HCO_3)Cl]^-$, which mimic bicarbonate or chloride, respectively, and are transported into cells by an anion exchanger (Clarkson, 1993). Cadmium and barium ions are similar to calcium ions, and manganese ions are similar to magnesium ions, which might deceive several specific biochemical mechanisms. Cadmium transport by voltage-dependent calcium channels has been reported in some cells (Clarkson, 1993). Trace manganese ions can displace essential magnesium ions in some enzymatic reactions (Dixon and Webb, 1979). Increased temperature, salinity, environmental pollution, or other stress factors enhance the probability of such mistakes, and as a result can alter shell composition. These alterations in the normal shell composition by trace metals are similar to those of vertebrate calcification, exhibiting the competitive mechanism of such accumulations (Sauer *et al.*, 1989). For this reason, it will be interesting to investigate calcium and magnesium selectivity during transport to developing shell walls and their interaction with matrix proteins by using fluorescent probes. It has been shown using a permeable Ca^{2+} probe (Indo-1 AM) and confocal microscopy that chondrocytes actively acquire Ca^{2+}, concentrate it in the cell periphery, and exfoliate it as Ca^{2+}-rich matrix vesicles (MV), a primary molecular machinery of bone calcification in vertebrates (Wuthier, 1993). During the second phase of MV-mediated mineralization, the hydroxyapatite crystals rupture the MV membrane and penetrate the bone matrix; their proteins regulate apatite deposition and growth (Wuthier, 1993; Anderson, 1995; Wu *et al.*, 1996; Boskey *et al.*, 1997; Kirsch *et al.*, 1997). The application of this methodology to foraminifera may also be effective.

The number, shape, and disposition of shell chambers in foraminifera is probably controlled by the cytoskeleton. It is known that the cytoskeleton determines the shape of cells and unicellular organisms, the structure of their surface, and the arrangement of daughter cells during the division of the molluskan zygote (Alberts *et al.*, 1983). Moreover, the cytoskeleton can pass some acquired peculiarities of the cell's shape and surface from mother to daughter cells (Alberts *et al.*, 1983). Various deformities or abnormalities in number, shape, and especially disposition of shell chambers were detected and described in fossil and modern benthic foraminifera (Boltovskoy and Wright, 1976; Haynes, 1981; Boltovskoy *et al.*, 1991; Stubbles, 1993; Yanko *et al.*, 1994, 1995, 1998, in press). These investigators noted that deformities might be a result of environmental stress or mechanical damage. The frequency of deformities increased in sites polluted by toxic metals or aromatic hydrocarbons and in sites with low salinity. Deformities are extremely variable and may be different from species to species in both calcareous and agglutinated foraminifera (Alve, 1991; Yanko *et al.*, 1994). A general classification of deformities could be based on the symmetry and arrangement of chambers in given species (Yanko *et al.*, 1994, 1995, 1998, in press; Geslin *et al.*, 1998; see also in this volume Geslin *et al.*, Ch. 9.

The shell wall structure and composition of the chambers with some abnormalities in hyaline foraminifera can also be altered. Yanko and Kronfeld (1992, 1993) found that deformed shells exhibit an increased Mg/Ca ratio. Examination of the protein-to-calcium ratio in the walls of normal and deformed tests of the foraminifer *Ammonia tepida*, using fluorescent probes for total protein (sulfoflavine) and bound calcium (chlorotetracycline), show that wall deformities caused by mechanical damage and subsequent regeneration contain more protein and less calcium than normal walls and wall deformities caused by pathological morphogenesis (Yanko *et al.*, 1995). Geslin *et al.* (1998) studied the ultrastructure of shell walls of normal and deformed foraminifera of *Ammonia* sp. with a scanning electron microscope, and detected a disorganization of calcite elements in the walls of deformed shells.

All of these data can be explained by the involvement of the cytoskeleton in abnormal shell development. Because the cytoskeleton controls both the shape of the cell and the transport of organelles or vesicular structures, injury to the cytoskeleton may alter the shape and arrangement of chambers and the transport of proteins and calcium to the shell wall. Therefore, experimental induction of shell deformities by mechanical damage and chemicals damaging the cytoskeleton, on the one hand, and examination of the cytoskeleton and connected structures in normal and deformed foraminifera from different clean and polluted habitats, on the other hand, could either corroborate or disprove this hypothesis. If the hypothesis is corroborated by such studies, the frequency of shell deformities could be used for pollution monitoring of different marine environments.

3.2. Environmental Quality, Antixenobiotic Defense Mechanisms and Sensitivity to Pollutants

As noted above, all marine environments, especially benthic epiphytic microenvironments, contain a large variety of natural xenobiotics produced by all members of the community. We assume that all uni- and multicellular organisms have created numerous multisubstrate antixenobiotic defense mechanisms during their coevolution that have allowed the organisms to protect themselves against these xenobiotics (Bresler *et al.*, 1990, 1998, in press; Bresler and Fishelson, 1994; Bresler and Yanko, 1995*a*). In this review, we have presented and analyzed direct and indirect evidence that benthic foraminifera also contain various antixenobiotic defense mechanisms, particularly export pumps, SATOA and MXRtr, and detoxifying enzymes. These defense mechanisms can also protect cells and organisms against some man-made xenobiotics that are chemically similar to those occurring in nature. However, several man-made xenobiotics can be converted by detoxifying enzymes into more toxic and dangerous compounds that can damage various cellular structures. Several natural and man-made xenobiotics in seawater can also irreversibly inhibit antixenobiotic export pumps (Williams and Jacobs, 1993; Smital and Kurelec, 1997). Thus, the invasion of man-made xenobiotics (pollutants) into a community not only creates an additional workload for antixenobiotic defense mechanisms, but also threatens to damage them. Thus pollution can eliminate specimens with slow antixenobiotic defenses and only the most protected specimens might survive in highly polluted sites. For example, we recently found that a population of the bivalve *Donax trunculus* from clean sites contained specimens with low and high MXRtr and SATOA activities in surface epithelia but populations from polluted sites contained only specimens with high MXRtr and SATOA activities (Bresler *et al.*, 1998, in press). Thus, in order to assess environmental quality at a station or community, we must first examine general antixenobiotic defense mechanisms (particularly export pumps) in key species and then compare them to the activity of the same mechanisms in the same species from a clean (reference) site(s). Alterations in the activity of antixenobiotic defense mechanisms may be the earliest means of detecting environmental pollutants.

High activity of antixenobiotic defense mechanisms and their cooperative work can protect the health of specimens even in the highly polluted environments. If the defense mechanisms cannot effectively eliminate and detoxify the penetrated xenobiotics, however, target metabolic reactions, functions, and cellular structures are damaged and different pathologies are induced. The study of marine mollusks from different clean and polluted sites along the Israeli Mediterranean shore, the Gulf of Eilat (Red Sea), and the Baltic and North Seas, shows a strong significant negative correlation (>0.9) between MXRtr activity and the frequency of signs of genotoxicity or frequency and expression of pathological alterations (Bresler *et al.*, 1998, in press). Thus, environmental pathology and genotoxicity can be mediated by an inherent or induced failure of antixenobiotic defense mechanisms.

3.3. Mechanisms of Adaptation and Resistance to Pollutants

Pollution of marine water and sediment decreases foraminiferal diversity, i.e., only selected foraminiferal species and specimens survive under these environmental conditions (Yanko *et al.*, 1994, in press). It is known that some other marine species, particularly mollusks, can survive in polluted environments (Kurelec, 1992, 1995, 1997; Bresler and Fishelson, 1994; Fishelson *et al.*, 1996; Bresler *et al.*, 1998, in press). Also some pathogenic protozoa, insects, and even fish are resistant to man-made xenobiotics such as drugs, insecticides and pesticides (Frommel and Balber, 1987; Mouches *et al.*, 1987; Foote *et al.*, 1989; Fournier and Mutero, 1994; Roark and Brown, 1996).

There are three main mechanisms that make organisms more resistant to xenobiotics: (1) enhancement of antixenobiotic defense mechanism(s); (2) elimination of targets for its action; and (3) selection of more protected genotypes (Bresler *et al.*, 1990, 1998; Kurelec, 1992, 1995, 1997; Fournier and Mutero, 1994). Antixenobiotic defense mechanisms can enhance their activity by genetic mechanisms (induction, overexpression, or amplification of corresponding genes) or some cytoplasmic posttranslational mechanisms. For example, amplification and overexpression of MXRtr genes was found to produce an excess of this transporter and create resistance of tumor cells and protozoans to drugs, and probably the resistance of some aquatic invertebrates to pollutants (Frommel and Balber, 1987; Foote *et al.*, 1989; Kurelec, 1992; Gottesman and Pastan, 1993). Overproduction of detoxifying esterases produced resistance of some insects to organophosphorus insecticides (Mouches *et al.*, 1987). As noted, many detoxification enzymes are inducible. Thus, organisms can temporarily enhance activity of their antixenobiotic defense mechanisms in order to protect themselves against excessive pollutants in the environment.

Posttranslational activity of membrane export pumps can also be modulated by protein kinase C. Thus, MXRtr can change its functional activity within broad limits by alteration of its activity and concentration in membranes.

Modification of the target enzyme, acetylcholinesterase, can be a mechanism of resistance to several insecticides that inhibit acetylcholinesterase (Fournier and Mutero, 1994). Environmental selection by pollutants also was shown in mollusks and fish (Roark and Brown, 1996; Fishelson *et al.*, 1996; Bresler *et al.*, 1998, in press). Thus, it is very important to study the adaptive responses of antixenobiotic defense mechanisms in different benthic foraminifera, particularly indications of phenotypic selection.

4. Conclusions

The nondestructive cytophysiological and cytochemical examination of living foraminifera using biophysical methods, fluorescent probes, markers, fluorogenic substrates, and fluorescent microscopy makes it possible to visualize and study metabolic reactions, enzyme activity, and transport processes; specific subcellular structures or molecular complexes, and their functions; DNA, RNA, and total protein and lipid content; and the concentration of primary inorganic ions in the cytoplasm or cell compartments. These methodologies allow us to detect various antixenobiotic defense mechanisms previously unknown in foraminifera and to study their roles in response to polluted environments. All these techniques permit a new approach to foraminiferal ecology: the complex study of antixenobiotic defense mechanisms and their interrelation with natural and man-made xenobiotics. The methodologies in combination with conventional microscopic examination make it possible to objectively assess foraminiferal health and detect alterations caused by external (environmental) stresses, thereby resulting in a new early warning monitoring system. Finally, the proposed methodology and approach may be a bridge connecting cytology, molecular biology, ecology, and foraminiferal paleoecology.

ACKNOWLEDGMENTS: The authors thank Mr. Sven T. Hombach, University of North Dakota, Grand Forks, North Dakota, for valuable comments in editing this paper. We are grateful to Dr. H. Peter Hombach, the Chief Scientist and Chairman of the Board of BioDigestor Technologies, Inc., in Grand Forks, for his remarks and constructive criticism. We sincerely thank Dr. Lev Fishelson (Department of Life Sciences, Zoological Faculty, Tel Aviv University), Dr. Ronald E. Martin (Department of Geology, University of Delaware) and an anonymous reviewer for reviewing this manuscript. This study was supported mainly by a grant from the EC (AVICENNE Program, AVI CT92-0007).

References

Agosin, M., Naquira, C., Paulin, J., and Capdevila, J., 1976, Cytochrome P450 and drug metabolism in *Trypanosoma cruzi*: Effects of phenobarbital, *Science* **194**:195–197.

Ahern, T. J., Allan, G. G., and Medcalf, D. G., 1980, New bromoperoxidase of marine origin, *Biochim. Biophys. Acta* **616**:329–339.

Alberts, B., 1998, The cell as a collection of protein machines: Preparing the next generation of molecular biologists, *Cell* **92**:291–294.

Alberts, B., Bray, D., Lewis, J., Raff, M., Roberts, K., and Watson, J. D., 1983, *Molecular Biology of the Cell*, Garland, New York–London.

Albone, E. S., 1984, *Mammalian Semiochemistry: The Investigation of Chemical Signals between Mammals*, John Wiley & Sons, New York, pp. 266–280.

Alve, E., 1991, Benthic foraminifera in sediment cores reflecting heavy metal pollution in Sorfjord, Western Norway, *J. Foram. Res.* **21**:1–19.

Anderson, H. C., 1995, Molecular biology of matrix vesicles, *Clin. Orthop.* **314**:266–280.

Anderson, O. R., 1988, *Comparative Protozoology: Ecology, Physiology, Life History*, Springer-Verlag, Heidelberg.

Anderson, O. R., and Bé, A. W. H., 1976, A cytochemical fine structure study of phagotrophy in a planktonic foraminifer, *Hastigerina pelagica* (d'Orbigny), *Biol. Bull. Mar. Biol. Lab. Woods Hole* **151**:437–449.

Anderson, O. R., and Lee, J. J., 1991, Cytology and fine structure, in: *Biology of Foraminifera* (J. J. Lee and O. R. Anderson, eds.), Academic Press, London, pp. 7–40.

Anderson, O. R., and Tuntivate-Choy, S., 1984, Cytochemical evidence for peroxisomes in planktonic foraminifera, *J. Foram. Res.* **14**:203–205.

Arillo, A., and Melodia, F., 1990, Protective effect of fish mucus against Cr (VI) pollution, *Chemosphere* **20**:397–402.

Avila, C., 1995, Natural products of Opistobranch mollusks: A biological review, *Oceanogr. Mar. Biol. Ann. Rev.* **33**:487–559.

Baba, H., Oishi, R., and Saeki, K., 1988, Enhancement of flood–brain barrier permeability to sodium fluorescein by stimulation of μ-opioid receptors in mice, *Naunyn-Schmiedeberg's Arch. Pharmacol.* **337**:423–428.

Barbier, M., 1976, *Introduction a l'ecologie chimique*, Masson, Paris, New York, Barcelona, Milan.

Berk, S. G., and Roberts, R. O., 1998, Development of a protozoan chemoattraction inhibition assay for evaluating toxicity of aquatic pollutants, in: *Microscale Testing in Aquatic Toxicology: Advances, Techniques, and Practice* (P. G. Wells, K. Lee, and C. Blaise, eds.), CRC Press, Boca Raton, pp. 337–348.

Bernhard, J. M., and Bowser, S. S., 1996, Novel epifluorescence microscopy method to determine life position of foraminifera in sediments, *Micropaleontology* **15**:68.

Bernhard, J. M., Newkirk, S. G., and Bowsen, S. S., 1995, Towards a non-terminal viability assay for foraminiferan protists, *J. Euk. Microbiol.* **42**:357–367.

Bernhard, J. M., Gupta, B. K. S., and Borne, P. F., 1997, Bentic foraminiferal proxy to estimate dysoxic bottom-water oxygen concentrations: Santa Barbara Basin, US Pacific continental margin, *J. Foram. Res.* **27**:301–310.

Bird, S., and Lloyd, J. B., 1995, Mechanism of lysosome rupture by dipeptides, *Cell Biochem. Funct.* **13**:79–83.

Boltovskoy, E., and Wright, R., 1976, *Recent Foraminifera*, W. Junk, The Hague.

Boltovskoy, E., Scott, D. B., and Medioli, F. S., 1991, Morphological variations of benthic foraminiferal tests in response to changes in ecological parameters: A review, *J. Paleontol.* **65**:175–185.

Boskey, A. L., Boyan, B. D., and Schwartz, Z., 1997, Matrix vesicles promote mineralization in a gelatin gel, *Calcif. Tissue Int.* **60**:309–315.

Bowser, S. S., 1985, Invasive activity of *Allogromia* pseudopodial networks: Skyllocytosis of gelatin/agar gel, *J. Protozool.* **32**:9–12.

Bowser, S. S., and McGee-Russell, S. M., 1982, EM cytochemical study of phagotrophy in the benthic foraminifera, *Allogromia* sp., *J. Protozool.* **29**:474–475.

Bowser, S. S., McGee-Russell, S. M., and Rieder, C. L., 1985, Digestion of prey in foraminifera is not anomalous: A correlation of light microscopic, cytochemical and HVEM technics to study phagotrophy in two Allogromiids, *Tissue and Cell* **17**:823–839.

Bowser, S. S., Travis, J. L., and Rieder, C. L., 1988, Microtubules associate with actin-containing filaments as discrete sites along the ventral surface of *Allogromia* reticulopods, *J. Cell Sci.* **89**:297–307.

Boyle, E., 1981, Cadmium, zinc, copper and barium in foraminiferal tests, *Earth Planet. Sci. Lett.* **53**:11–35.

Boyle, E., 1986, Paired carbon isotope and cadmium data from benthic foraminifera: Implications for changes in oceanic phosphorus, oceanic circulation, and atmospheric carbon dioxide, *Geochim. Cosmochim. Acta.* **50**:265–276.

Bradbury, M., 1979, *The Concept of a Blood—Brain Barrier*, John Wiley and Sons, Chichester.

Brattsten, L. B., 1979, Biochemical defense mechanisms in herbivores against plant allelochemicals, in: *Herbivores: Their Interaction with Secondary Plant Metabolites* (G. A. Rosental and D. H. Janzen, eds.), Academic Press, New York, pp. 199–270.

Brattsten, L. B., 1986, Fate of ingested plant allelochemicals in herbivorous insects, in: *Molecular Aspects of Insect–Plant Associations* (L. B. Brattsten and S. Ahmad, eds.), Plenum Press, New York, pp. 211–255.

Bresler, V., 1989, Organisms defend themselves against pollution, *Sci. Life* (Moscow) 7:21–25.

Bresler, V. M., Belyaeva, E. A., and Mozhaeva, M. G., 1990, A comparative study on the system of active transport of organic acids in Malpighian tubules of insects, *J. Insect. Physiol.* **36**:259–270.

Bresler, V. M., and Fishelson, L., 1994, Microfluorometrical study of bioaccumulation of benzo[a]pyrene and marker xenobiotics in the bivalve *Donax trunculus* from clean and polluted sites along the Mediterranean shore of Israel, *Dis. Aquat. Org.* **19**:193–202.

Bresler, V. M., and Nikiforov, A. A., 1981, *Transport of Organic Acids across Plasma Membranes*, Nauka, Leningrad (in Russian).

Bresler, V. M., and Yanko, V. V., 1995*a*, Chemical ecology: A new approach to study living benthic epiphytic foraminifera, *J. Foram. Res.* **25**:267–279.

Bresler, V. M., and Yanko, V. V., 1995*b*, Acute toxicity of heavy metals for benthic epiphytic foraminifera *Pararotalia spinigera* (Le Calvez) and influence of seaweed-derived DOC, *Environ. Toxicol. Chem.* **14**:1687–1695.

Bresler, V. M., Bresler, S. E., and Nikiforov, A. A., 1975, Structure and active transport in the plasma membrane of the tubules of frog kidney, *Biochim. Biophys. Acta.* **406**:526–537.

Bresler, V. M., Bresler, S. E., Kazbekov, E. N., Nikiforov, A. A., and Vasilyeva, N., 1979, On the active transport of organic acid (fluorescein) in the choroid plexus of rabbit, *Biochim. Biophys. Acta* **550**:110–119.

Bresler, V. M., Mozhaeva, M. G., and Belyaeva, E. A., 1985, A comparative study of the system of active transport of organic acids in Malpighian tubules of the tropical cockroach, *Blaberus giganteus, Comp. Biochem. Physiol.* **80A**:393–397.

Bresler, V. M., Valter, S. N., Isaev-Ivanov, V. V., Kazbekov, E. N., Kleiner, A. R., Orlov, Y. N., Ostapenko, I. A., Sukhodolova, A. T., and Fomichev, V. M., 1989, The influence of the lipid bilayer phase transition on the PAH carrier and activity of alkaline phosphatase in brush border membrane vesicles from normal and mutant rats, *Biochim. Biophys. Acta.* **982**: 288–294.

Bresler, V. M., Fishelson, L., and Abelson, A., 1998, Anti-xenobiotic defense mechanisms and environmental health, *35th CIESM Congr. Proc.* **35(2)**:338–339.

Bresler, V., Bissinger, V., Abelson, A., Dizer, H., Sturm, A., Kratke, R., Fishelson, L., and Hansen, P.-D., in press, Marine mollusks and fish as biomarker of pollution stress in littoral regions of the three seas: Red Sea, Mediterranean and North Sea. *Helgol. Meeresun.*

Bretscher, A., 1991, Microfilament structure and function in the cortical cytoskeleton, *Ann. Rev. Cell Biol.* **7**:337–374.

Brown, D. J., Clark, G. C., Gardner, G. R., and van Beneden, R. J., 1996, Identification of dioxin-specific binding proteins in marine bivalves, *Mar. Environ. Res.* **42**:7–12.

Burton, G. A., Jr., 1991, Assessing the toxicity of freshwater sediments (annual review), *Environ. Toxicol. Chem.* **10**:1585–1627.

Clarkson, T. W., 1993, Molecular and ionic mimicry of toxic metals, *Ann. Rev. Pharmacol. Toxicol.* **32**:545–571.

Cotran, R. S., Kumar, V., and Robbins, S. L., 1989, *Robbins Pathologic Basis of Disease, 4th Ed.*, W. B. Saunders, Philadelphia.

Debenay, J. P., Guillou, J. J., and Lesourd, M., 1996, Colloidal calcite in foraminiferal tests: Crystallization and texture of the test, *J. Foram. Res.* **26**:277–288.

DeLaca, T. E., Karl, D. M., and Lipps, J. H, 1981, Direct use of dissolved organic carbon by agglutinated benthic foraminifera, *Nature* **289**:287–289.

Deguchi, Y., Nozawa, K., Yamada, S., Yokoyama, Y., and Kimura, R., 1997, Evaluation of brain distribution and blood–brain barrier efflux transport of probenecid in rats by microdialysis: Possible involvment of the monocarboxylic acid transport system, *J. Pharmacol. Exp. Ther.* **280**:551–560.

Delaney, M. L., Bé, A. W., and Boyle, E., 1985, Li, Sr, Mg and Na in foraminiferal calcite shells from laboratory culture, sediment traps, and sediment, *Geochim. Cosmochim. Acta* **49**:1327–1341.

Di Virgilio, F., Steinberg, T. H., Swanson, J. A., and Silverstein, S. C., 1988, Fura-2 secretion and sequestration in macrophages: A blocker of organic anion transport reveals that these processes occur via a membrane transport system for organic anions, *J. Immunol.* **140**:915–919.

Di Virgilio, F., Steinberg, T. H., and Silverstein, S. C., 1990, Inhibition of Fura-2 sequestration and secretion with organic anion transport blockers, *Cell Calcium* **11**:57–62.

Dixon, M., and Webb, E. C., 1979, *Enzymes, 3rd Ed.*, Longman, London.

Faulkner, D. J., 1984*a*, Marine natural products: Metabolites of marine algae and herbivorous marine mollusks, *Nat. Prod. Res.* **1**:251–280.

Faulkner, D. J., 1984*b*, Marine natural products: metabolites of marine invertebrates, *Nat. Prod. Res.* **1**:551–598.

Fenical, W., 1982, Natural products chemistry in the marine environment, *Science* **215**:223–228.

Fishelson, L., Bresler, V., and Abelson, A., 1996, The use of marine mollusks and other animals for biomarker-based pollution monitoring in coastal areas, in: *MARS1-Biological Indicators of Natural and Man-made Changes in Marine and Coastal Waters*, Progr. Abstr. 2nd Workshop, Ministry of Science, pp. 7–11, October 27–28, Jerusalem.

Foote, S. J., Thompson, J. K., Cowman, A. F., and Kemp, D. J. 1989, Amplification of the multidrug resistance gene in some chloroquine-resistant isolates of *P. falciparum*, *Cell* **57**:921–930.

Fontana, A., Ciavatta, M. L., Gavagnin, M., and Cimino, G., 1998, Highly specialized detectors of bioactive molecules from marine environment, *35th CIESM Congr. Proc.* **35**(2):316–319.

Fournier, D., and Mutero, A., 1994, Modification of acetylcholinesterases as a mechanism of resistance to insecticides, *Compar. Biochem. Physiol.* **108C**:19–31.

Frommel, T. O., and Balber, A. E., 1987, Flow cytofluorometric analysis of drug accumulation by multidrug-resistant *Trypanosoma brucei brucei* and *T. b. rhodesiense*, *Mol. Biochem. Parasitol.* **26**:183–191.

Galgani, P., Cornwall, R., Toomey, B. H., and Epel, D. D., 1996, Interaction of environmental xenobiotics with a multixenobiotic defense mechanism in the bay mussel *Mytilus galloprovincialis* from the coast of California, *Environ. Toxicol. Chem.* **15**:325–331.

Geiselman, J. A., and McConnell, O. J., 1981, Polyphenols in brown algae *Fucus vesiculosus* and *Ascophyllum nodosum*: Chemical defenses against the marine herbivorous snail, *Littorina littorea*, *J. Chem. Ecol.* **7**:1115–1133.

Geslin, E., Debeney, J.-P., and Lesourd, M., 1998, Abnormal wall textures and test deformation in *Ammonia* (hyaline foraminifera), *J. Foram. Res.* **28**:148–156.

Gilron, G. L., and Lynn, D. H., 1998, Ciliated protozoa as test organisms in toxicity assessments, in: *Microscale Testing in Aquatic Toxicology: Advances, Techniques, and Practice* (P. G. Wells, K. Lee, and C. Blaise, eds.), CRC Press, Boca Raton, pp. 323–336.

Gottesman, M. M., and Pastan, I., 1993, Biochemistry of multidrug resistance mediated by the multidrug transporter, *Ann. Rev. Biochem.* **62**:385–428.

Gschwend, P. M., MacFarlane, L. K., and Newman, K. A., 1985, Volative halogenated organic compounds released to seawater from temperate marine macroalgae, *Science* **227**:1033–1035.

Guengerich, F. P., 1990, *Mammalian Cytochromes P450, Vol. 1*, CRC Press, Boca Raton, pp. 2–54.

Hahn, M. E., Poland, A., Glover, E., and Stegeman, J. J., 1992, The Ah receptor in marine animals: Phylogenetic distribution and relationship to cytochrome P450 IA inducibility, *Mar. Environ. Res.* **34**:87–92.

Hamer, D., 1986, Metallothionein, *Ann. Rev. Biochem.* **55**:913–951.

Handschumacher, R. E., Harding, M. W., Rice, J., Drugge, R. J., and Speicher, D. W., 1984, Cyclophilin: A specific cytosolic binding protein for cyclosporin A, *Science* **226**:544–547.

Harborne, J. B., 1993, *Introduction to Ecological Biochemistry, 4th Ed.*, Academic Press, London, New York and Paris.

Haugland, R. P., 1996, *Handbook of Fluorescent Probes and Research Chemicals, 6th Ed.*, Molecular Probes, Eugene, Ore.

Haynes, J. R., 1981, *Foraminifera*, John Wiley & Sons, New York.

Hemleben, C., Anderson, O. R., Berthold, W. U., and Spindler, M., 1986, Calcification and chamber formation in foraminifera—a brief overview, in: *Biomineralization in Lower Plants and Animals*, Systematics Association Special Volume 30 (B. S. C. Leadbeater and R. Riding, eds.) Clarendon Press, Oxford, pp. 237–249.

Higa, T., Fujiyama, T., and Scheuer, P. J. 1980, Halogenated phenol and indole constituents of acorn worms, *Compar. Biochem. Physiol.* **65B**:525–530.

Higgins, C. F., 1992, ABC transporters: From microorganisms to man, *Ann. Rev. Cell Biol.* **8**:67–113.

Hinton, D. E., Baumannn, P. C., Gardner, G. R., Hawkins, W. E., Hendricks, J. D., Murchelano, R. A., and Okihiro, M. S., 1992, Histopathologic biomarkers, in: *Biomarkers: Biochemical, Physiological and Histological Biomarkers of Anthropogenic Stress* (R. J. Hugget, R. A. Kimmerle, R. M. Mehrle, and Bergman H. L., eds.), Lewis, London, pp. 155–209.

Hollebone, B. R., Brownlee, L. J., Davis, D., Michelin, N., and Pardy, D., 1995, A mechanistic structure–activity relationship for hepatic polysubstrate monooxygenase, *Environ. Toxicol. Chem.* **14**:29–41.

Hoyt, M. A., Hyman, A. A., and Buhler, M., 1997, Motor proteins of the eukaryotic cytoskeleton, *Proc. Natl. Acad. Sci. USA* **94**:12747–12749.

Jakob, I., Hauser, I. A., Thevenod, F., and Lindemann, B., 1998, MDR1 in taste buds of rat vallate papilla: Functional, immunohistochemical, and biochemical evidence, *Am. J. Physiol. (Cell Physiol.)* **274**:C182–C191.

Kagi, J. H. R., and Schaffer, A., 1988, Biochemistry of metallothionein, *Biochemistry* **27**:8509–8515.

Kashman, Y., and Rudi, A., 1998, Recent marine nitrogen-containing metabolites, *35th CIESM Congr. Proc.* **35**(2):320–323.

Keppler, D., Leier, I., and Jedlitschky, G., 1997, Transport of glutathione conjugates and glucuronides by the multidrug resistance proteins MRP1 and MRP2, *Biol. Chem.* **378**:787–791.

Khan, M. A. Q., Kamal, A., Wolin, R. J., and Runnels, J., 1972, *In vivo* and *in vitro* epoxidation of Aldrin by aquatic food chain organisms, *Bull. Environ. Contam. Toxicol.* **8**:219–228.

King, G. M., 1986, Inhibition of microbial activity in marine sediment by a bromophenol from a hemichordate, *Nature* **323**:257–259.

Kirsch, T., Nah, H.-D., Shapiro, I. M., and Pacifici, M., 1997, Regulated production of mineralization-competent matrix vesicles in hypertrophic chondrocytes, *J. Cell Biol.* **137**:1149–1160.

Kotyk, A., and Janacek, K., 1977, *Membrane Transport: An Interdisciplinary Approach*, Plenum Press, New York and London.

Kurelec, B., 1992, The multixenobiotic resistance mechanism in aquatic organisms, *Crit. Rev. Toxicol.* **22**:23–43.

Kurelec, B., 1995, Inhibition of multixenobiotic resistance mechanism in aquatic organisms: Ecological consequences, *Sci. Total Environ.* **171**:197–204.

Kurelec, B., 1997, Pumping out: The first line cellular defense to water pollutants in aquatic organisms, *Toxicol. Ecotoxicol. News* **4**:104–109.

Langer, M. R., 1992, Biosynthesis of glycosaminoglycans in foraminifera: A review, *Mar. Micropaleontol.* **19**:245–255.

Langer, M. R., and Gehring, C. A., 1993, Bacteria farming: A possible feeding strategy of some smaller, motile foraminifera, *J. Foram. Res.* **23**:40–46.

Lea, D., and Boyle, E., 1989, Barium content of benthic foraminifera controlled by bottom-water composition, *Nature* **338**:751–753.

Lea, D., and Boyle, E., 1991, Barium in planktonic foraminifera, *Geochim. Cosmochim. Acta* **55**:3321–3331.

Lee, J. J., and Anderson, O. R., 1991, *Biology of Foraminifera*, Academic Press, London, San Diego and New York.

Lindroth, R. L., 1988, Adaptations of mammalian herbivores to plant chemical defense, in: *Chemical Mediation of Coevolution* (K. C. Spencer, ed.), Academic Press. San Diego, pp. 415–445.

Lipps, J. H., 1983, Biotic interactions in benthic foraminifera, in: *Biotic Interactions in Recent and Fossil Benthic Communities* (M. J. S. Tevesz and P. L. McCall, eds.), Plenum Press, New York, pp. 331–376.

Ma, L., Cantley, L. C., Janmey, P. A., and Kirschner, M. W., 1998, Corequirement of specific phosphoinositides and small GTP-binding protein Cdc42 in inducing actin assembly in Xenopus egg extracts, *J. Cell Biol.* **140**:1125–1136.

MacIntyre, A. C., and Cutler, D. J., 1988, The potential role of lysosomes in tissue distribution of weak bases, *Biopharm. Drug Dispos.* **9**:513–526.

Meyers, T. R., and Hendricks, J. D., 1984, Histopathology, in: *The Fundamentals of Aquatic Toxicology: Methods and Applications* (G. M. Rand and S. R. Petrocelli, eds.), Hemisphere, New York, pp. 283–334.

Miller, D. S., Fricker, G., and Drewe, J., 1997, *P*-glycoprotein-mediated transport of a fluorescent rapamycin derivative in renal proximal tubules, *J. Pharmacol. Exp. Ther.* **282**:440–444.

Millot, C., Millot, J.-M., and Morjani, H., 1997, Characterization of acidic vesicles in multidrug-resistant and sensitive cancer cells by Acridine Orange staining and confocal microspectrofluorometry, *J. Histochem. Cytochem.* **45**:1255–1264.

Montellano de, P. R. O. (ed.), 1995, *Cytochrome P450: Structure, Mechanism, and Biochemistry, 2nd Ed.*, Plenum Press, New York.

Moriyama, Y., Manabe, T., Yoshimori, T., Tashiro, Y., and Futai, M., 1994, ATP-dependent uptake of anti-neoplastic agents by acidic organelles, *J. Biochem. (Tokyo)* **115**:213–218.

Mouches, C., Magnin, M., Berge, J. B., de Silvestri, M., Beyssat, V., Pasteur, N., and Georghiou, G. P., 1987, Overproduction of detoxifying esterases in organophosphorus insecticide resistant Culex mosquitoes and their presence in other insects, *Proc, Natl. Acad. Sci.* USA **84**:2113–2116.

Murphy, S., Drotar, A.-M., and Fall, R., 1982. Biotransformation of the fungicide pentachloronitrobenzene by *Tetrahymena thermophila, Chemosphere* **11**:33–39.

Murray, J. W., 1991, *Ecology and Paleoecology of Benthic Foraminifera*, Longman Scientific and Technical, London.

Nebert, D. W., and Jansen, N. M., 1979, The Ah-locus: genetic regulation of the metabolism of carcinogens, drugs and other environmental chemicals by cytochrome P450 mediated monooxygenases, *Crit. Rev. Biochem.* **6**:401–437.

Neyfakh, A. A., Budnenko, V. E., and Chen, L. B., 1991, Efflux-mediated multidrug resistance in *Bacillus subtilis*: Similarities and dissimilarities with the mammalian system. *Proc. Natl. Acad. Sci. USA.* **88**:4781–4785

Palenik, B., Zafirion, O. C., and Morel, F. M. M., 1987, Hydrogen peroxide production by a marine phytoplanktoner, *Limnol. Oceanogr.* **32**:1365–1369.

Pancer, Z., Gershon, H., and Rinkevich, B., 1993, DNA cloning of a putative prochordate FK506-binding protein, *Biochem. Biophys Res. Comm.* **197**:973–977.

Parke, D. V., 1971, *The Biochemistry of Foreign Compounds*, Pergamon Press, Oxford.

Paulsma, C. C., Bosma, P. J., Zaman, G. J. R., Bakker, C. T., Otter, M., Scheffer, G. L., Scheper, R. J., Borst, P., and Oude Elferink, R., 1996, Congenital jaundice in rats with a mutation in a multidrug resistance-associated protein gene, *Science* **71**:1126–1128.

Pearce, A. G. E., 1968, *Histochemistry: Theoretical and Applied, 3rd Ed*, Churchill, London.

Perdue, E. M., and Gjessing, E. T. (eds.), 1990, *Organic Acids in Aquatic Ecosystems*, Wiley Interscience, Chichester.

Piccinni, E., 1989, Response to heavy metals of uni- and multicellular organisms: Homologies and analogies, *Bull. Zool.* **56**:265–271.

Piccinni, E., Irato, P., Cavallini, L., and Ammermann, D., 1992, Effects of cadmium in *Stylonychia lemnae, Stylonychia notophora* and *Oxytricha granulifera*: Isolation of a cadmium-binding protein, *J. Protozool.* **39**:589–593.

Polek, B., Godocikova, J., Batora, R., and Ursinvova, M., 1993, Cu-binding proteins of digestive tract of *Galleria mellonella* caterpillars (Lepidoptera; Pyralide), *Biologia* **48**:631–636.

Poot, M., Hudson, F. N., Grossmann, A., Rabinovitch, P. S., and Kavanagh, T. J., 1996, Probenecid inhibition of fluorescence extrusion after MCB-staining of rat-1 fibroblasts, *Cytometry,* **23**:78–81.

Pratt, J. R., Mochan, D. G., and Bowers, N. J., 1998, Ciliate microbiotest applications, in: *Microscale Testing in Aquatic Toxicology: Advances, Techniques, and Practice* (P. G. Wells, K. Lee, and C. Blaise, eds.), CRC Press, Boca Raton, pp. 349–357.

Proksch, P., 1994, Defensive role for secondary metabolites from marine sponges and sponge-feeding nudibranchs, *Toxicol.* **32**: 639–655.

Prosser, C. L., 1973, *Comparative Animal Physiology, 3rd Ed.*, Saunders, Philadelphia.

Ragan, M. A., and Glombitza, K., 1986, Phlorotannins, brown algal polyphenols, in: *Progress in Phycological Research*, Vol. 4 (F. E. Round and D. J. Chapman, eds.), Biopress, Bristol, pp. 129–241.

Rapoport, S. I., 1976, *Blood–Brain Barrier in Physiology and Medicine*, Raven Press, New York.

Rice, E. L., 1985, *Pest Control with Nature's Chemicals: Allelochemics and Pheromones in Gardening and Agriculture*, University of Oklahoma Press, Norman.

Roark, S., and Brown, K., 1996, Effects of metal contamination from mine tailings of allozyme distributions of populations of Great Plains fishes, *Environ. Toxicol. Chem.* **15**:921–927.

Robbins, L. L., and Brew, K., 1990, Proteins from the organic matrix of core-top and fossil planktonic foraminifera, *Geochim. Cosmochim. Acta.* **54**:2285–2292.

Robbins, L. L., and Healy-Williams, N., 1991, Toward a classification of planktonic foraminifera based on biochemical, geochemical, and morphological criteria, *J. Foram. Res.* **21**:159–167.

Roesijadi, G., 1996, Metallothionein and its role in toxic metal regulation, *Comp. Biochem. Physiol.* **133B**:117–123.

Rupp, G., Bowser, S. S., Mannella, C. A., and Rieder, C. L., 1986, Naturally occurring tubulin-containing paracrystals in *Allogromia*: Immunocytochemical identification and functional significance, *Cell Motil. Cytoskel.* **6**: 363–375.

Sarmento, A., Borges, N., and Azevedo, I., 1991, Adrenergic influence on the control of blood–brain barrier permeability, *Naunyn-Schmiedeberg's Arch. Pharmacol.* **343**:633–637.

Sauer, G. R., Adkisson, H. D., Genge, B. R., and Wuthier, R. E., 1989, Regulatory effect of endogenous zinc and inhibitory actions of toxic metal ions on calcium accumulation by matrix vesicles *in vitro, Bone Miner.* **7**:233–244.

Schreiber, S., Becker, K., Bresler, V., and Fishelson, L., 1997, Dietary L-carnitine protects the gills and skin of guppies (*Poecilia reticulata*) against anionic xenobiotics, *Compar. Biochem. Physiol.* **117C**:99–102.

Segner, H., Marthaler, R., and Linnenbach, M., 1988, Growth, aluminium uptake and mucous cells morphometrics of early life stages of brown trout, *Salmo trutta*, in low pH water, *Environ. Biol. Fish* **21**:153–159.

Smital, T., and Kuzelec, B., 1997, The concentrations of inhibitors of MXR mechanism in natural water, *Environ. Toxicol. Chem.* **16**:2164–2170.

Sirotnak, E. M., Yang, C.-H., Mines, L., Orbé, E., and Biedler, J. L., 1986, Markedly altered membrane transport and intracellular binding of Vincristine, *J. Cell Physiol.* **126**:266–274.

Stathuplos, L., and Tuross, N., 1994, Proteins and DNA from modern planktonic foraminifera, *J. Foram. Res.* **24**:49–59.

Stenseth, K., and Thyberg, J., 1989, Monesin and chloroquine inhibit transfer to lysosomes of endocytosed macromolecules in cultured mouse peritoneal macrophages, *Eur. J. Cell. Biol.* **49**:326–333.

Stoddart, D. M., 1990, *The Scented Ape*, Cambridge University Press, Cambridge.
Stubbles, S. J., 1993, Recent benthic Foraminiferida as indicators of pollution in Restronguet Creek, Cornwall. *Proc. Ussher Soc.* **8**: 200–204.
Supriyono, A., Schwarz, B., Wray, V., Witte, L., Mueller, W. E. G., Van Soest, R., Sumaryono, W., and Proksch, P., 1995, Bioactive alkaloids from the tropical marine sponge *Axinella cartere, Zeits. Naturforsch. B: Biosci.* **50**:669–674.
Sweet, D. H., Wolff, N. A., and Pritchard, J. B., 1997, Expression cloning and characterization of ROAT1, the basolateral organic anion transporter in rat kidney, *J. Biol. Chem.* **272**:30088–30095.
Teeyapont, R., Woerdenbag, H. J., Kreis, P., Hacker, J., Wray, V., Witte, L., and Proksch, P., 1993, Antibiotic and cytotoxic activity of brominated compounds from the marine sponge *Verongia aerophoba, Zeits. Naturforsch. B: Biosci.* **48**:939–945.
Ter Kuile, B., 1991, Mechanisms for calcification and carbon cycling in algal symbiont-bearing foraminifera, in: *Biology of the Foraminifera* (J. J. Lee and O. R. Anderson, eds.), Academic Press, London, San Diego, and New York, pp. 73–89.
Theiler, R., Coo, J. C., and Hager, L. P., 1978, Halohydrocarbon synthesis by bromoperoxidase, *Science* **202**:1094–1096.
Tiberghien, F., and Looz ,F., 1996, Ranking of P-glycoprotein substrates and inhibitors by calcein AM fluorometry screening assay, *Anti-Cancer Drug* **7**:568–574.
Towe, K., and Cifelli, R., 1967, Wall ultrastructure in the calcareous foraminifera: Crystallographic aspects and a model for calcification, *J. Paleontol.* **41**:742–762.
Travis, J. L., and Bowser, S. S., 1986, A new model of reticulopodal motility and shape: Evidence for a microtubule-based motor and an actin skeleton, *Cell Motil. Cytoskel.* **6**:2–14.
Travis, J. L., and Bowser, S. S., 1990, Microtubule-membrane interactions *in vivo*: Direct observation of plasma membrane deformation mediated by actively bending microtubulas. *Protoplasma* **154**:184–189.
Travis, J. L., and Bowser, S. S., 1991, The motility of foraminifera, in: *Biology of Foraminifera* (J. J. Lee and O. R. Anderson, eds.), Academic Press, London, pp. 91–155.
Travis, J. L., Kenealy, J. F. X., and Allenm R. D., 1983, Studies on the motility of the foraminifera: II. The dynamic microtubular cytoskeleton of the reticulopodial network of *Allogromia laticollaris, J. Cell Biol.* **97**:1668–1676.
Trick, C. G., Andersen, R. J., Gillam, A., and Harrison, P. J., 1983, Prorocentrin: An extracellular siderophore produced by the marine dinoflagellate *Prorocentrum minimum, Science* **219**: 306–308.
Triebskorn, R., Kohler, H. R., Zahn, T., Vogt, G., Ludwig, M., Rumpf, S., Kratzmann, M., Alberti, G., and Storch, V., 1991, Invertebrate cells as targets for hazardous substances, *Z. Angew. Zool.* **78**:277–287.
Van Alstyne, K. L., 1988, Herbivore grazing increases polyphenolic defenses in the intertidal brown alga *Fucus distichus, Ecology* **69**:655–663.
Van Pee, K. H., and Lingens, F., 1985, Purification of bromoperoxidase from Pseudomonas aureofaciens, *J. Bacteriol.* **161**:1171–1175.
Villalobos, A. R., Dunnick, C. A., and Pritchard, J. B., 1996, Mechanism mediating basolateral transport of 2,4-dichlorophenoxyacetic acid in rat kidney, *J. Pharmacol. Exp. Ther.* **278**:582–589.
Walker, D. A., Linton, A. E., and Schaffer, C. T., 1974, Sudan Black B: A superior stain to Rose Bengal for distinguishing living from non-living foraminifera, *J. Foram. Res.* **4**:205–215.
Weber, D., Shaw, C. F., and Petering, D. H., 1988, Properties of structure and function of cadmium-binding proteins/peptides from *Euglena gracilis, Mar. Environ. Res.* **24**:159–162.
Weiner, S., and Erez, J., 1984, Organic matrix of the shell of the foraminifer, *Heterostegina depressa, J. Foram. Res.* **14**:206–212.
Weiss, S. J., Test, S. T., Eckmann, C. M., Roos, D., and Regiandi, S., 1986, Brominated oxidants generated by human eosinophils, *Science* **234**:200–203.
Welnhofer, E. A., and Travis, J. L., 1996, *In vitro* microtubule dynamics during experimentally induced conversion between tubulin assembly states in *Allogromia laticollaris, Cell Motil. Cytoskel.* **34**:181–189.

Wester, P., and Vos, J., 1994, Toxicological pathology in laboratory fish: An evaluation with two species and various environmental contaminants, *Ecotoxicology* **3**:21–44.

Williams, A. B., and Jacobs, R. S., 1993, A marine natural product, patellamide D, reverses multidrug resistance in a human leukemic cell line, *Cancer Lett.* **71**:97–102.

Whittaker, R. H., and Feeny, P. P., 1971, Allelochemicals: Chemical interactions between species, *Science* **171**:757–770.

Woodin, S. A., Walla, M. D., and Lincoln, D. E., 1987, Occurrence of brominated compounds in soft-bottom benthic organisms, *J. Exp. Mar. Biol. Ecol.* **107**:209–217.

Wotton, R. S., 1988, Dissolved organic material and trophic dynamics, *BioScience* **38**: 172–178.

Wu, L. N., Genge, B. R., Sauer, G. R., and Wuthier, R. E., 1996, Characterization and reconstitution of the nucleational complex responsible for mineral formation by growth plate cartilage matrix vesicles, *Connect. Tissue Res.* **35**:309–315.

Wuthier, R. E., 1993, Involvement of cellular metabolism of calcium and phosphate in calcification of avian growth plate cartilage, *J. Nutr.* **123**(Suppl):301–309.

Yanko, V., and Kronfeld, J., 1992, Low and high magnesian calcitic tests of benthic foraminifera chemically mirror morphological deformations, *IV Int. Congr. Paleoceanogr., Kiel*, p. 308.

Yanko, V., and Kronfeld, J., 1993, Trace metal pollution affects the carbonate chemistry of benthic foraminiferal shell, *Israel Soc. Ecol. Environ. Qual. Sci., 24th Ann. Meet.*, Tel Aviv University, Tel Aviv, Israel, p. 46.

Yanko, V., Kronfeld, J., and Flexer, A., 1994, Response of benthic foraminifera to various pollution sources: Implications for pollution monitoring, *J. Foram. Res.* **24**:1–17.

Yanko, V., Arnold, A., and Bresler V., 1995, Morphological abnormalities in foraminiferal tests: Response to heavy metal pollution along the Israeli coast, eastern Mediterranean, GSA Annual Meeting, Prog. Abstr., Vol. 27, p. A244.

Yanko, V., Ahmad, M., and Kaminski, M. 1998, Morphological deformities of benthic foraminiferal tests in response to pollution by heavy metals: Implications for pollution monitoring, *J. Foram. Res.* **28**:177–200.

Yanko, V., Arnold, A., and Parker, W., in press, The effect of marine pollution on benthic foraminifera, in: *Modern Foraminifera* (B. Sen Gupta, ed.), Kluwer Academic, Dordrecht.

Yanovich, S., and Taub, R. B., 1983, Differences in daunomycin retention in sensitive and resistant P388 leukemic cells as determined by digitized video fluorescence microscopy, *Cancer Res.* **43**:4167–4171.

Yawetz, A., and Agosin, M., 1979, Epoxide hydrase in *Trypanosoma cruci* epimastigotes, *Biochim. Biophys. Acta.* **585**:210–219.

Yawetz, A., and Agosin, M., 1980, Glutathione-*S*-transferase and drug metabolism in *Trypanosoma cruci: In vivo* and *in vitro* formation of thioethers, *Compar. Biochem. Physiol.* **66C**:265–267.

Yawetz, A., and Agosin, M., 1981, Purification of the glutathione-*S*-transferase of *Trypanosoma cruci, Compar. Biochem. Physiol.* **68B**:237–243.

IV

Disturbance and Recovery Through Time

Chapter 11

Use of Arcellacea (Thecamoebians) to Gauge Levels of Contamination and Remediation in Industrially Polluted Lakes

R. TIMOTHY PATTERSON and ARUN KUMAR

1. Introduction

Arcellaceans (thecamoebians) are freshwater microscopic protozoans, similar to amoebae, that form agglutinated tests, or shells. Occasionally they also occur in brackish water (<5‰) environments (Todd and Brönniman, 1957; Haman, 1982; Hayward *et al.* 1996). Arcellacean distributional studies have been carried out over the past 150 years, mainly in lakes from Europe and North America (see references in Ogden and Hedley, 1980; Tolonen, 1986).

R. TIMOTHY PATTERSON and ARUN KUMAR • Department of Earth Sciences, and Ottawa-Carleton Geoscience Centre, College of Natural Sciences, Carleton University, Ottawa K1S 5B6, Canada.

Environmental Micropaleontology, Volume 15 of *Topics in Geobiology*, edited by Ronald E. Martin. Kluwer Academic/Plenum Publishers, New York, 2000.

These studies were mainly of a reconnaissance nature, primarily concerned with determining the ranges of various species (see references in Medioli and Scott, 1983, 1988, and a complete thecamoebian bibliography at http://meguma.earthsciences.dal.ca/~fmedioli/intro.html).

During the past 15 years the focus of Arcellacean research has shifted to more of a concentration on their potential use as paleoenvironmental indicators. Their high abundance, often several hundred per cubic centimeter, and excellent preservation potential in lacustrine sediments have made them useful in the reconstruction of late Quaternary–Holocene paleoenvironments (Scott and Medioli, 1983; Patterson *et al.*, 1985; Medioli and Scott, 1988; McCarthy *et al.*, 1995; Ellison, 1995). Arcellaceans have a generation time of only a few days. They also develop distinct morphologies in response to environmental stress. These traits have made them excellent indicators of various natural, chemically polluted, and rehabilitated subenvironments within lakes affected by industrial and mining pollution (Asioli *et al.*, 1996; Patterson *et al.*, 1996; Kumar and Patterson, 1997; Reinhardt *et al.*, 1998).

2. Taxonomy and Biology

"Thecamoebians" are an artificial polyphyletic group of protozoans that includes the Order Arcellinida Kent, 1880, Subclass Filosa Leidy, 1879 and Order Gromida Claparède and Lachmann, 1859 (Loeblich and Tappan, 1964). Thecamoebians are commonly referred to in the literature as agglutinated rhizopods or testate amoebae. They are characterized by the presence of pseudopodia of variable nature, an amoeboid sarcodine cell, and a very simple saclike or caplike test. Only a small portion of the order Arcellinida, the arcellaceans, are common in the fossil record with the majority of thecamoebians being of little paleolimnological interest (Medioli and Scott, 1983; Fig. 1).

The test of thecamoebians can be secreted by the organisms themselves (autogenous tests), be proteinacous, siliceous, or rarely, calcareous. Most of the arcellacean subgroup build their tests by agglutinating foreign particles (xenosomes) in autogenous cement to form xenogenous tests. The cement in these tests is usually composed of mucopolysaccharide, highly resistant to most biological and chemical destructive agents. This test type is common in the fossil record. The nature of xenosomes is entirely controlled by the composition of the substrate, and may consist of sand grains and/or diatom frustules (Medioli *et al.*, 1990). They have a long, albeit scanty, fossil record that goes back to the Carboniferous period (Thibaudeau and Medioli, 1986; Thibaudeau *et al.*, 1987; Wightman *et al.* 1994) and is mainly from lakes (Medioli and Scott, 1988) or peat deposits (Warner, 1990; Warner and Charman, 1994; Warner and Bunting, 1996). Their small size (60–300 μm, few species are smaller than 60 μm) and high abundance make it simple to collect statistically significant populations, even from small samples.

Arcellaceans generally reproduce asexually, producing clonal offspring. Sexuality is rare but not totally absent (Volkanov, 1962*a*,*b*, 1966; Mignot and Raikov, 1992). This reproductive strategy has led to serious taxonomic problems within the group. They have usually been grouped into specific units on the basis of test morphology (Medioli and Scott, 1983; Medioli *et al.*, 1987), despite the fact that distinct environmentally controlled arcellacean morphotypes, or strains, occur abundantly in nature. Most systematic work has thus focused on creating new "species" of local interest. The resultant nomenclatural confusion has led to two extreme taxonomic approaches: (a) that all arcellaceans belong to the same species (Wallich, 1864), or (b) that every morphotype that occurs more than a few times represents a genuine species (see the discussion in Medioli and Scott, 1983; Medioli *et al.* 1987; Ogden and Hedley, 1980). Medioli *et al.* (1990) stated that this second approach to arcellacean taxonomy has progressed to such a degree that the vast majority of species used in the literature are so oversplit that they have become useless for paleoecological purposes. As a possible solution for the taxonomic difficulties with the group, Medioli and Scott (1983) proposed that arcellacean species be considered as widely variable groups that, collectively, for any given wild population, accommodate 75% or more of the entire population. Although this species concept is subjective, it does make allowance for a substantial amount of observed morphological instability and does not preclude the identification of informally designated paleoenvironmentally useful infrasubspecific strains (Asioli *et al.* 1996; Reinhardt *et al.*, 1998). This infrasubspecific level classification is quite useful for delineating large populations of arcellaceans into ecophenotypes without describing new species (Fig. 1).

3. Paleoecological Utility

Arcellaceans are found in most areas where there is sufficient moisture (e.g., soils, mires, peat bogs, freshwater to brackish ponds and lakes). Most of the lacustrine taxa prefer oligotrophic lakes with mildly acidic water and are found in reduced numbers in eutrophic lakes (Medioli *et al.*, 1990). Ecological factors controlling the distribution pattern of arcellaceans include: dissolved oxygen content, dystrophy-grade of the lake (C/N ratio in sediment which depends upon the existence of humus compounds), pH, the grain size of sediment, and the existence of *Sphagnum* carpet around the lake (Ellison, 1995; Tolonen, 1986).

A number of studies have used arcellaceans to identify paleoenvironmental parameters (e.g., paleohydrological changes within lakes, lake level changes, paleoclimatological changes) in late Quaternary to Recent sediments (Haman, 1982; Scott and Medioli, 1983; Patterson *et al.*, 1985; Tolonen, 1986; Honig and Scott, 1987; Medioli and Scott, 1983, 1988; Medioli *et al.*, 1985, 1987, and 1990, Collins *et al.*, 1990, McCarthy *et al.*1995; and Ellison, 1995;

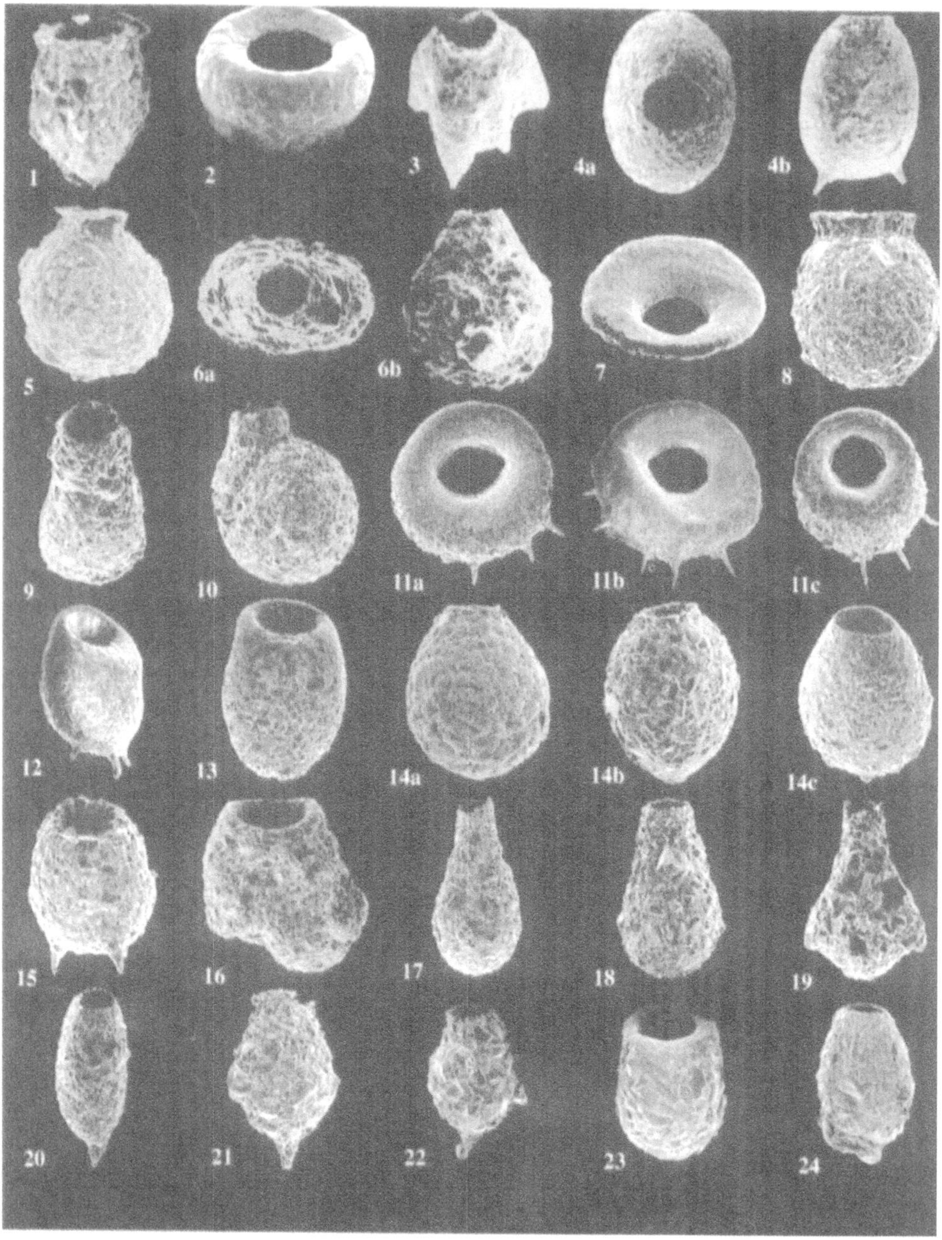

FIGURE 1. Common arcellacean species and strains commonly found throughout North America. For more information on the parameters indicated by these strains see Reinhardt *et al.*, 1998. A complete key to their identification is provided in Kumar and Dalby, 1998: (1) *Difflugia bacillariarum* Perty 1849 (size 100 m, specimen from Lake Erie). (2) *Centropyxis aculeata* Ehrenberg 1832, strain "discoides" Reinhardt *et al.* 1998 (size 160 μm, specimen from Lake Sentani, Irian Jaya, Indonesia). (3) *Difflugia fragosa* Hempel 1898 (size 156 μm, specimen from Lake Erie). (4) *Difflugia bidens* Penard 1902 (size 160 μm, specimen from Swan Lake, north of Toronto, Ontario). (5) *Difflugia urens* Patterson *et al.* 1985 (size 176 μm, specimen from Midway

Asioli *et al.*, 1996). Changes in arcellacean faunas also provide strong evidence of environmental changes in modern and late Quaternary peat deposits (Warner, 1990; Warner and Charman, 1994; and Warner and Bunting, 1996).

Arcellaceans are among the few benthic organisms that preserve well even in low pH freshwater sedimentary environments. Shells of other benthic organisms, such as mollusks and ostracodes, dissolve in the low-pH conditions typical of lake environments after the organisms die (Medioli and Scott, 1988). Other taxa commonly found in samples with arcellaceans, such as pollen grains and diatoms, are not indicative of the lake-bottom environment.

4. Analytical Methods

4.1. Collection and Preparation

A major advantage of working with arcellaceans lies with the very simple and inexpensive sampling and processing methodology. Methodologies for processing arcellaceans are described in detail in Medioli and Scott (1988), Medioli *et al.* (1990), and McCarthy *et al.* (1995). Our processing methods, outlined below, are similar but simpler (Patterson *et al.* 1996; Reinhardt *et al.*, 1998).

Lake, Nova Scotia). (6) *Pontigulasia compressa* Carter 1864 (size 126 μm, specimen from Lake Erie). (7) *Arcella vulgaris* Ehrenberg 1830 (size 124 μm, specimen from Crosswise Lake near Cobalt, Ontario). (8) *Difflugia urceolata* Carter 1864, strain "urceolata" Reinhardt et al. 1998 (size 160 m, specimen from Peterson Lake near Cobalt, Ontario). (9) *Lagenodifflugia vas* Leidy 1874 (size 90 μm, specimen from Peterson Lake near Cobalt, Ontario). (10) *Lesquerausia spiralis* Ehrenberg 1840 (size 93 μm, specimen from Peterson Lake near Cobalt, Ontario). (11) *Centropyxis aculeata* Ehrenberg 1832 strain "aculeata" Reinhardt *et al.* 1998 (size 135 μm, specimen from Crosswise Lake near Cobalt, Ontario). (12) *Centropyxis constricta* Ehrenberg 1843, strain "constricta" Reinhardt et al. 1998 (size 150 μm, specimen from Crosswise Lake near Cobalt, Ontario). (13) *Centropyxis constricta* Ehrenberg 1843 strain "aerophila" Reinhardt et al. 1998 (size 108 μm, specimen from Peterson Lake near Cobalt, Ontario). (14) *Cucurbitella tricuspis* Carter 1856 (size 95 μm, specimens from Peterson Lake near Cobalt, Ontario). (15) *Difflugia corona* Wallich 1864 (size 120 μm, specimen from Peterson Lake near Cobalt, Ontario). (16) *Difflugia globula* Ehrenberg 1848 (size 225 μm, specimen from Crosswise Lake near Cobalt, Ontario; the species name "*globulus*" was changed to "*globula*" by Kumar and Dalby, 1998). (17) *Difflugia oblonga* Ehrenberg 1832 strain "oblonga" Reinhardt *et al.* 1998 (size 104 μm, specimen from Crosswise Lake near Cobalt, Ontario). (18) *Difflugia oblonga* Ehrenberg 1832 strain "tenuis" Reinhardt *et al.* 1998 (size 120 μm, specimen from Crosswise Lake near Cobalt, Ontario). (19) *Difflugia oblonga* Ehrenberg 1832 strain "triangularis" Reinhardt *et al.* 1998 (size 255 μm, specimen from Crosswise lake near Cobalt, Ontario). (20) *Difflugia protaeiformis* Lamark 1816 strain "acuminata" Reinhardt *et al.* 1998 (size 98 μm, specimen from Crosswise Lake near Cobalt, Ontario). (21) *Difflugia protaeiformis* Lamark 1816 strain "amphoralis" Reinhardt *et al.* 1998 (size 75 μm, specimen from Peterson Lake near Cobalt, Ontario). (22) *Difflugia protaeiformis* Lamark 1816 strain "claviformis" Reinhardt *et al.* 1998 (size 75 μm, specimen from Peterson Lake near Cobalt, Ontario). (23) *Centropyxis constricta* Ehrenberg 1843 strain "aerophila" Reinhardt *et al.* 1998 (size 142 μm, specimen from Peterson Lake near Cobalt, Ontario). (24) *Difflugia oblonga* Ehrenberg 1832 strain "glans" Reinhardt *et al.* 1998 (size 146 μm, specimen from Lake Sentani, Irian Jaya, Indonesia).

A small boat provides an adequate sample collection platform in small lakes and ponds. We usually use an Eckman box corer but any similar sampling device is suitable. We also find that an inexpensive commercial sonar device (fish finder) equipped with a bottom hardness indicator is helpful in determining appropriate sample sites. Not only do these sonars measure water depth but with practice they can be used to distinguish rocky, muddy, or sandy substrates. Only muddy sites are sampled. Winnowed sandy substrates may have small allochthonous arcellacean communities and rocky substrates are normally barren. At each sampling station we also measure a variety of physical parameters (e.g., water depth, sedimentology, pH, water temperature, and oxygen concentration) to assist us in interpreting our results. We also usually determine the geographic location of each sampling station using a global positioning system (GPS) unit. In very small lakes siting on nearby landmarks may provide sufficient positioning accuracy.

For distributional studies the upper 2–3 mm of sediment (arcellaceans live at the sediment–water interface (Medioli and Scott, 1988; Medioli *et al.* 1990) from each Eckman grab are removed to obtain a 1-cc subsample for micropaleontological analysis. To avoid decay, all samples are treated with isopropyl alcohol and refrigerated immediately after collection. The samples are sieved through a 1000-μm screen to remove coarse organics, then through a 55-μm screen to retain arcellaceans and remove silts and clays.

We usually do not use biological stains (e.g., Rose Bengal) to detect living protoplasm. Owing to the rapid generation time for these arcellaceans, total (living and dead) populations provide a better estimate than living populations of seasonal standing crop (Scott and Medioli, 1980*a*).

Owing to the generally large arcellacean populations counted in most samples, these organisms lend themselves particularly well to statistical analysis (see Fishbein and Patterson, 1993, for recommended strategies). All micropaleontological samples are subdivided into aliquots for quantitative analysis using a wet splitter (as described by Scott and Hermelin, 1993). The subsamples may be examined either wet or dry depending on their organic content. Wet examination of organic-rich lacustrine sediments is necessary because in dried samples the vegetable debris mats together, making identification and quantitative analysis of arcellaceans nearly impossible. Wet aliquots are then examined under a binocular microscope and a statistically significant number of specimens are counted for each sample (see Patterson and Fishbein, 1989 for details).

4.2. Geochemical Analysis

The surface area of most Eckman box samplers is sufficiently large that the same grab can provide enough material for geochemical as well as micropaleontological analysis. We recommend analyzing sediment pore water rather than obtaining a bulk geochemical analysis. This is because the chemical makeup of compounds and elements found in pore water are in forms that can be generally directly ingested or absorbed by most organisms, including arcellaceans (Luoma, 1983; Campbell, 1995). The compounds and elements of the clays, silts, and other solids as recorded by a bulk geochemical analysis are in forms not generally available to organisms and may suggest erroneous correlations. We recommend that samples be analyzed using the Environmental Protection Agency's method 3051: microwave digestion for inductively coupled plasma atomic emission spectroscopy (ICP) analysis (United States Environmental Protection Agency, 1990). Additional analyses may also be carried out to measure arsenic and selenium levels using graphite furnace atomic absorption spectroscopy (GFAAS). Cold-vapor atomic absorption is used to analyze mercury levels in the samples as other metals in the ICP analysis obscure the characteristic wavelength of this element. All of these methodologies are readily available through most commercial environmental analysis firms.

5. Case Studies

Human settlement and tourism have exerted environmental stress on lacustrine ecosystems worldwide by causing eutrophication and acidification of lake waters and their bottom sediments. Severe environmental damage has also been caused to lakes by dumping of industrial pollutants and mine tailings. We will discuss application of arcellacean micropaleontological analysis in three industrially polluted lakes in northeastern Ontario, Canada (Fig. 2A). Two of these, Peterson and Crosswise lakes, are situated in the Cobalt region of northern Ontario. The third, James Lake, lies near the town of Temagami, Ontario. The results of these studies demonstrate the potential use of arcellaceans as cost-effective tools in monitoring the effects and rates of remediation in industrially polluted lakes.

5.1. Cobalt Area, Northeastern Ontario

Silver was discovered at Mile 103 of the Northern Ontario Railroad during the summer of 1903, and by 1911 silver production at Cobalt Camp exceeded 30,000,000 oz. (850,000,000 g) a year. In 8 years the boomtown of Cobalt had risen out of the forest to become the world's largest silver producer. Silver production, and the town's fortunes, began to tail off by the 1930s and by 1993 there were no active silver mines in the area (Murphy, 1977; Barnes, 1986; Dumaresq, 1993). An unfortunate legacy of the mining boom is the high level of environmental contamination of many area lakes and streams. These contaminants, primarily arsenic and mercury with significant amounts of nickel, cobalt, and cyanide, pose a signicant health risk for the 10,000 people still in the area.

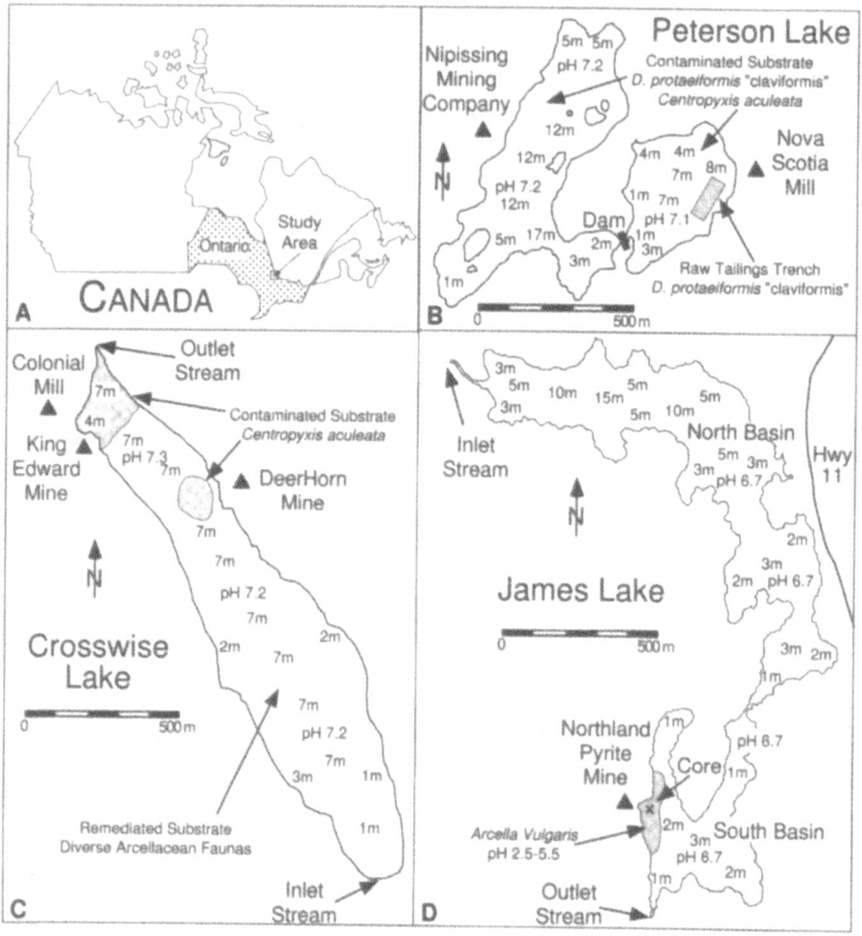

FIGURE 2. (A) Locations of studied lakes in northeastern Ontario. (B) Map of Peterson Lake in the Cobalt area showing general pH, bathymetry, location of mines and mills, as well as areas inhabited by various arcellacean faunas, particularly the *Difflugia protaeiformis* "claviformis" dominated fauna found in the highly contaminated excavated trench. (C) Map of Crosswise Lake in the Cobalt area showing general pH, bathymetry, location of mines and mills as well as areas inhabited by various arcellacean faunas, particularly areas of the lake showing clear indication of remediation. (D) Map of James Lake in the Temagami area showing general pH, bathymetry, location of the pyrite mine, core site, and areas inhabited by the low pH indicating *Arcella vulgaris* fauna.

During mining operations waste rock and fine mill tailings were dumped in the most convenient low-lying areas, usually lakes (Murphy, 1977; Barnes, 1986; Dumaresq, 1993). The silver in the ores was associated with arsenic minerals, most of which went into the tailings, and ultimately into area lakes. Huge quantities of mercury and cyanide were used in the ore-milling process. These dangerous contaminants found their way into the tailings and lakes as well (Dumaresq, 1993).

5.1.1. Major Contaminants in the Cobalt Area

The ores in the Cobalt area contain arsenic as sulfide, sulfarsenide, and arsenide minerals. In aerobic waters, arsenate is the most common arsenic-bearing solute, whereas under anaerobic conditions arsenite is common. Arsenic does not go readily into solution in the pore waters of lake sediments because it is adsorbed onto iron and manganese oxyhydroxides (Cullen and Reimer, 1989). Owing to its complex chemistry, arsenic transformations are common, resulting from changes in pH, Eh, temperature, and biological activity. Lower forms of aquatic animal life and bottom-feeding fish tend to accumulate the greatest amounts of arsenic. Arcellaceans, which are benthic organisms near the bottom of the food chain, are thus excellent indicators of arsenic contamination.

Mercury occurs in solid solution with native gold and silver (Berry *et al.* 1983) and with several sulfide minerals, particularly tetrahedrites (NRCC, 1978). Since mercury was used as an amalgam in the recovery of gold and silver, it is a very common metal in the mine tailings. Both mercury and methylmercury are available for organisms to take up and thus bioconcentration occurs, but bacterially mediated methylation is the most important bioconcentration process. The rate of methylation is pH-dependent, with rates in lake water at pH 4.5 being seven times faster than at pH 8.5 (D'Itri, 1991). Once formed, methylmercury is readily absorbed by organisms and its excretion is very slow (biological half-life of mercury in fish is 2 years; D'Itri, 1991; CCREM, 1987). In natural aerobic waters, elemental mercury is eventually oxidized and removed from the water column by sorption onto suspended and bottom sediments (CCREM, 1987). Nonetheless, mercury has been shown to be bioconcentrated in ecosystems, and higher trophic consumers contain higher mercury concentrations than organisms lower in the food chain (Cuthbert, 1992). Therefore arcellaceans may not be very sensitive indicators of mercury contamination.

5.1.2. Peterson Lake

Peterson Lake (Fig. 2B) is subdivided into an eastern and a westem basin by a man-made dam. The larger western basin of Peterson Lake (2.3 km^2) has a thermocline at a depth of about 8 m and has three distinct bathymetric areas: (1) a shallow weed-filled southern end; (2) a small shallow bay near the dam and; (3) a deep basin in the middle. The smaller eastern basin (0.8 km^2) has a flat bottom, the result of nearly 400,000 tons of tailings deposited from the nearby Nova Scotia Mill between 1910 and 1918. Additional tailings spilled into the lake when a tailings dam adjacent to the Nova Scotia Mill site broke. In 1965, the eastern basin was drained and nearly 55,000 tons of tailings were removed. In the nearshore area these tailings were dredged to a depth of 9.0 m. Tailings continue to migrate into the eastern basin from leaks in containment dams adjacent to the Nova Scotia Mill site (Dumaresq, 1993; Wallis, 1993; Patterson *et al.*, 1996, 1997).

The pH is slightly alkaline, ranging between 7 and 8. Arsenic levels measured from the substrate in the eastern basin range up to 8300 ppm and mercury levels reach 4.89 ppm, both well above acceptable levels (CCREM, 1987; Dumaresq, 1993; Patterson *et al.*, 1996, 1997; Reinhardt *et al.*, 1998).

5.1.3. Crosswise Lake

Crosswise Lake (Fig. 2C) has an elongate shape in a north–south direction with an area of approximately 3.2 km^2 (Patterson *et al.*, 1996). Fine tailings from five different mines were dumped into the northern portion of this lake from 1905 to 1970. This dumping shortened the lake by about 300 m and reduced its depth from about 14 to 7 m. The shallowing of the lake has produced a very flat bottom and destroyed the thermocline, in strong contrast to Peterson Lake. As in Peterson Lake the pH ranges between 7 and 8, and oxygen solubility is 11.0 ppm at the surface and 8.0 ppm at the bottom (Dumaresq, 1993; Patterson *et al.*, 1996).

Previous studies identified mercury and arsenic as the major pollutants. Concentrations for arsenic and mercury range up to 7100 ppm and 5.7 ppm, respectively, in the substrate. Arsenic values of up to 27.25 ppm were measured in the lake water (Dumaresq, 1993; Patterson *et al.*, 1996; Reinhardt *et al.*, 1998). Unfortunately, no mercury values for lake water were obtained. Pore-water arsenic and mercury concentrations from the uppermost layer of the bottom sediment, where most benthic biological activity occurs, were also not obtained. However, unpublished measurements of heavy metal concentrations in nearby James Lake indicate that, as would be expected, pore-water values are invariably much higher than in the water column. To put these levels of contamination in perspective, natural freshwater arsenic concentrations are typically less than 2.0 ppm and the maximum acceptable concentration in drinking water and for aquatic life is 50.0 ppm. Background concentrations of mercury in Canadian freshwater sources are usually about 0.05 ppm, the maximum acceptable concentration in drinking water is 1.0 ppm, while the maximum acceptable concentration for aquatic life is 0.1 ppm (CCREM, 1987).

5.1.4. Arcellacean Assemblages in Crosswise and Peterson Lakes

Several distinct arcellacean assemblages are recognized in sediment samples from these lakes and correlate well with various distinct polluted and remediated environments. The hallmark of the most polluted environments is a dramatic reduction in diversity of arcellacean species or "strains," with one or two of them clearly dominant. Arcellacean infrasubspecific strains sometimes discriminate among environments better than species units (Patterson *et al.*, 1996; Reinhardt *et al.*, 1998), and their use is recommended when studying lake microenvironments, pollutants, and rates of lake remediation. For example, the most highly contaminated substrate in these lakes, in the excavated trench in Peterson Lake, is dominated by *Difflugia protaeiformis* "claviformis" (Fig. 2B). This strain is opportunistic and able to thrive in areas where high levels of pollutants (e.g. Hg, As, Cd, Cr, Cu, Pb) in these lakes would preclude most species. This relationship has also been observed in Italian lakes (Asioli *et al.*, 1996).

Other important indicators of hostile conditions include the centropyxids, particularly the various strains of *Centropyxis aculeata* (Fig. 2B,C). Centropyxid species are capable of withstanding a variety of hostile conditions better than most other arcellacean species. These conditions include cold temperatures (Decloitre, 1956), low salinities (<5%) Decloître, 1953; Scott and Medioli, 1980*b*; Patterson *et al.*, 1985; Honig and Scott, 1987), low nutrient levels, oligotrophic conditions (Schönborn, 1984), and sites heavily contaminated by mercury and arsenic (Patterson *et al.*, 1996). Hostile conditions in substrates dominated by centropyxids and *Difflugia protaeiformis* "claviformis" are further indicated by the generally low diversity (Shannon Diversity Index (SDI) values of <0.5; defined in Fig. 3) and abundance (between 30 and 150 specimens/cc). These faunas are dominated by only one or two taxa (species and/or strains) with most species being rare. In contrast, healthy arcellacean faunas usually have SDI values greater than 2.5 and abundances of over 500 specimens/cc. As in most stable climax communities, there is an equitable distribution of species in these healthy environments with none overwhelmingly dominating the fauna. Various strains of *Difflugia oblonga* typically characterize these assemblages.

In addition to providing easy recognition of contaminated areas of these lakes arcellacean assemblages provide data on rates of substrate remediation (Patterson *et al.*, 1996; Reinhardt *et al.*, 1998). The remediation of lakes contaminated by heavy metals presents unique problems. Unlike organic pollutants, such as polychlorinated biphenyls (PCBs) and benzene compounds that can be broken down, metal pollutants are common in, or quickly revert to, their natural state (e.g., stable mineral form) without losing their toxicity. Thermodynamic processes dictate that stable minerals will not break down naturally to other less toxic compounds. Possible solutions that have been suggested include the conversion of toxic metals to harmless compounds so that they would not be bioavailable. Bacterial processes can accomplish this by either reducing or oxidizing the contaminants (S_2^-, C_2O_3, OH^-) (Wildeman *et al.*, 1994). Unfortunately, mine tailings sites such as those at Crosswise and Peterson lakes often contain such a large amount of heavy metal pollutants that these processes are not practical and other remediation methods need to be explored.

In Crosswise and Peterson lakes, we observed natural remediation taking place as evidenced by the return of vegetation and "normal" arcellacean faunas in parts of the lakes (Patterson *et al.*, 1996; Reinhardt *et al.*, 1998). The return of vegetation accelerates the rate of natural remediation by stabilizing cover material that in effect "caps" the tailings. For this process to occur, the pH must be relatively neutral (Forstner and Wittmann, 1981). As the lakes in the Cobalt area are alkaline, this was not an issue, but in order to remediate

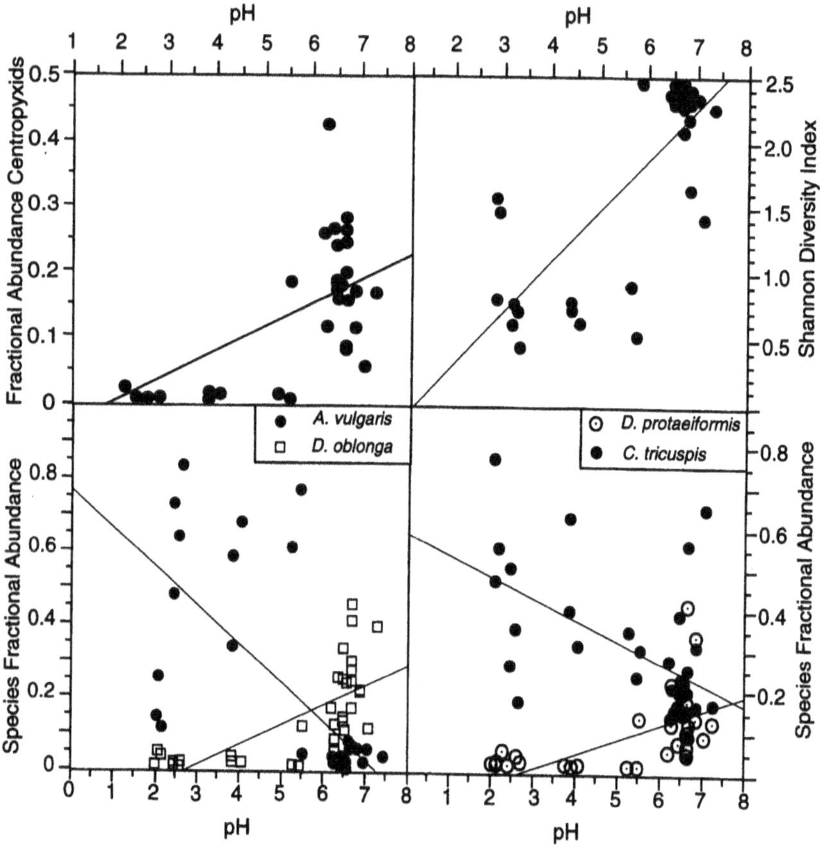

FIGURE 3. Relative proportion of various taxa, number of species, and SDI in relation to pH in James Lake. The linear least-squares method was used to determine the best fit for the linear curves using Deltagraph 4.0 (Deltapoint, 1996). The SDI $H(S) = -\Sigma p_i^* \ln(p_i)$, where p_i is the proportion of the ith species (and/or strains) in the assemblage. SDI is a better measure of diversity than numbers of species because it also takes into account the relative proportions of species (equitability) in the population.

acidic lakes, they must be made pH neutral (e.g., by adding large quantities of lime).

An excavated trench in the eastern portion of Peterson Lake represents a subenvironment that is devoid of vegetation and is the most heavily polluted site in the lakes examined (Fig 2B). Interestingly it was at this location that artificial remediation was attempted. The dredging of about 55,000 tons of sediment in 1965 disturbed the tailings that had already settled, destabilizing the pollutants and creating a subenvironment of pure tailings in contact with the lake. This procedure was not attempted in Crosswise Lake or in the western portion of Peterson Lake. In these areas the process of natural remediation is well underway and a vegetative layer is forming, or has formed, creating a barrier between the tailings and the environment (Fig. 2C). Remediation is progressing particularly well in Crosswise Lake. Despite being floored by over 7 m of contaminated tailings, the natural capping process has resulted in development of near normal arcellacean faunas in all areas except those adjacent to mine and mill operations active up until 1970.

Successful lake remediation in these and similarly polluted lakes is thus best achieved by leaving the tailings undisturbed to be buried naturally or by speeding the process by addition of an allochthonous sediment cap. In neutral pH settings such as those found in the Cobalt area lakes, our results suggest that only a thin cap (a few mm thick) of natural sediment and vegetative cover is required to be effective. As our arcellacean faunal analysis indicates, dredging of tailings only serves to nullify any natural remedial effects that have already occurred. In addition, when tailings are removed, a new location must be found for them, thus moving the problem elsewhere rather than solving it.

5.2. James Lake, Temagami Area, Northeastern Ontario

James Lake is a narrow, mesotrophic; C-shaped lake located along Highway 11 north of Temagami (Fig. 2D). The lake covers an area of 45.3 hectares, and is elongated in a north–south direction. The lake is fed by an inlet stream at the north end and drained by an outlet stream at the south end. A narrow region divides the lake into north (80%) and south (20%) basins. The southern basin is shallow, reaching a maximum depth of only 4.0 m. The northern basin is deeper with a maximum depth of 15.0 m (Fig. 2D). The lake is stratified in the northern basin, with both oxygen levels and temperature dropping significantly below 5.0 m depth. During the summer, temperature and oxygen levels in the upper epilimnion are 25°C and 9.0 mg/liter, respectively. Temperature and oxygen concentration drops to 10°C and 2.0 mg/liter, respectively, in the lower hypolimnion.

The Keewatin age volcanic rocks surrounding the lake are quite rich in pyrite. Pyrite was discovered in lenses within soft green schists in 1903. The Northland Pyrite Mine Co. was operational on the southwest shore of the lake from February 1906 to March 1911, shipping more than 38,000 tons of pyrite to Cobalt, a short distance to the north. Pyrite was used for making sulfuric acid for the milling of silver ore. Most waste rock (about 3,500 m^3 containing 25% pyrite) from this mine was dumped on the southwest lakeshore. Rainwater percolating through the waste rock becomes acidified and has contributed to acidification of the adjacent lake water and bottom sediments. The water flow leaches sulfates derived from the pyrite and pyrrhotite minerals. In areas near the waste rock pile sulfate concentrations in the sediment are extremely high, up to 7500 μg/g in places. The interaction between these sulfate ions and hydrogen ions from water produces sulfuric acid. In addition, localized bacterial reduction of some sulfate to H_2S may contribute to development of a toxic benthic environment for many aquatic invertebrates (Environment Canada, 1979). At the times of sampling there was a gradual change in acidity from a low of pH 2.0 in some bottom sediments adjacent to the waste rock piles to almost neutral conditions (pH 6.8) in more distant areas of the southern basin of the lake. The situation of the outlet stream adjacent to the mine site and the overall north–south flow of water in the lake help maintain this gradient.

Several metals, most notably iron and aluminum, are being leached out of the waste rock. Sediment pore water iron levels vary from a low of 1.52 mg/liter in the northern basin to a high of 11,800 mg/liter near the waste rock pile. Iron concentration in lake water varies from 0.09 mg/liter in the northern basin to 2.4 mg/liter near the waste rock pile (Gale, 1990). During freshet and particularly during spring turnover, the pH of the lake water rises to nearly neutral values throughout the lake (Gale, 1990). When the acidic, metal-laden water near the waste dump mixes with the neutral water of the rest of the lake the metal precipitates out as iron hydroxide (FeOH). All measured pore-water and lake-water iron concentrations were found to be well in excess of the maximum value (0.3 mg/liter) set by provincial drinking-water guidelines.

Iron is so plentiful in the environment that very high levels often accumulate in invertebrates. Since it is an essential trace element, a certain amount of bioconcentration can occur with little ill effect (Vymazal, 1984; Tessier *et al.*, 1984). The low drinking-water guideline standard is based more on aesthetics than on health concern. The taste of iron can readily be detected at 1.8 mg/liter and high concentrations also lead to staining of laundry and plumbing and massive growth of bacteria in water systems (Moore, 1991). The recommended daily intake is 10 mg for men and 18 mg for women. The daily intake of iron from drinking water containing 0.3 mg Fe/liter would be only 0.6 mg. Both Gale (1990) and ourselves observed a large number of vertebrates (fish and amphibians) in the lake, although mostly in the eastern part of the southern basin. Although guidelines for the protection of aquatic life range from 0.3 to 1.0 mg/liter, the tolerance of these organisms seems to be much higher ($>$10 mg/liter; Moore, 1991).

Aluminum concentrations in pore water varied from 0.19 to 415 mg/liter near the waste rock pile. Aluminum values in the lake water itself varied from 0.05 mg/liter in the northern basin to 0.24 mg/liter near the waste rock pile (Gale, 1990). Aluminum is not essential for survival but is found in virtually all plant and animal species. In higher pH regimes aluminum complexes into relatively stable complexes (Plankey and Patterson, 1987, 1988). However, in lower pH environments ($pH < 5.5$), such as found in the southern basin of James Lake, it mobilizes into biologically usable forms (Burrows, 1977). The guidelines for control of aluminum are highly variable and reflect differing opinions on the hazard posed by aluminum in drinking water, but 0.2 mg/liter seems to be the maximum allowable concentration agreed to by most agencies (Moore, 1991). The guidelines for the protection of aquatic life in Canada and several European nations are 0.005 mg/liter at $pH < 6.5$ and 0.1 mg/liter at $pH > 6.5$.

5.2.1. Arcellacean Assemblages in James Lake

James Lake provides an ideal laboratory for assessing the sensitivity of arcellaceans to industrial pollutants (Kumar and Patterson, 1997). The flow of lake water from north to south, exiting near the pollution point source at the old mine site, has created habitats that range from unimpacted in the northern basin to extremely contaminated near the mine site itself. Furthermore, the low-pH environment at James Lake provides an instructive contrast to the near neutral pH lakes of the Cobalt area. While both areas are contaminated by toxic heavy metals, the different acidity levels are reflected in the respective arcellacean faunas.

Arcellacean faunas in James Lake mirror the highly contaminated low pH areas quite well. The highly iron- and aluminum-contaminated, acidic areas of the lake (pH 2.0–5.5) adjacent to the mine waste rock pile are characterized by low-diversity faunas with SDI values of generally less than 1.0 and an average of less than six species per sample. In higher-pH areas (pH 6.5–7.5), where the level of contamination was low the SDI increased dramatically to 1.5–2.5 with most values greater than 2.0 (Fig. 3).

The distribution of individual taxa in James Lake indicate that pH, rather than iron and/or aluminum, may be the dominant factor controlling arcellacean faunas in the lake. *Arcella vulgaris* is the dominant species (90–100%) in the most contaminated areas at less than pH 5.5, and forms less than 5% of the total assemblage (or is totally absent) in the less-contaminated regions at pH between 6.5 and 7.5 (Fig. 3). *A. vulgaris* is an important component of arcellacean faunas in boggy ponds in the Arctic and further south. The low pH typical of these ponds has preadapted this species to dominate similar low-pH environments. Other indications that pH may be the dominant controlling factor on arcellacean distribution is the notable absence of opportunistic centropyxid taxa. In higher-pH environments in the Cobalt area species such as *Centropyxis aculeata* dominate contaminated substrates (Patterson *et al.*, 1996; Reinhardt *et al.*, 1998). Also notably missing from the lower-pH environments of this lake are any strains of *Difflugia protaeiformis*, although the species is often abundant in portions of James Lake with pH of 6.5–7.5, and in higher-pH and highly contaminated areas of Peterson and Crosswise lakes.

In the higher pH environments of the lake there is a greater abundance and diversity of arcellacean species and phenotypes, although, as found in most northern lakes, various strains of *Difflugia oblonga* usually dominate (25–50%; Fig. 3). Other taxa, such as *Difflugia corona*, *Lesquerasia spiralis*, *Difflugia urceolata*, and *D. protaeiformis*, also occur significantly (10–50%) at pH 6.5–7.5, and do not occur in this lake at pH $<$5.5. The highly variable proportion of the other arcellacean taxa recorded in higher-pH areas of James Lake reflects the presence of a variety of subenvironments not discussed here (Fig. 3; Patterson and Kumar, unpublished data).

5.2.2. Paleolimnological Analysis at James Lake

A major impetus for research in the James Lake area was concern by local area cottagers and wilderness outfitters that ongoing contamination from the site was having an adverse effect on the environment. Although it is obvious that mining activity has had a deleterius effect on the lake, some key information must be obtained before an effective ecosystem management plan for James Lake can to be put in place. This information includes knowledge of baseline conditions and natural variability, identification of the time when conditions in the lake first began to change, and a determination of possible outcomes (Ford, 1988). This information includes a temporal component and thus requires long-term data so that realistic targets for remediation efforts can be set, anthropogenic activity can be discerned and measured, and future scenarios inferred (Likens, 1988; Elliot, 1990; Smol, 1992). For a situation like that at James Lake, aquatic ecosystem managers generally choose from four main sources of data to address the objectives above. These include direct historical measurements, space-for-time substitution (i.e., comparing chemis-

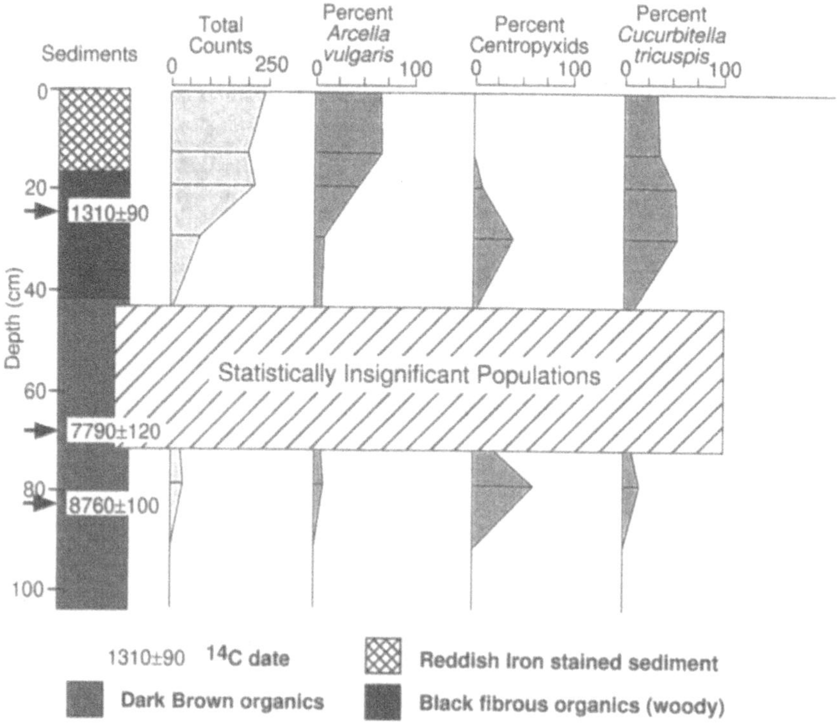

FIGURE 4. Sedimentology, biostratigraphy, and ^{14}C dates from a core collected in James Lake adjacent to the waste rock pile. The ^{14}C date was obtained from a core collected for geochemical analysis (Mason, 1998) less than 1 m away.

try and biota in similar but unaffected lakes), computer models based on empirical or dynamic data, and paleolimnological constructions (Smol, 1992).

At James Lake, and with most lakes, direct historical measurements are rarely available for the time frame of interest. Owing to the unusual configuration of the lake and the position of the pollution source with regard to the outlet, a comparative approach was possible within the lake itself. Both geochemical analysis and the observed arcellacean faunas indicate a highly contaminated area adjacent to the old mine site, and it is obvious that mining activity has seriously impacted the lake. However, according to the requirements for effective aquatic management, a determination of baseline conditions must be made prior to any remedial action.

Paleolimnological methods are the best approach for determining baseline conditions. Unfortunately, except where the sediment sink is large and stable, geochemical methods do not always provide an accurate depiction of the historical record (Horowitz, 1991; Bethke, 1996). Preservable micropaleontological bioindicators such as arcellaceans are thus the best tools to make a temporal assessment of the history of pollution and/or remediation of the contaminated area. Microfossils, except in areas of excessive bioturbation, do not get mixed and archive data on scales varying from millennia to single seasons provide invaluable information to ecosystem managers.

Our paleolimnological analysis is based on a 1-m core obtained from the lake-bottom sediments adjacent to the waste rock site (Fig. 2D). The upper 30 cm of the core are dominated by *Arcella vulgaris* faunas in both a black organic-rich unit and iron-stained sediment (top 18 cm), indicating deposition under low pH and probably highly contaminated conditions (Fig. 4). Lower intervals of the core are dominated by centropyxids indicating environmentally stressed but higher pH conditions. The high abundances of the seasonally floating arcellacean *Cucurbitella tricuspis* in most samples is the result of current transport unrelated to substrate conditions at the core site (Schönborn, 1984). Prior to obtaining ^{14}C dates it was assumed that the rise of the *A. vulgaris* dominated fauna coincided with the initiation of mining activity and lake acidification. As the average sedimentation rate in lakes varies between 1 and 5 mm/year (Förstner and Wittmann, 1983; Kukal, 1971) we assumed that the interval dominated by *A. vulgaris* was deposited after mine activity began. However, seven ^{14}C dates obtained from three cores, one collected less than 1 m from the one studied here, indicates that sedimentation rates in the area of the lake adjacent to the mine site have been very low for at least the past 9000 years (Mason, 1998). Low-pH, contaminated conditions have thus characterized the site for over 2000 years, and this site was a stressed environment for several thousand years prior to that. Thus natural acidification of the site, from sulfates leaching from naturally exposed large pyrite veins, began long before any anthropogenic contribution. This is not an unusual occurrence as any area characterized by metal-bearing formations will have values of those metals occurring at elevated levels (see Förstner and Wittmann, 1983 for examples). Detection of these elevated levels is a major exploration tool used by mineral exploration companies searching for ore. However, this reality is often unknown to environmental scientists, who have not had geological training, and who may assume that any elevated contaminant levels in an area is anthropogenic. In the case of James Lake it seems that the most grievous damage caused by the mining activity concerns aesthetics and safety, the result of huge piles of unsightly waste rock, deep shafts, and fallen head frames. Although mining may have exacerbated pollution in an already highly contaminated and acidified environment, the pollution was clearly present already. Geochemical analysis of cores in the area provide corroborative evidence that elevated levels of iron, aluminum, and sulfate have existed for thousands of years at this site (Mason, 1998).

6. Summary

Arcellaceans are sensitive to a whole host of natural and anthropogenic environmental factors including pH and heavy metal contamination, and as such are excellent bioenvironmental indicators. Because of their asexual reproductive mode the generation of environmentally induced phenotypes is particularly useful in identifying various chemically polluted and remediated benthic environments in lakes. Unlike other lacustrine bioindicators, such as mollusks and ostracodes, arcellaceans preserve well in the generally lower pH conditions found in lakes and thus are useful to track paleolimnological conditions on scales varying from millennia to single seasons. They are also very abundant with several hundred specimens being found per cubic centimeter, making them ideal for statistical analysis.

ACKNOWLEDGMENTS: NSERC Research Grant OGP 0041665 TO RTP supported this research. We gratefully acknowledge the assistance of students in the Department of Earth Sciences Environmental Geology Field Camp (67.285) for their help in the field.

References

Asioli, A., Medioli, F. S., and Patterson, R. T., 1996, Thecamoebians as a tool for reconstruction of paleoenvironments in some Italian lakes in the foothills of the southern Alps (Orta, Varese and Candia), *J. Foram. Res.* **26**(3):248–263.

Barnes, M., 1986, Fortunes in the ground:Cobalt, Porcupine and Kirkland Lake, *Boston Mills Press*, Erin, Ontario, 263 pp.

Berry, L. G., Mason, B., and Dietrich, R. V., 1983, *Mineralogy*, W. H. Freeman, San Francisco, 561 p.

Bethke, C. M., 1996, *Geochemical Reaction Modelling*, Oxford University Press, New York, 397 pp.

Burrows, W. D., 1977, Aquatic aluminum: Chemistry, toxicology and environmental prevalence, *CRC Crit. Rev. Environ. Contr.* **7**:167–216.

Campbell, P. G. C., 1995, Interactions between trace metals and aquatic organisms: A critique of the free-ion activity model, in: *Metal Speciation and Bioavailability in Aquatic Systems* (A. Tessier and D. R. Turner eds.), John Wiley & Sons, New York.

CCREM, 1987, Canadian water quality guidelines:Task Force on Water Quality Guidelines of the Canadian Council of Resource and Environment Ministers.

Claparède, E., and Lachmann, J., 1859, E'tudes sur les infusoires et les rhizopodes, Tome I (2me partie): Me'moires de l'Institut National Genevois, Vol. 6 (for 1858), pp. 261–242.

Collins, E. S., McCarthy, F. M., Medioli, F. S., Scott, D. B., and Honig, C. A., 1990. Biogeographic distribution of modern thecamoebians in a transect along the eastern North American coast, in: *Paleoecology, Biostratigraphy, Paleoceanography and Taxonomy of Agglutinated Foraminifera* (C. Hemleben, M. A. Kaminski, W. Kuhnt, and D. B. Scott, eds.) NATO Advanced Study Institute Series, Series C, *Mathematical and Physical Sciences,* Vol. 327, pp. 783–791.

Cullen, W. R., and Reimer, K. J., 1989, Arsenic speciation in the environment. *Chem. Rev.* **89**:713–764.

Cuthbert, I. D., 1992, A regional assessment of mercury contamination in yellow perch (*Percaf-lavescens*) in Ontario: Implications for common loons (*Gavia immer*), in: *Proc. 18th Ann. Aquatic Toxicology Workshop,* September 30–October, 3 1991. Ottawa, Ontario (A. J. Niimi and M. C. Taylor, eds.), pp. 87–97.

Decloître, L., 1953, Recherches sur les Rhizopodes thecamoebiens d'A. O. F. (Suite). Faune du Lac Tamna (Senegal), *Bull. Inst. Franc, ais Afri. N.,* No. 31:249.

Decloître, L., 1956, Les thecamoebiens de l'Eqe (Groenland): Expeditions Polaires Francaises Missions Paul-Emile Victor VIII, *Actualites Scientifiques et Industrielles* 1242, 100 pp.

Deltapoint, 1996. Deltagraph User's Guide. Monterey, Cal.

D Itri, E. M., 1991, Mercury contamination—what we have learned since Minimata, *Environ. Monitor. Assess.* **19**:165–182

Dumaresq, C. G., 1993, The occurrence of arsenic and heavy metal contamination from natural and anthropogenic sources in Cobalt area of Ontario. Unpub. M.Sc. thesis, Carleton University and Ottawa, Carleton Geoscience Centre, Ottawa. 326 pp.

Elliot, J. M., 1990, The need for long-term investigation in ecology and the contributions of the Freshwater Biological Association, *Freshwater Biol.* **23**:1–5.

Ellison, Robert L., 1995, Paleolimnological analysis of Ullswater using testate amoebae, *Paleolimnology* **13**:51–63.

Environment. Canada, 1979, *Water Quality Source Book.*

Fishbein, E., and Patterson, R. T., 1993, Error-weighted maximum likelihood (EWML): A new statistically based method to cluster quantitative micropaleontological data, *J. Paleontol.* **67**(3):475–485.

Ford, J., 1988, The effects of chemical stress on aquatic species composition and community structure, in: *Ecotoxicology: Problems and Approaches* (S. A. Lewin, M. A. Harwell, J. R. Kelly, and K. D. Kimball, eds.), Springer-Verlag, New York, pp. 99–144.

Forstner, O., and Wittmann, G. T. W., 1983, Heavy metals in industrial and domestic emuents, in: *Metal Pollution in the Aquatic Environment,* Springer-Verlag, Heidelberg, Ch. G2, pp. 340–351.

Gale, P., 1990, James Lake, Northland Pyrite Mine. Ontario Department of the Environment Technical Memorandum, Sudbury, Ontario, 15 pp.

Haman, D., 1982, Modern Thecamoebinids (Arcellinida) from the Belize Delta, Louisiana, *Trans. Gulf Coast Assoc. Geol. Soc.* **32**:353–376.

Hayward, B. W., Grenfell, H., Cairns, G., and Smith, A., 1996, Environmental controls on benthic foraminiferal and thecamoebian associations in a New Zealand tidal inlet, *J. Foram. Res.* **26**(2):150–171.

Honig, C. A., and Scott, D. B., 1987, Post-glacial stratigraphy and sea-level change in southwestern New Brunswick, *Can. J. Earth Sci.* **24**:354–363.

Horowitz, A. J., 1991, *A Primer on Sediment-Trace Element Chemistry, 2nd Rev. Ed.* Lewis, Chelsea, Mich., 136 pp.

Kent, W. S., 1880, *A Manual of the Infusoria, Including a Description of All Known Flagellate, Ciliate and Tentaculiferous Protozoa, British and Foreign, and an Account of the Organization and Affinities of the Sponges,* Vol. 1, D. Bogue, London, 472 pp.

Kukal, Z., 1971, *Geology of Recent Sediments*, Academic Press, London, 490 pp.

Kumar, A., and Dalby, A. P., 1998, Identification key for Holocene lacustrine arcellacean (thecamoebian) taxa, *Palaeontol. Electron.* **1**(1):39.

Kumar, A., and Patterson, R. T., 1997, Arcellaceans (Thecamoebians) as proxies of lake bottom acidity: An example from James Lake, northeastern Ontario. Geological Association of Canada/Mineralogical Association of Canada Annual Meeting, May 19–21, 1997 (Abstr.), p. A-150.

Leidy, J., 1879, Fresh water rhizopods of North America. United States Geological Survey of the Territories, Report 12, 1–324.

Likens, G. E. (ed.), 1988, *Long-Term Studies in Ecology: Approaches and Alternatives*, Springer-Verlag, New York, 214 pp.

Loeblich, A. R., Jr., and Tappan, H., 1964, Sarcodina, chicfly "Thecamoebians" and Foraminiferida, in: *Treatise on Invertebrate Paleontology, Part C. Protista II*, Vol.1 (R. C. Moore, ed.), Geological Society of America and University of Kansas Press, Lawrence, pp. 1–510.

Luoma, S. N., 1983, Bioavailability of trace metals to aquatic organisms—A review, *Sci. Total Environ.* **28**:1–22.

Mason, G., 1998, A study of the effects of past mining activities on James Lake sediment chemistry, Carleton University Department of Environmental Science, Unpublished Honours Thesis, 53 pp.

McCarthy, F. G., Collins, E. S., McAndrews, J. H., Kerr, H. A., Scott, D. B. and Medioli, F. S., 1995. A comparison of post glacial Arcellacean ("Thecamoebian") and pollen succession in Atlantic Canada, illustrating the potential of arcellaceans for paleoclimatic reconstruction, *J. Paleontol.* **69**(5):980–993.

Medioli, F. S., and Scott, D. B., 1983. Holocene Arcellacea (Thecamoebians) from eastern Canada, Cushman Found. *Foram. Res. Spec. Publ. No. 21*, 63 pp.

Medioli, F. S., and Scott, D. B., 1988, Lacustrine thecamoebians (mainly Arcellaceans) as potential tools for paleolimnological interpretations, *Paleogeogr. Paleoclim. Paleoecol.* **62**:361–386.

Medioli, F. S., Petrucci, F., and Scott, D. B., 1985. Sulla presenza di Tecamebe in campioni carotati sul fondo del Lago di Garda (missione Picard, sett. 1981), *Civiltà Gardesana: L'ambiente Lago*, **3**:83–103.

Medioli, F. S., Scott, D. B., and Abbott, B. H., 1987, A case study of protozoan interclonal variability: Taxonomic implications, *J Foram. Res.* **17**:28–47.

Medioli, F. S., Scott, D. B., Collins, E. S., and McCarthy, F. M. G., 1990. Fossil thecamoebians: Present status and prospects for the future, in: *Paleoecology, Biostratigraphy, Paleoceanography and Taxonomy of Agglutinated Foraminifera* (C. Hemleben, M. A. Kaminski, W. Kuhnt, and D. B. Scott, eds.), NATO Advanced Study Institute Series, Series C. Mathematical and Physical Sciences, Vol. 327, pp. 813–840.

Mignot, J. P., and Raikov, I. B., 1992, Evidence for meiosis in the testate amoeba *Arcella*, *J. Protozool.* **39**:287–289.

Moore, J. W., 1991, *Inorganic Contaminants of Surface Water: Research and Monitoring Priorities*, Springer-Verlag, New York, 334 pp.

Murphy, J. P., 1977, *Yankee Takeover of Cobalt:* Highway Bookshop Press, Cobalt, Ontario, p. 200.

NRCC, 1978, Effects of arsenic in the Canadian environment:National Research Council of Canada, No. 15391, Associate Committee on Scientific Criteria for Environmental Quality, Ottawa, Ontario, Canada, p. 349.

Ogden, C. G., and Hedley, R. H., 1980, *An Atlas of Fresh Water Testate Amoeba. British Museum (Natural History)*, Oxford University Press, Oxford, 222 pp.

Patterson, R. T., and Fishbein, E., 1989, Re-examination of the statistical methods used to determine the number of point counts needed for micropaleontological quantitative research. *J. Paleontol.* **63**:245–248.

Patterson, R. T., MacKinnon, K. D., Scott, D. B., and Medioli, F. S., 1985, Arcellaceans (Thecamoebians) in small lakes of New Brunswick and Nova Scotia: Modern distribution and NHolocene stratigraphic changes, *J. Foram. Res.* **15**:14–137.

Patterson, R. T., Barker, T., and Burbridge, S. M., 1996, Arcellaceans (Thecamoebians) as proxies of arsenic and mercury contamination in northeastern Ontario lakes, *J. Foram. Res.* **26**(2):172–183.

Patterson, R. T., Kumar, A., and Dumaresq, C., 1997, Arcellaceans: Shelled microinvertebrates as cost-effective tools for monitoring current and historical environmental trends, 24th Ann. Aquatic Toxicity Workshop, Niagara Falls, Ontario, October 19–22, 1997 (Abstr.).

Plankey, B. J., and Patterson, H. H., 1987. Kinetics of aluminum fulvic acid complexation in acidic waters, *Env. Sci. Tech.* **21**:595–601.

Plankey, B. J., and Patterson, H. H., 1988, Effect of fulvic acid on the kinetics of aluminum fluoride complexation in acidic waters, *Env. Sci. Tech.* **22**:1454–1456.

Reinhardt, E. D., Dalby, A., Kumar, A., and Patterson, R. T., 1998, Utility of arcellacean morphotypic variants as pollution indicators in mine tailing contaminated lakes near Cobalt, Ontario, Canada, *Micropaleontology* **44**(2):131–148.

Schönborn, W. S., 1984, Studies on remains of Testacea in cores of the Great Woryty Lake (NE-Poland), *Limnologica (Berlin)* **16**:185–190.

Scott, D. B., and Hermelin, J. O. R., 1993, A device for precision splitting of micropaleontological samples in liquid suspension. *J. Paleontol.* **67**:151–154.

Scott, D. B., and Medioli, F. S., 1980*a*. Living vs. total foraminiferal populations; Their relative usefulness in paleoecology. *J. Paleontol.* **54**:814–831.

Scott, D. B., and Medioli, F. S., 1980*b*, Post-glacial emergence curves in the maritimes determined from marine sediments in raised basins, *Proc. Coastlines '80*, National Science and Engineering Research Council, pp. 428–449.

Scott, D. B., and Medioli, F. S., 1983. Testate rhizopods in Lake Erie: Modern distribution and stratigraphic implications, *J. Paleontol.* **57**:809–820.

Smol, J. P., 1992, Paleolimnology: An important tool for effective ecosystem management, *J. Aquat. Ecosys. Health* **1**:49–58.

Tessier, A., Campbell, P. G. C., Auclair, J. C., and Bisson, M., 1984, Relationships between the partitioning of trace metals in sediments and their accumulation in the tissues of the fresh water mollusc *Elliptio complanata* in a mining area, *Can. J. Fish. Aquat. Sci.* **41**:1463–1472.

Thibaudeau, S. A., and Medioli, F. S., 1986, Carboniferous thecamoebians and marsh foraminifera: New stratigraphic tools for ancient paralic deposits (Abstr), Geological Society of America, Annual Meeting, San Antonio, p. 771.

Thibaudeau, S. A., Medioli, F. S., and Scott, D. B., 1987, Carboniferous marginal-marine rhizopods: A morphological comparison with recent correspondents (Abstr.), Geological Society of America, Annual Meeting, Phoenix, p. 866.

Todd, R., and Brönnimann, P., 1957, Recent foraminifera and thecamoebina from the eastern Gulf of Paria, *Cushman Found. Foram. Res. Spec. Publ. No. 3*, 43 pp.

Tolonen, K., 1986, Rhizopod analysis, in: *Handbook of Holocene Paleoecology and Paleohydrology* (B. E. Berglund, ed.), John Wiley & Sons, New York, pp. 645–666.

United States Environmental Protection Agency, 1990, Test Methods for Evaluating Solid Waste. 3rd Ed. SW-846, 3051-1–3051-13.

Volkanov, A., 1962*a*. Uber de Kopulation der Testaceen (Rhizopoda-Testacea), *Rep. Bulgarian Acad. Sci.* **15**(3):305–308.

Volkanov, A., 1962*b*. *Paraquadrula madarica* n sp. (Rhizopoda-Testacea), *Rep. Bulgarian Acad. Sci.* **15**(4):423–426.

Volkanov, A., 1966, Uber die Fortpflanzung der Testaceen (Rhizopoda-Testacea): Bulgarian Acad. Sci., *Bull. Inst. Mus. Zool. (Sofia)*, **22**:5–49, 12 plates.

Vymazal, J., 1984, Short-term uptake of heavy metals by periphyton algae, *Hydrobiologia* **119**:171–179.

Wallich, G. C., 1864, On the extent and some of the principal causes of structural variation among the difflugian rhizopods, *Ann Mag. Nat. Hist. Series 3* **13**:215–245.

Wallis, S. D., 1993, Sonar profiling: A subbottom study of Gillies, Peterson and Crosswise lakes, Ontario. Unpublished B.Sc. thesis, Carleton University, Ottawa, Ontario, p. 40.

Warner, B., 1990, Methods in Quaternary Ecology #5: Testate Amoebae (Protozoa). *Geosc. Can.* **15**(4):251–260.

Warner, B., and Bunting, M. J., 1996, Indicators of rapid environmental change in northern peat lands, in: *Geoindicators, Assessing Rapid Environmental Changes in Earth Systems* (A. R. Berger and W. J. Iams, eds.), A. A. Balkema, Rotterdam, pp. 235–246.

Warner, B. G., and Charman, D. J., 1994, Holocene changes on a peat land in northwestern Ontario interpreted from testate amoebae (Protozoa) analysis. *Boreas* **23**:270–279.

Wightman, W. G., Scott, D. B., Medioli, F. S., and Gibling, M. R., 1994, Agglutinated foraminifera and thecamoebians from the Late Carboniferous Sydney coalfield, Nova Scotia: Paleoecology, paleoenvironments and paleogeographical implications, *Palaeogeogr. Palaeoclim. Palaeoecol.* **106**:187–202.

Wildeman, T. R., Updegraff, D. M., Reynolds, J. S., and Bolis, J. L., 1994, Passive bioremediation of metals from water using reactors or constructed wetlands, in: *Emerging Technology for Bioremediation of Metals* (J. L. Means and R. E. Hinchee, eds.), CRC Press, Boca Raton, pp. 13–25.

Wilkinson, L., 1992. *SYSTAT: Statistics Version 5. 2 Ed.*, SYSTAT, Inc., Evanston, Ill. p. 724.

Chapter 12

Sedimentary Diatoms and Chrysophytes as Indicators of Lakewater Quality in North America

SUSHIL S. DIXIT and JOHN P. SMOL

1. Introduction

Water resources are being increasingly affected by anthropogenic activities. However, long-term data are required to track shifts in environmental variables and to characterize the trajectories of aquatic ecosystems accurately. Unfortunately, historical water quality data are often lacking, and so environmental monitoring programs must either continue for many years before meaningful trends can be inferred or else indirect proxy techniques, such as paleolimnological approaches (Smol, 1992; Charles *et al.*, 1994; Anderson and Battar-

SUSHIL S. DIXIT and JOHN P. SMOL • Paleoecological Environmental Assessment and Research Laboratory (PEARL), Department of Biology, Queen's University, Kingston, Ontario K7L 3N6, Canada.

Environmental Micropaleontology, Volume 15 of *Topics in Geobiology*, edited by Ronald E. Martin. Kluwer Academic/Plenum Publishers, New York, 2000.

bee, 1994), must be used. This chapter summarizes some of the ways that we have used diatom and chrysophyte paleoindicators to assess changes in water quality in North American lakes.

Effective implementation of environmental policies and laws require government agencies to address such questions as: What is the current status of our aquatic environment? Has it changed over the last 100–150 years owing to anthropogenic activities? How has it changed, and to what extent? Is the aquatic environment getting better or worse, or is it in steady state? Have recent restoration efforts had any effects? Without historical data, it is not possible to characterize preindustrial or precultural environmental conditions with which current conditions can be compared or to assess the extent of environmental change/damage. Paleolimnological approaches can be effectively melded with management programs to provide some of these missing data and address many pressing management questions (Smol, 1992).

As noted above, frequent resampling protocols over long time frames are rarely viable options in large-scale environmental monitoring programs because of logistic and economic constraints. Although tradeoffs have to be made when optimizing temporal versus spatial coverage, a selected number of sites can be monitored intensively to answer specific questions. Alternatively, we can select an indicator group that will provide both spatial and temporal coverage of environmental trends in a larger number of sites. Many biological indicators are available to track environmental changes; however, when both spatial and temporal coverage is required, the choice is somewhat limited. Ideally the indicator should: (1) be simple; (2) be able to identify the extent and rate of degradation of water quality; (3) be regionally applicable; (4) provide a significant degree of confidence and estimates of low error; (5) relate to biotic integrity; (6) detect short-term trends; and (7) be able to monitor the recovery and effectiveness of control or regulatory measures.

Among the biological indicators preserved in lake sediments, diatom valves and chrysophyte scales often form the mainstay of most paleolimnological environmental assessments (Dixit *et al.*, 1992*a*; Smol, 1995). Both these indicators have been used to provide both short- and long-term environmental assessment data (Dixit *et al.*, 1992*a,b*, 1999; Charles *et al.*, 1994; chapters in Stoermer and Smol, in press; see also in this volume Rosenfeld *et al.*, Ch. 7; Schornikov, Ch. 8; Patterson and Kumar, Ch. 11; Dale, Ch. 13; Alve, Ch. 14; Ishman, Ch. 16; van der Zwaan, Ch. 17).

Diatoms are single-celled microscopic algae belonging to the class Bacillariophyceae. They are characterized by a sculptured siliceous (SiO_2) cell wall (Fig. 1). Each cell wall is made up of two halves called valves that may be held together by beltlike structures called girdle bands. The taxonomy of diatoms is based primarily on valve morphology and ultrastructure. Because many thousands of diatom taxa exist, taxonomists require significant training before they are sufficiently skilled to undertake detailed environmental assessments. However, as has been shown in several large-scale paleolimnological programs (e.g., Charles and Whitehead, 1986; Battarbee *et al.*, 1990), standardized taxonomic protocols can be implemented and followed.

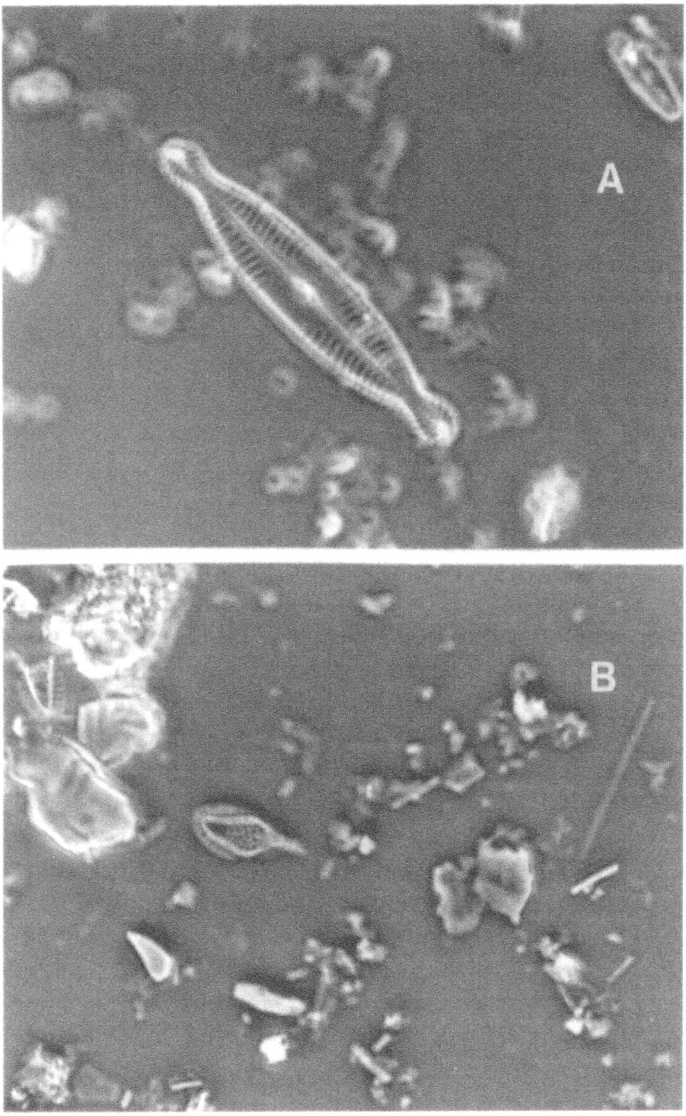

FIGURE 1. Light micrographs of diatom *Pinnularia braunii* (A) and chrysophyte scale of *Mallomonas pseudocoronata* (B).

Diatoms exhibit a variety of life strategies in aquatic environments. For example, planktonic taxa occupy open-water pelagic regions, whereas benthic, periphytic assemblages live attached or associated with diverse substrates, such as aquatic macrophytes (epiphytic), mud (epipelic), rocks (epilithic), and sand (epipsammic). Pelagic sediments typically integrate diatoms from the many habitats present in a lake, and so a deep-water sediment sample, often taken from near the center of a lake, can provide an integrated (space-

and time-averaged) assessment of diatoms from various habitats (Charles *et al.*, 1991).

Diatoms possess many other important characteristics that make them effective bioindicators (summarized in Dixit *et al.*, 1992*a*). For example, valves are generally well preserved and present in very high numbers in lake sediments. Because diatoms have rapid immigration and replication rates, they respond quickly to changes in the aquatic environment (Dixit *et al.*, 1992*b*). Changes in diatom assemblages also correspond to shifts in other biotic communities, such as other algae, macrophytes, zooplankton, and fish (Charles *et al.*, 1994).

Chrysophyte taxa belonging to the classes Chrysophyceae and Synurophyceae are characterized by an external covering of siliceous scales (Siver, 1991). Similar to diatom valves, chrysophyte scales are species-specific and possess taxonomically unique shapes and ornamentations. The use of scaled chrysophytes in paleolimnological investigations has lagged behind the use of diatoms, as research on sedimentary chrysophyte scales began in earnest only in the early 1980s (e.g., Smol 1980; reviewed in Smol, 1995). Because scaled chrysophytes are exclusively euplanktonic, they track changes occurring in the pelagic region closely (Dixit *et al.*, 1989*a*), whereas diatoms track changes occurring in both pelagic and littoral environments. In paleolimnological studies, chrysophytes have not only provided complementary data to diatom-based paleoenvironmental assessments (Charles and Smol, 1988; Dixit *et al.*, 1992*b*), but they have also provided additional information, such as early evidence of detrimental effects and/or recovery from anthropogenic activities (Dixit *et al.*, 1989*a*). Because of their planktonic nature and because they often bloom in spring, chrysophytes may respond more quickly than diatoms to some environmental stresses, such as acidification (Dixit *et al.*, 1992*b*).

All chrysophyte taxa are characterized by the endogenous formation of a siliceous resting stage or stomatocyst. These cysts, which are often highly ornamented and appear to be species-specific (Duff *et al.*, 1995), also possess the necessary characteristics to become excellent paleoindicators (Smol, 1995). However, since their applications to monitoring programs are still in the development stages, we will restrict our examples in this chapter to the use of diatom valves and chrysophyte scales.

Diatoms and chrysophytes have now been used in many different types of environmental assessments. This chapter summarizes some of our recent North American work.

2. Field and Laboratory Methods

A schematic diagram of the overall paleolimnological approach, as it applies to algal microfossils, is shown in Fig. 2. Because diatom valves and chrysophyte scales share many common features, they can be studied using the same field and laboratory procedures.

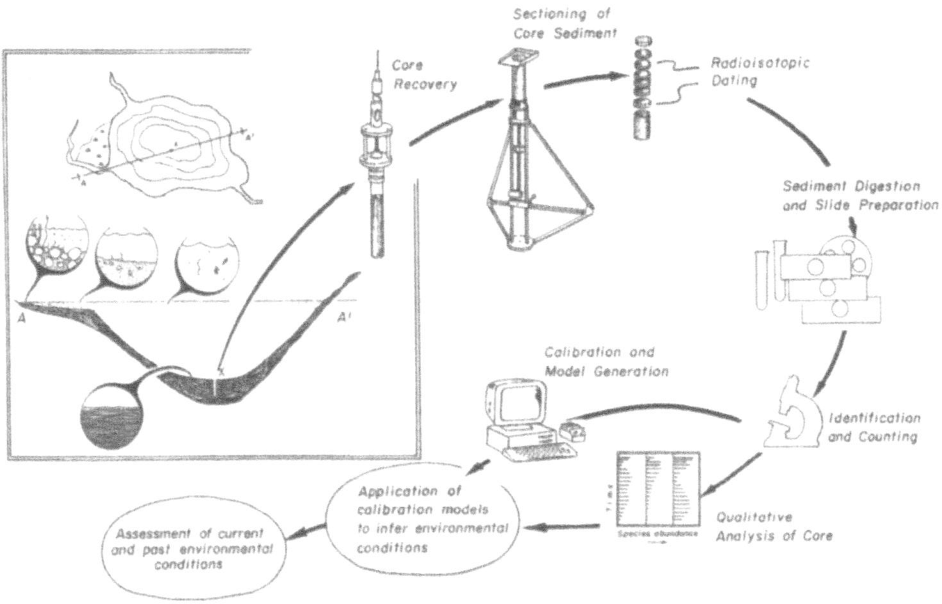

FIGURE 2. A schematic diagram showing inputs of microfossils from various habitats to the lake bottom and various steps taken in a paleolimnological investigation (reprinted with permission from Dixit *et al.*, 1992*a* © 1992 American Chemical Society).

Sediment cores spanning a lake's recent history, which are typically about 50 cm long, and potentially contain the last few hundred years of environmental history, are obtained near the deepest part of the lake. Although a variety of corers are available for recovering profiles spanning recent (i.e., 10 to 200 years) environmental histories, gravity corers [e.g., modified Glew (1989) corer] or a frozen-crust corer (e.g., Huttunen and Meriläinen, 1978) are most widely used. If longer-term profiles are required, such as recovering the entire postglacial history of a lake, piston corers [e.g., Livingstone (1955) corer] are typically used. For close-interval work, which is usually required for many environmental assessments, the sediment layers are carefully extruded out of the coring tube in specified increments using, e.g., a Glew (1988) extruder. The thickness of the sediment slice is set by the investigator to meet the time resolution required for the study. In oligotrophic and/or acidic lakes, cores sectioned at 0.25-cm intervals have provided fine-resolution (e.g., 2- to 3-year increments) assessments of recent environmental change (Dixit *et al.*, 1989*a*; Cumming *et al.*, 1994), whereas in more productive lakes, a 0.25-cm slice may represent sediment accumulated in less than a year. In some high-elevation and arctic lakes, a 0.25 cm sediment section may represent many years of sediment accumulation. Core sectioning techniques are also available to obtain annual or even seasonal scale changes in some lake environments (Simola, 1977).

A critical component of any detailed paleolimnological assessment is geochronology, or the dating of the various sediment sections. Sediment accumulation rates can only be calculated once the sediment increments are dated. The ^{210}Pb radioisotope technique has been commonly used for dating the last approximately 100 to 150 years of sediment accumulation (Appleby *et al.*, 1990; Binford, 1990). This time scale represents a period when industrial and cultural activities greatly increased in North America. ^{137}Cs and ^{241}Am isotopes can also be used to check the accuracy of ^{210}Pb dates (Appleby *et al.*, 1990). Chronostratigraphic markers, such as pollen (e.g., *Ambrosia*), charcoal, soot particles, and other techniques are also used for dating and correlation.

Siliceous algal microfossils are separated and cleaned from the organic matrix by digesting the sediment with strong acids, followed by repeated washes in distilled water (Battarbee, 1986). Subsamples of the resulting siliceous slurry, which contains the cleaned diatom valves and chrysophyte scales, are then dried onto coverslips and mounted on glass slides using a permanent mounting medium with a high refractive index, such as Naphrax®. Identifications are made to the lowest taxonomic level possible (e.g., variety) at a magnification of 1000 × or higher. Typically, a minimum of 500 valves and 500 scales are counted, along transects of the microscope slide, for each sample interval.

3. Qualitative Assessment of Water Quality

3.1. Indicator Taxa

The distributions of some diatom and chrysophyte taxa are sufficiently specific that they can be used qualitatively as indicator taxa for certain environmental variables, such as lakewater pH, nutrients, or salinity (Stoermer and Smol, 1999; Smol, 1995). For example, Dixit *et al.* (1993) found that many diatom taxa in 71 Adirondack Park (New York) lakes were only commonly found within certain pH ranges. From the relationships between the percent abundance of selected taxa and lakewater pH (Fig. 3), it was evident that taxa such as *Fragilaria acidobiontica, Navicula tenuicephala, Eunotia exigua,* and *Tabellaria quadriseptata* were indicative of acidic waters, whereas *Asterionella formosa, Cyclotella comta,* and *Fragilaria brevistriata* were common in alkaline waters. *Cyclotella stelligera* occurred at a wider pH range with the highest abundances at pH ~ 7.

Chrysophyte taxa have also shown similar trends for lakewater pH in many regions of the United States and Canada (Dixit *et al.*, 1989*b*, 1990; Cumming *et al.*, 1992*a*). For example, in northern New England lakes (Dixit *et al.*, 1990), we observed that taxa such as *Mallomonas hindonii, M. hamata,* and *Chrysodidymus synuroides* are more common in acidic waters, whereas *M. caudata* and *M. pseudocoronata* are more common in alkaline waters (Fig. 4). *Chrysosphaerella longispina* occurred at a narrow pH range of 6 to 7.

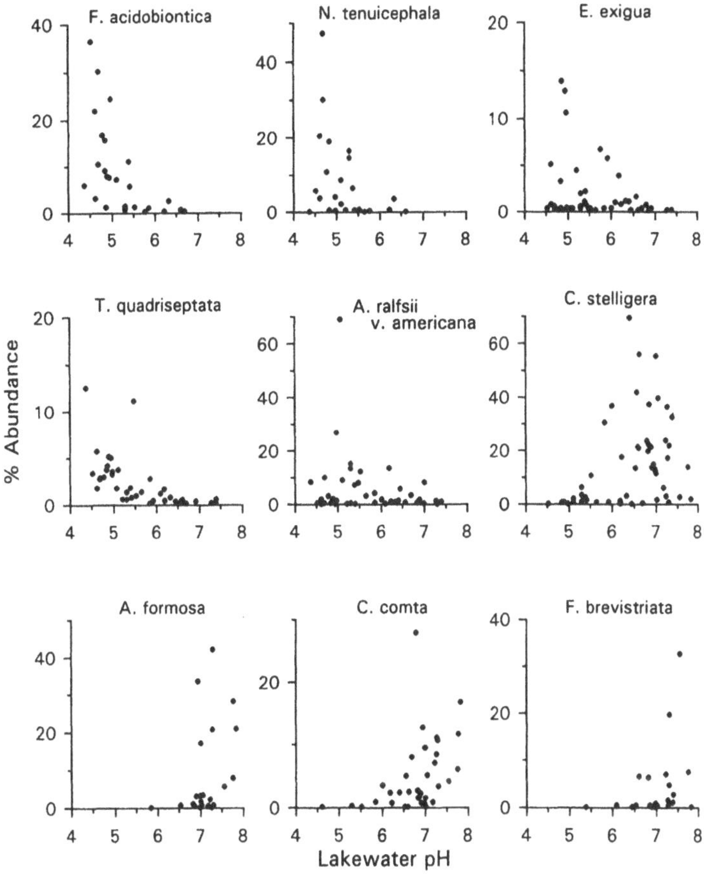

FIGURE 3. Percent distribution of selected diatom taxa vs. measured lakewater pH of 71 Adirondack lakes.

In low-pH waters, metal levels can also influence chrysophyte distributions (Dixit *et al.*, 1989*b*). In both the New England and Adirondack regions, scales of *M. hindonii* were shown to be excellent indicators of acidic waters with high aluminum concentrations. A similar indicator status for this taxon was found in the acidic and metal-contaminated lakes of Sudbury, where scales of *Synura echinulata* were additionally shown to characterize acidic lakes that have high copper and nickel concentrations (Dixit *et al.*, 1989*b*).

The presence of indicator taxa at specific core depths can provide some initial water quality assessments. As a simple example, we present some of the diatom species changes that occurred in the last ca. 150-year history of Little Round Lake, a small presently meromictic (chemically stratified) lake in southeastern Ontario (Smol *et al.*, 1983; Dixit *et al.* 1992a). Prior to the arrival of European settlers (pre-1850), the Little Round Lake diatom flora was dominated by oligotrophic-indicating *Cyclotella* spp. (Fig. 5), suggesting that the lake was unproductive. With the arrival of European settlers to the

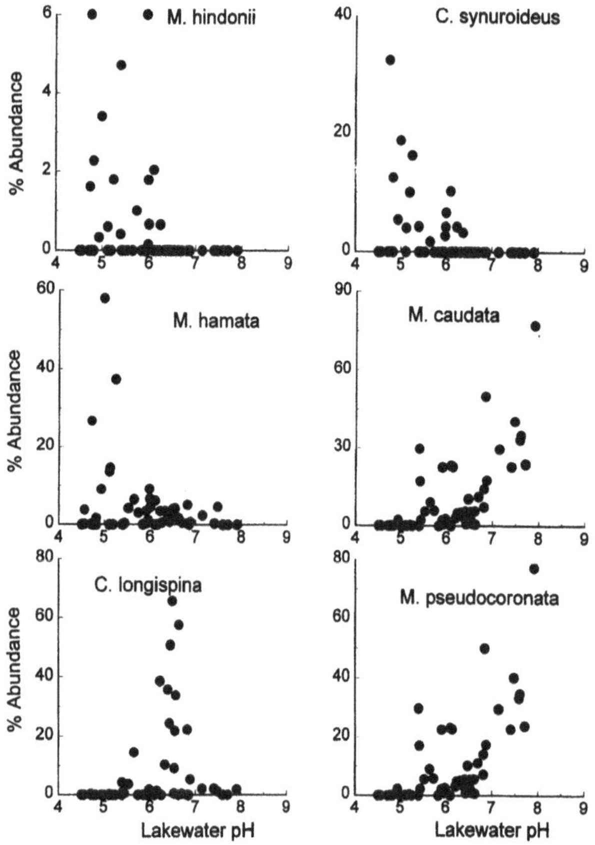

FIGURE 4. Percent distribution of selected chrysophyte taxa vs. measured lakewater pH of 58 northern New England lakes (modified from Dixit *et al.*, 1990).

drainage basin in the 1860s, distinct species changes occurred. Initially, *Cyclotella* spp. were replaced by more mesotrophic-indicating *Synedra* taxa (1860s in Fig. 5). Species successions continued to taxa characteristic of eutrophic waters, such as small *Stephanodiscus* species in the early 1900s, coinciding with the period of more intense agriculture in the area. By the 1970s, the return of many *Cyclotella* species indicates that the lake has returned to a more oligotrophic state, which coincides with the cessation of most cultural disturbances in the watershed.

3.2. Indicator Assemblages

Although individual taxa can provide some information on changes in environmental quality, it is clearly more powerful to use many taxa to make environmental inferences. Moreover, assessments of biological integrity

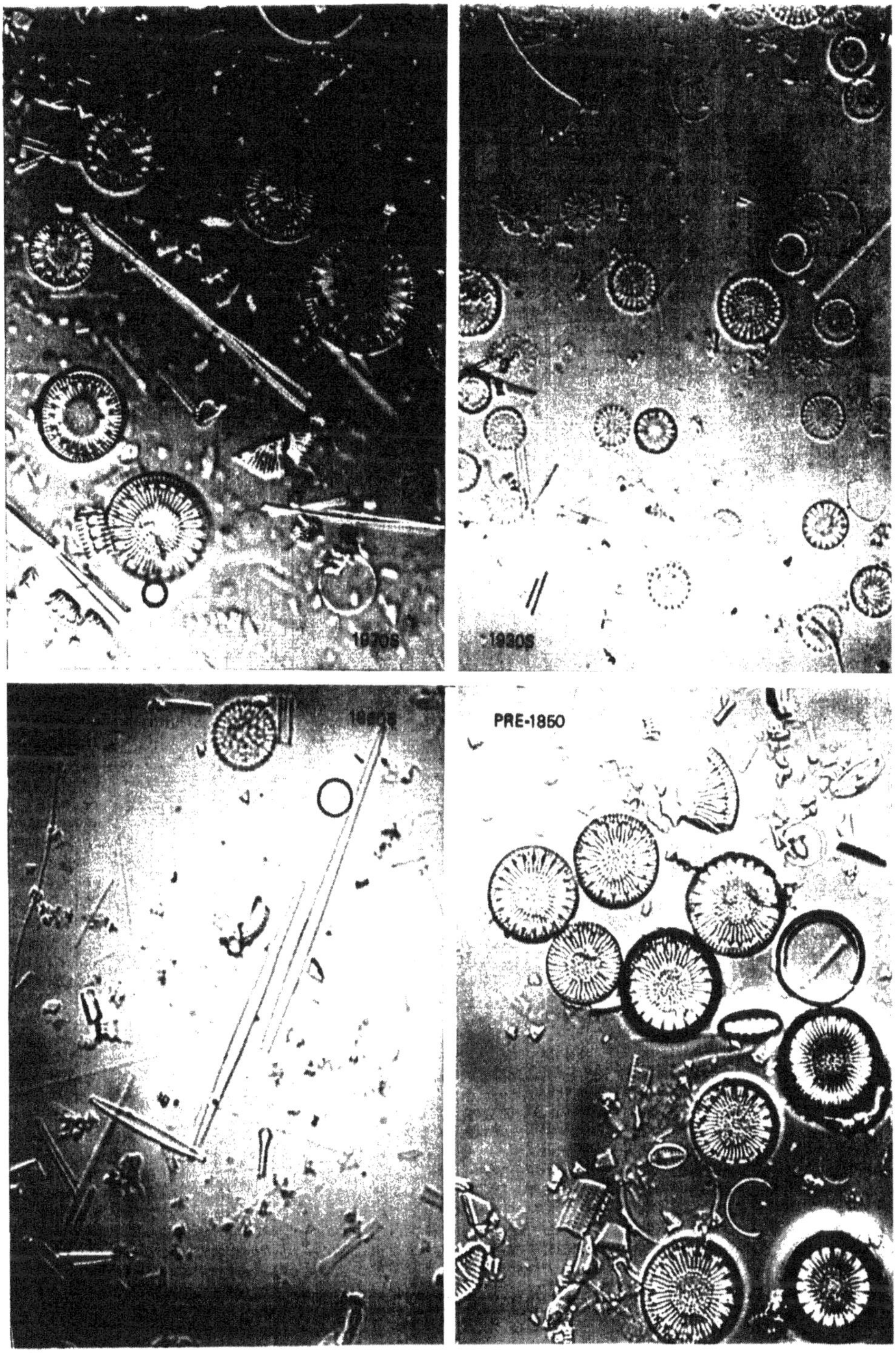

FIGURE 5. Light micrographs of diatom assemblages at different sediment levels of Little Round Lake, Ontario, showing response of assemblages to water quality through time (reprinted with permission from Dixit *et al.*, 1992*a* © 1992 American Chemical Society).

(Hughes, 1995) require data on assemblage changes and not just individual indicator taxa.

As early as 60 years ago, Hustedt (1939) defined five pH indicator diatom assemblages, namely, acidobiontic, acidophilous, indifferent (or circumneutral), alkaliphilous, and alkalibiontic. These assemblages could be clearly delineated in the same 71 Adirondack lakes referred to earlier (Fig. 6). Acidobiontic assemblages were common in lakes with pH<5.5, whereas acidophilous diatoms dominated in lakes with a pH range of 5.0–6.5. Circumneutral diatoms occurred in higher abundances in circumneutral lakes (pH 6.5–7.5), and alkaliphilous diatoms occurred mainly in alkaline waters.

Distinct diatom assemblages also exist along, e.g., trophic and salinity gradients. Dixit and Smol (1994) identified strong relationships between lakewater phosphorus and oligotrophic, mesotrophic, and eutrophic indicator assemblages for 66 calibration lakes in the northeastern United States (Fig. 7). Not surprisingly, a major portion of the diatom taxa in low-phosphorus (<10 μg/liter) lakes was mainly comprised of taxa characterized as oligotrophic diatoms. Mesotrophic diatom assemblages generally occurred in high abundances in mesotrophic lakes (phosphorus 10–30 μg/liter). In more productive lakes (phosphorus >30 μg/liter), eutrophic diatom assemblages were most common. These data suggest that it would be possible to monitor trophic state changes, at least in a qualitative way, by monitoring shifts in diatom assemblages.

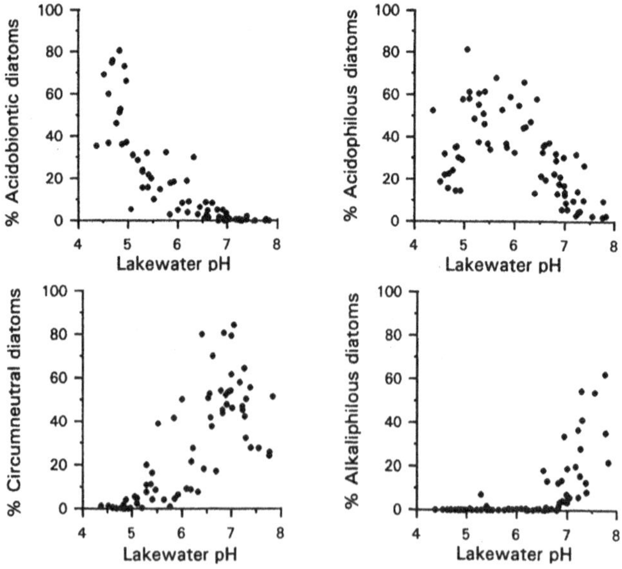

FIGURE 6. Percent distribution of pH indicator diatom assemblages plotted vs. the measured lakewater pH of 71 Adirondack lakes.

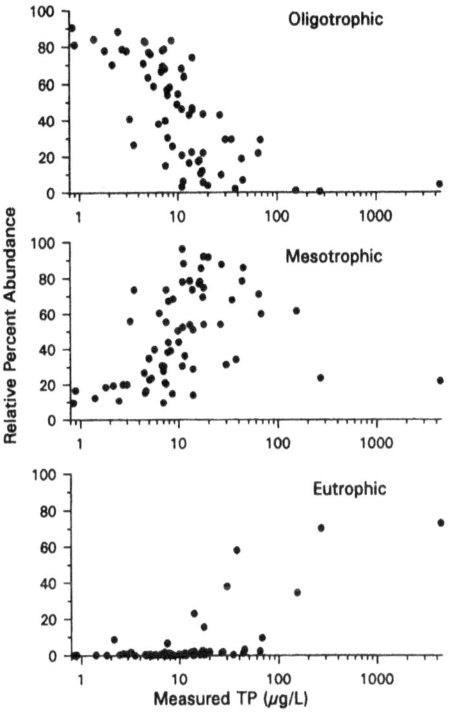

FIGURE 7. Percent distribution of trophic indicator diatom assemblages from surface sediments of 64 EMAP lakes plotted vs. measured lakewater total phosphorus (modified from Dixit and Smol, 1994).

4. Developing Surface Sediment Calibration (Training) Data Sets

In our view, the most effective means of quantitatively determining the relationship between water quality and the distributions of diatom and chrysophyte taxa in a large suite of lakes is to use so-called "surface sediment calibration sets" or "training sets." The overall concepts are very simple. Indicators (such as diatoms or chrysophytes) are analyzed from the surface sediments (e.g., the top 1 cm or 0.5 cm, representing the last few years of sediment accumulation) of a suite of lakes (usually 50 or more in number) for which measured lakewater quality data are also available. Statistical techniques are then used to relate the species assemblages to the environmental variables that are shown to influence the species composition. Transfer functions (see below) can then be constructed for the strongest environmental variables. Charles and Smol (1994) review the development of surface sediment calibration sets, and Birks (1995) summarizes the statistical techniques that are often used in these analyses. Below we summarize some of these quantitative approaches, with examples of some of our North American work.

4.1. Multivariate Ordination Analysis

Powerful multivariate statistical techniques allow paleolimnologists working with surface sediment calibration sets to determine which environmental variables influence species distributions, and how and with what confidence we can infer these variables from species composition (Birks, 1995). For example, canonical correspondence analysis (CCA); (Ter Braak, 1987), a direct gradient analysis technique, has been widely used to explore the relationship between environmental variables and species distributions.

As one example, we present a CCA ordination (Fig. 8) developed to examine the relationship between water chemistry characteristics and diatom taxa identified from the surface sediments of 64 northeastern U.S. lakes, which were sampled as part of U.S. EPA's Environmental Monitoring and Assessment Program—Surface Waters (EMAP-SW) in 1991 (Dixit and Smol, 1994). Forward selection and Monte Carlo permutation tests (available in the computer program CANOCO, Ter Braak, 1988) were used to identify which measured variables explained the maximum amount of variation in the species data.

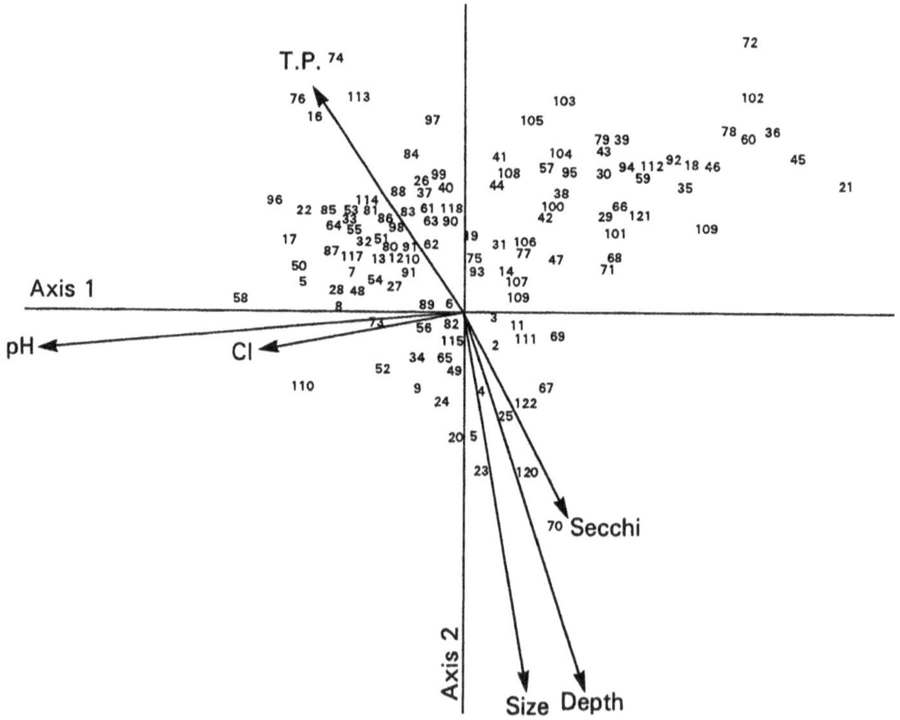

FIGURE 8. Canonical correspondence analysis biplot showing the significant environmental variables (as arrows) and diatom species scores of 122 common taxa found in surface samples of 64 EMAP lakes (T.P. = total phosphorus) (from Dixit and Smol, 1994).

On the environmental–species ordination (Fig. 8), six significant ($p<0.05$) environmental variables are marked by arrows, whereas species are shown as numbers. Environmental arrows point in the direction of maximum change of that variable, and the length of the arrow is a measure of the amount of variance a particular environmental variable captures. In this example (Fig. 8), the contributions of the significant variables in explaining the species data were, in descending order, pH (23.5%), sampling depth (14.8%), phosphorus (7.8%), Secchi depth (7.8%), chloride (6.7%), and lake size (6.2%). The species–environmental correlations for CCA axes 1 (0.92) and 2 (0.83) were high and statistically significant ($p \leqslant 0.01$). The relative influence of selected environmental variables to the CCA axes can be further evaluated by examining their canonical coefficients and approximate *t*-tests, and by intraset correlations between axes and environmental variables (Dixit and Smol, 1994).

Because the position of each taxon approximates its weighted-average optimum relative to other taxa on the CCA biplot, it is possible to approximate the indicator value of various taxa (Fig. 8). For example, the taxa positioned toward the right of the biplot are indicators of low lakewater pH, whereas the taxa toward the left are indicators of high pH. Simultaneous influences of other environmental variables can also be estimated. For example, taxa common in waters of high phosphorus (and low Secchi depths) are positioned on the upper left quadrant of axis 1, whereas taxa commonly found in waters of low phosphorus (and high Secchi depths) are positioned on the lower right quadrant.

CCA can also be used to examine temporal data from more than one trophic level. For example, Kingston *et al.* (1992) used diatom assemblages to characterize fishery resource health in Adirondack lakes. These were the same Adirondack lakes that had been studied for acidification trends using algal microfossils (Charles *et al.*, 1990) and invertebrate indicators (i.e., *Chaoborus* mandibles), which were used to track past fish populations in the same lakes (Uutala, 1990). Kingston *et al.* (1992) showed that recent fishery declines in soft-water Adirondack lakes could be explained in part by major increases in inferred lakewater monomeric aluminum, which were related to anthropogenic acidification.

4.2. Inference Models (Transfer Functions)

Over the last decade, there have been major advances in the development of algal-based inference models for several important water quality variables. Initial efforts were mainly centered on lake acidification research (reviewed in Battarbee *et al.*, 1999), but more recently predictive models have been developed for trophic state (Hall and Smol, 1992; Bennion *et al.*, 1996; Dixit *et al.*, 1999) and climate-related variables, such as lakewater salinity (Wilson *et al.*, 1996), temperature (Pienitz *et al.*, 1995), and other variables (chapters in Stoermer and Smol, 1999).

Weighted averaging regression and calibration (WA) approaches, which can be implemented using the computer program WACALIB (Line *et al.*, 1994), are highly effective for developing inference models (Birks, 1995). WA assumes that unimodal relationships more closely approximate the distribution of species to environmental variables. This assumption is generally true if the environmental gradient is long enough. Statistical techniques are available to assess these relationships (Birks, 1995). The general premise with WA is that at a given environmental condition, such as at pH = 5.2, taxa with pH optima near this value will most likely be encountered. Thus, an optimum for a taxon would be an average of all the values of lakes in the calibration set in which that taxon occurred, weighted by its relative abundance (regression step). Computed optima of various taxa are then used to infer any particular variable by taking an average of the taxa abundances in a fossil sample, each weighted by its optimum (calibration step).

In computing weighted averages, tolerance values of individual taxa can be also used (Birks, 1995), and taxa that have narrow tolerances can be given more weight in the inference functions than taxa that have wider tolerances (Birks *et al.* 1990).

In WA, averages are taken twice: first in the regression step and again in the calibration step. This results in shrinkage of the range of inferred values, which is normally corrected by applying a linear deshrinking ("classical regression") or an "inverse regression" approach (Birks, 1995) using the WACALIB program. Classical regression is generally preferable in developing WA models (Birks *et al.*, 1990; Birks, 1995).

In the calibration step, it is generally assumed that the species assemblage present in a sample is also well represented in the calibration set. However, inferences may become less reliable as one goes downcore if the fossil diatom assemblages diverge from the floristic composition of the surface sediment analogues (i.e., the calibration set samples). Various analogue assessment techniques can be used to determine the representativeness of the downcore diatom assemblage as compared to the assemblages used in the calibration data set (Birks *et al.*, 1990).

We again draw on examples of inference models developed for the Adirondack lakes (Cumming *et al.*, 1992*a*; Dixit *et al.*, 1993). Powerful diatom inference models were constructed for lakewater pH, acid neutralizing capacity (ANC), dissolved organic carbon (DOC), and monomeric aluminum from the Adirondack Park calibration lakes (Fig. 9). The strengths of these types of models have, in the past, been assessed by: (1) the correlation coefficient between the measured and inferred value; (2) the apparent root mean squared error (RMSE) of prediction; and (3) the bootstrapped RMSE of prediction (Birks *et al.*, 1990). $RMSE_{boot}$, and other similar techniques, such as jackknifing, provide the most realistic error estimates (Birks, 1995). Bootstrapping is a computer-intensive resampling procedure wherein a subset of calibration samples that is the same size as the original calibration set is selected, at random, with replacement. The remaining unselected samples form an independent test set. After a large number of bootstrap cycles (usually 1000), a new

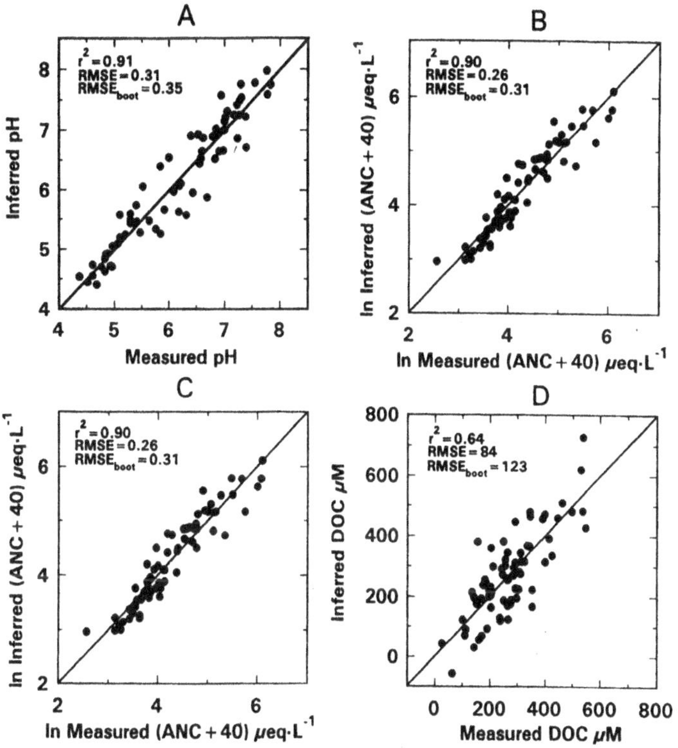

FIGURE 9. The relationship between measured and diatom-inferred pH (A), acid neutralizing capacity (B), monomeric aluminum (C); and dissolved organic carbon (D) values for the Adirondack Park calibration lakes (modified from Dixit *et al.*, 1993).

RMSE$_{boot}$ is obtained. The bootstrap error can be split into an estimate of error unique to each sample (s_{i1}), and a constant error (s_2) for the entire dataset, largely representing the inherent natural variation in the dataset. Bootstrap error estimations can be easily made using the WACALIB program (Line *et al.*, 1994).

The relationship between measured and diatom-inferred pH for the Adirondack calibration set (Fig. 9) indicates that the pH model is strong ($r^2 = 0.91$; RMSE$_{boot}$ = 0.35 of a pH unit). Not surprisingly, the ANC model is also strong ($r^2 = 0.90$ and has a low RMSE$_{boot}$. The relationship between measured and inferred monomeric aluminum is also robust, and, as with the pH and ANC models, most of the error using this model can be accounted for by inherent natural variation in the dataset ($s_{i1} = 0.11$ and $s_2 = 0.43$). The model for DOC is moderately strong ($r^2 = 0.64$), and the errors ($s_{i1} = 59$ μM, $s_2 = 108$ μM) are becoming more evenly distributed.

Using chrysophyte assemblages from the same Adirondack lakes, Cumming *et al.* (1992*a*) also developed significant predictive models to infer lakewater pH and monomeric aluminum concentrations (Fig. 10). Their model evaluations were also based on correlation coefficients and the RMSE of prediction derived from bootstrapping.

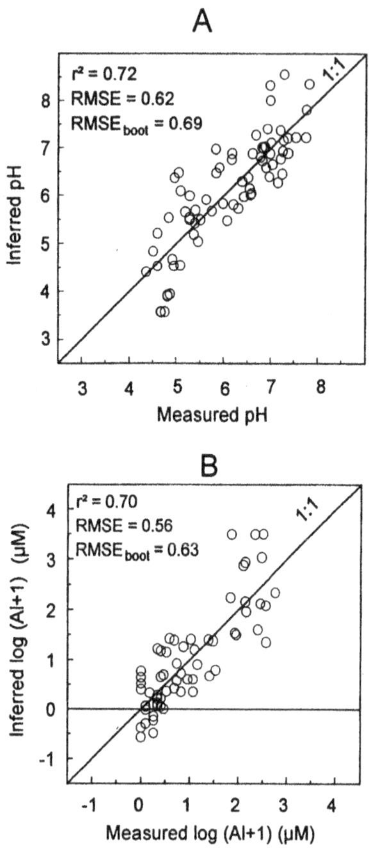

FIGURE 10. The relationship between measured and chrysophyte-inferred pH (A), and monomeric aluminum (B) from the Adirondack Park calibration lakes (modified from Cumming *et al.*, 1992*a*).

In addition to lakewater pH, diatom and chrysophyte models for aluminum have provided useful information on trends in past lakewater aluminum changes in Adirondack lakes and fisheries damage (Kingston *et al.*, 1992). These inferences are especially important because monomeric aluminum is toxic to biota at various levels in the food chain (Baker and Christensen, 1991; Gensemer, 1991). Meanwhile, the DOC inference model has provided valuable data on historical changes in DOC, especially to separate "natural" versus "anthropogenic" acidification (Cumming *et al.*, 1992*b*), and to evaluate chemical models (Sullivan *et al.*, 1992). We have recently developed inference models for phosphorus and chloride to assess trophic and other watershed-activity-related changes in lakes in the northeastern United States (Dixit *et al.*, 1999).

New statistical approaches are always being developed (Birks, 1995). Because the weighted-averaging partial-least-square (WA-PLS) approach uses residual structure in species data, it may improve on some WA inferences (Ter Braak and Juggins, 1993).

5. Historical Assessments of Environmental Change

As an example of how algal microfossils can be used to quantitatively infer past water quality changes, we present one study from Sudbury, Ontario (Dixit *et al.*, 1992*b*). In the Sudbury region of Canada, the metal smelting industry had emitted large amounts of SO_2 and toxic metals, leading to acidification and metal contamination of a large number of lakes (Keller *et al.*, 1986). Baby Lake (46°24′ N, 80°52′ W) is a small (12-hectare) lake located about 1 km southwest of the Coniston Smelter, which was closed in 1972. Vegetation and soils are absent in the drainage basin because in the past very high emissions of SO_2 and metals killed local forests and any available wood was used as fuel in nearby roast beds (Hutchinson and Havas, 1986). A 26-cm long, high-resolution sediment core was taken from the lake in 1987, and the results of analyses are summarized below.

The profiles of the common diatom taxa in Baby Lake recorded two distinct patterns (Fig. 11A). During the pre-1940 period, circumneutral and/or alkaline taxa (*Asterionella formosa, Achnanthes minutissima, Cyclotella ocellata, C. stelligera,* and *Tabellaria flocculosa* strain IIIp) dominated the diatom community, but thereafter several acid-indicating and metal-tolerant taxa (e.g., *Frustulia rhomboides* v. *saxonica, Pinnularia hilseana,* and many *Eunotia* species) increased in abundance. The second major change commenced after ca. 1970 with a return of taxa that were common in pre-1940 sediments (Fig. 11A).

Profiles of the common chrysophyte taxa in Baby Lake also recorded two major shifts in species composition since about 1940 (Fig. 11B). Prior to the

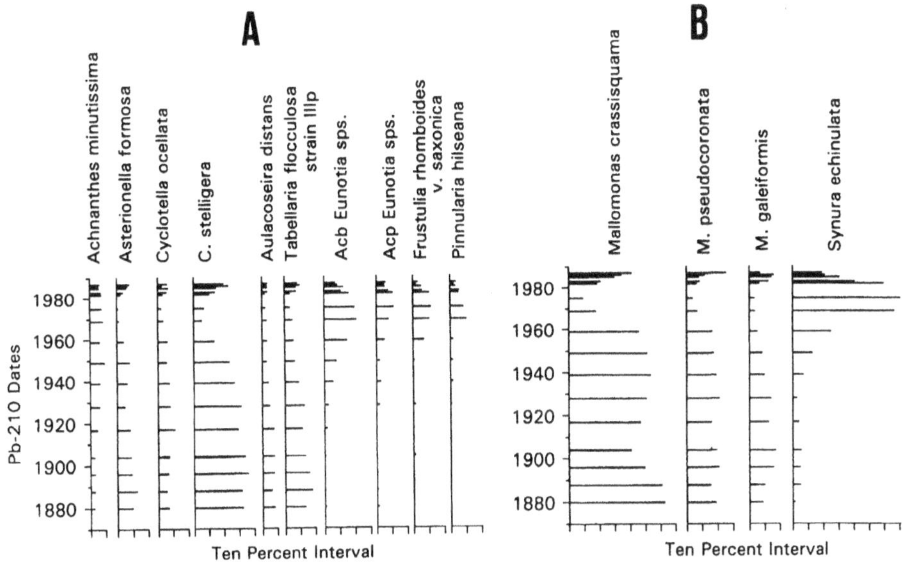

FIGURE 11. Stratigraphic distribution of common diatom (A) and chrysophyte (B) taxa from Baby Lake, Sudbury (modified from Dixit *et al.*, 1992*b*).

1940s, the chrysophyte flora was dominated by *Mallomonas crassisquama, M. pseudocoronata*, and *M. galeiformis*. The first major shift started after ca. 1940, when these taxa declined in abundance and *Synura echinulata* began to dominate the chrysophyte population. In the Sudbury region, *S. echinulata* is characteristic of acidic waters that have high copper and nickel concentrations (Dixit *et al.*, 1989*b*). *S. echinulata* declined in abundance during the second assemblage shift, which occurred between the early 1970s and 1987, whereas circumneutral/alkaline taxa increased in relative abundance.

Weighted-averaging models, developed from a 72-lake surface sediment calibration set in the Sudbury region (Dixit *et al.*, 1989*b*, 1991), were then applied to the Baby Lake fossil diatom and chrysophyte assemblage data. With the use of these transfer functions, the diatom- and chrysophyte-inferred pH histories indicate that the preimpact lakewater pH ranged between 6.4 and 6.8 (Fig. 12A). Acidification commenced about 1950 and continued until about 1975. The lowest chrysophyte-inferred pH of 4.2 agrees closely with the lakewater pH measured by government scientists at that time (Fig. 12A). Both indicators have shown marked recovery in lakewater pH in Baby Lake since about 1975, albeit the response in diatom assemblages was somewhat slower than among chrysophytes.

Chrysophyte inferences of pH often predate changes inferred from diatoms, suggesting that the former are "early warning indicators" of limnological change. As noted earlier, this pattern is probably due to the ecological characteristics of these two algal groups. Many chrysophytes are vernal bloomers and thus may reflect spring water chemistry values more closely than diatoms. This has important environmental implications because many aquatic organisms are sensitive to pH depressions associated with the spring

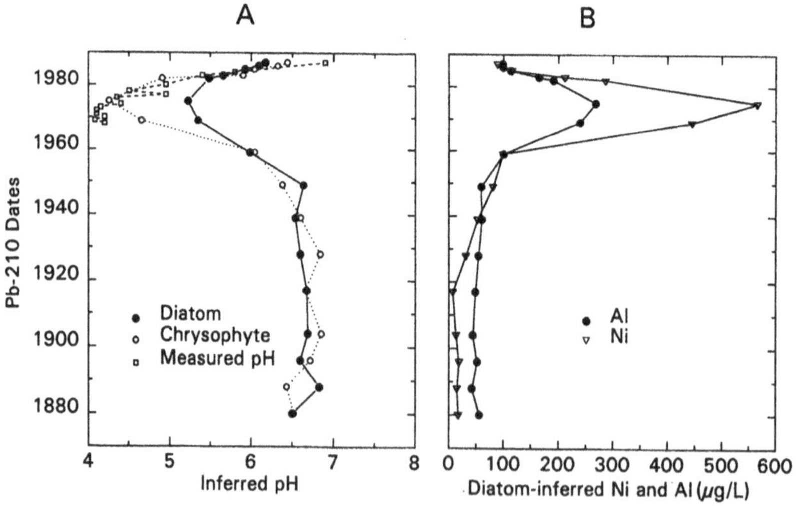

FIGURE 12. Diatom and chrysophyte-inferred pH (A) and diatom-inferred aluminum and nickel (B) reconstructions for Baby Lake (modified from Dixit *et al.*, 1992*b*).

snow-melt (Baker and Christensen, 1991). Diatoms tend to be abundant throughout spring and summer and appear to reflect an average water chemistry for the growing season (Dixit *et al.*, 1992*a*). The close relationship between chrysophyte-inferred and measured pH recovery demonstrates that chrysophytes provide high-resolution data for short-term water quality trends.

Diatom-inferred trends in aluminum and nickel concentrations indicate that metal concentrations in Baby Lake have also increased during this century (Fig. 12B). The inferred trend in the nickel profile suggests that nickel levels started to increase after 1930, whereas lakewater aluminum concentrations started to increase about 1950. These differences are likely due to differences in the source of nickel and aluminum inputs (Dixit *et al.*, 1992*b*). Nickel inputs to the lake increased soon after the smelter started operation, whereas aluminum levels only increased after acidification. Although nickel concentrations started to increase about 1930, the largest increase occurred between 1960 and the early 1970s, coinciding with the peak metal mining and smelting at the nearby smelter (Keller *et al.*, 1986). Contemporary with the pH recovery, inferred aluminum and nickel levels also declined immediately after the closure of the Coniston Smelter in 1972.

The paleolimnological study of Baby Lake clearly shows that acidification was caused by sulfur deposition from the smelter. The recent pH recovery was due mainly to the closure of the nearby smelter in 1972, as well as reduced sulfur dioxide emissions from the other smelters that are still in operation in the Sudbury region (Dixit *et al.*, 1992*b*). The data strongly suggest that, in addition to establishing trends for the last 150 years or so, paleolimnological techniques can be successfully used to detect changes in lakewater quality that have occurred over relatively short time spans (i.e., 5 to 10 years).

6. Regional Assessments

In addition to simply tracking environmental trends in individual lakes, it is possible to use sedimentary diatoms and chrysophytes to characterize reference (or precultural) conditions in a large suite of lakes and to assess the extent, magnitude, and spatial pattern of change in lakewater quality as a result of anthropogenic activity on a regional scale. The approach that we present here was developed for the PIRLA-II project (Charles and Smol, 1990) to estimate regional changes in the chemistry of Adirondack lakes (Cumming *et al.*, 1992b). There were major concerns that, during this century, a large number of water bodies in the Adirondacks had been adversely affected by acidic deposition. However, because of the lack of long-term monitoring data, paleolimnological techniques had to be employed to infer these data.

In the PIRLA-II project, 37 low-ANC Adirondack lakes were statistically selected, and diatom and chrysophyte assemblages were analyzed from the top samples (0–1 cm slice), representing present-day conditions, and from the bottom samples (generally from >25 cm core depth), representing preindustrial conditions. This so-called "top/bottom approach" is a relatively rapid way to assess overall patterns of change for a large suite of lakes. Inferences of current and preindustrial water chemistry variables were based on the WA inference models presented earlier in Figs. 9 and 10.

The above study showed that: (1) during preindustrial times, very few Adirondack lakes were naturally very acidic; (2) assessments of changes in lakewater chemistry from preindustrial times to the present showed that recent, widespread, chronic acidification has occurred in most lakes that have at present measured pH $<$ 6.0; (3) historical changes in DOC were small, with no consistent trends, suggesting that acidification is due mainly to atmospheric deposition of strong mineral acids; and (4) the recently acidified lakes are in the southwestern parts of the Adirondacks, an area that is geologically especially sensitive to acidification, receives the highest annual rainfall, and has the highest mineral acid deposition in the region (Cumming *et al.*, 1992*b*). The extrapolation of these data to a target population of 675 Adirondack lakes indicated that about 80% of the lakes with current pH $<$ 5.2, and 30–45% of the lakes with a current pH range between 5.2 and 6.0 have undergone declines in pH and increases in monomeric aluminum concentrations since preindustrial times.

The use of algal microfossils for making regional assessments was further developed during U.S. EPA's nationwide Environmental Monitoring and Assessment Program–Surface Waters (EMAP-SW) project (Dixit *et al.*, 1999). This monitoring program was designed to address a number of issues, such as: (1) to catalogue the present extent and geographical distribution of lakes and their current ecological conditions; (2) to determine the proportion of lakes that are degrading or recovering; (3) to determine where, and at what rates, these changes are occurring; and (4) to identify the likely causes of these changes (Larsen *et al.*, 1991). Sedimentary diatoms were selected as one of the five biological indicators (i.e., fish, zooplankton, macrobenthos, diatoms, and riparian birds) in EMAP-SW. The program adapted a probability-based sampling design, so that lake-specific data can be projected to the entire population of lakes in the country or a region (Paulsen *et al.*, 1991).

In EMAP-SW, the diatom assemblages preserved in the "top" and "bottom" sediment samples (similar to PIRLA-II) of 238 lakes and reservoirs in the northeastern United States were studied to determine how this region and various ecoregions have responded to human activities over the last 150 years or so (Dixit *et al.*, 1999). Based on the diatoms preserved in the surface sediments of these lakes, Dixit *et al.* (1999) showed that the distribution of species was closely related to several environmental variables, but primarily to lakewater pH, total phosphorus, and chloride.

Using the transfer functions developed from the above calibration set, population estimates of historical changes in water quality (from the top/bottom analyses of the same sediment cores) were made for all lakes and reservoirs of the northeastern United States, and also for the Adirondacks, New England Uplands, and Coastal Lowlands/Plateau ecoregions. These data indicated that the extent of cultural impact has been quite variable among the ecoregions, with marked water quality deterioration occurring in hundreds of water bodies. Chloride and phosphorus levels have especially increased in sites that currently have high concentrations, and low pH sites have become more common in all three ecoregions.

Figure 13 presents a summary of the EMAP-SW target population data for inferred total phosphorus. Although there are more eutrophic lakes in the northeast now than in the past, the data also suggest that there were some naturally productive lakes in the region. On a regional basis, the percentages of oligotrophic (TP < 10 μg/liter), mesotrophic (TP 10–30 μg/liter), and eutrophic (TP > 30 μg/liter) lakes have not changed much for both natural lakes and reservoirs (Fig. 13). Nevertheless, by examining the data for individual ecoregions, distinct patterns emerge. Despite distinct pH-related changes in the Adirondacks, this region has not been severely affected by nutrient enrichment since preindustrial times. In the New England Uplands, the population of eutrophic lakes has increased (from 2 to 4%) and oligotrophic lakes declined from 41 to 38%, whereas the population of mesotrophic lakes has remained relatively unchanged (57–58%). The reservoirs of this ecoregion have responded somewhat differently, i.e., the population of eutrophic sites has declined from 22 to 9%, and both oligotrophic and mesotrophic reservoirs have increased from 12 to 17% and 65 to 75%, respectively.

The diatom data also inferred that an increasing number of Coastal Lowlands/Plateau lakes have undergone nutrient enrichment (Fig. 13). Among natural lakes, the proportions of eutrophic and oligotrophic lakes increased from 6 to 8% and 48 to 63%, respectively, whereas the population of mesotrophic lakes has declined from 46 to 29% since preindustrial time. Although 38% of the reservoirs were eutrophic in the past, the percentage of eutrophic reservoirs has increased to 46% at present, and the numbers of both oligotrophic and mesotrophic reservoirs have declined.

7. Conclusions

In the absence of long-term water quality data, the rapidly growing field of algal-based paleolimnology offers many important opportunities for environmental monitoring and assessments. Sedimentary diatom and chrysophyte assemblages have been repeatedly shown to be reliable indicators of environmental change, and can be used to compare present-day water quality with natural ("reference" or "background") conditions. These data are much in demand for many lake management issues.

Many advances have been made over the last decade in the development of biological-based paleolimnological approaches, and their applications to problems of lake management. Although many more exciting advances can be expected in the next few years, these techniques are now sufficiently robust and tested that they can be effectively integrated, in a cost-effective manner, into ongoing water quality monitoring programs.

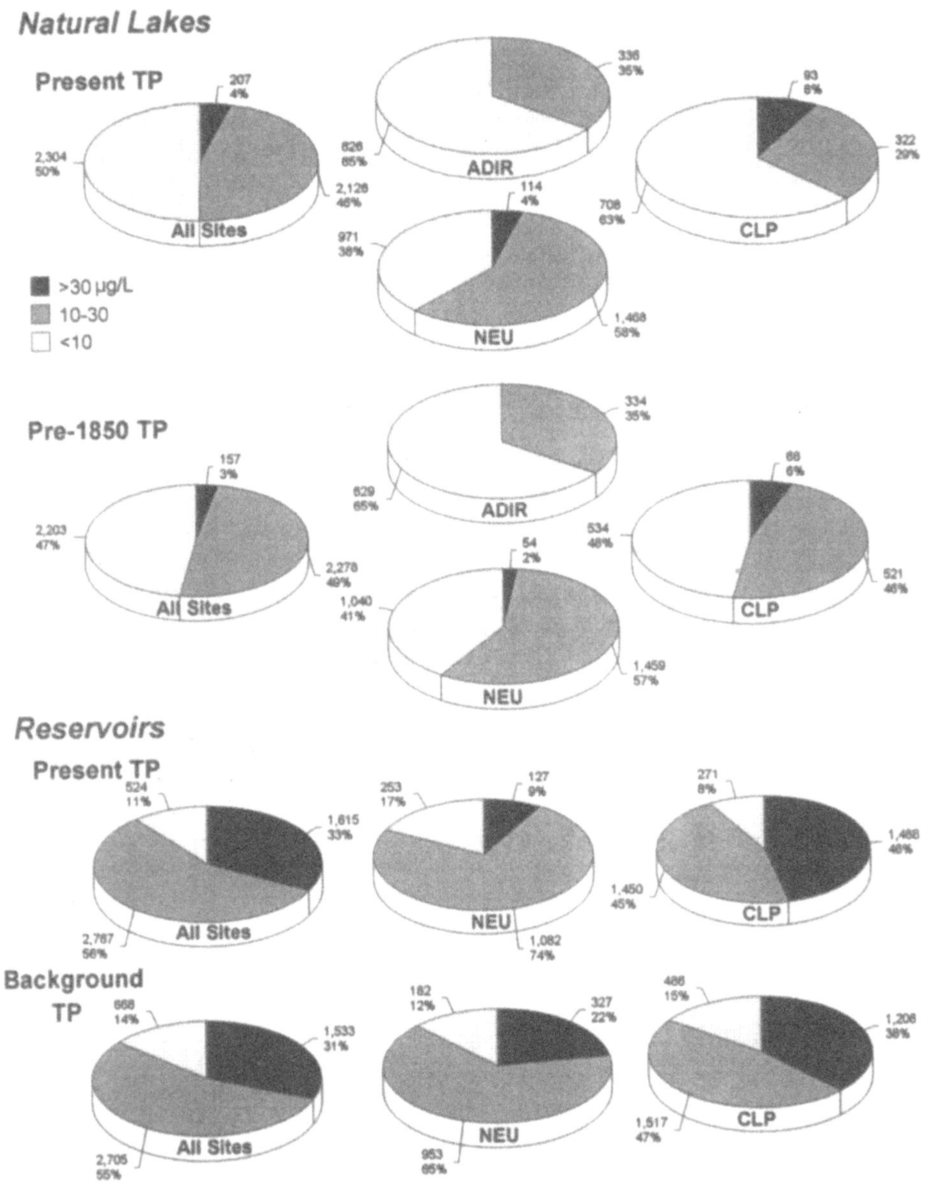

FIGURE 13. The target populations for oligotrophic (TP < 10 μg/liter), mesotrophic (TP 10–30 μg/liter), and eutrophic (TP > 30 μg/liter) lakes and reservoirs at present and during "background" (preimpact) periods for the entire northeastern region, and for three ecoregions in the northeastern United States (reprinted with permission from Dixit *et al.*, 1999 NRC Research Press). For the natural lakes, the background is pre-1850, whereas for the reservoirs the background represents the time when their bottom samples were deposited (i.e., the time the reservoir was formed; dates vary depending on the reservoir) (TP = Total phosphorus; ADIR = Adirondacks; NEU = New England Uplands; CLP = Coastal Lowlands and Plateau).

ACKNOWLEDGMENTS: The preparation of this paper was funded in part by the U.S. Environmental Protection Agency through a cooperative agreement (CR-821898) with Queen's University. Additional funding for research reported on in this chapter was provided by grants from the Natural Sciences and Engineering Research Council of Canada. The paper has not been subjected to the EPA's peer and administrative review.

References

Anderson, N. J., and Battarbee, R. W., 1994, Aquatic community persistence and variability: A paleolimnological perspective, in: *Aquatic Ecology: Scale, Pattern and Process* (P. S. Giller, A. G. Hildrew, and D. G. Raffaelli, eds.), Blackwell Scientific Publications, Oxford, pp. 233–259

Appleby, P. G., Richardson, N., Nolan, P. J., and Oldfield, F., 1990, Radiometric dating of the United Kingdom SWAP sites, *Phil. Trans. R. Soc. Lond. B* **327**:233–238.

Baker, J. P., and Christensen, S. W., 1991, Effects of acidification on biological communities in aquatic ecosystems, in: *Acidic Deposition and Aquatic Ecosystems: Regional Case Studies* (D. F. Charles, ed.), Springer-Verlag, New York, pp. 83–106.

Battarbee, R. W., 1986, Diatom analysis, in: *Handbook of Holocene Palaeoecology and Palaeohydrology* (B. E. Berglund, ed.), John Wiley & Sons, Chichester, pp 527–570.

Battarbee, R. W., Mason, J., Renberg, I., and Talling, J. F., (eds.), 1990, *Palaeolimnology and Lake Acidification*, The Royal Society, London.

Battarbee, R. W., Charles, D. F., Dixit, S. S., and Renberg, I., 1999, Diatoms as indicators of surface water acidity, in: *Applications and Uses of Diatoms* (E. F. Stoermer and J. P. Smol, eds.), Cambridge University Press, New York and Cambridge, pp. 85–127.

Bennion, H., Anderson, N. J., and Juggins, S., 1996, Predicting epilimnetic phosphorus concentrations using an improved diatom-based transfer function and its application to lake eutrophication management, *Environ. Sci. Technol.* **30**:2004–2007.

Binford, M. W.,1990, Calculation and uncertainty analysis of ^{210}Pb data for PIRLA project lake sediment cores, *J. Paleolim.* **3**:253–267.

Birks, H. J. B., 1995, Quantitative palaeoenvironmental reconstructions, in: *Statistical Modeling of Quaternary Science Data*, Technical Guide 5 (D. Maddy and J. S. Brew, eds.), Quaternary Research Association, Cambridge, pp. 161–254.

Birks, H. J. B., Line, J. M., Juggins, S., Stevenson, A. C., and ter Braak, C. J. F., 1990, Diatoms and pH reconstructions, *Phil. Trans. Roy. Soc. Lond. B* **327**:263–278.

Charles, D. F., and Smol, J. P., 1988, New methods for using diatoms and chrysophytes to reconstruct past lakewater pH, *Limnol. Oceanogr.* **33**:1451–1462.

Charles, D. F., and Smol, J. P. 1990. The PIRLA II project: regional assessment of lake acidification trends. *Verh. Int. Verein. Limnol.* **24**:474–480.

Charles, D. F., and Smol, J. P., 1994, Long-term chemical changes in lakes: Quantitative inferences using biotic remains in the sediment record, in: *Environmental Chemistry of Lakes and Reservoirs*, Advances in Chemistry Series 237 (L. Baker, ed.), American Chemical Society, Washington, D. C., pp. 3–31.

Charles, D. F., and Whitehead, D. R., 1986, Paleoecological investigation of recent lake acidification, Report EA-4906, Electric Power Research Institute, Palo Alto, Cal.

Charles, D. F., Binford, M. W., Furlong, E. T., Hites, R. A., Mitchell, M. J., Norton, S. A., Oldfield, F., Paterson, M. J., Smol, J. P., Uutala, A. J., White, J. R., Whitehead, D. R., and Wise, R. J., 1990, Paleoecological investigation of recent lake acidification in the Adirondack Mountains, N. Y., *J. Paleolim.* **3**:195–241.

Charles, D. F., Dixit, S. S, Cumming, B. F., and Smol, J. P., 1991, Variability in diatom and chrysophyte assemblages and inferred pH: paleolimnological studies of Big Moose L., N. Y., *J. Paleolim.* **5**:267–284.

Charles, D. F., Smol, J. P., and Engstrom, D. R., 1994, Paleolimnological approaches to biomonitoring, in: *Biological Monitoring of Aquatic Systems* (S. Loeb and A. Spacie, eds.), Lewis Press, Ann Arbor, pp. 233–293.

Cumming, B. F., Smol, J. P., and Birks, H. J. B., 1992*a*, Scaled chrysophytes (Chrysophyceae and Synurophyceae) from Adirondack drainage lakes and their relationship to environmental variables, *J. Phycol.* **28**:162–178.

Cumming, B. F., Smol, J. P., Kingston, J. C., Charles, D. F., Birks, H. J. B., Camburn, K. E., Dixit, S. S., Uutala, A. J., and Selle, A. R., 1992b, How much acidification has occurred in Adirondack region (New York, USA) lakes since pre-industrial times?, *Can. J. Fish. Aquat. Sci.* **49**:128–141.

Cumming, B. F., Davey, K. A., Smol, J. P., and Birks, H. J. B., 1994, When did acid-sensitive Adirondack lakes (New York, U.S.A.) begin to acidify and are still acidifying?, *Can. J. Fish. Aquat. Sci.* **51**:1550–1568.

Dixit, S. S., and Smol, J. P., 1994, Diatoms as indicators in the Environmental Monitoring and Assessment Program-Surface Waters (EMAP-SW), *Environ. Monit. Assess.* **31**:275–306.

Dixit S. S., Dixit A. S., and Smol J. P., 1989*a*, Lake acidification recovery can be monitored using chrysophycean microfossil, *Can. J. Fish. Aquat. Sci.* **46**:1309–1312.

Dixit, S. S., Dixit, A. S., and Smol, J. P., 1989*b*, Relationship between chrysophyte assemblages and environmental variables in seventy-two Sudbury lakes as examined by canonical correspondence analysis (CCA), *Can. J. Fish. Aquat. Sci.* **46**:1667–1676.

Dixit, S. S., Smol, J. P., Davis, R. B., and Anderson, D. S., 1990, Utility of scaled chrysophytes in predicting lakewater pH in northern New England lakes, *J. Paleolimnol.* **3**:269–286.

Dixit, S. S., Dixit, A. S., and Smol, J. P., 1991, Multivariable environmental inferences based on diatom assemblages from Sudbury (Canada) lakes, *Freshwater Biol.* **26**:251–265.

Dixit, S. S., Smol, J. P., Kingston, J. C., and Charles, D. F., 1992a, Diatoms: Powerful indicators of environmental change, *Environ. Sci. Technol.* **26**:22–33.

Dixit, A. S., Dixit, S. S., and Smol, J. P., 1992b, Algal microfossils provide high temporal resolution of environmental trends, *Water Air Soil Pollut.* **62**:75–87.

Dixit, S. S., Cumming, B. F., Birks, H. J. B., Smol, J. P., Kingston, J. C., Uutala, A. J., Charles, D. F., and Camburn, K. E., 1993, Diatom assemblages from Adirondack lakes (New York, U.S.A.) and the development of inference models for retrospective environmental assessment, *J. Paleolim.* **8**:27–47.

Dixit, S. S., Smol, J. P., Charles, D. F., Hughes, R. M., and Collins, G. B., 1999, Assessing water quality changes in the lakes of the northeastern United States using sediment diatoms, *Can. J. Fish. Aquat. Sci.* **56**:131–152.

Duff, K., Zeeb, B., and Smol, J. P., 1995, *Atlas of Chrysophycean Cysts*, Kluwer Academic, Dordrecht.

Gensemer, R. W., 1991, The effect of pH and aluminum on the growth of the acidophilous diatom *Asterionella ralfsii* var. *americana*, *Limnol. Oceanogr.* **36**:123–131.

Glew, J. R., 1988, A portable extruding device for close interval sectioning of unconsolidated core samples, *J. Paleolim.* **1**:235–239.

Glew, J. R., 1989, A new trigger mechanism for sediment samplers, *J. Paleolim.* **2**:241–243.

Hall, R. I. and Smol, J. P., 1992, A weighted-averaging regression and calibration model for inferring total phosphorus concentration from diatoms in British Columbia (Canada) lakes, *Freshwater Biol.* **27**:417–434.

Hughes, R. M., 1995, Defining acceptable biological status by comparing with reference conditions, in: *Biological Assessment and Criteria: Tools for Water Resource Planning and Decision Making* (W. S. Davis and T. P. Simon, eds.), Lewis Press, Boca Raton, pp. 31–47.

Hustedt, F., 1939, Systematische und ökologische Untersuchungen über die Diatomeen-Flora von Java, Bali und Sumatra, *Arch. Hydrobiol (Suppl.)* **16**:1–155, 264–394.

Hutchinson, T. C. and Havas, M., 1986, Recovery of previously acidified lakes near Coniston, Canada following reductions in atmospheric sulphur and metal emissions, *Water Air Soil Pollut.* **28**:319–333.

Huttunen, P., and Meriläinen, J., 1978, New freezing device providing large unmixed sediment: samples from lakes, *Ann. Bot. Fennici* **15**:128–130.

Keller, W., Pitblado, J. R., and Conroy, N., 1986, Water quality improvements in the Sudbury Ontario, Canada area related to reduced smelter emissions, *Water Air Soil Pollut.* **31**:765–774.

Kingston, J. C., Birks, H. J. B., Uutala, A. J., Cumming, B. F., and Smol, J. P., 1992, Assessing trends in fishery resources and lakewater aluminum from paleolimnological analyses of siliceous algae, *Can. J. Fish. Aquat. Sci.* **49**:116–127.

Larsen, D. P., Stevens, D. L., Selle, A. R., and Paulsen, S. G., 1991, Environmental monitoring and assessment program-surface waters: A northeast lake pilot, *Lake Reserv. Manage.* **7**:1–11.

Line J. M., Ter Braak, C. J. F., and Birks, H. J. B.,1994, WACALIB version 3.3—a computer program to reconstruct environmental variables from fossil assemblages by weighted averaging and to derive sample-specific errors of prediction, *J. Paleolim.* **10**:147–152.

Livingstone, D. A., 1955, A lightweight piston sampler for lake deposits, *Ecology* **36**:137–139.

Paulsen, S. G., Larsen, D. P., Kaufmann, P., Whittier, T., Baker, J. R., Peck, D. V., McGue, J., Stevens, D., Stoddard, J., Hughes, R., McMullen, D., Lazorchak, J., Kinney, W., *et al.*, 1991, EMAP-Surface Waters Monitoring and Research Strategy—Fiscal Year 1991. U. S. Environmental Protection Agency. EPA/600/3-91/022.

Pienitz, R., Smol, J. P., and Birks, H. J. B., 1995, Assessment of freshwater diatoms as quantitative indicators of past climate change in the Yukon and Northwest Territories, Canada, *J. Paleolim.* **13**:21–49.

Simola, H., 1977, Diatom succession in the formation of annually laminated sediment in Lovojarvi, a small eutrophicated lake, *Ann. Bot. Fennici* **14**:143–148.

Siver, P. A., 1991, *The Biology of Mallomonas*, Kluwer Academic, Dordrecht.

Smol, J. P., 1980, Fossil synuracean (Chrysophyceae) scales in lake sediments: a new group of paleoindicators, *Can. J. Bot.* **58**:458–465.

Smol, J. P., 1992, Paleolimnology: an important tool for effective ecosystem management, *J. Aquat. Ecosyst. Health* **1**:49–58.

Smol, J. P., 1995, Application of chrysophytes to problems in paleoecology, in: *Chrysophyte Algae: Ecology, Phylogeny and Development* (C. Sandgren, J. P. Smol, and J. Kristiansen, eds.), Cambridge University Press, Cambridge, pp. 303–329.

Smol, J. P., Brown, S. R., and McNeely, R. N., 1983, Cultural disturbances and trophic history of a small meromictic lake from central Canada, *Hydrobiologia* **103**:125–130.

Stoermer, E. F., and Smol, J. P., (eds.), 1999, *The Diatoms: Applications for the Environmental and Earth Sciences,* Cambridge University Press, Cambridge.

Sullivan, T. J., Turner, R. S., Charles, D. F., Cumming, B. F., Smol, J. P., Schofield, C. L., Driscoll, C. T. Cosby, B. J., Birks, H. J. B., Uutala, A. J., Kingston, J. C., Dixit, S. S., Bernert, J. A., Ryan, P. F. and Marmorek, D. R., 1992, Use of historical assessment for evaluation of process-based model projections of future environmental change—lake acidification in the Adirondack Mountains, New-York, U.S.A., *Environ. Pollut.* **77**:253–262.

Ter Braak, C. J. F., 1987, CANOCO—A FORTRAN program for canonical community ordination by [partial] [detrended] [canonical] correspondence analysis, principal components analysis and redundancy analysis (version 2.0), TNO Institute of Applied Computer Science, Statistical Department Wageningen, 6700 AC Wageningen, The Netherlands.

Ter Braak, C. J. F., 1988, Partial canonical correspondence analysis, in: *Classification and Related Methods of Data Analysis* (H. H. Bock, ed.), North-Holland, Amsterdam, pp. 551–558.

Ter Braak, C. J. F., and Juggins, S., 1993, Weighted averaging partial least squares regression (WA-PLS): An improved method for reconstructing environmental variables from species assemblages, *Hydrobiologia* **269/270**:485–502.

Uutala, A. J., 1990, Chaoborus (Diptera: Chaoboridae) mandibles—Paleolimnological indicators of the historical status of fish populations in acid-sensitive lakes, *J. Paleolim.* **4**:139–151.

Wilson, S. E., Cumming, B. F., and Smol, J. P., 1996, Assessing the reliability of salinity inference models from diatom assemblages: An examination of a 219-lake data set from western North America, *Can J. Fish. Aquat. Sci.* **53**:1580–1594.

Chapter 13

Dinoflagellate Cysts as Indicators of Cultural Eutrophication and Industrial Pollution in Coastal Sediments

BARRIE DALE

1. Introduction

This chapter summarizes the first attempts to use dinoflagellate cysts as indicators of eutrophication and industrial pollution. This is new work, published just recently, and though so far based on only a few studies it shows a clear potential for development as a robust working method for environmental science. In contrast to almost all other microfossil groups included in

BARRIE DALE • Department of Geology, University of Oslo, Blindern, N-0316 Oslo, Norway.

Environmental Micropaleontology, Volume 15 of *Topics in Geobiology*, edited by Ronald E. Martin. Kluwer Academic/Plenum Publishers, New York, 2000.

this book, the cysts used here are acid-resistant and therefore not subject to the dissolution that sometimes affects the mineralized shells of foraminifers and diatoms. Furthermore, they provide environmental signals from the first level of the food web—primary production.

1.1. Dinoflagellate Cysts

Dinoflagellates constitute one of the major groups of phytoplankton, present in all the main aquatic ecosystems (freshwater, brackish, and marine). They are an ancient group of organisms, displaying a broad range of biological complexity (Taylor, 1987), and they have been the subject of increasing scientific interest lately as one of the primary groups producing harmful algal blooms (Reguera *et al.*, 1998). In addition to the biflagellate motile stage characteristically found in the plankton, many species also produce nonmotile resting cysts that accumulate in bottom sediments (Dale, 1983). Dinoflagellate resting-cysts (called cysts from here on in the interest of brevity) are most likely hypnozygotes produced in the sexual life cycle. The cells include large quantities of food storage products (starch and lipids), facilitating a resting period of up to 10 years at least. This allows the cysts to function as: (1) protection during periods of adverse conditions (including overwintering at higher latitudes); (2) effective agents of dispersion (including via ships ballast tanks); and (3) seed beds for initiating plankton blooms.

Many cysts are covered by cell walls that are resistant to physical, chemical, and biological attack, thus allowing fossilization. These include species with calcareous and siliceous walls (the latter known only from a few fossil forms), but most are acid-resistant palynomorphs (i.e., they contain sporopollenin-like material termed dinosporin). Acid-resistant cysts constitute one of the main groups of microfossils used routinely in biostratigraphy of sediments from Mesozoic to Pliocene ages, e.g., in the oil industry (Stover *et al.*, 1996). They have also proved increasingly useful as indicators of climatically-produced short-term environmental change in more recent sediments (Dale and Nordberg, 1993; Dale and Fjellså, 1994; Dale, 1996, Thorsen and Dale, 1998; see also in this volume Dixit and Smol, Ch. 12; Alve, Ch. 14; Ishman, Ch. 16; van der Zwaan, Ch. 17).

1.2. Background to the Work Reported Here

The rationale for attempting to use cysts in environmental studies is the same as that for other microfossils. In many cases of public concern over environmental deterioration, the central question raised, "To what extent is this a result of human influence?," has proved difficult to answer. This is because problems such as eutrophication and pollution involve changes over time; understanding the problem therefore depends on sufficient time series observations. Ideally, these would cover both the amount of pollution (e.g., nutrients from sewage outfalls) and the effects on the ecosystem (e.g., levels of dissolved oxygen in the water and amounts of primary productivity). They would also cover sufficient time to permit an assessment of natural background levels of variation prior to human impact.

In reality, these ideals are seldom met, and the general lack of adequate time-series data has prompted the need for *retrospective* methods for assessing the extent and timing of human impact. The recovery of environmental information archived in bottom sediments of aquatic systems has long been recognized as one of these methods, and both geochemical signals (Walsh *et al.*, 1981) and microfossils have been used successfully (see, e.g., many references elsewhere in this volume). However, it is important to realize the limitations of any one geochemical parameter or microfossil signal. Within the oil industry, *integrating signals* from various geochemical and microfossil data has proved most useful for interpreting paleoenvironments. The future strategy for similar applications in environmental science should also involve developing as many such methods as possible, and integrating the signals they provide.

The work reported here was started because cysts, as the fossilized remains of a major group of phytoplankton, offered an obvious potential for recording quantitative and qualitative changes induced by cultural eutrophication. Furthermore, global modeling of recent cysts in marine to brackish systems (Wall *et al.*, 1977; Dale, 1996) showed them to be sensitive indicators of various environmental factors (e.g., water temperature, salinity, and nutrient levels), suggesting that they may also provide useful signals for tracing marine pollution.

Results are summarized from two case studies from the southeastern coast of Norway: cultural eutrophication in the Oslofjord and marine pollution in the Frierfjord. It should be noted that these examples are from marine systems; there are also freshwater cysts, but the extent to which they fossilize and may provide environmental signals is unknown.

1.3. Materials and Methods

Both case studies employed the same type of material and the same methods. Bottom sediments were cored using conventional gravity corers, and 1- or 2-cm slices were taken out for samples. The outer ring of sediment from each slice was cut off and discarded to avoid contamination owing to sediment sliding along the core barrel, leaving a sample from the center of about 2 cm^3. Samples were oven-dried at 50°C, weighed, and prepared using standard palynological methods (similar to those of Wood *et al.*, 1996): (1) cold 10% HCl to dissolve carbonates; (2) warm (60°C) 40% HF to dissolve silicates; (3) sieving to retain the >25-μm fraction, and finally (4) an aliquot portion of this was mounted on microscope slides for cyst analysis using light microscopy. Results were expressed both quantitatively (cysts/g dry sediment) and qualitatively (percentage composition of the cyst assemblage).

Duplicate cores and samples were taken for dating by the ^{210}Pb method (Pheiffer Madsen and Sorensen, 1979).

2. Cultural Eutrophication in the Oslofjord

We have recently put together the results from several studies from the Oslofjord, Norway, over the past few years, which demonstrate that cysts may be useful indicators of cultural eutrophication in the inner part of the fjord (Dale *et al.*, 1999). These results are summarized in the form of a generalized model for comparison with subsequent attempts to test for cultural eutrophication elsewhere.

2.1. Cultural Eutrophication

Eutrophication is an increase in the rate of supply of organic matter to an ecosystem (Nixon, 1995), and cultural eutrophication is when this increase results from human activities (e.g., sewage outfalls). Cultural eutrophication was considered to be a problem largely affecting freshwater systems, but has been the subject of increasing concern among marine scientists, especially since the Group of Experts on the Scientific Aspects of Marine Pollution (GESAMP, 1991) identified it as "... one of the causes of immediate concern in the marine environment."

It is generally accepted that cultural eutrophication has produced harmful effects on marine ecosystems in different parts of the world (Nixon, 1990), particularly the devastating effects of hypoxia on benthos and demersal fish stocks (e.g., Rosenberg *et al.*, 1990). There is even a controversial hypothesis to the effect that it is causing a global epidemic of harmful algal blooms (Smayda, 1990).

Conditions in the inner Oslofjord are particularly favorable for the development of cultural eutrophication. The city of Oslo is adjacent to the two innermost basins of the fjord (Vestfjord and Bunnefjord, both about 160 m deep), which are separated from the outer fjord by a narrow channel (Drobak Sound), with a maximum sill depth of only 19.5 m (Fig. 1). Circulation, exchange, and renewal of water is therefore severely restricted (Gade, 1968), and additional nutrient-loading from the city would thus tend to be retained largely within the inner fjord, leading to a potential to fertilize increased primary production.

This appears to have been the case, with increased amounts of sewage discharge to the fjord from a rapidly growing population causing substantial negative consequences for water quality. Oslo experienced larger-scale industrial development starting about the middle of the 1800s. This was accompanied by almost exponential growth in the population, which grew from around 40,000 to 470,000 by the 1960s and remained at about this level through the 1980s. As in other examples, there are no consistent, systematic,

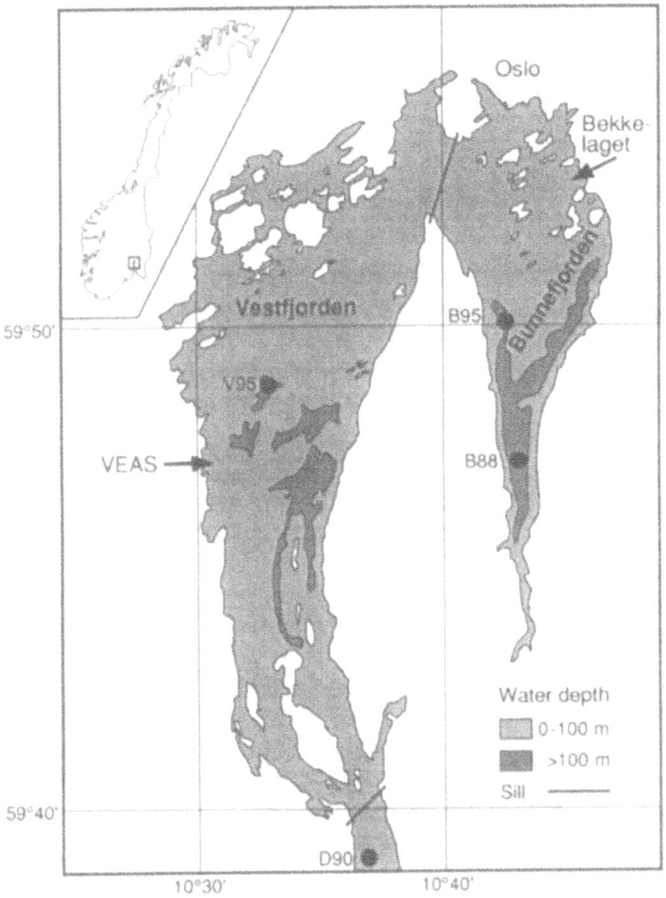

FIGURE 1. Map of the inner Oslofjord showing the location of cores studied for cultural eutrophication (modified from Dale *et al.*, 1999). Note the location of two main sewage effluent discharges: Bekkelaget (major discharge until 1980s) and VEAS (a new and cleaner treatment plant that replaced the older ones around Oslo from the 1980s onward).

long-term data documenting the development of cultural eutrophication, but there is a great deal of substantiate evidence (summarized in Table 1).

The inner Oslofjord represents an excellent case for investigating possible retrospective signals of cultural eutrophication, since: (1) cultural eutrophication occurred early (its maximum development was probably in the 1940s to 1950s); (2) it was studied early on by marine scientists from most of the main relevant disciplines (water chemistry, both phyto- and zooplankton; fisheries, and benthos); and (3) as a result of early warnings from the scientists, serious attempts were made to clean up sewage discharge (particularly in the 1970s and 1980s), such that there is a sufficiently long record of possible environmental recovery to be found in the sediments.

TABLE 1. Summary of Evidence for Cultural Eutrophication in the Inner Oslofjord

Time period	Evidence	Reference
1917	High phytoplankton populations	Gaarder and Gran, 1927
1933–1934	Pronounced effect on phytoplankton from pollution	Braarud and Bursa, 1939
1950–1951	Deep waters of innermost Bunnefjord void of oxygen-H_2S from bottom up to 75 m	Beyer and Foyn, 1951
1960s	Reports from comprehensive research projects on the Oslofjord and its pollution problem	Ruud, 1968*a*,*b*; Beyer, 1968; Munthe-Kaas, 1968
mid-1970s on	Negative effects reduced through improved sewage treatment	Magnusson and Rygg, 1988

2.2. The Eutrophication Signal

The four cores studied from the inner Oslofjord (Fig. 1) all contained ample cyst records (a total of 32 species) covering both the development of eutrophication and the cleanup phases, allowing us to assess possible signals. The results (diagrammatically represented in Fig. 2) strongly suggest a eutrophication signal comprising the following two elements:

1. Increasing total cysts/g sediment (up to about double), considered to reflect the increase in total primary production through eutrophication, including cyst-forming dinoflagellates.
2. A notable increase in one particular species, *Gonyaulax polyedra*, accounting for most of the total increase in cysts/g. This is also a proportional increase, raising the species from a minor element of preindustrial assemblages (<5%) to dominance during eutrophication (around 40–50%).

The signal is considered to be particularly robust, since the clear trends accompanying the buildup of eutrophication are reversed to around preindustrial levels following cleanup efforts in the late 1980s, especially in the Bunnefjord. The timing of this corresponds well with the opening of a new and cleaner treatment plant for most of the city's sewage, with an outfall in the Vestfjord (VEAS, Fig. 1).

The importance of *G. polyedra* in the signal is interpreted as reflecting the fact that this is a typical bloom species (Lewis and Hallet, 1997) in summer plankton of the Oslofjord. In its natural state, the system is nutrient-limited (Paasche and Erga, 1988), such that spring phytoplankton growth is unlimited by nutrients, whereas the summer phytoplankton is usually limited by nutrients remaining after a spring bloom (predominantly of diatoms). *G. polyedra* seems to have been able to exploit the added nutrients from human impact,

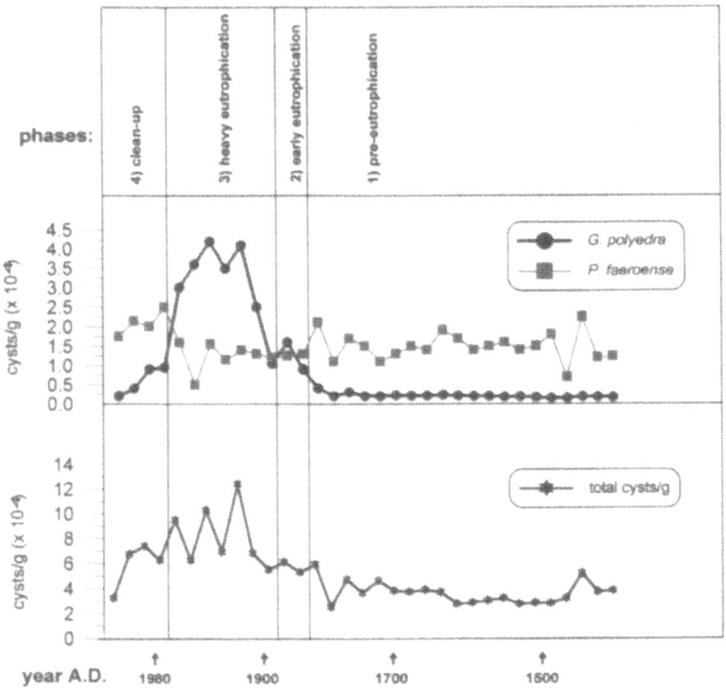

FIGURE 2. Schematic diagram summarizing the main elements of the dinoflagellate cyst signal for eutrophication in the Oslofjord (Dale *et al.*, 1999): increased total cysts/g dry sediment (roughly double) and a marked increase in *G. polyedra*, while the population of *P. faeroense* remained fairly constant.

allowing it to develop a larger population than would otherwise have been possible with the naturally limited nutrients. This interpretation is supported by the fact that *Peridinium faeroense* (a typical species in the spring plankton of Oslofjord, and therefore presumably not nutrient-limited) maintained about the same population throughout the development of cultural eutrophication (Fig. 2).

3. Industrial Pollution in the Frierfjord

Two cores were studied from a fjord system in southwestern Norway to investigate the potential for using cysts as indicators of industrial pollution in the marine environment (Saetre *et al.*, 1997). This work was carried out in cooperation with Dr. E. Alve, who had previously studied benthic foraminifera from this area (see the Acknowledgments, here, and Ch. 14 by Alve in this volume).

3.1. Hydrography and Pollution

The location of the cores is shown in Fig. 3: Core 1 from the heavily polluted Frierfjord, and core 2 from the adjoining relatively unpolluted Brevikfjord. Hydrography of the Frierfjord (maximum depth 98 m) is largely controlled by a relatively shallow sill (maximum depth 23 m) at its southern entrance. This restricts water circulation, particularly in the periodically stagnant deeper waters (estimated retention time of 1.5–2.5 years); the intermediary waters (8–30 m below surface) are renewed on average about once a month (Molvaer *et al.*, 1979). The River Skien produces a 2- 8-m-thick brackish surface layer that flows quickly through the fjord. There is more circulation in the adjoining Brevikfjord, where deeper waters are renewed completely or partially about twice a year.

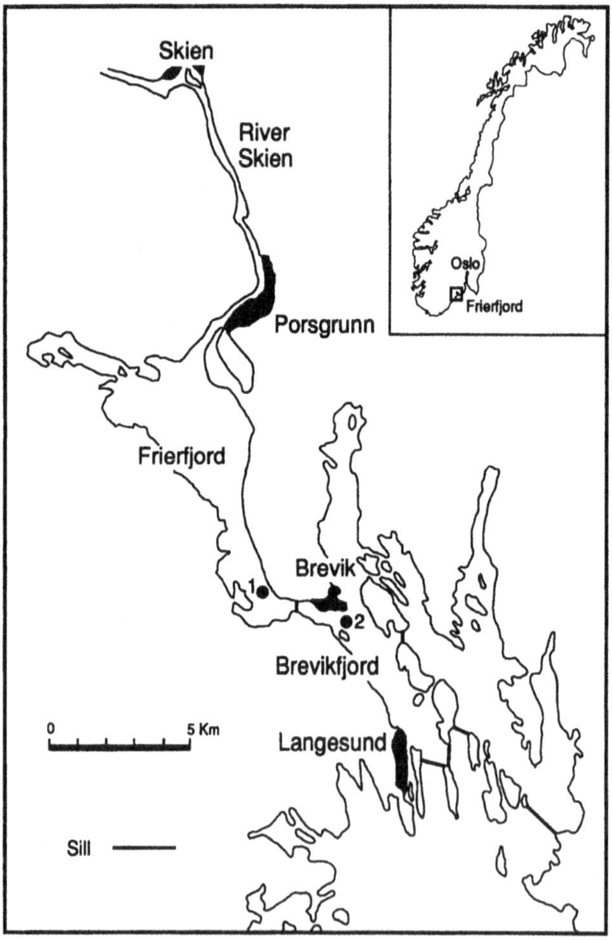

FIGURE 3. Map of the Frierfjord region showing the location of cores studied for dinoflagellate cyst signals: core 1 (=core 28 of Alve, Ch. 14, this volume) in the heavily polluted Frierfjord, and core 2 (=core 30 of Alve, Ch. 14, this volume) from the relatively unpolluted Brevikfjord.

The region around Frierfjord developed into an industrial center during the 19th and 20th centuries, and populations in the towns of Skien and Porsgrunn increased from about 4000 in 1800 to about 27,000 in 1960. The River Skien and the Frierfjord received increasing amounts of sewage effluents and industrial waste, which eventually had a detrimental effect on the fjord environment. Heavy industrial pollution included very slightly soluble material (e.g., wood fiber and particular inorganic material), and toxins such as heavy metals, chlorinated hydrocarbons, and polycyclic aromatic hydrocarbons (PAH) (Molvaer *et al.*, 1979). Rapid consumption of oxygen from remineralization of the organic matter, combined with restricted water circulation, contributed to the anoxic or critically low oxygen levels regularly recorded from deep water in Frierfjord.

As with the Oslofjord, the possibility of a long period of environmental recovery makes the Frierfjord a particularly good site for testing for useful signals from microfossils. During 1974–1977, pronounced restrictions imposed on industrial discharges substantially reduced pollution levels (e.g., heavy metals, chlorinated hydrocarbons, and PAH in fish and mussels decreased). This allowed us to study possible responses to pollution reflected in changes in the cyst record by comparing the records in the polluted Frierfjord (station 1, Fig. 3) with those of the adjoining, relatively unpolluted Brevikfjord (station 2) before and during heavy pollution and after the cleanup.

Dating suggested that the 35-cm core from Frierfjord represented roughly the last 300 years (Fig. 4). Our study (Saetre *et al.*, 1997) included several types of geochemical data (not repeated here): water content of sediment, loss of weight on ignition, total organic carbon analysis, and CHN analysis to investigate amounts and types of organic matter, and analysis for six metals: Cu, Zn, Cd, Ni, Fe, and Mn. The geochemical data agreed with both the dates and the known history of pollution, thus helping to establish a framework for interpreting the cyst data. This suggested that contaminants started to increase around 1836, with an initial peak about 1878, reaching a final peak around 1960 and then leveling off, before decreasing from about 1974 onward.

3.2. Cyst Signals from the Frierfjord

Cyst concentrations were generally low compared with other fjord environments, but a total of 22 cyst species was recovered, of which 4 were dominant. The results suggest possible signals reflecting industrial pollution (summarized in Fig. 4) comprising the following elements:

1. Decreasing total cysts/g sediment (by as much as two-thirds), considered to reflect, at least in part, deteriorating environmental conditions for many cyst-forming dinoflagellates.
2. A series of proportional changes within the dominant species, involving systematic shifts corresponding to progressive stages of pollution (1–4 in Fig. 4).

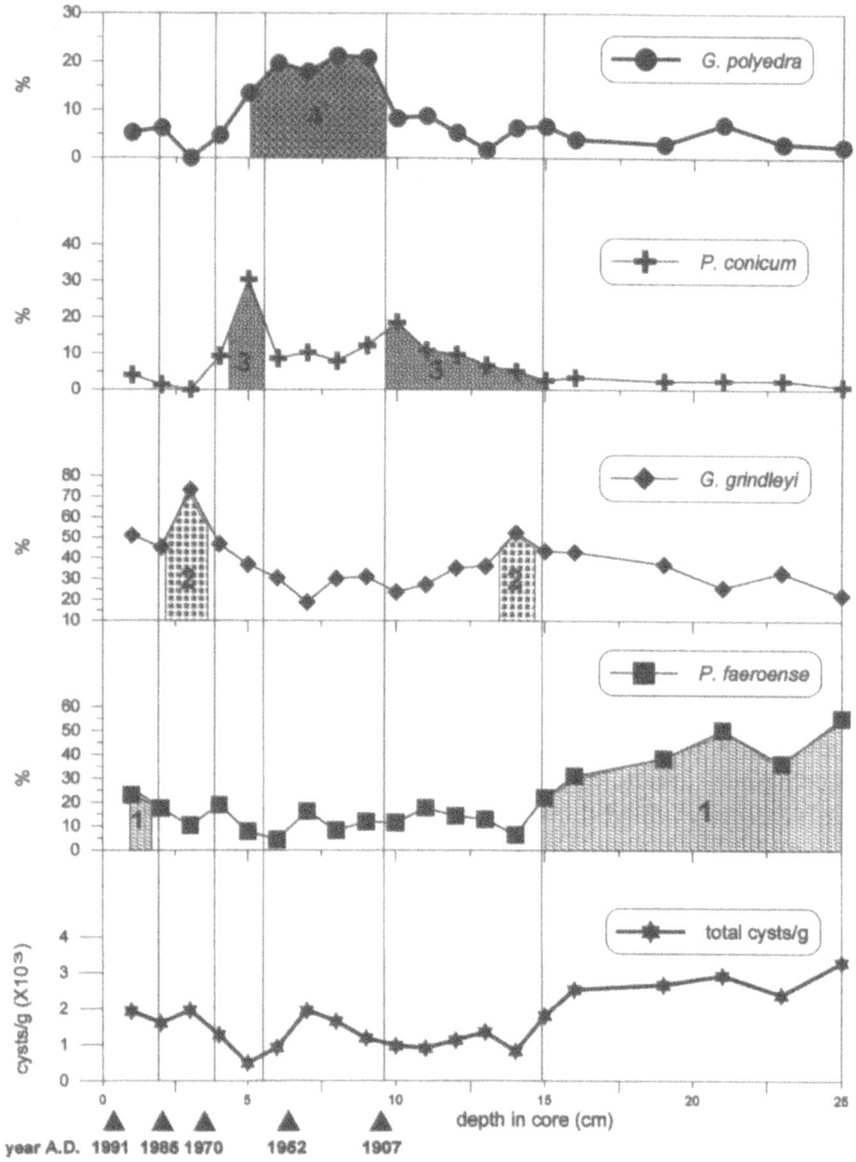

FIGURE 4. Summary of the main elements of the dinoflagellate cyst signals for pollution from core 1, Frierfjord (data from Saetre *et al.*, 1997). A sequence of four phases of pollution (1–4) are suggested, with two peaks (Phase 3) in the late 1800s and the 1960s, with a reversal after substantial cleanup of the environment starting in the 1970s.

These signals are considered to be fairly robust, since the main elements accompanying the presumed increase in pollution are reversed around the 1970s, after which time pollution was substantially reduced. The ecological preferences of the four dominant species are reasonably well known (Dale, 1996), which allows us to suggest possible interpretations for their proportional changes. The first change recorded (therefore a possible response to pollution) is the notable decrease in *P. faeroense* (from >50% to >30%). Two alternative interpretations may be suggested: either this is a species that is sensitive to very early stages of pollution (industry began with a saw mill in the 16th century), in which case environmental recovery has not yet reached the stage allowing it to reestablish previous levels of dominance; or this may be a climatic signal (this is a colder water species and phase 1 here probably corresponds to the warming at the end of the Little Ice Age).

The next change (phase 2) involves a proportional increase in *Gonyaulax grindleyi*. This is the most cosmopolitan species in present-day cyst assemblages, occurring in polar to equatorial regions, and abundance "spikes" are characteristic of a wide variety of environmental changes (Dale, 1996). In the Frierfjord, this is interpreted as probably reflecting early changes resulting from pollution.

Phase 3 is characterized by the proportional increase of *Protoperidinium conicum*. This is also a cosmopolitan species, but almost never accounts for more than 1–2% of the assemblage elsewhere (with the exception of upwelling areas, discussed below). Its increase here, from this "normal" background level to two peaks (around 20 and 30%, respectively) corresponding roughly to the peaks of contamination, is particularly noteworthy. Unlike the other species shown here, *P. conicum* is heterotrophic, feeding on phytoplankton. There are at least two possible explanations for its increase: (1) that a reduction in light penetration in surface waters (reduced Secchi depth) caused by heavy contamination favored heterotrophs that are independent of light; or (2) alternatively, there may have been a pollution-induced shift in phytoplankton composition toward more suitable prey for *P. conicum*.

Phase 4 is very similar to the eutrophication signal from the Oslofjord, involving at least a proportional increase in *G. polyedra*. Indeed, this may reflect true cultural eutrophication, since it occurs in a period around the early to mid-1900s, when large amounts of poorly treated sewage effluent were released into the system. This possibility is further supported by a corresponding overall increase in total cysts/g sediment, though true cyst production is difficult to assess given the strong possiblity of "dilution" by increased sedimentation of waste particulate matter (discussed below).

4. Discussion

4.1. Comparison between the Signals of Eutrophication and Industrial Pollution

The two investigations summarized here allow us to suggest possible signals for tracing the history of eutrophication and industrial pollution. The main elements of these appear somewhat opposite, the former being characterized by increased total cysts/g sediment accompanied by massive increase in one autotrophic species and the other showing decreased total cysts/g and at least the proportional increase of one heterotrophic species. However, whereas it is reasonable to interpret the eutrophication signal as an increase in cyst production (since there is no suggestion of major changes in the rather stable sedimentation rates in the Oslofjord), interpreting cyst production in the Frierfjord is more uncertain.

In the Frierfjord, the reduced cysts/g could reflect lower production (at least of cyst-forming dinoflagellates) owing to the toxic effects of chemical pollution; reduction of the euphotic zone owing to light-reducing waste particulate matter; or increased sedimentation by waste material (even though core 1 was taken well away from the river that transported much of the waste, there was a noticeable increase of, e.g., cellulose particles in the more polluted sediments). Most likely it reflects a combination of all these factors. Nevertheless, this is considered to be an important element of the pollution signal, especially since equivalent sediments in the relatively unpolluted core 2 (identified by similar, but much less pronounced chemical signals) showed *increasing* cysts/g (from ~2000/g to >3500/g).

The cyst record from core 2 (not repeated here) also supported the increase of, especially, *P. conicum* as an important element of the pollution signal, since no such increase was seen (all samples included <4%). Otherwise, the record in core 2 showed a weak trend toward decreasing *P. faeroense*, similar to, but much weaker than in core 1 (supporting the climatic interpretation?). The only clear trend in core 2 was an increase, particularly in the upper 10 cm, in *G. grindleyi* (from about 40 to almost 60%), probably reflecting overall environmental change (possibly due to pollution/eutrophication, but if so then at lower levels not producing distinguishable signals).

4.2. Comparison with Other Cyst Signals

The main cyst signals reflecting environmental factors described prior to the work summarized here include: climatic (surface water temperatures), oceanic vs. coastal waters, salinity, and nutrients (e.g., upwelling) (Dale, 1996). In assessing signals of cultural eutrophication or industrial pollution, we need to consider other possible causes, and two of the above signals are of particular interest for comparison. Climate has changed during the time period of interest here (Jones, 1994), with a general warming since the end of the Little Ice Age. *G. polyedra* is a useful warm-water indicator in southern Scandinavian waters (Dale, 1996). It could therefore have been interpreted as reflecting climatic warming if it had only increased in the Oslofjord record. However, since it shows a marked *decrease* since the 1980s during continued climatic *warming*, the suggested eutrophication signal remains distinctive.

Various aspects of the upwelling signal are pertinent to the work summarized here. Cyst assemblages in recent sediments associated with upwelling in many parts of the world have been studied in order to establish their possible use in tracing the development of these natural sources of nutrient enrichment. The general signal obtained suggests no particular increase in total cysts/g (possibly even reduction) and a heavy dominance of heterotrophic species (Dale, 1996). This is because other phytoplankton groups (notably diatoms) are more successful competitors than autotrophic dinoflagellates in the presence of an ample supply of appropriate nutrients. Thus, cyst concentrations tend to be low (cysts/g sediment) owing to fewer autotrophic dinoflagellates, including the cyst-forming species, while the heterotrophic species (e.g., feeding on diatoms) dominate the cyst assemblages.

There are clear parallels between the upwelling signal and the pollution signal obtained from Frierfjord; certain cyst assemblages from upwelling areas are characterized by dominance of the same species as in core 1 from the Frierfjord: *P. conicum* (17–44% off Mauritania, 20–30% along the Chilean coast, 19–40% off Iceland, and 14–53% round the Faeroe Islands) (Dale, 1996). From this perspective, the pollution signal from Frierfjord may be interpreted as reflecting a similar shift in relative amounts of various groups within the plankton: reduced amounts of autotrophic dinoflagellates and at least a proportional shift to heterotrophs (especially *P. conicum*). However, the main difference is still that the shift in assemblages in the Frierfjord appears to be closely associated with pollution, both in timing and location (since it is not reflected in core 2), and therefore may prove useful for tracing pollution in coastal zones, where the signal cannot be confused with upwelling. However, Thorsen and Dale (1997) showed a marked increase in *P. conicum* in supposedly eutrophicated sediments in a fjord near Bergen, on the western coast of Norway. Since contamination in this fjord is considered to be mainly from domestic sewage, the question remains as to what extent this record reflects more toxicity (as in the Frierfjord?) than suspected or ultimately represents support for an alternative eutrophication signal.

There are similar parallels between the cyst signal obtained from some waters influenced by upwelling areas and the eutrophication signal from the Oslofjord. Shelf waters shoreward of major upwelling regions, both off California and off northwestern Africa, are enriched by nutrients from the upwelling further offshore, though at substantially lower levels. As in the eutrophication signal from the Oslofjord, cyst assemblages from these shelves are strongly dominated by *G. polyedra*, in this case a signal of natural eutrophication.

5. Future Work

The two cases presented here may be used as models against which the results from other similar studies can be compared. Future work should include documenting cyst records for the past several hundred years from other examples of cultural eutrophication and industrial pollution, as well from uncontaminated (or relatively uncontaminated) sites. The most informative comparisons presumably will come from studies carried out in climatic regimes similar to those described here from southwestern Norway, since the overall assemblage composition will be similar. On the other hand, the species forming the main elements of these signals are widespread and fairly cosmopolitan, and it will be interesting to see how broadly applicable these signals prove to be.

The discussion here strongly suggests that the signal obtained from a record of cultural eutrophication in the Oslofjord is most likely a signal for eutrophication in general. As such, it opens up the possibility of developing "state-of-the-environment" assessment with respect to eutrophication that could be useful for obtaining baseline values against which future trends could be assessed. For example, this could help to indicate the need for improved cleanup of sewage discharges or to demonstrate environmental recovery following cleanup (interestingly, the cyst record from the core from the Vestfjord, nearest to the new sewage treatment plant VEAS suggested that recovery there had not progressed as far as in the Bunnefjord, though more detailed work is needed to substantiate this).

The possible signal from one of the most polluted fjords in Norway, the Frierfjord, raises several questions: To what extent is the apparent reduction in total cysts/g reflecting reduced cyst production (e.g., due to toxicity) or dilution by added sediment from pollution? Has pollution caused a shift within the plankton away from autotrophic dinoflagellates toward other groups, including prey for the heterotrophic species? To what extent have there been toxic effects on the phytoplankton—or does the signal reflect a different form of nutrient enrichment (somehow analogous to offshore upwelling)? Answering these, and other questions raised here, and integrating the cyst signals with those from other microfossils, should prove an exciting challenge to a new and developing field within the environmental sciences.

However, results just published from cyst records in two cores from the heavily polluted Yokohama Port, Tokyo Bay, Japan, Matsuoka (1999) illustrates the need to differentiate between industrial pollution and eutrophication. The study area in Yokohama Port is in many ways similar to the Frierfjord, a sedimentary basin that has received heavy pollution from both sewage and industrial waste, including massive river transport of particulate matter into the system. The main signals from the cysts are also similar, with Frierfjord and at least Matsuoka's station 1 clearly showing increased heterotrophic influence with pollution. Matsuoka interprets this as a eutrophication signal in Yokohama Port, but the possible influence of toxic waste or reduced light from particulate matter emphasized in the Frierfjord study is not discussed. An increase in diatom biomass is a common response to nutrient loading where sufficient silica is present (e.g., Conley *et al.*, 1993). Therefore, such cases may be expected to produce a shift to more heterotrophic dinoflagellates that feed on diatoms, as postulated by Matsuoka (1999), but as yet the examples from Yokohama Port and Frierfjord cannot be said to demonstrate this unequivocally since eutrophication as a factor cannot be separated from heavy industrial pollution. The Nordåsvannet, discussed in the previous section here, with increased occurrence of the diatom *Skeletonema costatum* and cysts of a heterotrophic dinoflagellate corresponding to increased nutrient loading, may represent an example of this type of eutrophication signal (as discussed by Thorsen and Dale, 1998, p. 440).

6. Taxonomic Note

The taxonomy of dinoflagellates is characterized by frequent name changes, owing in part to the problems caused by a long history until recently of separate classification systems for motile stages and cysts (discussed fully in Dale, 1983). A full discussion of these problems would be out of place here, but suffice it to say that species names used here follow the more conservative philosophy suggested by Dale (1983) aimed at more easily retaining ecological information. The following list of synonyms (cyst-based names marked *) should help the reader relate this work to other literature:

- *Gonyaulax polyedra* (=*Lingulodinium machaerophorum**; *Lingulodinium polyedrum*)
- *Gonyaulax grindleyi* (=*Operculodinium centrocarpum**; *Protoceratium reticulatum*)
- *Peridinium faeroense* (=*Pentapharsodinium dalei*)
- *Protoperidinium conicum* (=*Selenopemphix quanta**)

ACKNOWLEDGMENTS: The work summarized here is from Dale *et al.*, 1999 and Saetre *et al.*, 1997; while these studies are discussed from my own current perspective, the major contributions of coauthors are greatfully acknowledged. Thanks are also due to Dr. Jens Skei and other colleagues at NIVA for various forms of support for the eutrophication studies in the Oslofjord, and to the Nansen Fund for financing the dating of the cores. I thank Dr. Elisabeth Alve, University of Oslo (with financial support from the Norwegian Research Council) for cooperating on the Frierfjord work—providing samples, dates for core 1, and sharing her data on weight loss on ignition, TOC, and water content [see also her chapter (Ch. 14) in this volume].

References

Beyer, F., 1968, Zooplankton, zoobenthos, and bottom sediments as related to pollution and water exchange in the Oslofjord, *Helgol. Wiss. Meeresun.* **17**:496–509.
Beyer, F., and Foyn, E., 1951, Surstoffmangel I Oslofjorden. En kritisk situasjon for fjordens dyrebestand, *Naturen* **75**(10):289–306.
Braarud, T., and Bursa, A., 1939, The phytoplankton of the Oslofjord 1933–34, *Hvalr. Skrif.* **19**:1–63.
Conley, D. J., Schelske, C. L., and Stoermer, E. F., 1993, Modification of the biogeochemical cycle of silica with eutrophication, *Mar. Ecol. Prog. Ser.* **101**:179–192.
Dale, B., 1983, Dinoflagellate resting cysts: "benthic plankton," in: *Survival Strategies of the Algae* (G. A. Fryxell, ed), Cambridge University Press, New York, pp. 69–136.
Dale, B., 1996, Dinoflagellate cyst ecology: Modelling and geological applications, in *Palynology: Principles and Applications* (J. Jansonius and D. C. McGregor, eds), American Association of Stratigraphic Palynologists Foundation, Publishers Press, Salt Lake City, pp. 1249–1276.
Dale, B., and Nordberg, K., 1993, Possible environmental factors regulating prehistoric and historic "blooms" of the toxic dinoflagellate *Gymnodinium catenatum* in the Kattegat-Skagerrak region of Scandinavia, in: *Toxic Phytoplankton Blooms in the Sea* (T. J. Smayda and Y. Shimizu, eds.), Elsevier, Amsterdam, pp. 53–57.
Dale, B., and Fjellså, A., 1994, Dinoflagellate cysts as Paleoproductivity indicators: State of the art, potential, and limits, in: *Carbon Cycling in the Glacial Ocean: Constraints on the Ocean's Role in Global Change* (R. Zahn, T. F. Pedersen, M. A. Kaminski, and L. Labeyrie, eds.), Springer-Verlag, Berlin, pp. 521–537.
Dale, B., Thorsen, T. A., and Fjellså, A., 1999, Dinoflagellate cysts as indicators of cultural eutrophication in the Oslofjord, Norway, *Est. Coast. Shelf Sci.* **48**:371–382.
Gaarder, T., and Gran, H. H., 1927, Investigations of the production of plankton in the Oslo Fjord, *Rapp. Proces-Verbaux Reun. Cons. Int. Explor. Mer.* **42**:1–48.
Gade, H. G., 1968, Horizontal and vertical exchange and diffusion in the water masses of the Oslofjord, *Helgol. Wiss. Meeresun.* **17**:462–475.
GESAMP, 1991, *The State of the Marine Environment*, Blackwell, Oxford.
Jones, P. D., 1994, Hemispherical surface air temperature variations: A reanalysis and update to 1993, *J. Clim.* **7**:1794–1802.
Lewis, J., and Hallet, R., 1997, *Lingulodinium polyedrum* (*Gonyaulax polyedra*) a blooming dinoflagellate, *Oceanogr. Mar. Biol. An Ann. Rev,.* **35**:97–161.
Magnusson, J., and Rygg, B., 1988, Eutrofisituasjonen i ytre Oslofjord. Rapport 338/88 Norsk institutt for vannforskning (NIVA), Oslo, Norway.
Matsuoka, K., 1999, Eutrophication process recorded in dinoflagellate cyst assemblages: A case of Yokohama Port, Tokyo Bay, Japan, *Sci. Total Environ.* **231**:17–35.
Molvaer, J., Bokn, T., Kirkerud, L., Kvalvågnes, K., Nielsen, G., Rygg, B., and Skei, J., 1979, Resipientundersogelse av nedre Skienselva, Frierfjorden og tilleggende fjordområder. Report no. 0-70111, NIVA, Oslo, Norway, 252 pp.
Munthe-Kaas, H., 1968, Surface pollution and light extinction in the Oslofjord, *Helgol. Wiss. Meeresun.* **17**:476–488.
Nixon, S. W., 1990, Marine eutrophication: A growing international problem, *Ambio* **19**:101.
Nixon, S. W., 1995, Coastal marine eutrophication: A definition, social causes, and future concerns, *Ophelia* **41**:199–219.
Paasche, E., and Erga, S. R., 1988, Phosphorus and nitrogen limitation of phytoplankton in the inner Oslofjord (Norway), *Sarsia* **73**, 229–243.
Pheiffer Madsen, P., and Sørensen, J., 1979, Validation of the lead-210 dating method, *J. Radioanal. Chem.* **54**:39–48.
Reguera, B., Blanco, J., Fernández, Ma, L., and Wyatt, T., eds., 1998, *Harmful Algae*. Xunta de Galicia and Intergovernmental Oceanographic Commission of UNESCO.
Rosenberg, R., Elmgren, R., Fleischer, S., Jonsson, P., Persson G., and Dahlin, H., 1990, Marine eutrophication case studies in Sweden, *Ambio.* **19**:102–108.

Ruud, J. T., 1968*a*, Introduction to the studies of pollution in the Oslofjord, *Helgol. Wiss. Meeresunt.* **17**:455–461.

Ruud, J. T., 1968*b*, Changes since the turn of the century in the fish fauna and the fisheries of the Oslofjord, *Helgol. Wiss. Meeresunt.* **17**:510–517.

Smayda, T., 1990, Novel and nuisance phytoplankton blooms in the sea: Evidence for a global epidemic, in: *Toxic Marine Phytoplankton* (E. Graneli, B. Sundstrom, L. Edler, and D. A. Anderson, eds.), Elsevier, New York, pp. 29–40.

Stover, L. E., Brinckhuis, H., Damassa, S. P., de Verteuil, L., Helby, R. J., Monteil, E., Partridge, A. D., Powell, A. J., Riding, J. B., Smelror, M., and Williams, G. L., 1996, Mesozoic–Tertiary dinoflagellates, acritarchs and prasinophytes, in: *Palynology: Principles and Applications* (J. Jansonius and D. C. McGregory, eds.) American Association of Stratigraphic Palynologists Foundation, Publishers Press, Salt Lake City, pp. 641–750.

Saetre, M. L. L., Dale, B., Abdullah, M. I., and Saetre, G.-P., 1997, Dinoflagellate cysts as possible indicators of industrial pollution in a Norwegian fjord, *Mar. Environ. Res.* **44**(2):167–189.

Taylor, F. J. R., ed., 1987, *The Biology of Dinoflagellates, Botanical Monographs*, Vol. 21, Blackwell Scientific, Oxford.

Thorsen, T. A., and Dale, B., 1997, Dinoflagellate cysts as indicators of pollution and past climate in a Norwegian fjord, *Holocene* **7**(4):433–446.

Thorsen, T. A., and Dale, B., 1998, Climatically influenced distribution of *Gymnodinium catenatum* during the past 2000 years in coastal sediments of southern Norway, *Palaeogeogr. Palaeoclim. Palaeoecol.* **143**:159–177.

Wall, D., Dale, B., Lohmann, G. P., and Smith, W. K., 1977, The environmental and climatic distribution of dinoflagellate cysts in modern marine sediments from regions in the North and South Atlantic Oceans and adjacent seas, *Mar. Micropaleontol.* **2**:121–200.

Walsh, J., Prmuzic, E. T., and Whitledge, T. E., 1981, Fate of nutrient enrichment of continental shelves as indicated by the C/N of bottom sediments, in: *Ecohydrodynamics* (J. C. J. Nihoul, ed.), Elsevier, Amsterdam, pp. 13–50.

Wood, D. G., Gabriel, A. M., and Lawson, J. C., 1996, Palynological techniques—processing and microscopy, in: *Palynology: Principles and Applications* (J. Jansonius and D. C. McGregor, eds.), American Association of Stratigraphic Palynologists Foundation, Publishers Press, Salt Lake City, pp. 29–50.

Chapter 14

Environmental Stratigraphy

A Case Study Reconstructing Bottom Water Oxygen Conditions in Frierfjord, Norway, over the Past Five Centuries

ELISABETH ALVE

ELISABETH ALVE • Department of Geology, University of Oslo, Blindern, N-0316 Oslo, Norway.

Environmental Micropaleontology, Volume 15 of *Topics in Geobiology*, edited by Ronald E. Martin. Kluwer Academic/Plenum Publishers, New York, 2000.

1. Introduction

Many silled fjords have a natural potential for developing dysoxic and even anoxic bottom water conditions. This often occurs in microtidal silled fjords with an estuarine circulation pattern, where surface or transitional low-density water extends down to sill depth and thereby inhibits frequent renewals of the denser deep basin water. The oxygen regimes in such fjords are very sensitive to increased fluxes of organic carbon whether due to natural or anthropogenic causes.

Particularly during the last century, many coastal areas and fjords have served as repositories for different kinds of human-induced organic material and of nutrients. The major sources have been municipal effluents (e.g., sewage), effluents from paper and pulp mill industries, drainage from farmlands, and in situ fish farm operations. In several areas, these discharges, either directly or indirectly, have increased the oxygen consumption both in the overlying water masses and in the seabed sediments. A fundamental ecological question, and one of concern to environmental management, is: Has the overall position of the redox-boundary within the sediments or in the water column been elevated owing to anthropogenic discharges, or is it positioned in accordance with the natural hydrographical, geochemical, and geomorphological setting of the area?

Ideally, hydrographical and geochemical time-series data extending more than 100 years or at least several decades back in time would help to clarify this, and biological time series would provide information about possible negative impacts on the ecosystem. However, such long time series are virtually nonexistent for most areas, and the only way of providing information about the recent past environmental conditions is through interpretation of the geological record, i.e., environmental stratigraphy. In other words, environmental stratigraphy is reconstruction of recent environmental history through detailed paleontological, sedimentological, and geochemical analysis of sediment cores. Sedimentology and geochemistry provide data on the fluxes and composition of both inorganic and organic material. However, it is only by studying benthic organisms that are preserved in the sediments that it is possible to reconstruct the oxygen conditions of past bottom water and sediment pore water. An ideal group for those studies is the benthic foraminifera, which are sensitive indicators of dysoxic/anoxic and other environmental conditions (see Murray, Ch. 1, in this volume for a discussion of environmental change using benthic foraminifera as proxies). Such background information will also be useful in the future, as it will provide a baseline for comparison to establish to what extent governmental regulations on anthropogenic discharges have had a positive effect on the benthic communities (for discussion, see Alve, 1995*a*).

To date, relatively few investigations have used environmental stratigraphy to evaluate the benthic ecological health of possibly polluted areas (e.g., Ellison *et al.*, 1986; Nagy and Alve, 1987; Alve, 1991; Schafer *et al.*, 1991;

Barmawidjaja *et al.*, 1995; Blackwelder *et al.*, 1996; Sen Gupta *et al.*, 1996), but there is a growing international interest in the subject.

The present study is an example of how environmental stratigraphy can provide otherwise inaccessible information about changing bottom water oxygen conditions in a Norwegian fjord over the past 500 years (see also in this volume Dixit and Smo, Ch. 12; Dale, Ch. 13; Ishman, Ch. 16; Van der Zwaan, Ch. 17).

2. Area Description and Discharge History

2.1. General Setting

Frierfjord is about 10 km long and represents one of the innermost parts of the Grenlandsfjord system in southern Norway (Fig. 1). It can be divided into a major basin, representing the widest and deepest (maximum depth about 98 m) northern part of the fjord, and a narrower and shallower (maximum depth about 50 m) southern part, which is separated from the seaward

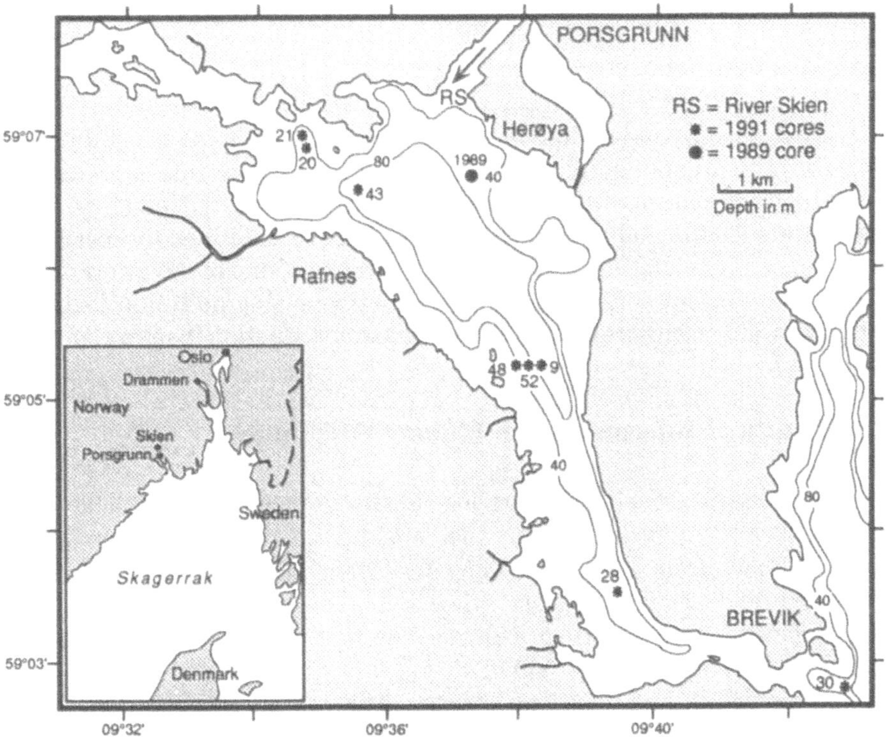

FIGURE 1. Map of Frierfjord showing location of stations.

fjord system by a sill at 23 m water depth at Brevik. The river Skien enters Frierfjord from the northeast and the runoff varies from 50 to 700 m^3/sec with an annual average of 270 m^3/sec (Molvaer, 1980). The resulting brackish surface water (salinity <10%) usually has a thickness of 3–6 m. Below the halocline, intermediate water masses extend down to about 25–35 m, whereas more stable waters (salinity 33.0–34.5‰; temperature 6.1–6.5°C) occupy the deeper areas (Molvaer *et al.*, 1979; Molvaer, 1980). The brackish surface water has an average residence time of about 1–4 days and that of the intermediate water masses is on the order of days to weeks (Rygg *et al.*, 1987). Moderate deep-water renewals generally take place in January to May each year (Molvaer, 1980) but efficient deep water renewals, at depths greater than about 50 m, occur only once every 1 to 3 years (Rygg *et al.*, 1987). The area is microtidal, with a tidal range of about 20 cm.

At water deeper than 20 m clay and silt make up 95% of the sediments on average (Naes and Oug, 1991) but a major (unquantified) portion of the >63 μm-fraction consists of sand-sized fecal pellets. About 50% of the river-derived fine sediments (<63 μm) settles in the fjord during low freshwater supply (<150 m^3/sec). During greater supplies (near the annual average), more sediment is introduced into the fjord but, because of the short residence time of the brackish water, about 90% of the sediment is transported out over the sill (Molvaer *et al.*, 1979). Some of the nearshore, industrialized areas in the inner Frierfjord and parts of the river were dredged 8–10 times between 1974 and 1986 (Rygg *et al.*, 1987) and the sediments were dumped in the central part of the fjord. The most extensive dumping activities were in 1974/75 and in 1986, involving about 600,000 m^3 and 120,000 m^3 of sediment, respectively.

Blooms of phytoplankton were not recorded during investigations in the 1970s despite exceptionally high nutrient concentrations; this is probably due to the brackish nature of the surface water, its reduced transparency, and the short residence time, all representing unfavorable conditions for marine algae (Molvaer *et al.*, 1979). Dinoflagellate cyst analysis of one of the Frierfjord cores used in the present study indicates that the increasing pollution load on the fjord has had a negative effect on dinoflagellate production (Saetre *et al.*, 1997).

2.2. Historical Information on Effluent Discharges

Frierfjord has received anthropogenically induced biodegradable organic material for several centuries: initially bark and wood fibers from water-driven sawmills, later from pulp and paper industries, and over the last century, increasing amounts of more easily degradable, organic material and nutrients from domestic sewage and agriculture. The first water-driven sawmills were established along the River Skien during the first half of the 16th century (Tønnesen, 1956). In 1585/1586, at least 30 sawmills were active in the area, and by the end of the century, it was the largest wood-exporting district in the country. By 1620, the number of sawmills had increased to 64. The discharges of wood fibers, bark, etc., caused shallowing of the water depth in Skien

Harbor and created problems for ships as early as the middle of the 17th century. Finally, it became impossible for ships to sail up to the sawmills, so during the 1750s the harbor was dredged and renovated. Yet, the problems persisted. During the 1700s, the area was the third most important wood-exporting area in Norway after Christiania (Oslo) and Drammen (Tønnesen, 1956).

From about 1860, the sawmills were no longer dependent on water power, as they were driven by steam, and the water power was used in pulp production. By 1875, three pulp mills were established near Skien (the first in 1870) and several steam saws and other industries were established along the river. Since then, industrial development has been extensive and the fjord has received increasing amounts of effluents from pulp and paper industries, and, during the 20th century, from fertilizer and chemical industries, smelting plants, and domestic sewage. Legislative limitations on the discharges were introduced during the middle of the 1970s. During the 1980s, the discharges of organic material were approximately halved compared to values for the 1970s (Rygg *et al.*, 1987).

The human population in Skien, Porsgrunn, and the surrounding areas increased from about 1600 in 1661 to about 5300 in 1801 (Tønnesen, 1956, pp. 121, 403). In 1920, it was 25,045 (Tønnessen, 1957, p. 540) and by the middle of the 1980s the population had increased to about 90,000 (Rygg *et al.*, 1987), implying a drastic increase in the quantity of domestic sewage over the last century. Major sewage treatment did not start until 1991/1992 (i.e., when the last set of cores in the present study were collected).

2.3. Records of Oxygen Conditions until 1990

In this paper, the phrase "oxygen depletion" is used to indicate both instances of anoxia (absence of detectable dissolved oxygen) and dysoxia (<1 ml O_2/liter), following Bernhard (1996).

Only a few records concerning environmental conditions in Frierfjord are available from before 1974, when comprehensive environmental monitoring was initiated. In 1898, Johan Hjort reported that there were no organisms living on the bottom sediments of Frierfjord and that the sediments consisted of black mud and rotting substances with a strong smell of H_2S (Hals, 1968). In both April and May 1924, the oxygen concentration was less than 1 ml/liter at water depths greater than 60–70 m (Dannevig, 1930). In June 1933, anoxic conditions (with H_2S) occurred at depths greater than 60–70 m and the upper 21 cm of a 62-cm-long sediment core collected at 90-m water depth the same year consisted of black sediments smelling of H_2S, which rested on "slicky clay with a little sand" (Strøm, 1936, pp. 40, 60, 64). A ^{210}Pb-dated core collected from the same deep basin and dated in 1975 gave an estimated sedimentation rate of 1.8 mm/year (Molvaer *et al.*, 1979). According to this dating, the area where Strøm's core was collected (90 m water depth) started to accumulate black sediments around 1820.

TABLE 1. Records of H_2S in the Water Column at a Station Midway between Cores 1989 and 48 (Fig. 1)

Date	H_2S records at >50 m	H_2S records at >80 m	Source
Late winter 1974	×	×	Molvaer, 1980
October 1975–March 1977	×	×	Molvaer, 1980
Late 1978–spring 1980	×	×	Knutzen *et al.*, 1983
Summer 1980–spring 1981	×	×	Knutzen *et al.*, 1983
December 1981–spring 1984	×	×	Knutzen *et al.*, 1983; Rygg *et al.*, 1984
January 1986		×	Rygg *et al.*, 1987
October 1986		×	Rygg *et al.*, 1987
Throughout 1987		×	Rygg *et al.*, 1988
December 1987	×	×	Rygg *et al.*, 1988
January–March 1988	×	×	Molvaer, 1992
December 1988–January 1989		×	Molvaer, 1992
April–May 1989		×	Molvaer, 1992

Comprehensive investigations of the oxygen conditions in Frierfjord were carried out by the Norwegian Institute for Water Research (NIVA) during the years 1974–1990. Anoxia was frequently recorded at greater than 50 m water depth from 1974 to 1984 and at depths greater than 80 m until May 1989 (see Table 1 for summary of anoxic records).

3. Material and Methods

The present study is based on nine short (33 to 52-cm) sediment cores collected by means of a light gravity corer (6.7-cm inner diameter). One core was collected during a preliminary survey of the area in 1989 (core 1989), whereas the rest were collected in February (core 9), May (cores 20, 21, 28, and 30), September (core 43), and November (cores 48 and 52), 1991 (Fig. 1, Table 2). Except for core 30, which was collected just outside the sill for reasons of comparison, all the cores are from within the fjord proper. The cores were sectioned in slices 1- or 2-cm-thick on board the ship and frozen to preserve them for subsequent sedimentological and micropaleontological analyses.

During the February and May 1991 cruises, bottom water samples were collected at most stations for dissolved oxygen analysis by transferring 250 ml of the seawater immediately overlying the sediment in the core liner (~0–15 cm above the sediment/water interface) to glass bottles for standard Winkler titrations.

In the laboratory, subsamples were taken (while the sediments thawed) for determination of water content (calculated as percent of wet sample), total

TABLE 2. Details of Stations and Cores, Dissolved Oxygen Concentrations (ml/liter) in Bottom Water Collected in February and May 1991 and Number of Samples Analyzed for Benthic Foraminifera (N.D. = no data)

Core No.	Core length (cm)	Latitude (°N)	Longitude (°E)	Water depth (m)	Oxygen (ml/liter)		Number of foraminifera samples
					February	May	
1989	36	—	—	50	N.D.	N.D.	13[a]
30	38	59°02.85′	09°42.70′	65	N.D.	6.13	9[a]
28	35	59°03.62′	09°39.10′	50	2.28	5.61	7[a]
48	33	59°05.42′	09°37.75′	50	2.59	4.98	21[a]
52	35	59°05.38′	09°37.95′	70	1.51	4.94	0
9	51	59°05.34′	09°38.12′	90	1.12	N.D.	23[a]
43	52	59°06.68′	09°35.35′	93	N.D.	N.D.	24
20	44	59°06.86′	09°35.30′	72	N.D.	4.99	18[a]
21	41	59°06.95′	09°35.21′	50	N.D.	4.71	18

[a]Stained surface samples.

organic carbon (TOC) analysis, and/or ignition loss, acid soluble zinc analysis (cores 28 and 52), and microfossil analysis. These samples were weighed, frozen, freeze-dried, and weighed again. The samples from cores 1989 and 9 were dried at 40°C instead of freeze-dried because of logistical problems. The remaining part of the surface samples from six cores (Table 2) were gently shaken while thawed in a mixture of 70% ethanol and Rose Bengal stain (1 g/liter) and left for at least 1 hr prior to processing. All microfossil samples were washed through a 125-μm sieve and dried (40°C). Samples rich in organic matter were spread over a large area while drying (Lehmann and Röttger, 1997) to prevent them from forming a hard block. The >125-μm-fractions were carefully dry-sieved through a 500-μm sieve (to separate material held together by organic matter; the >500-μm-fraction was checked for foraminifera), and subsequently floated in tetrachloroethylene (C_2Cl_4, density 1.62) owing to abundant fecal pellets, which are frequently found in many fjordic environments (e.g., Alve, 1991). The >125-μm-fraction (rather than the commonly used >63-μm-fraction) was used for the microfossil analysis in order to exclude at least some of the wood fibers, which were extremely abundant in the upper 10–20 cm of most cores. This choice of size fraction caused the loss of many small individuals (particularly *Stainforthia fusiformis*, Alve, unpublished data from similar environments) but as the loss should be consistent throughout the dataset, faunal comparisons can still be performed within and between cores. When possible, about 300 dead benthic foraminifera were picked from each of the 133 analyzed samples and all live (stained) foraminifera were picked in the stained samples. The species diversity refers to the number of species recorded per counted number of specimens. Additionally, the number of juvenile bivalves and gastropods per counted number of dead foraminifera were recorded. All abundance data refer to number of tests per gram of dry sediment. No correction was made for the salt content of the sediment, as it was found to have an insignificant effect on the numerical results.

Cores 21 and 28 were ^{210}Pb-dated at the Force Institutes, Division of Isotope Technic and Analysis, Denmark, by the CRS-method (Pheiffer Madsen and Sørensen, 1979). The TOC content and ignition loss were determined by the Leco combustion method (Leco Industrial Furnace). For the zinc analyses, 4–6 g of wet sediment was dispersed in 20 ml of 1M HCl and shaken for 20 min at room temperature; the filtrate was then diluted to 100 ml and analyzed on a Perkin Elmer Model 503 atomic absorption spectrophotometer with a detection limit of 1 ppm. Zinc is a trace metal that often shows elevated concentrations in contaminated and polluted sediments (e.g., Skei, 1996).

4. Results

Owing to the large number of analyzed cores, diagrams are provided only for selected datasets.

4.1. Dating

The ^{210}Pb-datings gave linear accumulation rates of 0.13 cm/year (core 21) and 0.17 cm/year (core 28). Based on the variations in water content and ignition loss, the reports from the Force Institutes noted changes in the sedimentation patterns at times corresponding to core depths around 13 cm (core 21) and 12 cm (core 28). The change at 13 cm in core 21 was assigned an estimated age of 135 years (=1856).

4.2. Dissolved Oxygen Concentrations

In February 1991, the dissolved oxygen concentrations ranged from 2.59 ml/liter at 50 m water depth to 1.12 ml/liter at 90 m (Table 2). In May 1991, the bottom water oxygen concentrations were greater than 4.7 ml/liter at all investigated stations.

4.3. Sediment Characteristics

The lower parts of core 30 (65 m water depth) just outside the sill and core 28 (50 m) from the southern part of the fjord consisted of light gray, homogeneous sediments. The lower parts of all other cores had medium gray, homogeneous sediments (Figs. 2–4). The color of cores 30 and 28 darkened gradually upcore to medium gray and that of the other cores to dark gray at core depths where the organic carbon and the water content start to increase (see below). Cores 30 and 28 had light-brown surface sediments of various thicknesses (2 cm and 0.4 cm, respectively), whereas all other cores had black, laminated surface sediments with a smell of H_2S at, or just below, the sediment/water interface. The black sediments had a soupy appearance, sometimes with brownish fecal pellets and light gray, fluffy sediment aggregates in the topmost veneer. The thickness of the black surface sediments varied between cores but increased consistently in cores from increasing water depths. The laminated nature was clearly seen by gently breaking the sediment slices in two, perpendicular to the bedding, after they had been freeze-dried. All three cores from the central part of the fjord (cores 48, 52, and 9) had a layer of light gray sediments within the black zone, at a core depth of about 2 to 4 cm.

The background (i.e., deeper core intervals with minimal changes) TOC values were in the range 0.6–0.8% in all cores except core 28 (1%). The background water content was about 30% in cores 20, 21, 48, and 52, it was 30–40% in cores 30 and 43, and somewhat higher in core 28 (40–45%). Both the TOC and the water content showed maxima in the upper parts of all the cores but the values decreased or became constant near the sediment surface. The increase started at about 14–17 cm in most cores, except cores 20 and 43, where it started at about 27 and 34 cm, respectively. Core 30 (collected just

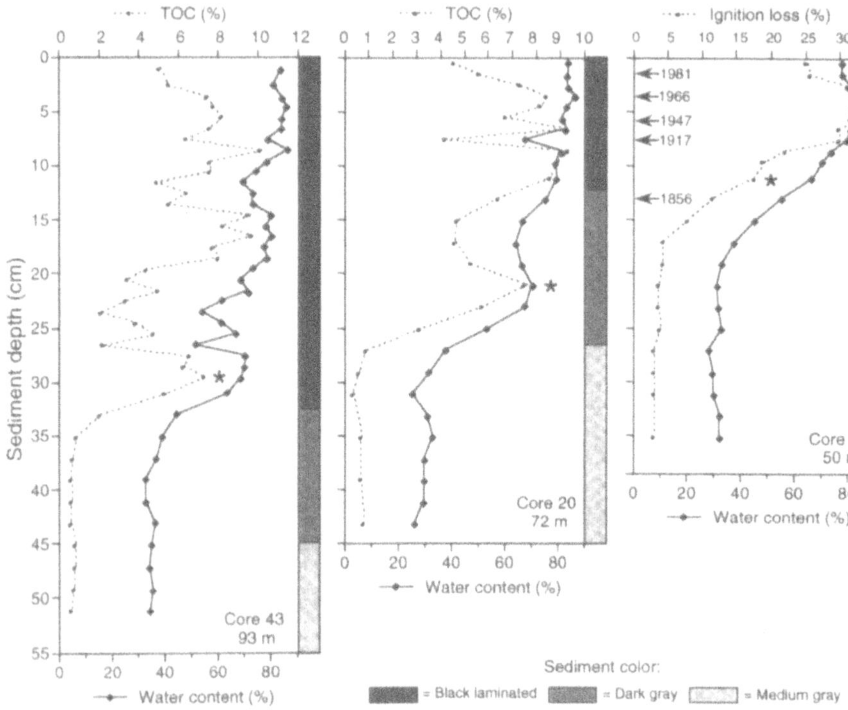

FIGURE 2. Sediment characteristics of cores 43, 20, and 21. Dates on y-axis, core 21, obtained by ^{210}Pb-datings (star = year 1875).

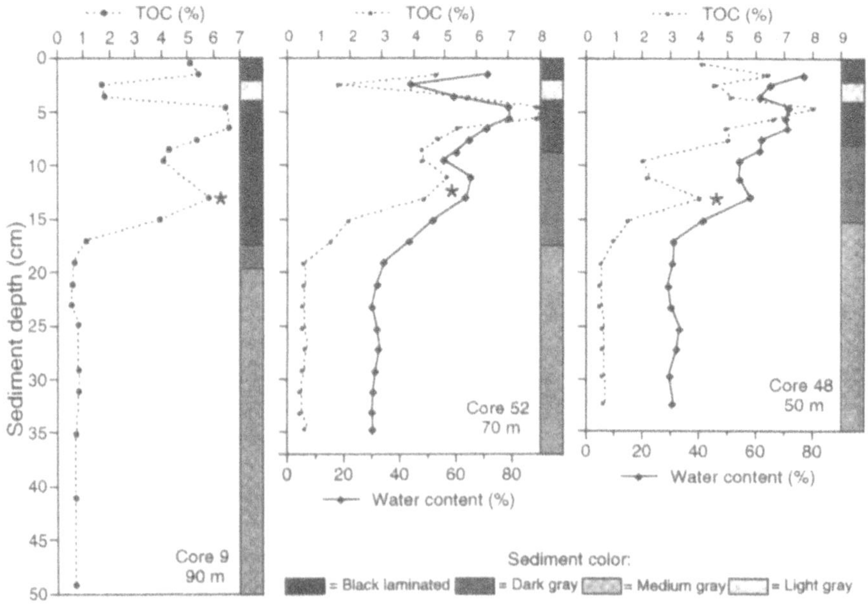

FIGURE 3. Sediment characteristics of cores 9, 52, and 48 (star = year 1875).

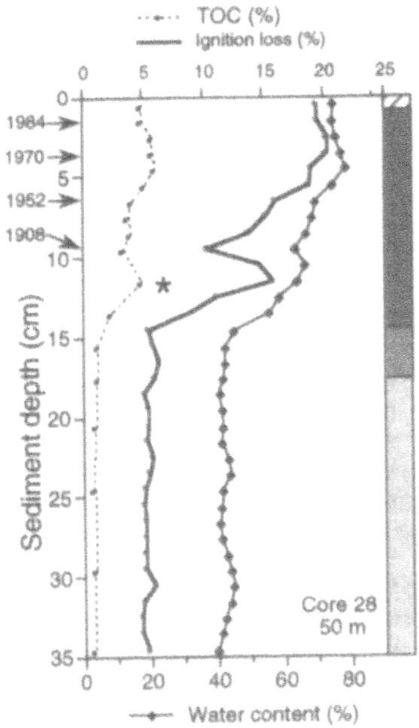

FIGURE 4. Sediment characteristics of core 28. Dates on *y*-axis, core 28, obtained by ^{210}Pb-datings (star = year 1875). For sediment color legend, see Fig. 3, and hatched area at the top = brown color.

outside the sill) had a subsurface TOC maximum of 2.7% at 1–2 cm depth, and a surface value of 2.3%. All other cores had about 5% TOC in the surface sediments, whereas the subsurface maximum values varied from 6% in the southern part of the fjord (core 28) to 10% in the northern part (core 43). There was a highly significant positive correlation between the TOC and ignition loss values in core 28 (referred to as core 1 in Saetre *et al.*, 1997). Except for the upper 2–3 cm, wood chips were present in the >500-μm-fraction at intermediate and shallower core depths.

The background values of acid-soluble zinc were 61–89 ppm and 32–62 ppm at greater than 14 cm in cores 28 and 52, respectively. The values increased dramatically upcore and culminated at 3–4 cm with maximum values of 509 (core 28) and 555 ppm (core 52) before they decreased to 315 and 378 ppm in the surface sediments, respectively.

4.4. Benthic Foraminifera

4.4.1. Living (Stained) Assemblages

No living foraminifera were recorded in the surface sample (0–1 cm) of core 20. The living abundance ranged from 2 individuals/g sediment (core 48) to 21/g sediment (core 9) in the other surface (0–2 cm) samples from within Frierfjord and they were strongly dominated by *Stainforthia fusiformis* (Williamson) which comprised 96% in core 1989, 81% in core 48, 99% in core 9, and 52% in core 28. *Bulimina marginata* d'Orbigny and *Leptohalysis scottii* (Chaster) were the second and third most abundant species. The number of species was 2 in cores 1989 and 9, 6 in core 48, and 13 in core 28. Core 30 from outside the sill had 27 individuals/g sediment and 32 species, of which *Globobulimina auriculata* (Bailey) (25%), *Nonionellina labradorica* (Dawson) (20%), *B. marginata* and *Nonionella turgida* (Williamson) (both 13%) were the most abundant, whereas *S. fusiformis* was rare (2%).

4.4.2. Dead (Unstained) Assemblages

Assemblages dominated by *Eggerelloides scaber* (Williamson) occurred at 10–12 cm in core 43, and at 2–4 cm in core 9 (Fig. 5) and core 48. Associated species included *Ammoscalaria runiana* (Heron-Allen and Earland), *Miliammina fusca* (Brady), *Balticammina pseudomacrescens* Brönnimann, Lutze and Whittaker, *Jadammina macrescens* (Brady), *Haynesina germanica* (Ehrenberg),

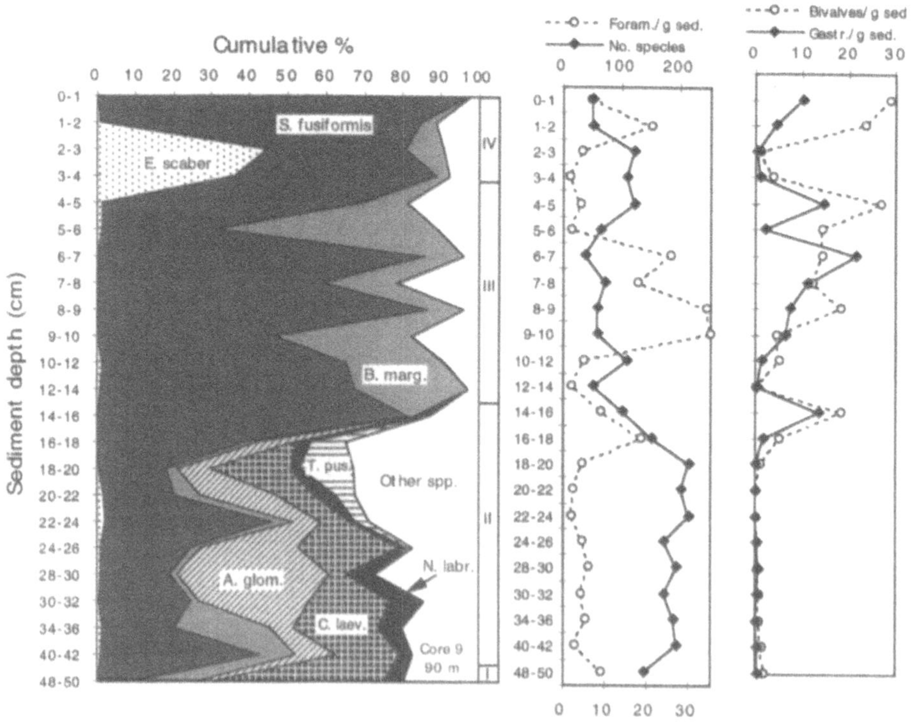

FIGURE 5. Cumulative percentages of characteristic foraminiferal species, number of foraminiferal species, and numerical densities of foraminifera and juvenile bivalves and gastropods, core 9. Note the compressed vertical scale below 26 cm core depth. Time periods in Roman numerals.

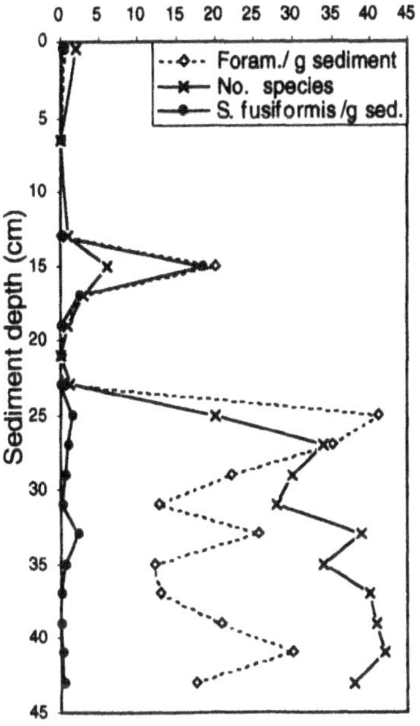

FIGURE 6. Number of foraminiferal species and numerical density of benthic foraminifera and of *Stainforthia fusiformis* in core 20, 72 m water depth.

Elphidium williamsoni Haynes, and *Ammonia beccarii* (Linné). These are typical shallow water species in southern Norway (Alve and Murray, 1999), and most probably they originated from the previously mentioned extensive dredging and dumping activities in 1974/75.

Cores 20 and 43 (72 and 93 m water depth, northwest part of the fjord) showed comparable faunal patterns and had abundant foraminiferal assemblages below 24 cm (Fig. 6) and 32 cm, respectively. *Spiroplectammina biformis* (Jones and Parker) dominated at 25 cm (60%) and 27 cm (28%) in core 20 and had its maximum relative abundance at 24–25 cm in core 43. Otherwise no single species dominated these lower core intervals but *Trochamminopsis pusillus* (Höglund) (mean = 21%, SD = 7, $n = 17$), *Cassidulina laevigata* d'Orbigny (mean 17%, SD = 8, $n = 17$), and *Adercotryma glomeratum* (Brady) (mean = 14%, SD = 9, $n = 17$) were abundant. The numerical density was relatively stable (mean = 17 tests/g sediment, SD = 6, $n = 14$) below 28 cm (core 20) and 38 cm (core 43). Above this, both cores showed peak abundances (max. 41/g sediment at 25 cm, core 20; 82/g sediment at 35 cm, core 43) followed by a faunal collapse where the abundance fell to zero. At shallower core depths, benthic foraminifera were rare (<1 test/g sediment) in core 20 except for a nearly monospecific *Stainforthia fusiformis* assemblage at 14–18 cm with a maximum abundance of 20 tests/g sediment. Except for 2–3 tests/g sediment (mostly *S. fusiformis*) in the surface 5 cm, core 43 had a similar foraminiferal peak (max. 45 tests/g sediment) at 21–25 cm within the otherwise nearly barren (<1 test/g sediment) core interval above 32 cm. The number of species was stable in the lower parts of the cores (mean = 36, SD = 4, $n = 14$) and decreased upcore. With one exception in core 43 (8–19 species at 21–25 cm) and one exception in core 20 (4 species at 14–18 cm), either none or only a few species were present above 24 (core 20) and 32 cm (core 43).

The deepest sample in core 9 (90 m, central part of the fjord) was dominated by *Cassidulina laevigata* (45%) with abundant *Bulimina marginata* (17%) and a minimum value of 10% for *Stainforthia fusiformis* (Fig. 5). At 41–18 cm either one of these species or *Adercotryma glomeratum* dominated. *Stainforthia fusiformis* dominated with abundant *B. marginata* above 18 cm. *Leptohalysis gracilis* (Kiaer) and *Trochamminopsis pusillus* were abundant only at 16–20 cm (max. 19 and 14% respectively) and *Spiroplectammina biformis* (max. 13%) at 9–16 cm. The mean numerical density was 31 tests/g sediment (SD = 16, $n = 9$). Below 18 cm, it showed a first peak (131 tests/g sediment) at 17 cm, a second and main peak at 9–10 cm (250 tests/g sediment), and a third peak (150 tests/g sediment) at 1–2 cm. The number of species was stable (mean = 26, SD = 3, $n = 9$) below 18 cm but decreased drastically upcore, where it fluctuated between 5 and 15.

The lower parts of cores 1989, 21, and 48 (all 50 m, northern and central fjord areas) were dominated by *Cassidulina laevigata* (mean = 43%, SD = 16, $n = 18$) up to 20, 20, and 18 cm, respectively (except one sample in core 21, Fig. 7). Its abundance decreased dramatically upcore and it was not recorded in the surface 2, 6, and 5 cm of the three cores, respectively. Other common

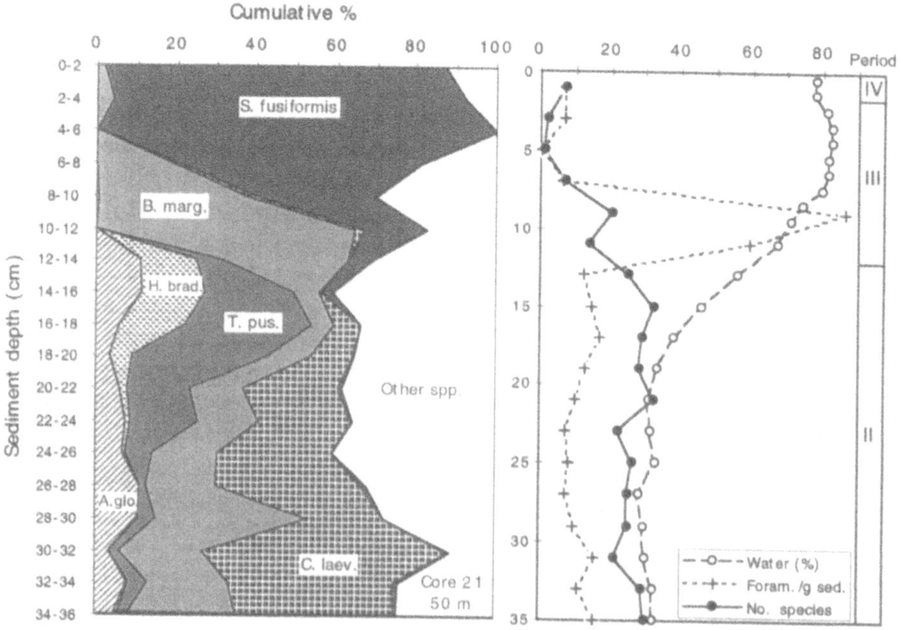

FIGURE 7. Cumulative percentages of characteristic foraminiferal species, number of foraminiferal species, numerical densities of foraminiferal tests, and water content, core 21.

species in the lower core intervals were *Bulimina marginata*, *Adercotryma glomeratum*, and *Trochamminopsis pusillus* but their abundances also decreased dramatically upcore, and only *B. marginata* was recorded (1–2%) in the upper 2 cm. *Spiroplectammina biformis* was generally rare but had maximum abundances of 51–55% at 12–16 cm in core 48 and 6–7% at 10–18 cm in core 21. *Stainforthia fusiformis* made up less than 4% of the assemblages below 14 cm in cores 1989 and 21 and below 20 cm in core 48. It showed a marked increase upcore and strongly dominated the surface 2 cm in core 48 (87%), the surface 4 cm in core 1989 (72–92%), and the surface 6 cm in core 21 (86–100%). The numerical densities of tests below 10 cm (cores 1989 and 48) and 12 cm (core 21) showed stable background values with a mean of 13 tests/g sediment (SD = 5, $n = 31$). Above this, the values increased abruptly in all cores to maxima of 86 (at 8–10 cm, core 21), 125 (at 6–7 cm, core 48), and 37 tests/g sediment (at 4–6 cm, core 1989) before they fell to less than 1 at 4–6 cm and at 1–2 cm in cores 21 and 48, respectively, and to 13 tests/g sediment at 2–4 cm in core 1989. The surface sediments had values of 2, 7, and 33 tests/g sediment in cores 48, 21, and 1989, respectively. Core 48 showed a generally decreasing number of species from 46 in the deepest sample to a minimum of 2 at 1–2 cm, and 3 species in the surface 0–1 cm. Cores 21 and 1989 had a more stable number of species (mean = 26, SD = 4, $n = 22$) below 12 and 6 cm, respectively. Above these levels, the values decreased to 7 species in the surface 1989 sample. Core 21 had a minimum number of 2 species at 2–4 cm but 7 species in the surface sample.

Core 28 differed from the other 50-m cores in being dominated by *Bulimina marginata* below 3 cm, except at 10–12 cm where *Spiroplectammina biformis* dominated (40%). *Cassidulina laevigata*, *Nonionellina labradorica*, and *Hyalinea balthica* (Schröter) were commmon below 20 cm, where the number of species was 40–45. The relative abundance of *Stainforthia fusiformis* increased from less than 3% in the lower part of the core to 13% at 6–7 cm, through 35% at 3–4 cm (33 spp.), and reached a maximum of 71% at 1–2 cm, where the number of species reached a minimum of 17. The surface 0–1 cm had 57% *S. fusiformis*, 15% *B. marginata* and 25 species. The numerical density of tests was an order of magnitude higher in samples from the surface 4 cm (250 tests/g sediment) compared to those below 20 cm.

Core 30 (65 m outside the sill), showed minor faunal variations. *Bulimina marginata* (mean = 28%, SD = 2, $n = 9$) was the most abundant species with subsidiary *Cassidulina laevigata*, *Hyalinea balthica*, and *Cibicides lobatulus* (Walker and Jacob). There was no consistent pattern of variation in the numerical density (mean = 1170 tests/g sediment, SD = 300, $n = 9$), but the number of species increased from 30 in the deepest to 41 in the two shallowest samples.

4.5. Juvenile Bivalves and Gastropods

Juvenile bivalve and gastropod shells were rare in the four cores from 50 m water depth (generally <0.6 shells/g sediment). Minor peak densities were recorded only in core 21 (bivalves: 2.3; gastropods: 1.6, at 6–7 cm core depth) and core 48 from the central part of the fjord (bivalves: 5.6 at 4 cm; gastropods: 3.5 at 5 cm). However, such shells showed abundance peaks in the upper halves of cores 20, 9, and 43 from the deeper (>70 m) parts of the fjord (Fig. 5). Core 20 had a maximum numerical density of 35 bivalves/g sediment. In the lower core halves (deeper than 26, 18, and 38 cm in the three cores respectively), these shells were rare and there were always fewer than 1.5 shells/g sediment.

5. Discussion

5.1. Intercore Age Correlation

Because bioturbation by macrofauna probably has been minimal or nonexistent owing to oxygen depletion during deposition of the sediments, where TOC values are strongly elevated relative to background levels (Pearson and Rosenberg, 1976; discussion in Alve, 1995*a*), it is likely that the chronological succession of events is well preserved in the sediments.

In addition to the ^{210}Pb-dating, sediment and faunal data have been used to correlate between and, indirectly, date all except two cores. These are core 1989, where no sedimentological data are available, and core 30 outside the sill, which did not show clear patterns in the parameters on which the correlations are based. The other seven cores show a pronounced increase in both the organic carbon and water content at specific sediment depths, with a first peak (indicated by a * in Figs. 2–4), followed by a decrease. It is reasonable to assume that the increase and the peak represent synchronous events within the fjord. This is also strongly supported by the fact that the peak lies within the narrow core interval where *Spiroplectammina biformis* has its maximum abundance in all seven cores. The ages of these two levels (first increase and peak) are calculated by assuming that the sedimentation rate between the levels of the two oldest ^{210}Pb-dated samples is representative for greater core depths. Following this method, the initial increase, which starts at 16 cm in core 21 and at 14 cm in core 28, corresponds to the years 1822 and 1835, respectively, and the year 1830 has been chosen to define the initial increase. The first peak is located at 11.5 cm in core 28 and at 11 cm in core 21. The calculated ages of the peak in the dated cores are 1879 and 1871 and the year 1875 has been used to define the age of the peak. After inferring that the increase in organic carbon started in 1830 and that the first peak occurred in 1875, the approximate age of the pre-1875 levels has been calculated, indicating that only core 9 penetrated sediments deposited as early as in the 15th century, whereas all other cores penetrated back into 17th-century sediments.

Based on the previously mentioned historical information about industrial development in the area, the time span covered by the cores has been divided into four periods (Table 3). The establishment of water-driven

TABLE 3. Time Periods and Summary of the Interpretations of Environmental Changes in Frierfjord over the Last Five Centuries

Year	Time period	Interpretations of environmental changes in Frierfjord		
		Northwest part	Central part	Southern part
1991	IV Pollution abatement	Occasional anoxic surface sediments at >50 m and predominant anoxia at >70 m	Occasional anoxic surface sediments at >50 m and frequent anoxia at 90 m	Slightly improved oxygen conditions in surface sediments at 50 m
1975	III Modern industry	Increasing frequency of anoxic bottom waters at >50 m. Anoxia dominates at >70 m	Increasing deterioration of oxygen conditions in surface sediments at 50 m with occasional anoxia near the end of the period. Frequent anoxia at 90 m	Depleted oxygen conditions in surface sediments at 50 m
1870	II Sawmill	Impoverished oxygen conditions in surface sediments at 50 m during the latter part of the period, and throughout the period at depths >70 m. Occasional anoxic bottom water at >70 m by the end of the period	Impoverished oxygen conditions in surface sediments at >50 m during the latter part of the period and throughout the period at >90 m. Occasional anoxia at >90 m near the end of the period	Slightly impoverished oxygen conditions in surface sediments at 50 m throughout the period
1550	I Preindustry	Well-oxygenated conditions throughout the water column	Well-oxygenated conditions throughout the water column	Well-oxygenated conditions throughout the water column

sawmills proceeded slowly in the first half of the 16th century and did not accelerate until the second half of the century. Consequently, the year 1550 has been used to define the border between "the preindustrial period," Period I, and "the sawmill period," Period II. The establishment of "modern" industries started around 1870 and, consequently, this year has been used to define the border between "the sawmill period" and "the modern industrial period," Period III. Post-1975 has been denoted "the pollution abatement period," Period IV, because governmental restrictions on industrial effluents were introduced during the middle of the 1970s.

5.2. Faunal Basis for Environmental Interpretations

Detailed information about the ecology of the living foraminifera and taphonomic processes that modify the original faunal composition is a prerequisite for using their fossil remains as a basis for reliable paleoenvironmental interpretations. Our understanding of the biology of foraminifera inhabiting oxygen depleted environments is still poorly understood but important contributions have been made during the last decade (for reviews, see Sen Gupta and Machain-Castillo, 1993; Bernhard, 1996). Concerning their preservation potential in the sedimentary record, Denne and Sen Gupta (1989) concluded that taphonomic processes are least active in areas of rapid sedimentation (where the bioturbated zone is thin) and under oxygen-deficient conditions. Consequently, there is reason to believe that the preservation potential is satisfactory in the Frierfjord sediments and that reliable interpretations can be made from the fossil foraminiferal assemblages.

The environmental interpretations in the present study are based primarily based on the following faunal characteristics: (1) assemblage composition, (2) species diversity, (3) numerical density of tests, and (4) recolonization properties. As a preparation for the following interpretation of the Frierfjord data, many of the examples given below are from southern Norwegian waters.

1. *Cassidulina laevigata* is known to dominate in areas with well-oxygenated bottom water (see references in Murray, 1991). In the middle and outer Oslo Fjord, Norway, it was the most abundant species in surface samples collected at about 100 m in 1961–1963 (Risdal, 1964) in areas where the bottom water has an annual range of 4.7–6.0 ml O_2/liter (median dissolved oxygen concentration between 1933 and 1989, Baalsrud and Magnusson, 1990). In the same general area, and additionally in inner Oslo Fjord, it was a common species (5–20%) in samples dominated by *Bulimina marginata* at 25–110 m water depth (Risdal, 1963, 1964). In the innermost Oslo Fjord however, *C. laevigata* was rare or absent in areas where oxygen depletion is common (October median bottom water $[O_2] = 1.0–1.2$ ml/liter at 50–90 m from 1973 to 1982, Magnusson *et al.*, 1996). These samples were dominated by *B. marginata* and more than 20% of the assemblages were *Stainforthia fusiformis*. *S. fusiformis* is a typical opportunistic species (Alve, 1994) and may be the first and most successful colonizer of defaunated sediments

whether the defaunation is due to anoxia (Alve, 1995*b*) or to freezing of the sediments (Alve and Olsgard, 1999). This species also seems to flourish in sediments enriched by easily biodegradable organic material whether the bottom water is oxygen-depleted (Alve, 1990) or not (Alve and Murray, 1997). *Spiroplectammina biformis* is the most abundant and widely distributed species below the brackish surface layer in the organically enriched Drammensfjord, southern Norway (Alve, 1990). However, in sediments where the dissolved oxygen concentration in the water just above the sediment/water interface was less than 0.5 ml/liter (and occasionally anoxic) during most of the year, the only living (stained) individuals were *S. fusiformis* (Alve, 1995b).

2. High species diversity is generally characteristic of stable marine environments (e.g., Murray, 1991) and in the Oslo Fjord, there were always more than 20 species present in samples with common *Cassidulina laevigata* (Risdal, 1963, 1964). In contrast, the number of species was only 2–6 in strongly *Stainforthia fusiformis/Bulimina marginata*-dominated, black, organic-rich surface sediments from the innermost parts of the fjord (Risdal, 1963), which often experience several months of anoxia a year (Magnusson *et al.*, 1996). The same pattern was seen in the most oxygen depleted areas in Drammensfjord, Norway (Alve, 1990). It seems that if the number of species (per about 250 counted specimens) is not limited by other environmental factors (e.g., low salinity), a "general rule of thumb" is that the presence of more than about 15–20 species reflects well-oxygenated bottom water conditions. On the other hand, the presence of fewer than about 5 species (in dry-picked samples, see below) indicates stressful conditions, which might be due to a dominance of severe oxygen depletion.

3. Benthic foraminifera are not able to survive prolonged anoxia (e.g., Bernhard and Reimers, 1991; Alve, 1995*b*) but there are several lines of evidence suggesting that certain benthic foraminifera are microaerophiles and some are facultative anaerobes (review in Bernhard, 1996). For example, experimental studies have shown that *Stainforthia fusiformis*, *Bulimina marginata*, and *Adercotryma glomeratum* are able to survive more than 3 weeks of anoxia (Bernhard and Alve, 1996). Furthermore, other hard-shelled foraminifera were not significantly negatively affected when exposed to more than 2 months of anoxic bottom water conditions, whereas soft-shelled ones showed a significant numerical density decrease (Moodley *et al.*, 1997). I have never recorded soft-shelled benthic foraminifera in dry-picked subrecent assemblages, which implies that they do not have a good preservation potential and that we are left with the (seemingly less-sensitive) hard-shelled foraminifera for use in environmental stratigraphy. Even though many species are reported to survive several weeks of anoxia, behavioral experimental studies have shown that infaunal species (e.g., *S. fusiformis* and *B. marginata*) respond to oxygen depletion by migrating toward more oxygenated habitats and even become epifaunal when exposed to concentrations lower than 0.2 ml O_2/liter (Alve and Bernhard, 1995). Increased supply of labile organic material initially causes hypertrophic conditions and the faunal density increases several times compared to what is "normal" for the area (for examples and discussion, see

Alve, 1995*a*). This increase is accompanied by an elevation of the redox-boundary within the sediments and it can, in extreme cases, emerge into the water column. Consequently, excess supplies of labile organic material may cause a faunal collapse owing to the development of prolonged anoxia. Exceptionally high numerical densities (one to two orders of magnitude higher than in "normal" benthic environments) have also been reported in offshore, severely oxygen-depleted ($\leqslant$0.1 ml O_2/liter) environments (e.g., Phleger and Soutar, 1973; Bernhard and Reimers, 1991).

4. So far, we have only a poor understanding of the processes and biological dynamics of benthic foraminiferal dispersal and colonization patterns (for review, see Alve, 1999). However, it seems that recolonization of sediments that have experienced prolonged anoxia may take at least 1 year (e.g., Cato *et al.*, 1980; Alve, 1995*b*). On the other hand, a faster colonization of about 1 month was recorded at a well-oxygenated offshore dumpsite used for disposal of dredge spoil rich in organic material (Schafer, 1982).

Finally, it must be noted that most stratigraphical data represent an average of sediments deposited over several years. Consequently, some samples may represent one or several years of anoxia (no foraminifera living at the site) as well as oxic time intervals (with abundant foraminifera). More detailed information about the biology of foraminifera inhabiting anoxic sediments (e.g., survival time, possible reproduction) and their colonization pattern following prolonged anoxia is needed to improve our interpretations of the fossil record.

5.3. Environmental Conditions before the Establishment of Modern Industries: Periods I and II

The faunal composition (e.g., abundant *Cassidulina laevigata*) and high species diversity in the lower parts of cores 9, 20, and 43 suggest that the bottom water at depths greater than 70 m was generally well oxygenated during Period I and the first and middle part of Period II (Fig. 8). However, the fluctuating, but high, abundance of *Stainforthia fusiformis*, particularly in core 9 (Fig. 5), indicates that the sediments were enriched in organic material and that the redox-boundary was already close to the sediment/water interface during the early part of Period II. Such a slight organic enrichment is probably not detectable by TOC analyses after so many years owing to diagenetic decomposition. It may be speculated that the strongly increasing frequency of *S. fusiformis* from the oldest sample to the next one up in core 9 reflects an increased flux of organic material (decay products and bacteria associated with degrading wood fibers) from Period I to Period II owing to the establishment of the first sawmills. However, data from one sample are inadequate to draw any firm conclusions.

Diverse *Cassidulina laevigata*-dominated assemblages with only rare *Stainforthia fusiformis* were recorded in the lower parts of all cores from 50 m water depth. This implies that well-oxygenated conditions prevailed during

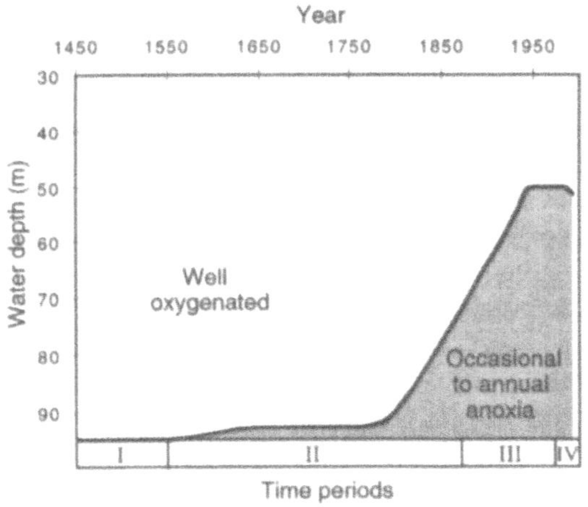

FIGURE 8. Schematic diagram illustrating how areas at successively shallower water depths have been affected by anoxia over the past five centuries.

Period I and that neither the surface sediments nor the bottom water at intermediate water depths, from north to south in the fjord, were negatively affected during the initial and middle parts of Period II.

Some faunal changes occurred in all cores during the latter part of Period II. *Trochamminopsis pusillus*, which is common throughout Period II, shows maximum abundances during the late 18th and early 19th centuries, *Cassidulina laevigata* shows a drastic reduction, and at water depths greater than 70 m, the species diversity decreases as the numerical density of foraminifera reaches the first of several peaks. Höglund recorded only a few specimens of *T. pusillus* in the Skagerrak in 1937 (Höglund, 1947, p. 202). However, by 1992/1993 it was one of the most characteristic and common species there, and increased nutrient supply over recent decades was suggested as a possible explanation for its more frequent occurrence (Alve and Murray, 1995). Overall, this indicates that the discharges from the sawmills caused drastic changes in the benthic environment at water depths of 50 m or more during the late 18th and early 19th centuries with a redox-boundary at or close to the sediment/water interface and probably occasional anoxia below 70 m.

The fact that these sediments were deposited before harmful and toxic chemical contaminants (e.g., heavy metals, PAHs, PCBs) were introduced to the fjord shows that decaying organic matter and oxygen depletion rather than other pollutants were the main causes for the faunal changes. However, it can be argued that climatic changes might have had an impact on the faunal composition. There are primarily two aspects to consider regarding this issue: (1) In northern Scandinavia, the summer air temperature after 1400 has fluctuated ±0.8°C relative to the 20th century (Briffa *et al.*, 1990, 1992), but it is not likely that such minor fluctuations had an impact on the benthic organisms at greater than 50 m water depth. (2) During the "Little Ice Age,"

Europe experienced increased precipitation accompanied by swollen rivers (Lamb, 1977, p. 452). However, even though increased precipitation and river runoff during that period had a negative effect on the deep water circulation in Drammensfjord (Alve, 1991), it is not likely that this (or the slightly changing air temperatures) had the same effect in Frierfjord because the sill is located at greater water depth, about 20 m below the present-day brackish surface layer. This would leave enough room for oxygenated coastal water to enter below the brackish layer, even if its thickness occasionally increased by a few meters.

5.4. The Modern Industrial Period: Period III

In open coastal areas where the bottom water is continuously circulated and renewed, conditions are not suitable for anoxia to develop in the water column as a response to high organic influx. However, in silled fjords and other areas with restricted deep-water renewals, excess input of oxygen-consuming organic material that accumulates in the sediments may cause the redox-boundary to be elevated above the sediment/water interface and into the water column if the oxygen consumption exceeds the supply between successive deep-water renewals.

At the transition to Period III, the faunal density peaks seen in the cores from greater than 70 m water depth in the northwest part of the fjord are immediately followed by a faunal collapse and the single abundance peak recorded higher up in both cores is strongly dominated by *Stainforthia fusiformis* (Fig. 6). This indicates that with one exception (i.e., the *S. fusiformis* peak), the surface sediments at more than 70 m in the northwest area have been dominantly anoxic since the modern industries were established (Fig. 8). The frequent occurrences of juvenile bivalves and gastropods, which probably died from anoxia at an immature stage, also support this view. The position of the single *S. fusiformis* peak above the initial TOC and water-content peaks (see stars in Fig. 2) strongly suggests that it represents the same event in both cores. The TOC and water-content values are at a minimum at core depths where *S. fusiformis* peaks. There are two possible explanations for this coincidence: (1) that the sediments are redeposited, or (2) that it represents several years of reduced organic input and/or more frequent deep-water renewals causing a reestablishment of oxic bottom water conditions, which allowed recolonization of a low diverse *S. fusiformis* assemblage.

There is no faunal or sedimentological evidence that the sediments in the *Stainforthia fusiformis* peak samples are redeposited and the remains of a flexible polychaete tube indicates *in situ* oxic conditions. Furthermore, the accompanying TOC and water-content minima are clearly seen in all investigated cores from within the fjord, even in core 28 from the southernmost area (Figs. 2–4), and this also fits well with the temporally increased species diversity in core 9. Consequently, there is strong evidence that the *S. fusiformis* peak seen at depths greater than 70 m in the northwest and the increased species diversity at 90 m in the southern part of the central basin reflect several years of well-oxygenated bottom waters during the last decade of the 1800s.

The first and, to my knowledge, only previous record of benthic foraminifera in Frierfjord is Kiaer (1900), who lists 21 species of which *Virgulina schreibersiana* was by far the most abundant. His sample, probably collected in the 1890s, was from about 86 m water depth and consisted of "drab coloured mud." According to the illustrations in Goës (1894, probably used by Kiaer for identifications) *V. schreibersiana* includes what we today know as *Stainforthia fusiformis*. This record fits well with the core data.

Except for the *Stainforthia fusiformis* peak, the near absence of foraminifera below 70 m in the northwest and the low diversity *S. fusiformis* assemblages at 90 m further south suggest that the dissolved oxygen in the bottom water generally was consumed between successive deep-water renewals. The different faunal developments in these two areas, although at similar water depths, are probably related to their different positions relative to the river mouth. The highest TOC values are recorded in the northwest cores, showing that this area was the most heavily affected by organic discharges. This is also indicated by the greater thickness of Periods III and IV sediments compared to other areas. It is reasonable to assume that the redox-boundary in these particularly organic-rich sediments was positioned near the sediment/water interface following the occasional deep-water renewals and, as the oxygen became exhausted, the boundary rose into the water masses. This would cause nearly permanently anoxic sediment pore water to prevail and thereby strongly reduce the possibilities for foraminiferal colonization of the sediments. On the other hand, the more southerly area, which is positioned where the oxygenated coastal water flows down into the deep basin during the renewals (Fig. 1), was not as directly affected by the river discharges. This is clearly reflected by the faunal composition (Fig. 5).

A faunal density peak, similar to those that occurred in the deeper parts of the basin during the first half of the 19th century, developed around 1910–1920 at 50 m in the northwest and two to three decades later at the same water depth further south. This reflects a progressive expansion of anoxia, basically within the sediments but also in the bottom waters, up to shallower water depths, and the faunal development shows that the process continued until 1974/1975 when the foraminiferal record was interrupted by the dumping of dredging spoils.

5.5. The Pollution Abatement Period: Period IV

The absence of both live and dead benthic foraminifera at depths greater than 70 m in northwest Frierfjord shows that the benthic environment in this area still suffered from anoxia when the cores were collected in 1991. The fact that the dumping activity took place in the mid-1970s, when the industrial discharges were at a maximum, makes it impossible to infer from the foraminifera whether or not prolonged anoxic conditions occurred below 50 m water depth in the southern part of the deep basin. However, the observation that only *Stainforthia fusiformis* and *Bulimina marginata* occur immediately above the dumped sediments indicates either that these two species were the sole survivors or that, for several years, they were the only species that had recolonized the dumped sediments. Independently of how they got there, these were the only species that were able to live in the area for several years after deposition of the dumped sediments. The predominance of *S. fusiformis* and the extremely low species diversities are in good agreement with the measured records of fluctuating oxic–anoxic conditions in the deep basin during the 1970s and 1980s (Table 1).

Although the faunal composition in the shallower, southern part of the fjord (core 28) changed during Period III, the faunal data do not indicate that this area had experienced prolonged periods of anoxia. However, the increased species diversity and reduced relative abundance of *S. fusiformis* toward the sediment surface might indicate that the reduced discharges have caused improved oxygen conditions in the surface sediments.

5.6. Fjords as Pollution Traps

There is a striking difference in the faunal development over the past few centuries in the Frierfjord cores as compared to the stable faunal composition throughout core 30 from just outside the sill. Furthermore, the concentrations of heavy metals in the surface 10 cm are two to three times higher in core 28 (the least affected of the Frierfjord cores) than in core 30 (cores 1 and 2, respectively, in Saetre *et al.*, 1997) and the TOC values are about twice as high. These fundamentally different faunal and geochemical characteristics in two areas that, geographically, are not far apart (Fig. 1) is an excellent example of how silled fjords act as pollution traps (e.g., Skei, 1996), leaving the more seaward areas less affected.

6. Summary and Conclusions

"Environmental stratigraphy" is used here as a term to describe the detailed study of recent to subrecent sediments for reconstructing possible environmental and ecological changes. As an example to illustrate its applicability, high-resolution studies of nine sediment cores (max. 52 cm long) from a Norwegian fjord are presented to demonstrate how detailed, and otherwise inaccessible, information about the successive development of anthropogenically induced oxygen depletion can be obtained and how silled fjords act as pollution traps.

Owing to natural processes, deep-water renewals in Frierfjord are limited and, consequently, the deep-water oxygen concentration is particularly sensitive to increased flux of biodegradable organic material. Micropaleontological and sedimentological data indicate that the sediment and bottom water oxygen conditions in Frierfjord have changed over the last four to five centuries. Intercore age correlations have been determined by ^{210}Pb-datings and synchronous TOC events. Based on the present results and historical information, the cores have been divided into four time periods: (I) the preindustrial period (sediments deposited before 1550); (II) the sawmill period (1550–1870); (III) the modern industrial period (1870–1975); and (IV) the pollution abatement period (post-1975). It is concluded that well-oxygenated conditions prevailed throughout the water column in the preindustrial period in all parts of the fjord. During the sawmill period, the oxygen conditions in the surface sediments were impoverished at water depths greater than 70 m. Near the end of the sawmill period, occasional anoxia developed at these depths in the northwest part of the fjord. At the same time, there was a drastic deterioration in the benthic environment below 50 m water depth both in the central and northern parts of the fjord. During the modern industrial period, nearly permanently anoxic conditions prevailed at depths greater than 70 m in the northwest, and there was an increasing frequency of anoxia below 50 m in the northwest and at 90 m water depth in the southern parts. Pollution abatement since the mid-1970s has slightly improved the oxygen conditions in the surface sediments, at least in the southernmost part of the fjord.

ACKNOWLEDGMENTS: I am most grateful to Tom Tønnesen, Eskil Odberg, and all the others who assisted during the cruises with the research vessels Bjørn Føyn and Trygve Braarud, to Inger Ann Hansen for carrying out the organic carbon and metal analyses, to Eigil Whist for drawing the map (Fig. 1), and to John Murray, Ron Martin, and the reviewers Joan M. Bernhard and Jean-Pierre Guilbault for constructive comments on the manuscript. The project was funded by the Norwegian Research Council for Science and the Humanities (NAVF).

References

Alve, E., 1990, Variations in estuarine foraminiferal biofacies with diminishing oxygen conditions in Drammensfjord, SE Norway, in: *Paleoecology, Biostratigraphy and Taxonomy of Agglutinated Foraminifera* (C. Hemleben, M. A. Kaminski, W. Kuhnt, and D. B. Scott, eds.), Kluwer, Dordrecht, pp. 661–694.

Alve, E., 1991, Foraminifera, climatic change and pollution: A study of Late Holocene sediments in Drammensfjord, SE Norway, *Holocene* **1**:243–261.

Alve, E., 1994, Opportunistic features of the foraminifer *Stainforthia fusiformis* (Williamson): Evidence from Frierfjord, Norway, *J. Micropalaeontol.* **13**:24.

Alve, E., 1995*a*, Benthic foraminiferal responses to estuarine pollution: A review, *J. Foram. Res.* **25**:190–203.

Alve, E., 1995*b*, Benthic foraminiferal distribution and recolonization of formerly anoxic environments in Drammensfjord, southern Norway, *Mar. Micropaleontol.* **25**:169–186.

Alve, E., 1999, Colonization of new habitats by benthic foraminifera: A review, *Earth-Sci. Rev.* **46**:167–185.

Alve, E., and Bernhard, J. M., 1995, Vertical migratory response of benthic foraminifera to controlled oxygen concentrations in an experimental mesocosm, *Mar. Ecol. Prog. Ser.* **116**:137–151.

Alve, E., and Murray, J. W., 1995, Benthic foraminiferal distribution and abundance changes in Skagerrak surface sediments: 1937 (Höglund) and 1992/93 data compared, *Mar. Micropaleontol.* **25**:269–288.

Alve, E., and Murray, J. W., 1997, High benthic fertility and taphonomy of foraminifera: A case study of the Skagerrak, North Sea, *Mar. Micropaleontol.* **31**:157–175.

Alve, E., and Murray, J. W., 1999, Marginal marine environments of the Skagerrak and Kattegat: A baseline study of living (stained) benthic foraminiferal ecology, *Palaeogeogr. Palaeoclimatol. Palaeoecol.* **146**:171–193.

Alve, E., and Olsgard, F., 1999, Benthic foraminiferal colonization in experiments with copper contaminated sediments, *J. Foram. Res.* **29**:186–195.

Baalsrud, K., and Magnusson, J., 1990, Eutrofisituasjonen i Ytre Oslofjord, Hovedrapport. NIVA-rapp. O-8907509/2480, 116 pp.

Barmawidjaja, D. M., van der Zwaan, G. J., Jorissen, F. J., and Puskaric, S., 1995, 150 years of eutrophication in the northern Adriatic Sea: Evidence from a benthic foraminiferal record, *Mar. Geol.* **122**:367–384.

Bernhard, J. M., 1996, Microaerophilic and facultative anaerobic benthic foraminifera: A review of experimental and ultrastructural evidence, *Rev. Paleobiol.* **15**:261–275.

Bernhard, J. M., and Alve, E., 1996, Survival, ATP pool, and ultrastructural characterization of benthic foraminifera from Drammensfjord (Norway): Response to anoxia, *Mar. Micropaleontol.* **28**:5–17.

Bernhard J. M., and Reimers, C. E., 1991, Benthic foraminiferal population fluctuations related to anoxia: Santa Barbara Basin, *Biogeochemistry* **15**:127–149.

Blackwelder, P., Hood, T., Alvarez-Zarikian, C., Nelsen, T. A., and McKee, B., 1996, Benthic foraminifera from the NECOP study area impacted by the Mississippi River plume and seasonal hypoxia, *Quat. Int.* **31**:19–36.

Briffa, K. R., Bartholin, T. S., Eckstein, D., Jones, P. D., Karlén, W., Schweingruber, F. H., and Zetterberg, P., 1990, A 1,400-year tree-ring record of summer temperatures in Fennoscandia, *Nature* **346**:434–439.

Briffa, K. R., Jones, P. D., Bartholin, T. S., Eckstein, D., Schweingruber, F. H., Karlén, W., Zetterberg, P., and Eronen, M., 1992, Fennoscandian summers from A.D. 500: temperature changes on short and long timescales, *Climate Dynam.* **7**:111–119.

Cato, I., Olsson, I., and Rosenberg, R., 1980, Recovery and decontamination of estuaries, in: *Chemistry and Biogeochemistry of Estuaries* (E. Olausson, and I. Cato, eds.), John Wiley & Sons Ltd., Chichester, pp. 403–440.

Dannevig, A., 1930, The Propagation of Our Common Fishes During the Cold Winter 1924. Investigations on the Norwegian Skagerrak Coast. *Rep. Norw. Fishery Mar. Invest.* **3**(10):1–133, A. S. John Grieg, Bergen.

Denne, R. A., and Sen Gupta, B. K., 1989, Effects of taphonomy and habitat on the record of benthic foraminifera in modern sediments, *Palaios* **4**:414–423.

Ellison, R. L., Broome, R., and Ogilvie, R., 1986, Foraminiferal response to trace metal contamination in the Patapsco River and Baltimore Harbour, Maryland, *Mar. Pollut. Bull.* **17**:419–423.

Goës, A., 1894, A Synopsis of the Arctic and Scandinavian Recent Marine Foraminifera hitherto discovered, *Kongl. Sv. Vet. Ak. Handl.* **25**:1–127.

Hals, H., 1968, *Eidanger Bygdehistorie. Bind 1: Eidangers historie til omkring 1845*, Porsgrunn, Skien, 463 pp.

Höglund, H., 1947, Foraminifera in the Gullmarfjord and the Skagerrak, *Zool. Bidrag Uppsala* **26**:1–328.

Kiaer, H., 1900, Synopsis of the Norwegian Marine Thalamophora, *Rep. Norw. Fishery Mar. Invest.* **1**:1–58, Kristiania.

Knutzen, J., Molvaer, J., Norheim, G., Rygg, B., and Skei, J., 1983, Grenlandsfjordene og Skienselva 1982, Overvåkingsrapport nr. 91/83, ISBN 82-577-0666-3, 49 pp.

Lamb, H. H., 1977, *Climate: Present, Past and Future*: 2. *Climatic History and the Future*, Methuen, London.

Lehmann, G., and Röttger, R., 1997, Techniques for concentration of foraminifera from coastal salt meadow deposits, *J. Micropalaeontol.* **16**:144.

Magnusson, J., Lømsland, E. R., and Johnsen, T., 1996, Overvåking av forurensningssituasjonen i indre Oslofjord 1995, Overvåkingsrapport nr. 661/96, ISBN 82-577-3027-0.

Molvaer, J., 1980, Deep water renewals in the Frierfjord—an intermittently anoxic basin, in: *Fjord Oceanography* (H. J. Freeland, D. M. Farmer, and C. D. Levings, eds.), Plenum Press, New York and London, pp. 531–537.

Molvaer, J., 1992, Undersøkelse av eutrofiering i Grenlandsfjordene 1988–89, Overvåkingsrapport nr. 474/91, ISBN 82-577-2044-5, 46 pp.

Molvaer, J., Bokn, T., Kirkerud, L., Kvalåvgnaes, K., Nilsen, G., Rygg, B., and Skei, J., 1979, Resipientundersøkelse av nedre Skienselva, Frierfjorden og tilliggende fjordområder, Rapport nr. 8, Sluttrapport, NIVA-rapport O-70111, 252 pp.

Moodley, L., van der Zwaan, G. J., Herman, P. M. J., Kempers, L., and van Breugel, P., 1997, Differential response of benthic meiofauna to anoxia with special reference to Foraminifera (Protista: Sarcodina), *Mar. Ecol. Prog. Ser.* **158**:151–163.

Murray, J. W., 1991, *Ecology and Palaeoecology of Benthic Foraminifera*, Longman, Harlow, 397 pp.

Nagy, J., and Alve, E., 1987, Temporal changes in foraminiferal faunas and impact of pollution in Sandebukta, Oslo Fjord, *Mar. Micropaleontol.* **12**:109–128.

Naes, K., and Oug, E., 1991, Sedimentenes betydning for forurensningstilstanden i Frierfjorden og tilgrensende områder, NIVA-Rapport O-895903/E-90406, ISBN 82-577-1885-8, 193 pp.

Pearson T. H., and Rosenberg, R., 1976, A comparative study of the effects on the marine environment of wastes from cellulose industries in Scotland and Sweden, *Ambio* **5**:77–79.

Pheiffer Madsen, P., and Sørensen, J., 1979, Validation of the lead-210 dating method, *J. Radioanal. Chem.* **54**:39–48.

Phleger, F. B., and Soutar, A., 1973, Production of benthic foraminifera in three east Pacific oxygen minima, *Micropaleontology* **19**:110–115.

Risdal, D., 1963, Foraminiferfaunaen i en del sedimentkjerner fra indre Oslofjord, *Nor. Geol. Unders.* **224**:5–90.

Risdal, D., 1964, Foraminiferfaunaenes relasjon til dybdeforholdene i Oslofjorden, med en diskusjon av de senkvartre foraminifersoner, *Nor. Geol. Unders.* **226**:5–142.

Rygg, B., Bjerkeng, B., Knutzen, J., Molvaer, J., and Norheim, G., 1984, Grenlandsfjordene og Skienselva 1983, Overvåkingsrapport nr. 161/84, ISBN 82-577-0824-0, 61 pp.

Rygg, B., Green, N., Molvaer, J., and Naes, K., 1987, Grenlandsfjordene og Skienselva 1986, Overåkningsrapport nr. 287/87, ISBN 82-577-1293-0, 91 pp.

Rygg, B., Green, N., Knutzen, J., and Molvaer, J., 1988, Grenlandsfjordene og Skienselva 1987, Overvåkingsrapport nr. 327/88, ISBN 82-577-1444-5, 72 pp.

Schafer, C. T., 1982, Foraminiferal colonization of an offshore dump site in Chaleur Bay, New Brunswick, Canada, *J. Foram. Res.* **12**:317–326.

Schafer, C. T., Collins, E. S., and Smith, J. N., 1991, Relationship of foraminifera and thecamoebian distributions to sediments contaminated by pulp mill effluent: Saguenay Fiord, Quebec, Canada, *Mar. Micropaleontol.* **17**:255–283.

Sen Gupta, B. K., and Machain-Castillo, M. L., 1993, Benthic foraminifera in oxygen-poor habitats, *Mar. Micropaleontol.* **20**:183–201.

Sen Gupta, B. K., Turner, R. E., and Rabalais, N. N., 1996, Seasonal oxygen depletion in continental-shelf waters of Louisiana: Historical record of benthic foraminifers, *Geology* **24**:227–230.

Skei, J. M., 1996, Trace metals in sediments in Norwegian coastal water and the Norwegian Trench—Anthropogenic signals in deposition areas, in: *Scientific Symposium on the North Sea Quality Status Report 1993, 18—21 April, 1994, Ebeltoft, Denmark* (J. Andersen, H. Karup, and U. B. Nielsen, eds.), pp. 94–100, Ministry of Environment and Energy, Danish Environmental Protection Agency, Copenhagen.

Saetre, M. M. L., Dale, B., Abdullah, M. I., and Saetre, G.-P., 1997, Dinoflagellate cysts as potential indicators of industrial pollution in a Norwegian fjord, *Mar. Environ. Res.* **44**:167–189.

Strøm, K. M., 1936, Land-locked waters, *Skrifter utgitt av det Norske Vitenskaps-akademi i Oslo.* 1. Matematisk-Naturvitenskapelig Klasse, Vol. 1, No. 7, pp. 1–85.

Tønnesen, J. N., 1956, *Porsgrunns Historie, Vol. 1, Fra Lasteplass til Kjøpstad 1576–1807*, Porsgrunn, Porsgrunn kommune i kommisjon hos H. J. Dyring, 632 pp.

Tønnesen, J. N., 1957, *Porsgrunns Historie, Vol. 2, Fra Trelast og Skipsfart til Industri 1807–1920*, Porsgrunn, Porsgrunn kommune i kommisjon hos H. J. Dyring, 658 pp.

Chapter 15

Foraminifera of Storm-Generated Washover Fans

Implications for Determining Storm Frequency in Relation to Sediment Supply and Barrier Island Evolution, Folly Island, South Carolina

SCOTT P. HIPPENSTEEL and RONALD E. MARTIN

1. Introduction

This study investigates the sediment source for washover fans and the use of natural (foraminifera) and artificial (glass bead) tracers to quantify deposition and mixing in back-barrier marsh environments along the South Carolina coast. Erosion and deposition along South Carolina's coast are processes of

SCOTT P. HIPPENSTEEL and RONALD E. MARTIN • Department of Geology, University of Delaware, Newark, Delaware 19716-2544.

Environmental Micropaleontology, Volume 15 of *Topics in Geobiology*, edited by Ronald E. Martin. Kluwer Academic/Plenum Publishers, New York, 2000.

growing interest. Recent storm events have demonstrated the economic effects of such natural agents and the need to better understand the frequency of major storms in relation to barrier-island migration, sediment supply, and evolution. Foraminiferal assemblages provide insight into back-barrier deposition rates, including potential storm-generated washover intervals, and may help to identify processes acting on an evolving barrier island in a transgressive setting.

2. Methods

2.1 Area Description

In the spring of 1994, nineteen 3-in.-diameter vibracores were taken from Folly Island, South Carolina, and the adjacent back-barrier marsh. The core locations were selected to determine the stratigraphy of the region around Folly Island (Harris *et al.*, 1995). Initial examination of cores found sand-rich intervals that were in many cases enriched in offshore foraminifera that represent discrete overwash fans. Four of these cores from differing depositional environments (low-marsh, high-marsh, immediate back-barrier) were chosen for further micropaleontological analysis of potential overwash fans (Fig. 1). Marsh subenvironments were recognized based on discussions with D.E. Krantz and M.S. Harris and our own observations of vegetation (e.g., occurrence of *Spartina alterniflora*; Letzsch and Frey, 1980), fiddler crab burrow density (Sharma *et al.*, 1987), and elevation; modern agglutinated

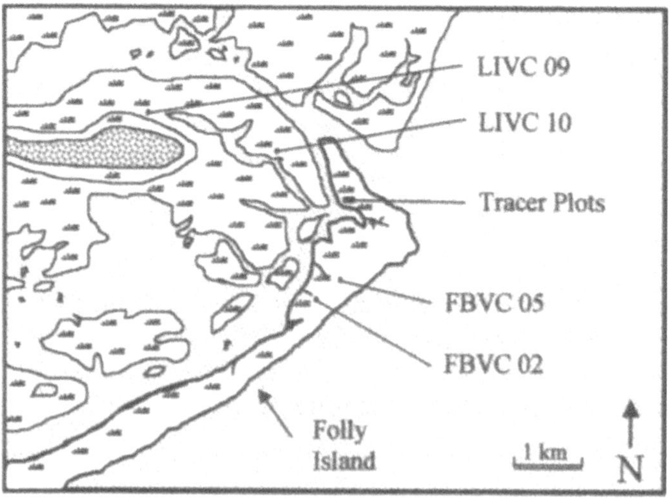

FIGURE 1. Location of the four cores and tracer plots used in this study. Approximate location of the study area is 32°66′N, 79°54′W. Charleston, Southern Carolina, is approximately 16 km North-Northwest of the study area.

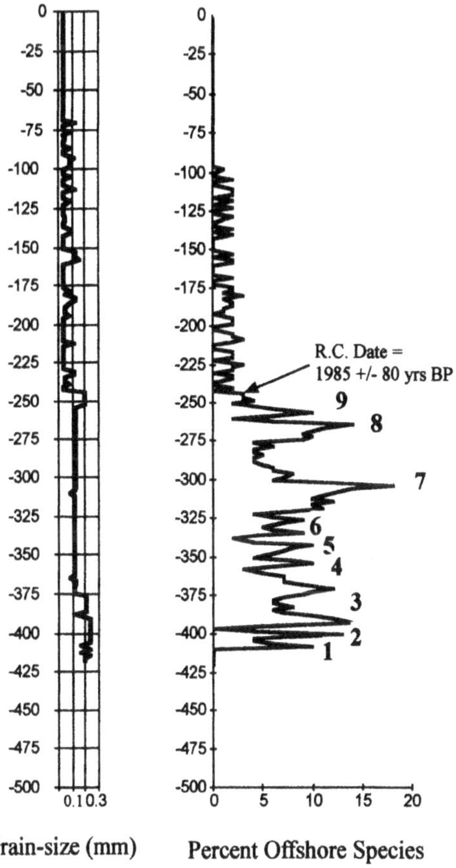

FIGURE 2. Comparison of the percent of offshore foraminifera and the grain size for FBVC 02. Nine washover intervals were identified in FBVC 02 (low marsh). A *Mulinia* (bivalve) shell taken from a depth of 2.45 m gave a radiocarbon (corrected AMS) age of 1985 ± 80 years BP.

foraminiferal assemblages (based on Goldstein and Harben, 1993) substantiated our field observations. Vibracore FBVC 02 (Folly Beach Vibracore 02) was taken from the margin of the Folly Island back-barrier environment (Figs. 1 and 2). Although FBVC 05 was taken farther from the barrier-island beach ridge than FBVC 02 (Figs. 1 and 3), it was recovered from the side of a small elevation that is higher relative to modern sea level. The core location is on the low-marsh/high-marsh border approximately 250 m northeast of FBVC 02 (M.S. Harris, 1994, per. comm.). LIVC 09 (Long Island Vibracore 09) was taken from a high-marsh subenvironment approximately 1.5 km northwest of FBVC 02 (Figs. 1 and 4). The bottom of the core penetrated the back-barrier of a Pleistocene barrier island (Harris *et al.*, 1995). Surface sediment was very sandy and, compared with the FBVC cores, there was little evidence of burrowing. LIVC 10 was taken along the outside of a marsh-creek meander in a low-marsh environment (Figs. 1 and 5). LIVC 10 had the lowest elevation and the most evidence of burrowing.

In May 1996, three 1 × 1 m tracer plots were selected to represent mixing characteristics for low-marsh, intermediate-marsh ("transitional"-marsh), and high-marsh subenvironments (Fig. 1). Agglutinated foraminifera, vegetation, and elevation (relative to mean-high water) were used to establish these subenvironments.

2.2. Sample Collection and Analysis

Sediment samples used for foraminiferal analysis were taken at 2-cm intervals throughout the vibracores. All samples were taken from the center of the cores to avoid potential smearing along the inner core wall. Ten grams of sediment were wet-sieved using distilled water and 0.5-mm and 1.0-mm screens and allowed to air-dry. Relative proportions of sand and mud and

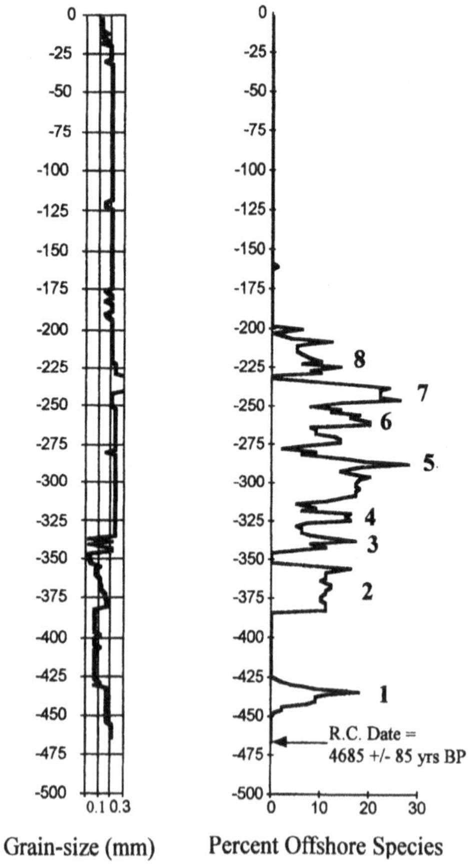

FIGURE 3. Comparison of the percent of offshore foraminifera and the grain size for FBVC 05. Eight washover intervals were identified in FBVC 05 (low-marsh/high-marsh transition zone). A *Mulinia* shell taken from a depth of 4.65 m gave a radiocarbon (corrected AMS) age of 4685 ± 85 years BP.

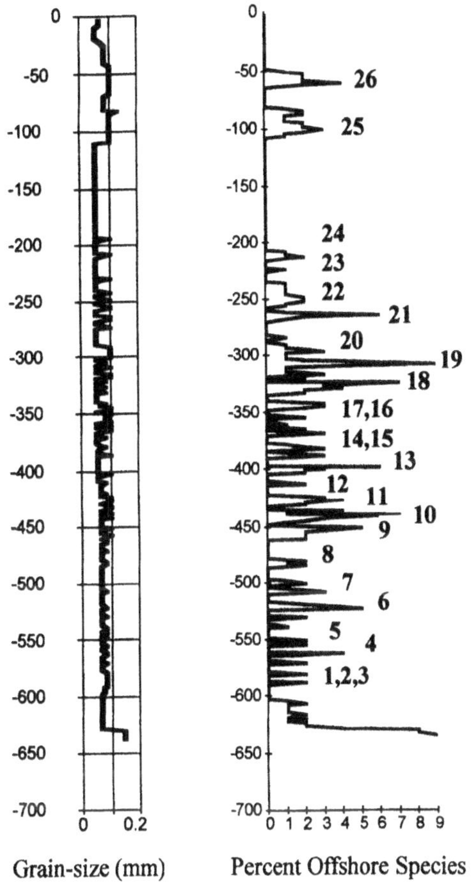

FIGURE 4. Comparison of the percent of offshore foraminifera and the grain size for LIVC 09. Twenty six washover intervals were identified in LIVC 09 (high-marsh/relic overwash fan); washover intervals are discrete and the mixing parameter was the lowest among the cores. Compare to Fig. 5.

average grain size were estimated based on sieving (Figs. 2–5; data available upon request). The remaining dry sediment was floated using CCl_4, and the foraminifera picked from the dried float and identified to genus; separation of tests between float, coarse fraction, and residue was excellent.

Foraminifera were divided into two groups: those indicative of offshore (modern or Oligo–Miocene) open-marine environments and those indicative of marsh/estuarine/nearshore environments ("background"). For the Folly Island cores these classifications included: (1) "Marsh/estuarine/nearshore genera": *Ammonia* spp. and *Elphidium* spp.; and (2) "Offshore genera": *Bolivina* spp., *Buccella* spp., *Bulimina* spp., *Buliminella* spp., *Cancris* spp., *Cibicides* spp., *Eponides* spp., *Fursenkiona* spp., *Hanzawaia* spp., *Nonionella* spp., *Quinqueloculina* spp., *Rosalina* spp., *Saracenaria* spp., *Siphogenerina* spp., *Stilostomella* spp., *Uvigerina* spp., *Virgulina* spp., and planktonic spp.

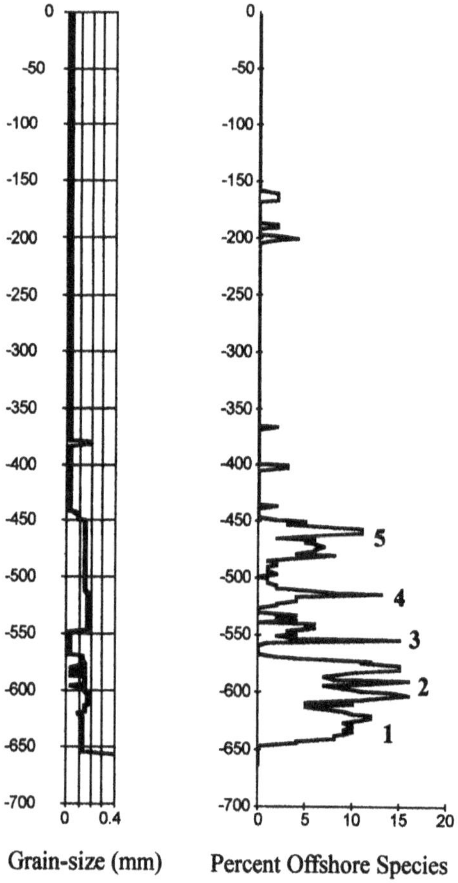

FIGURE 5. Comparison of the percent of offshore foraminifera and the grain size for LIVC 10. Five washover intervals were identified in LIVC 10 (low-marsh/creek-side); washover intervals are extensively smeared and the core had the highest mixing parameter. Compare to Fig. 4.

(primarily *Globigerina* spp.). Agglutinated foraminifera were not used in this study, as the vibracores had been allowed to dry and agglutinated marsh foraminifera that have been exposed to air or oxidizing bacteria tend to disintegrate via oxidation of their organic test cements (Scott and Medioli, 1986; Goldstein and Harben, 1993).

Two criteria were used to detect possible washover intervals: (1) percentage of offshore (modern or Oligo–Miocene) foraminifera, and (2) total number of calcareous foraminifera. In samples that contained fewer than ten foraminifera, the percentage of offshore species was considered to be zero.

2.3. Tracer Study

A 2-mm-thick layer of red sand (tracer) nearly identical in grain size to that of the calcareous foraminifera found in the washover intervals was spread on the plot surface; this thickness was chosen because it would yield an initial concentration of tracer (for the top 1 cm of sediment) of approximately 1,000–1200 grains/10 cm^3 (similar to the maximum concentration of calcareous foraminifera in washover intervals). Average plant density and burrow frequency were also measured for each plot.

Tracer was also planted in the subsurface in order to measure downward *and upward* mixing of sediment. Obviously, disturbance to the marsh was much greater for these plots than those of surface. A layer of tracer was buried at a depth of 5 cm in the low marsh to measure upward mixing and another layer of tracer was buried at 25 cm to measure mixing that may result from processes other than bioturbation (e.g., tidal flushing of pore waters and sediment compaction). If fiddler crabs have a maximum burrowing depth of 20 cm (Sharma *et al.*, 1987; personal observation), any mixing below this depth must result from processes such as these.

Coring of the plots was considered for sampling but this method would have caused compaction and smearing of sediment. Instead, immediately after planting beads, a plug of sediment 4 × 4 cm was cut from all plots to measure the initial (before mixing) concentration of tracer. Plugs were also cut to a depth of 15 cm at 50-day and 1-year intervals to measure the mixing of the tracer. After the plug was removed, it was carefully trimmed to 3.3 × 3.3 cm and cut at 1-cm intervals, which yielded 10 cm^3 samples for processing. All tracer concentrations in this study are presented as number of grains of tracer per 10 cm^3.

2.4. Calculation of Biodiffusion Coefficients

If potential storm-derived washover events and resulting offshore percentage peaks are considered to be instantaneous events, a single washover horizon is analogous to an "impulse" layer. This impulse layer, and the degree of upward and downward smearing of the event layers, can be used to calculate relative bioturbation rates (e.g., Berger and Heath, 1968; Cutler, 1993; Martin, 1993). Bioturbation rates for washover intervals were estimated using the method described by Guinasso and Schink (1975), who determined that mixing can be estimated by a dimensionless parameter *G*, where

$$G = D/m\nu \tag{1}$$

and *D* is the biodiffusion coefficient, *m* the mixing layer thickness, and ν the sedimentation rate. Calculation of the mixing parameter (*G*) gives an estimate of the sedimentation rate relative to mixing rates [see Martin, 1998, for criticisms of equation (1)]. In the specific case of the Folly Island back-barrier marsh, sedimentation rates vary but are usually quite low. Average sedimentation rates for similar environments range from 0.14–0.45 cm/year (South Carolina; Sharma *et al.*, 1987), 0.30–0.50 cm/year (Delaware; Church *et al.*,

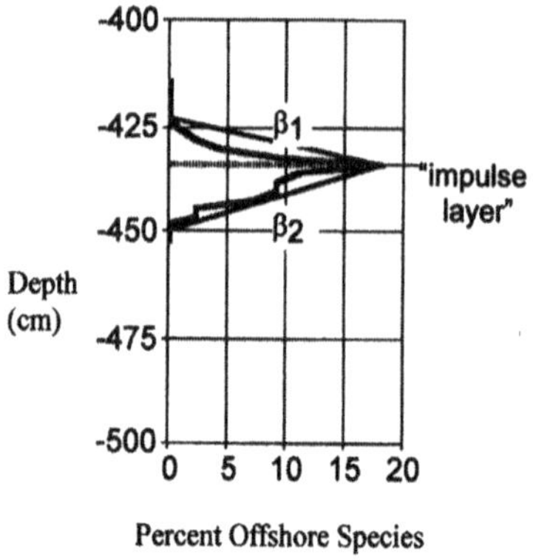

FIGURE 6. Best fit line for β_1 and β_2 using an offshore peak from FBVC 05. β_1 and β_2 were fitted both visually (G) and by simple linear regression (G'). The mixing parameter was then calculated for each pair of slopes (see Table 1).

1981, 1987), and 0.41–0.46 cm/year (Delaware, T. M. Church, University of Delaware, unpublished radiotracer data).

The mixing parameter was calculated for each washover layer from the equation (Officer and Lynch; 1983):

$$G = \{\beta_2 + [\beta_2^2 + \pi^2\beta_1(\beta_2 - \beta_1)]^{1/2}\}/2\pi^2\beta_1 \quad (2)$$

where β_1 and β_2 are the slopes of the percentage of offshore species plotted against depth (constant scale). For each washover, the slopes were estimated both visually (G) and by simple linear regression (G') to test visual estimates (Fig. 6). The original impulse (washover) was assumed to be represented by the maximum abundance of offshore foraminifera.

The mixing parameter (G), which was used to estimate mixing of washover intervals in vibracores (Hippensteel, 1995), could not be used in the tracer study. Because establishing a mixing parameter requires that the "impulse" layer (tracer) be smeared *above* the original layer, too little time would pass or too little sediment (several cm) would be deposited to allow significant upward mixing of the surface plots. Therefore, a different method of estimation for mixing was devised for the tracer studies. The simplest measure of mixing for a tracer layer would relate the maximum concentration of tracer to the amount of tracer that has been stratigraphically spread. In other words, an undisturbed tracer layer should have a distinct, thin peak with a relatively high concentration, whereas a tracer layer that has been significantly mixed is characterized by a smeared (damped) signal in which the maximum concentration has been mixed into adjacent layers of sediment. Consequently, the index of mixing (I_m) relates the maximum concentration of tracer to the amount of tracer that has been mixed into the overlying and underlying layers of sediment:

$$I_m = \frac{\int_{T_{d(\max)}}^{T_{d(\min)}} f(x)\,dx}{C_{\max}} \quad (3)$$

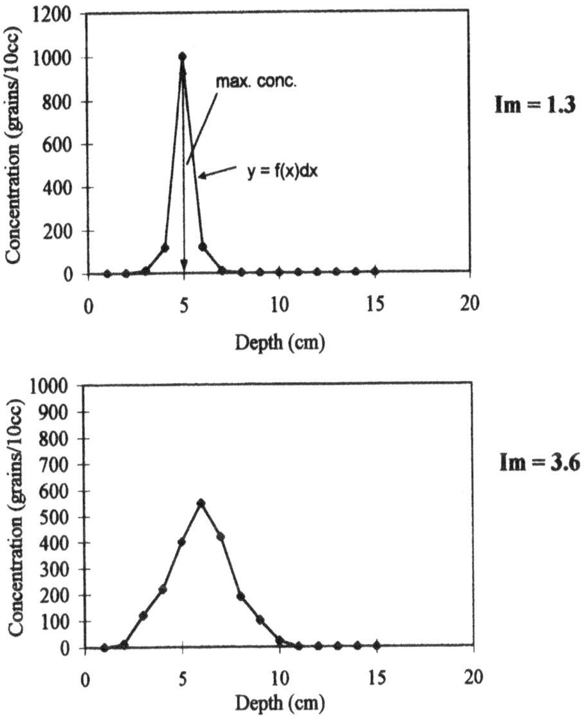

FIGURE 7. Relationship between I_m and mixing for hypothetical settings. The top plot has experienced little mixing and the tracer profile is distinct, whereas the bottom plot has been smeared into the overlying and underlying sediments.

where $T_{d(\min)}$ and $T_{d(\max)}$ are the minimum and maximum depths of the tracer, respectively, $f(x)dx$ is the curve created by the concentration profile, and $C_{\max}$ is the maximum concentration. This equation represents the integral of a curve created by the plot of tracer (vs. depth) divided by the maximum concentration of tracer. The higher the value of I_m, the more the tracer has been spread and the more mixing that has occurred (Fig. 7).

3. Results and Discussion

3.1. Foraminiferal Abundance

Total calcareous foraminiferal abundance varied from 0 to 1240 specimens per 10 g of sediment. The most abundant genus was *Elphidium* spp: in several samples more than 800 specimens of this taxon were present. Other taxa that exceeded 50 specimens in a single sample included *Ammonia beccarri*, *Uvigerina* spp., and planktonic spp. In no case, however, did any taxa other than *Elphidium* spp. exceed 100 specimens in a sample.

The following genera were found in the presumed washover intervals, listed from relatively most abundant to least abundant: *Uvigerina* spp., planktonic spp., *Nonionella* spp., *Bulimina* spp., *Bolivina* spp., *Siphogenerina* spp., *Stilostomella* spp., *Lenticulina* spp., *Cibicides* spp., *Hanzawaia* spp., *Cancris* spp., *Planularia* spp., and *Saracenaria* spp. (Hippensteel, 1995; Hippensteel and Martin, 1995*a, b*; data available upon request). Abundance of offshore taxa was higher in FBVC 02, FBVC 05, and LIVC 10 (all of which were located in low-marsh or transitional-marsh environments) than in LIVC 09 (high-marsh), in which only a maximum of 9% was reached (Fig. 4). The highest plotted peak in offshore percentage was 28% at 2.82 m in FBVC 05 (Fig. 3). In most instances, total foraminiferal abundance peaks coincided stratigraphically with abundance peaks of offshore taxa (Figs. 2–5).

3.2. Occurrence of Washover Intervals

Nine 10- to 20-cm thick washover intervals were identified in FBVC 02 below approximately 2.5 m, whereas marsh mud and associated calcareous foraminifera (such as *Ammonia* spp. and *Elphidium* spp.) dominated the upper 2.5 m (Fig. 2). Small subsidiary peaks in percentages of offshore foraminifera (<4 %) in marsh mud between 1.0 and 2.5 m could be the result of two factors: (1) bioturbation and upward reworking of sediments from the offshore peaks below 2.5 m, or (2) relatively complete mixing of frequent, smaller-scale washovers between 1.0 and 2.5 m. In either case, the volume of offshore sediments supplied to this area has apparently decreased through time. Subsidiary peaks within each of the presumed washover intervals (1–9) could also be the result of bioturbation.

FBVC 05 was taken from a higher elevation than FBVC 02. The low abundance of calcareous foraminifera in the upper 2.0 m of FBVC 05 is consistent with a modern high-marsh environment. In FBVC 02, on the other hand, the samples consistently revealed between 50 and 80 total calcareous tests per 10 g of sediment in the same interval (Figs. 3); the higher calcareous test abundances, combined with the lower offshore species component, suggest a low-marsh environment for FBVC 02.

Despite their differences, cores FBVC 05 and FBVC 02 are basically similar. Both cores have washover intervals with similar relative stratigraphic positions and are of roughly similar thicknesses (cf. Figs. 2 and 3). It is unlikely that the offshore percentage peaks in the two cores are from the same washover fan, as the cores were taken about 250 m apart; nevertheless, it is not unreasonable to conclude that washover sediments in these cores were deposited by the same major storms.

Radiocarbon (Corrected AMS) dating of an articulated *Mulinia* (bivalve) shell taken from a *mud interval* at a depth of 2.45 m in FBVC 02 gave an age of 1985 ± 80 years BP (Fig. 2). If this shell is considered to be *in situ* (and assuming no sediment compaction), the average sedimentation rate for the top 2.45 m of the core is approximately 0.12 cm/year. This value compares well with sedimentation rates (0.14-0.45 cm/year) based on radionuclides (^{210}Pb, ^{137}Cs, and ^{7}Be) reported by Sharma *et al.* (1987) for salt marshes near North Inlet, South Carolina.

Radiocarbon (Corrected AMS) dating of an articulated *Mulinia* shell from 4.65 m in FBVC 05 gave an age of 4685 ± 85 years BP (Fig. 3). If this shell is considered *in situ* (sediment compaction not considered), the average sedimentation rate for the core is approximately 0.10 cm/year. If the washover intervals are considered instantaneous (washover intervals were not considered when calculating the average sedimentation rate), the average sedimentation rate is approximately 0.50 cm/year. However, unlike the shell from FBVC 02, this shell was not taken from a mud layer, and therefore the possibility that it has been reworked is greater. Nevertheless, the calculated sedimentation rate for FBVC 05 is in general agreement with that of FBVC 02.

Core LIVC 09 (Fig. 4) was taken from a paleo-back-barrier environment that is classified as high marsh on the basis of its elevation and the plants and agglutinated-foraminifera present. The large number of thin overwash peaks may have resulted from either of two causes. First, because of its distance from the beach, offshore-sediment supply was perhaps significantly smaller. Second, if only overwash peaks with the highest percent offshore species are considered (> ~3%), the number of actual overwashes is reduced to nine or ten, which is in general agreement with other cores. Although the large number of smaller peaks could have been generated by bioturbation of thicker overwash intervals, bioturbation rates are substantially lower in the high marsh (Sharma *et al.*, 1987; Hippensteel and Martin, unpublished data from bead experiments at Bombay Hook National Wildlife Refuge, Delaware); moreover, each of the smaller apparent impulse layers correlates to a thin, discrete sand layer (Fig. 4), which also suggests that the peaks are in fact storm-derived.

LIVC 10 (Fig. 5) was taken from a low-marsh environment adjacent to a migrating tidal creek. Five washover intervals were identified below 4.5 m based on the occurrence of offshore species. Peaks 3, 4, and 5 in LIVC 10 (Fig. 5) may not be discrete washover intervals, but the result of reworking of sediment from layers 1 and 2. A large burrow at 6.10 m and smaller burrows throughout the core support this premise. Intervals 3 through 5 could be the result of reworking of sediments as the depositional environment shifted through time from high-marsh/back-barrier (sandy mud receiving offshore foraminifera) to low-marsh (mud). Note that the total foraminifera for intervals 3 through 5 are all below 100 whereas those of 1 and 2 are over 1000 (Fig. 5).

3.3. Bioturbation and Washover Recognition

Size-selective reworking of microfossils may smear abundance peaks, thereby obscuring recognition of washovers. Bioturbation is expected to be highest in low-marsh, muddy sediments, as bioturbator populations and burrow frequencies are typically higher in the low marsh (Sharma *et al.*, 1987).

TABLE 1. Comparison of Mean Mixing Parameter *G*, Depositional Environment from Which Cores Were Recovered, and Visual Evidence of Bioturbation[a]

Core	*G*/*G′*	Depositional environment	Visual evidence of bioturbation
FBVC 02	0.120 ± 0.156 0.108 ± 0.154	Low-marsh	None
FBVC 05	0.197 ± 0.319 0.195 ± 0.309	Low-marsh/high-marsh border	Minimal
LIVC 09	0.034 ± 0.048 0.032 ± 0.046	High-marsh	None
LIVC 10	0.523 ± 0.563 0.527 ± 0.585	Low-marsh/creek-side	Extensive

[a] *G* was calculated using slopes which were fitted visually, whereas *G′* was calculated using slopes which were fitted using simple linear regression.

The primary animal bioturbator at the core sites is the fiddler crab (Uca sp.), which favors a low-marsh environment (Deery and Howard, 1977). Bioturbation in marshes is not spatially continuous, however (Goldstein *et al.*, 1995*a*, *b*): dense halophyte roots may impede bioturbation by insects, crabs, and polychaetes, although the halophytes themselves may be major bioturbators (Basan and Frey, 1977; Frey and Basan, 1981; Sharma *et al.*, 1987; Goldstein *et al.*, 1995*a*, *b*). Nevertheless, plant density and burrow density are both highest in the low marsh, suggesting that the fiddler crabs' preference for muddier (and softer) low-marsh mud supersedes the deterrents to burrowing caused by increased *Spartina* populations. Grain size of the sediment may be the most significant factor determining burrow density, as burrows were observed to be as much as 5 cm deeper in low-marsh sediments than in high-marsh sediments. Variable nutrient content of the sediments may also be an important factor (Sharma *et al.*, 1987).

Although variable, the relationship between bioturbation and environment is demonstrated by the distribution of values of the mean mixing parameter (G, G'; Table 1). In most cases the mean mixing parameter also reflects visual evidence of bioturbation (Table 1). In LIVC 09, which was taken from a high-marsh environment, washover intervals showed little evidence of bioturbation ($G = 0.034$ and $G' = 0.032$); other evidence of bioturbation is absent from LIVC 09, except for a single burrow 0.8 m from the top of the core. In LIVC 10, which exhibits a higher degree of smearing ($G = 0.523$ and $G' = 0.527$), burrows are abundant, which is consistent with the low-marsh environment of LIVC 10. Although no physical evidence of burrowing was evident in FBVC 02, the mixing parameters ($G = 0.120$ and $G' = 0.108$) indicate moderate bioturbation, which is consistent with a low-marsh environment of FBVC 02. FBVC 05 was taken from the low-marsh/high-marsh transition zone and mixing parameters ($G = 0.197$ and $G' = 0.195$) are intermediate, which is consistent with a transitional environment with minimal

TABLE 2. Mixing Parameter (G) vs. Index of Mixing (I_m)

Core/plot averages	G	I_m
FBVC 02 (low-marsh)	0.114	10.6
FBVC 05 (intermediate marsh)	0.196	12.6
LIVC 09 (high-marsh)	0.033	5.0
LIVC 10 (low-marsh, stream side)	0.525	15.8
Low-marsh plot (1 year)	—	4.8
Intermediate-marsh plot (1 year)	—	3.8
High-marsh (1 year)	—	1.6

visual evidence of bioturbation (Table 1). One source of discrepancy between the mixing parameter and the subenvironments may be the assumption that marsh subenvironments remain constant both laterally and through time (i.e., throughout each core).

As with the previously measured washover intervals, mixing rates measured in the tracer study were higher in the low marsh and lower in the high marsh. Table 2 shows I_m for the four vibracores from Folly Island and the three experimental plots from this study. The mixing parameters are also presented for the vibracores, for which the Spearman correlation between G and I_m is very strong (0.981, $\alpha < 0.05$). Table 3 summarizes the concentration profiles for each experimental plot over the 50-day and 1-year intervals (Figs. 8–10). Table 4 relates I_m to physical evidence of mixing (burrow density) and plant density. Subenvironments with the highest number of burrows per square meter had the highest index of mixing. The I_m values for the surface plots are lower than in the vibracores, but the surface plots have been subjected to mixing for only 1 year, whereas the washovers were no doubt subjected to mixing for several decades or longer.

I_m values for surface plots will continue to increase, assuming that significant amounts of tracer are not removed from the plots via physical and biological agents. Interestingly, the highest percentage of tracer remaining after 1 year was found in the intermediate-marsh plot (75%). The low marsh (46% remaining) and high marsh (30% remaining) had the majority of tracer removed from the plots. The primary mechanism in removing the tracer from the plot was probably tidal currents (low marsh) and storms/wind (high

TABLE 3. Index of Mixing for Experimental Plots

Environment	I_m	Environment	I_m	Environment	I_m
Low (original)	1.0	Intermediate (original)	1.1	High (original)	1.4
Low (50-day)	2.8	Intermediate (50-day)	4.1	High (50-day)	2.1
Low (1-year)	4.8	Intermediate (1-year)	3.8	High (1-year)	1.6

TABLE 4. Comparison of Index of Mixing, Burrow Density, and Plant Density for the Three Experimental Plots[a]

Plots	I_m	Burrow density (burrows/m^2)	Plant density (*Spartina* spp.stems/m^2)
Low-marsh	4.8	88	52
Intermediate-marsh	3.8	22	40
High-marsh	1.6	16	45

[a]Densities are the average of measurements made in March 1997, May 1997, and March 1998.

marsh). Fiddler crabs may also have mixed some sediment outside the plots as mixing can be both vertical *and* horizontal: a plug taken 10 cm *outside* the low-marsh plot revealed a total concentration of 82 grains (~7% of the initial concentration of the plot).

Fiddler crabs appear to be responsible for most of the mixing within the plots down to about 20–25 cm. Mixing to this depth from pore-water movement or differential compaction may require much more than a single year to be recorded. Mixing by fiddler crabs at shallower depths may, however, be more difficult to measure than previously thought. Random-walk or simple-diffusion models fail to consider that crabs are piping sediment several centimeters upward from its original location while not adding tracer to the adjacent layers ("nonlocal mixing"; Fig. 11; see Martin, 1998, for discussion). Several burrows were observed with red sand scattered around the freshly excavated aperture. If such mixing were to occur in a discrete storm event that

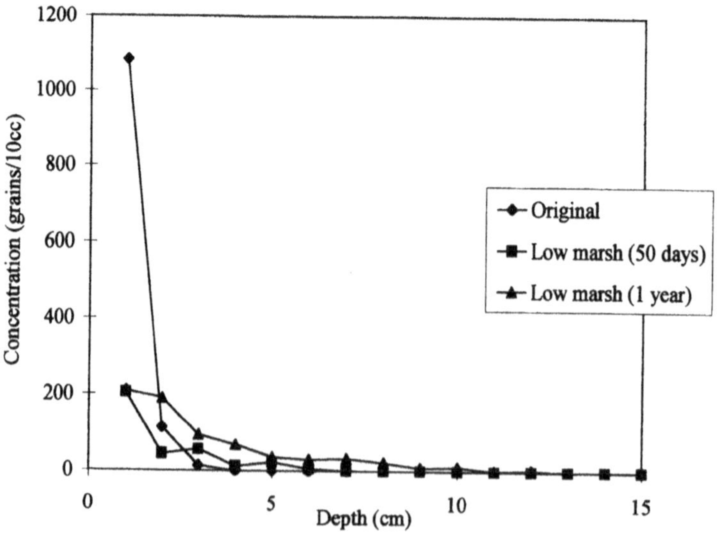

FIGURE 8. Mixing of the low-marsh tracer plot.

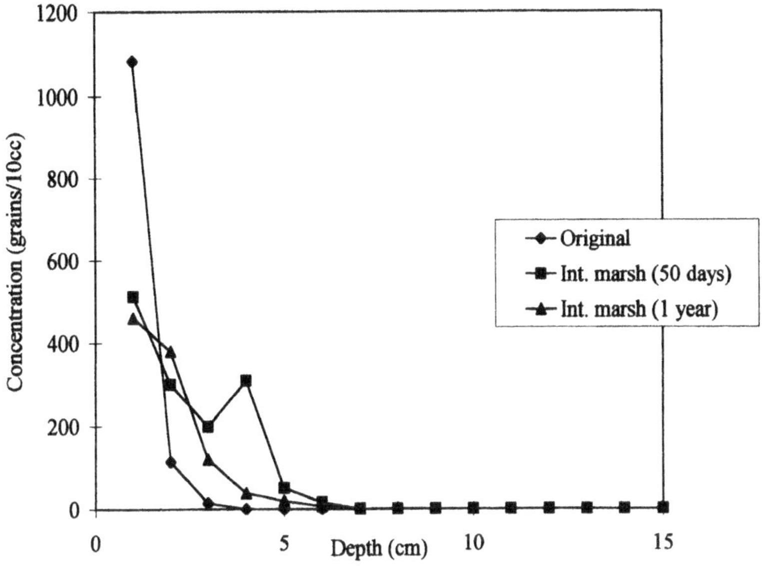

FIGURE 9. Mixing of the intermediate-marsh ("transitional marsh") tracer plot.

has been rapidly buried after initial mixing, it might appear as two separate events in a vibracore instead of a single mixed horizon.

For several reasons, it is difficult to estimate the sedimentation rate using the tracer layers. First, a single year is obviously a short time span in which to measure accumulation. Harrison and Bloom (1977), e.g., based their tracer-derived estimates on data collected over 10 years. Second, the profiles have

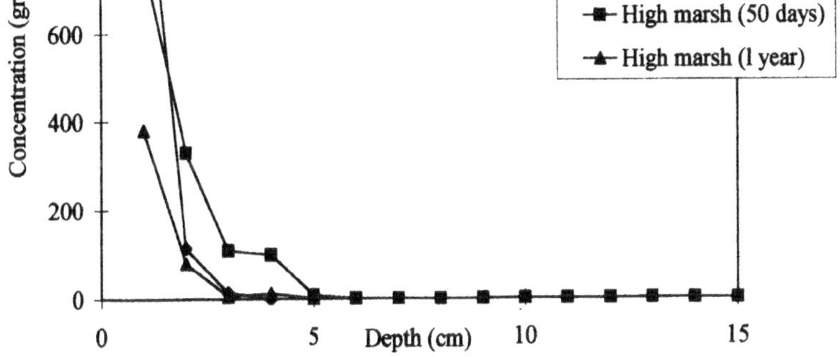

FIGURE 10. Mixing of the high-marsh tracer plot.

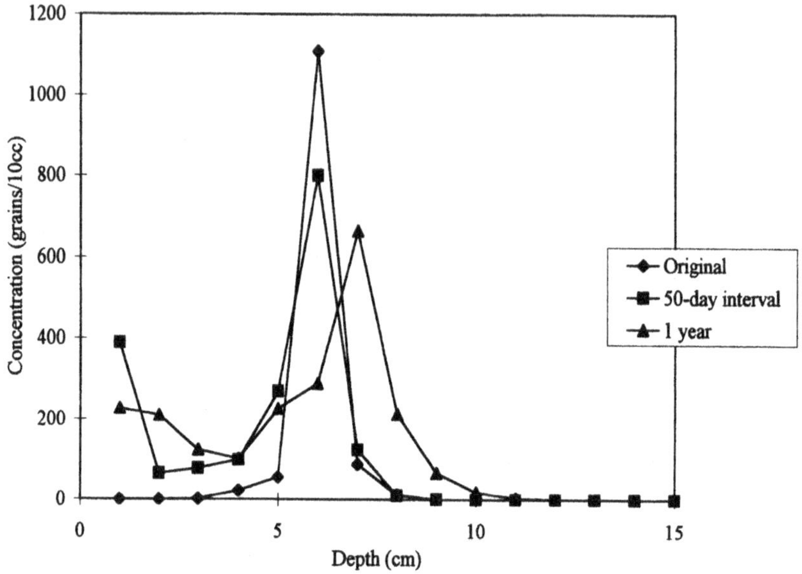

FIGURE 11. Mixing of the 5-cm (subsurface) tracer plot.

been mixed to the point that selecting the original depth of the peak concentration is quite difficult. The 5-cm deep plot is still distinct, but the sediment replaced above this plot probably suffered from at least a small degree of compaction after the initial disturbance (Fig. 11). Nevertheless, the maximum peak appears to have moved downward approximately 1 cm; since the sampling interval was only 1 cm, this is not a significant distance.

Given an average sedimentation rate of 0.25 cm/year and assuming that 50% of the tracer in the surface (0–1 cm of sediment) is removed per year and that 20% of the tracer is mixed into the overlying centimeter of sediment per year, nearly the entire original tracer horizon (>99%) will be destroyed (removed) before the horizon reaches a depth of 5 cm. Even if only 5% of the tracer is mixed into the overlying layer (an extremely conservative estimate based on our tracer study results), any evidence of the tracer will be removed from the plots before the horizon reaches a depth that will put it out of reach of the fiddler crabs (~20–25 cm). In essence, the fiddler crab mixing combined with the horizontal transport of sediment (wind, tides, etc.) will likely destroy any short-duration stratigraphic signal (e.g., seasonal reproductive pulses; small-to-moderate, *but not major*, washovers) in all subenvironments before it is preserved. Nevertheless, in high-marsh subenvironments (where mixing is minimal) the stratigraphic signal has a higher potential for preservation. The much greater thickness and coarseness of most Folly Island washovers probably increased the preservation potential of their signal by inhibiting mixing by fiddler crabs.

3.4. Foraminifera as Indicators of Sediment Supply

The premise that washover sediments are derived from offshore is supported by the similarity in foraminifera found in the cores and modern and Oligo–Miocene offshore species. Species in the paleowashover sediments from Folly Island and Miocene material dredged from Georgia tidal channels (Darby and Hoyt, 1964) and removed from cores from the North Carolina continental shelf (Snyder, 1988) include: *Cancris sagra* (d'Orbigny), *Hanzawaia concentrica* (Cushman), *Saracenaria senni* (Hedberg), *Siphogenerina transversa* (Cushman), *Stilostomella recta* (Palmer and Bermudez), and *Uvigerina calvertensis* (Cushman).

Theoretically, major storm events derive a larger supply of sediment from offshore and produce thicker washover intervals than do smaller storms. In all cores, presumed storm-layer (washover) thickness is highest in intervals of higher offshore percentages. The Spearman correlation between the two variables is relatively high ($\rho = 0.747$, $\alpha < 0.05$). The relationship between washover thickness and percent of offshore species is especially evident in the marsh-fringing core FBVC 05 (Fig. 3) and the paleo-back-barrier high-marsh core LIVC 09 (Fig. 4). Core FBVC 05 has fewer, but thicker, overwashes with offshore peaks approaching 30%, whereas LIVC 09 has more numerous, but thinner, washover intervals, and offshore peaks never exceed 10%.

4. Conclusions

1. The presence of Oligo–Miocene foraminifera in back-barrier cores from Folly Island indicates that much of the sediment for barrier-island formation is eroded from offshore. Sediment budgets for migrating barriers in this region must incorporate this potential sediment source.
2. Analysis of foraminiferal abundance peaks in back-barrier sediments demonstrates the potential overprint of bioturbation on the recognition of relatively thin vs. thick washovers (small vs. large-scale storms?). The four cores and the tracer plots sampled in this study reveal bioturbation rates consistent with depositional environment.
3. Documentation of the thickness and frequency of washovers within a sequence (sea-level)-stratigraphic framework would no doubt further reveal the long-term role of storms in barrier-island evolution and their potential effect on long-term anthropogenic coastal development in relation to coastal morphology. Such studies would also allow documentation of true thin washovers (small-scale storms?) vs. artifactual peaks produced by bioturbation.

ACKNOWLEDGMENTS: This study was funded by NSF Grant 9017864 and by a grant from the Cushman Foundation for Foraminiferal Research. All radiocarbon dates were supplied by John F. Wehmiller (Department of Geology, University of Delaware; processing by University of Arizona, Accelerator Mass

Spectrometer Facility, Tucson, Arizona). The cores used in this study were taken by David E. Krantz (U.S. Geological Survey, Dover, Delaware) and M. S. Harris (Department of Geology, University of Delaware). M. S. Harris also provided guidance concerning the stratigraphy and evolution of Folly Island. Thanks are due to Tim Patterson and Paul Gayes for reviewing the manuscript and supplying many constructive criticisms and helpful suggestions.

References

Basan, P. B., and Frey, R. W., 1977, Actual-palaeontology and neioichnology of salt marshes near Sapelo Island, Georgia. in: *Trace Fossils 2* (T. P. Crimes, Harper, P., and Harper, J. C, eds.), *Geol. J. Spec. Issue No. 9.*

Berger, W. H., and Heath, G. R., 1968, Vertical mixing in pelagic sediments, *J. Mar. Res.* **26**:134–143.

Church, T. M., Lord, C. J., and Somayajulu, B.L.K., 1981, Uranium, thorium and lead nuclides in a Delaware salt marsh sediment, *Est. Coast. Shelf Sci.* **13**:267–275.

Church, T. M., Biggs, R. B., and Sharma, P., 1987, The birth and death of salt marshes: Geochemical evidence for sediment accumulation and erosion, *EOS (Suppl.)* **68**:305.

Cutler, A. H., 1993, Mathematical models of temporal mixing in the fossil record, in: *Taphonomic Approaches to Time Resolution in Fossil Assemblages.* (S. M. Kidwell and A. K. Behrensmeyer, eds.), Paleontological Society Short Courses in Paleontology No. 6., Paleontological Society, Knoxville, Tennessee, pp. 169–187.

Darby, D. G., and Hoyt, J. H., 1964, An Upper Miocene fauna dredged from tidal channels of coastal Georgia, *J. Paleontol.* **38**(1):67–73.

Deery, J. R., and Howard, J. D., 1977, Origin and character of washover fans on the Georgia Coast, USA, *Trans. Gulf Coast Assoc. Geol. Soc.* **27**:261–270.

Frey, R. W., and Basan, P. D., 1981, Taphonomy of relict salt marsh deposits, Cabretta Island, Georgia, *Senckenb. Mar.* **13**:111–155.

Goldstein, S. T., and Harben, E. B., 1993, Taphofacies implications of infaunal foraminiferal assemblages in a Georgia salt marsh, Sapelo Island, *Micropaleontology* **39**(1):53–62.

Goldstein, S. T., Watkins, G. T., and Kuhn, R. M., 1995*a*. Microhabitats of salt marsh foraminifera: St. Catherines Island, Georgia, U.S.A. *Mar. Micropaleontol.* **26**:17–29.

Goldstein, S. T., Watkins, G. T., and Kuhn, R. M., 1995*b*. Microhabitats of salt marsh foraminifera: A case study from coastal Georgia, *Geol. Soc. Am. Prog. Abstr.* **27**:A29.

Guinasso, N. L., and Schink, D. R., 1975. Quantitative estimates of biological mixing rates in abyssal sediments, *J. Geophys. Res.* **80**:3032–3043.

Harris, M. S., Donovan-Early, P., Flocks, J., Gayes, P. T., Gielstra, N., Kindinger, J., Krantz, D. E., Nelson, D. D., York, L. L, 1995, Geologic framework influences on modern coastal morphology and processes along the South Carolina coast, *Geol. Soc. Am. Prog. Abstr.* **27**(6):78.

Harrison, E. Z., and Bloom, A. L., 1977, Sedimentation rates on tidal salt marshes in Connecticut. *J. Sed. Petrol.* **47**(4):1484–1490.

Hippensteel, S. P., 1995, Foraminifera as an indicator of overwash deposits and barrier island sediment supply, Folly Beach, South Carolina. M.S. Thesis, Department of Geology, University of Delaware, 128 pp.

Hippensteel, S. P., and Martin, R. E., 1995*a*. Foraminifera as an indicator of overwash deposits and barrier island sediment supply, Folly Beach, South Carolina, *SEPM Congr. Sed. Geol. Prog. Abstr.* **1**:69.

Hippensteel, S. P., and Martin, R. E., 1995*b*. Foraminifera as indicators of overwash deposits and barrier island sediment supply, Folly Beach, South Carolina, *Geol. Soc. Am. Prog. Abstr.* **27**(6):29.

Letzsch, W. S., and Frey, R. W., 1980. Deposition and erosion in a Holocene salt marsh, Sapelo Island, Georgia, *J. Sed. Petrol.* **50**(2):529–542.

Martin, R. E., 1993, Time and taphonomy: Actualistic evidence for time-averaging of benthic foraminiferal assemblages, In: *Taphonomic Approaches to Time Resolution in Fossil Assemblages,* (S. M. Kidwell and A. K. Behrensmeyer, eds.), Paleontological Society Short Courses in Paleontology No. 6., Paleontological Society, Knoxville, Tennessee, pp. 34–56.

Martin, R. E., 1998, *One Long Experiment: Scale and Process in Earth History*, Columbia University Press, New York, 262 pp.

Officer, C. B., and Lynch, D. R., 1983, Mixing, sedimentation rates and age dating for sediment cores. *Mar. Geol.* **46**:261–278.

Scott, D. B., and Medioli, F. S., 1986, Foraminifera as sea-level indicators, in: *Sea-Level Research: A Manual for the Collection and Evaluation of Data.* (O. van de Plassche, ed.), Geo Books, Norwich, England (IGCP Projects 61 and 200), pp. 435–455.

Sharma, P., Gardner, L. R., Moore, W. S., and Bollinger, M. S., 1987. Sedimentation and bioturbation in a salt marsh as revealed by ^{210}Pb,^{137}Cs, ^{7}Be studies, *Limnol. Oceanogr.* **32**(2):313–326.

Snyder, S. W., 1988, Micropaleontology of miocene sediments in the shallow subsurface of Onslow Bay, North Carolina Continental Shelf, *Cushman Found. Foram. Res. Spec. Publ. No. 25.*

Chapter 16

Benthic Foraminiferal Distributions in South Florida

Analogues to Historical Changes

SCOTT E. ISHMAN

1. Introduction

The South Florida ecosystem is complex and dynamic. The evolution of the ecosystem has been influenced by the influx of freshwater related to natural hydroperiods in the Everglades wetland, to hurricane events, and to sea-level rise, as well as to anthropogenic changes, such as alteration of the natural hydroperiod and changes in flow between Florida Bay and the Atlantic. Reduced fish and shellfish populations, altered seagrass densities and die-offs (Robblee *et al.*, 1991), and increased phytoplankton blooms show that the ecosystem has undergone significant change, the causes of which remain poorly understood (VanArman, 1984; Boesch *et al.*, 1993). While there have been detailed studies of aquatic animals and vegetation conditions, changes in the benthic community have not been as rigorously addressed.

Benthic foraminifera occur in a wide variety of marine and brackish environments (Murray, 1991), and their abundance and spatial and temporal

SCOTT E. ISHMAN • U.S. Geological Survey, 926A National Center, Reston, Virginia 20192. Present address: Department of Geology, Southern Illinois University, Carbondale, Illinois 62901.

Environmental Micropaleontology, Volume 15 of *Topics in Geobiology*, edited by Ronald E. Martin. Kluwer Academic/Plenum Publishers, New York, 2000.

distributions in marine sediments make them valuable paleoenvironmental indicators. Several studies of benthic foraminifera from Florida Bay sediments (Moore, 1957; Lynts, 1962; Phleger, 1966; Bock, 1971; Rose and Lidz, 1977; Lidz and Rose, 1989) showed abundant and diverse distributions related to a variety of environmental conditions. Analyses of downcore samples showed significant faunal changes, indicating fluctuations in ecosystem conditions (Brewster-Wingard *et al.*, 1998; Ishman *et al.*, 1998). The purpose of this chapter is to demonstrate the application of foraminiferal data to ecosystem reconstruction and the value of this method to land- and water-use managers (see also in this volume Ebrahim, Ch. 4; Hallock, Ch. 5; Eagar, Ch. 6).

2. Setting

Florida Bay (Fig. 1) represents a restricted platform interior environment that can be subdivided into a complex mosaic of subenvironments, including mudbank, open "lake" (Lidz and Rose, 1989), mangrove, and nearshore brackish. Oceanographic circulation and conditions (e.g., salinity, temperature, dissolved oxygen, and turbidity) within Florida Bay are controlled by its complex bathymetry (Prager *et al.*, 1996). Water depths in the eastern bay and Atlantic transition zone average 1.5 m and shallow to the northeast part of the bay. Circulation is restricted in the eastern part of the bay owing to its

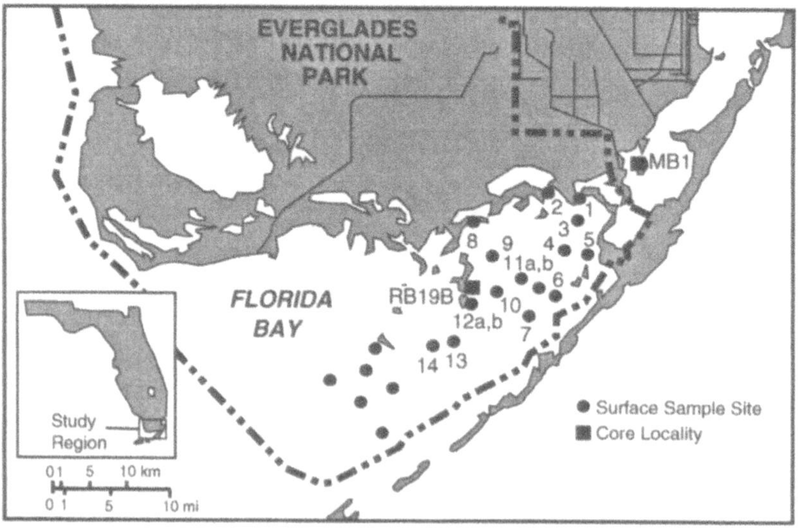

FIGURE 1. Map showing southern Florida study area, modern surficial sample localities (●), and core sites (■). RB19B refers to the core collected on Russell Bank and MB1 refers to the core collected in Manatee Bay.

enclosed nature, resulting in extremes in temperature and salinity (Boyer and Jones, 1996), with median temperatures between the years 1989 and 1995 at 28.6°C and median salinity at 30.6 parts per thousand (ppt) (Boyer and Jones, 1996). The western bay has an average depth of approximately 2.2 m and accommodates extensive exchange of water between the western bay and Gulf of Mexico water (Lee and Williams, 1996). This exchange results in higher median salinity (35.8 ppt) and lower median water temperature, 27.7°C (Boyer and Jones, 1996).

The complex paleobathymetry of Florida Bay is the result of erosion and deposition on the preexisting Pleistocene limestone topography. Ridges of *in-situ* peat and locally transported marl built during the Holocene (Wanless *et al.*, 1995; McPherson and Halley, 1996) are the primary features upon which the current mudbanks within Florida Bay are accumulating. Sedimentation on the mudbanks is largely controlled by sediment accumulation under baffling and trapping conditions, where seagrass is present and there is biogenic deposition of calcareous sands. Erosion on the north and east margins of the mudbanks and accumulation on the south and west sides result in a net western and southern migration of the mudbanks (Wanless *et al.*, 1995; Prager *et al.*, 1997).

3. Methodology

Surficial sediment samples were collected by hand using small-diameter (4-cm) push cores. The upper 10 cm from each push core was washed through nested 63-μm and 850-μm sieves to include infaunal species and account for time-averaging for comparison with downcore assemblages. Sediment greater than 63-μm and greater than 850-μm was dried at less than 50°C and picked for foraminifera. When possible, a total of 300 specimens were counted; otherwise the total sample was picked. The foraminifera were identified based on the classification of Loeblich and Tappan (1988), and species abundances were tabulated. The foraminiferal abundance data were standardized to relative abundance (percent) of the total sample assemblage (see Brewster-Wingard *et al.*, 1996). All quantitative analyses were conducted using the relative abundance data.

The modern foraminiferal data were quantified to determine the associations between faunas and specific environmental parameters using R- and Q-mode cluster analyses, respectively, and R-mode principal components analysis. When associations were identified, this information was used to evaluate downcore foraminiferal assemblages and interpret the paleoenvironmental conditions represented in the fossil assemblages.

Downcore samples were collected at 2-cm intervals and processed for foraminifera using the standard processing technique described for the surficial sediment samples.

4. Results

14.1. Benthic Foraminifera as Proxies for Salinity in Modern Sediments

A total of 16 surface sediment samples from Florida Bay (Fig. 1) were analyzed for benthic foraminifers and their abundances are reported in Brewster-Wingard *et al.* (1997). Preliminary results from these authors show a total of 29 frequently occurring species present in the ⩾63-μm size fraction. The benthic foraminiferal assemblages present are dominated by calcareous hyaline and miliolid forms. The dominant taxa include *Ammonia parkinsoniana, Elphidium galvestonense, E. delicatulum, Quinqueloculina bicostata, Q. bosciana, Q. poeyana,* and *Q. seminula.* Additional taxa present are *Miliolinella circularis, Rosalina floridensis, R. floridana,* and *Archaias angulatus.*

The R-mode cluster analysis of the benthic foraminiferal data identified three major clusters (Fig. 2a). Cluster 1 is composed of three species, *Ammotium* sp., *A. parkinsoniana,* and *E. galvestonense,* and represents the *Ammonia–Elphidium* assemblage. Cluster 2 includes *Triloculina* spp., *Archaias angulatus,* and *Trochammina conica,* all of which are more typical of the southeastern United States continental margin and reefs (Martin and Wright, 1988), and represents the *Archaias*–miliolid assemblage. Cluster 3 contains *Miliolinella circularis, Quinqueloculina* spp., and *Peneroplis proteus,* species associated with backreef or lagoonal environments with seagrass.

The Q-mode cluster analysis of the benthic foraminiferal data resulted in the definition of two predominant sample groupings (Fig. 2b). Cluster A contains samples from sites 4, 5, 7, 11 (July), 12 (July), and 13. The foraminiferal assemblages from these sites contain less than 10% *A. parkinsoniana* and more than 35% miliolids. Cluster B is composed of samples from sites 1, 2, 3, 6, 8, 9, 11 (February), and 12 (February). *A. parkinsoniana* and *E. galvestonense* comprise >50% of the assemblages from these samples.

R-mode principal components analysis of the data resulted in five principal components that explain 65.39% of the total variance in the data (Table 1). PC1 accounted for 18.64% of the variance. Benthic foraminiferal species with high PC1 scores ($> \pm 0.50$) include *A. parkinsoniana, E. galvestonense, M. circularis,* and *Quinqueloculina* spp. These taxa compose R-mode cluster 1. High PC2 scores were given to *Trochammina* spp., *Triloculina trigonula,* and *Quinqueloculina agglutinata.* Species with high PC3 scores were *Articulina mucronata, Sigmoilina antillarum, E. delicatulum,* and *Buccella hannai*; all are found in R-mode cluster 2. High PC4 scores were given to *Q. seminula, Bolivina transluscense, Nodobacularia cassis, Triloculina subrotunda,* and *A. angulatus. Rosalina floridensis, Trochammina* sp., and *Ammodiscus ensis* had high PC5 scores.

Trends observed in the Florida Bay benthic foraminiferal distributions indicate a strong association between foraminiferal distributions and salinity. *A parkinsoniana* and *E. galvestonense,* the primary taxa comprising cluster 1

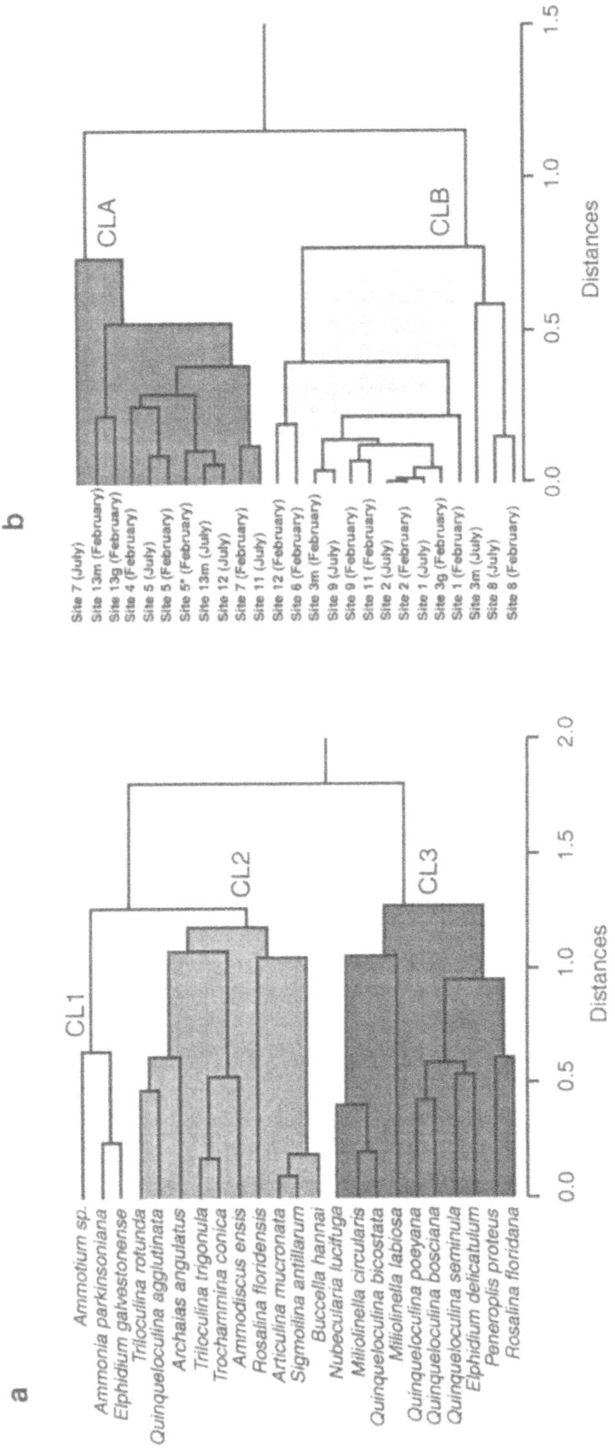

FIGURE 2. Dendrograms showing the results of R-mode (a) and Q-mode (b) cluster analyses.

TABLE 1. Results of R-Mode Principal Components Analysis: Varimax Rotation

	Principal components				
Species	1	2	3	4	5
Ammonia parkinsoniana	−0.897	0.225	−0.032	0.182	−0.115
Elphidium galvestonense	−0.864	−0.102	0.173	0.246	0.007
Miliolinella circularis	0.828	0.104	−0.007	0.199	0.134
Quinqueloculina bicostata	0.759	0.134	−0.102	0.421	0.225
Nubecularia lucifuga	0.709	−0.144	0.074	0.253	−0.052
Quinqueloculina bosciana	0.706	0.172	0.046	−0.307	−0.054
Quinqueloculina poeyana	0.685	0.005	0.21	−0.299	0.456
Elphidium delicatulum	0.598	0.19	−0.547	−0.329	0.071
Trochammina conica	−0.082	−0.941	−0.084	0.014	−0.047
Trochammina inflata	−0.082	−0.941	−0.084	0.014	−0.047
Triloculina trigonula	−0.08	−0.874	−0.003	−0.019	0.113
Quinqueloculina agglutinata	−0.263	−0.669	−0.043	0.34	0.102
Agglutinated sp.	0.055	−0.652	0.129	0.14	0.24
Ammodiscus ensis	0.258	−0.64	−0.038	0.07	−0.538
Articulina mucronata	0.019	0.084	−0.949	−0.209	0.054
Sigmoilina antillarum	−0.003	0.091	−0.929	0.002	−0.182
Buccella hannai	−0.038	−0.373	−0.899	0.019	−0.002
Quinqueloculina seminula	0.231	0.186	−0.036	−0.903	0.063
Bolivina transluscense	−0.123	0.046	0.052	−0.696	−0.354
Nodobaculariella cassis	0.05	0.007	−0.114	−0.624	0.09
Triloculina subrotunda	−0.039	−0.011	0.038	0.614	−0.055
Archaias angulatus	0.096	−0.427	0.123	0.503	0.148
Rosalina floridensis	−0.086	0.076	0.037	−0.082	−0.88
Trochammina sp.	−0.022	0.029	0.04	−0.046	−0.866
Miliolinella labiosa	0.06	−0.066	0.158	−0.082	0.414
R. floridana	0.437	0.204	0.112	−0.144	−0.287
Ammotium sp.	−0.482	0.232	0.012	0.128	0.062
Biloculina irregularis	0.057	0.027	0.231	−0.245	0.039
Peneroplis proteus	0.294	0.128	0.01	0.034	0.019
Percent of total variance explained	18.639	15.488	10.777	11.322	9.164

and having high negative PC1 scores, are common in brackish environments of the Gulf Coast region and Florida Bay (Phleger, 1951; Lynts, 1962; Poag, 1978), and represent the *Ammonia–Elphidium* foraminiferal association. The taxa *Q.* spp., *M. circularis, E. delicatulum*, and *A. angulatus* have high positive PC1 scores and occur in clusters 2 and 3. These species are associated more commonly with marine salinities in the Florida Bay region (Moore, 1957; Lynts, 1962; Poag, 1981) and represent the *Quinqueloculina–Miliolinella* and *Quinqueloculina–Archaias* foraminiferal associations. This indicates that clusters 1–3 and the PC1 grouping are controlled primarily by salinity. Linear regression analysis of species vs. salinity shows a strong correlation

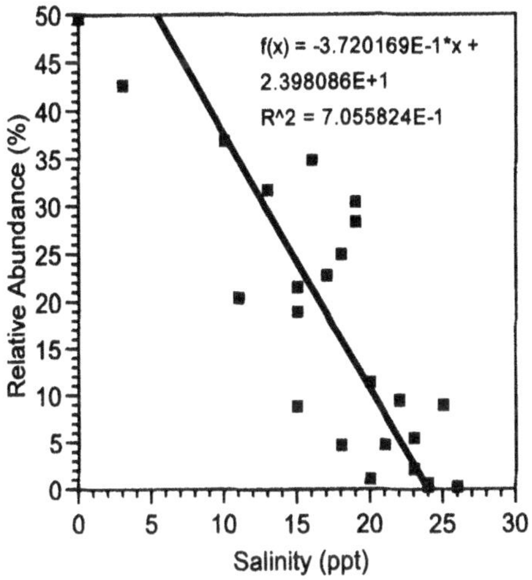

FIGURE 3. Plot of relative abundance of *Ammonia parkinsoniana* vs. salinity showing linear regression best fit.

($R^2 = 0.71$) between the relative abundance of *A. parkinsoniana* and salinity in modern Florida Bay (Fig. 3). *A. parkinsoniana* is most abundant (>20%) in Florida Bay where annual average salinity values are 18 ppt or less (oligohaline–mesohaline). A clear gradient in relative abundance of *A. parkinsoniana* exists with increasing salinity and is observed seasonally (Fig. 4). Also associated with low salinity in the Florida Bay and Gulf Coast regions is *E. galvestonense* (Lynts, 1962; Poag; 1981). At salinities greater than 18 ppt (polyhaline–euhaline), the relative abundance of *A parkinsoniana* drops significantly with *Quinqueloculina* spp. becoming the dominant taxon and *M. circularis, E. delicatulum,* and *A. angulatus* increasing in abundance. These results are consistent with observations of benthic foraminiferal distributions in the Gulf Coast where *A parkinsoniana* and *E. galvestonense* are associated with oligo- to polyhaline conditions and represent the *Ammonia–Elphidium* predominance facies (Poag, 1978, 1981). Lidz and Rose (1989) attributed the distribution of *A. beccarii ornata* (*A. parkinsoniana*) in Florida Bay to substrate and bay physiography. However, a clear trend in their data exists where the abundance of *A. beccarii ornata* decreases in regions where salinities are higher. In addition, the abundance of *Quinqueloculina* spp. and *A. angulatus* in higher-salinity environments is consistent with their distributions in the Gulf Coast region (Phleger, 1951; Poag, 1978, 1981), Florida Bay (Moore, 1957; Lynts, 1962, Phleger, 1966), and back-reef environments of the Florida Keys (Martin and Wright, 1988).

The other cluster and principal component groupings cannot be interpreted as clearly as cluster 1 and PC1. However, cluster 3 and PC3 contain the benthic foraminiferal species *Articulina mucronata, Buccella hannai,* and

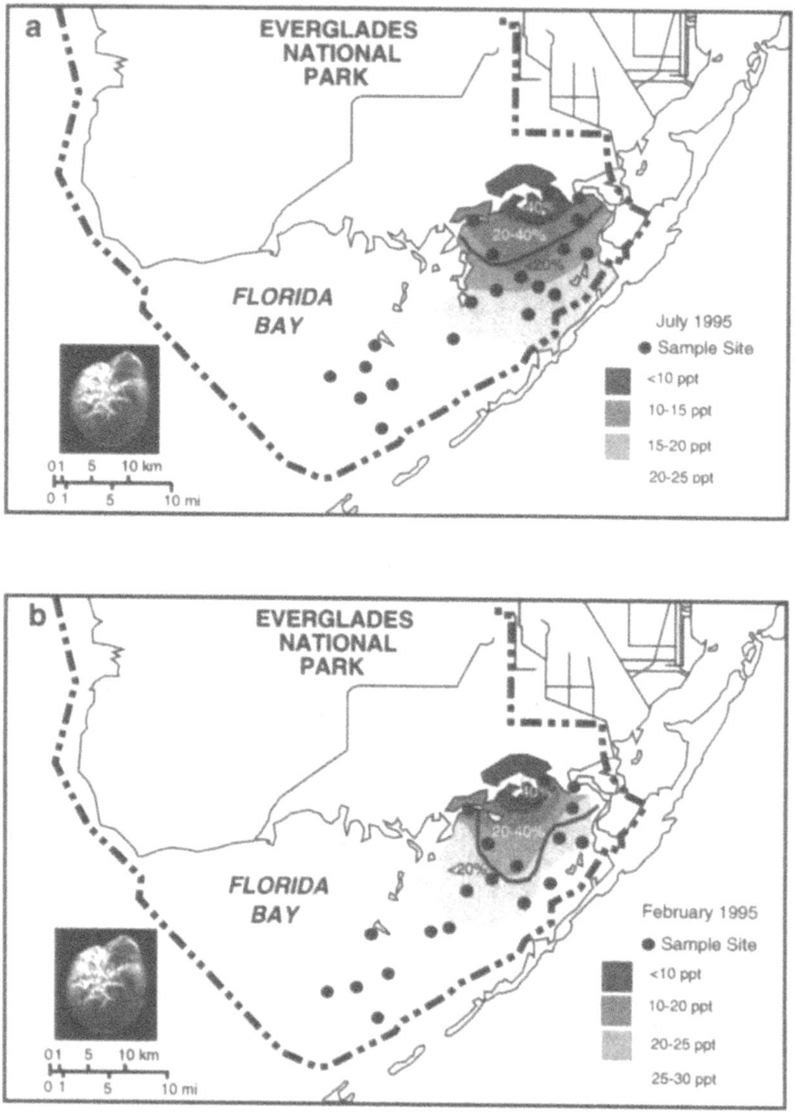

FIGURE 4. Relative abundance (%) distribution of *Ammonia parkinsoniana* in northeastern Florida Bay with respect to salinity in July (a) and February (b).

Sigmoilina antillarum, which are common to the southeastern United States continental shelf (Schnitker, 1971). These species occur primarily in the upper Keys region of the study area and are abundant in central Biscayne Bay (Bush, 1958; Brewster-Wingard *et al.*, 1996; Ishman *et al.*, 1997) and may indicate greater and more regular influx of Atlantic water into this region of Florida Bay.

4.2. Downcore Salinity Trends in Florida Bay

Results from two sediment cores, Russell Bank Core 19B (RB19B) and Manatee Bay Core 1 (MB1), show distinct faunal fluctuations, indicating salinity changes in the South Florida region. Core RB19B is 148 cm in length and was recovered from the central northeast region of Florida Bay (Fig. 1). Core MB1 is 120 cm in length and was recovered from Manatee Bay, which is located in Barnes Sound in the far southeastern part of Biscayne Bay (Fig. 1). Age control of the core material was determined by ^{210}Pb, ^{137}Cs, and the occurrence of exotic pollen in the cores. The ^{210}Pb/^{137}Cs record from core RB19B indicates an average accumulation rate of 1.22 cm/year (see Brewster-Wingard *et al.*, 1997, for details). If we assume a constant accumulation rate for the length of the core, the base of the core represents calendar year 1878. Age determination for core MB1 is based on the first occurrence of the exotic pollen *Casuarina*. The first notable occurrence of this taxon in the southern Florida region was documented in the early 1900s (Craighead, 1971), and it shows an accumulation rate of 0.80 cm/year for the upper 65 cm of the core. Linear extrapolation from the first occurrence of *Casuarina* to the base of the core yields an estimated age of AD 1848.

A total of 41 benthic foraminiferal species or species groups were identified from cores RB19B and MB1 (Brewster-Wingard *et al.*, 1997; Ishman *et al.*, 1998). The foraminiferal assemblages are dominated by calcareous benthic forms with the dominance patterns alternating between rotaliid forms (*Ammonia parkinsoniana, Elphidium galvestonense,* and *E. poeyanum*) and miliolids (*Miliolinella circularis, M. labiosa, Quinqueloculina bosciana, Q. seminula, Q. tenagos, Q. polygona, Q. poeyana,* and *Triloculina trigonula*). Other significant species include *Archaias angulatus, Peneroplis proteus, Rosalina floridana, Articulina mucronata,* and *Q. agglutinata.*

Three dominant foraminiferal associations are identified in the cores using principal components analyses: *Ammonia–Elphidium (A–E), Quinqueloculina–Miliolinella* (*Q*–M), and *Archaias–Articulina* (*A–A*) (Table 2). These associations are very similar to those described from modern Florida Bay and Biscayne Bay sediments and would appear to represent changes in salinity and seagrass conditions (Brewster-Wingard *et al.*, 1996; Ishman *et al.*, 1997). The fluctuations of these assemblages through time are much more apparent in core MB1, which is located in a more restricted environment than

TABLE 2. Benthic Foraminiferal Associations Observed in the Cores from Florida Bay with Their Significant Faunal Components

Ammonia–Elphidium (*A–E*)	*Quinqueloculina–Miliolinella* (*Q–m*)	*Archaias–Articulina* (*A–A*)
Ammonia parkinsoniana	*Quinqueloculina* spp.	*Archaias angulatus*
Elphidium galvestonense	*Miliolinella circularis*	*Articulina mucronata*
Elphidium poeyanum	*Miliolinella labiosa*	*Triloculina trigonula*
Elphidium gunteri	*Elphidium delicatulum*	*Rosalina floridensis*
Ammotium sp.	*Peneroplis proteus*	

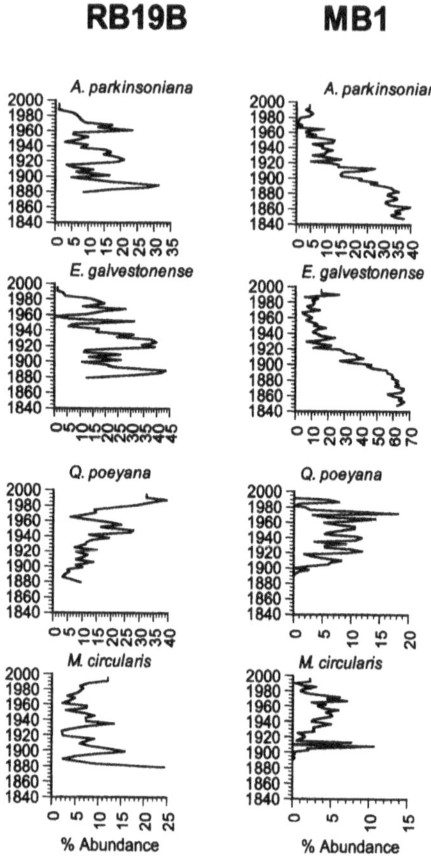

FIGURE 5. Downcore plots of the relative abundance of significant benthic foraminiferal taxa in cores RB19B and MB1.

RB19B (Fig. 1) and thus has a greater sensitivity to environmental fluctuations. At the MB1 site, assemblages in the mid- to late 1800s and earliest 1900s were dominated by the *Ammonia–Elphidium* association. The relationship of this assemblage with oligohaline to mesohaline conditions indicates salinities ranging from 5 to 18 ppt. A gradual salinity increase is shown throughout this interval by the progressive increase in the relative abundance of *E. galvestonense* and decrease in *A. parkinsoniana*. It also should be noted that epiphytic species, including the robust taxon *A. angulatus*, are rare or absent in the early part of the record from Florida Bay (Brewster-Wingard *et al.*, 1998; Ishman *et al.*, 1998), indicating seagrass abundance and densities lower than present day.

A significant faunal change occurs in the early 1900s with the introduction of the *Quinqueloculina–Miliolinella* assemblage. This assemblage indicates an increase in seagrass density, and polyhaline–euhaline salinity conditions ranging from 18 to 25 ppt. *A. parkinsoniana* becomes a much less dominant component of the benthic foraminiferal assemblage, and the relative abundance of *E. galvestonense* decreases along with a coincident increase in *Quinqueloculina* spp. and *M. circularis*. Other records from Florida Bay indicate significant changes in the early 1900s. Results from the stable isotopic analyses of a coral head from Lignumvitae Basin, Florida Bay, showed a positive shift in $\delta^{18}O$ occurring in the early 1900s (Swart *et al.*, 1996). Such a shift could indicate an increase in salinity, which is consistent with the faunal data from RB19B and MB1. Swart *et al.* (1996) also observed a shift in $\delta^{13}C$ and attributed this to carbon sequestration resulting from restricted circulation and reduced flushing of Florida Bay owing to the construction of the Flagler Railway. Similarly, freshwater delivery to Florida Bay as measured by fluorescence banding in a coral from Peterson Key Basin, Florida Bay, reflects a decrease in freshwater delivery to Florida Bay beginning in the early 1900s (Smith *et al.*, 1989).

The foraminiferal record shows that polyhaline-to-euhaline conditions persisted until about 1940. At that time the amplitude and frequency of foraminiferal faunal changes increased significantly. This also is reflected in other faunal groups (ostracodes and mollusks) in Florida Bay (Brewster-Wingard *et al.*, 1998; Ishman *et al.*, 1998), suggesting wide-ranging fluctuations in salinity conditions with the occurrence of periodic hypersaline conditions. Smith *et al.* (1989) cited a similar change in character in the fluorescence record from the Peterson Key Basin coral in which pre-1932, the fluorescence record showed a periodicity between 4 and 6 years, and post-1932, no periodicity could be identified. It is likely that these changes in the character of faunal patterns and fluorescence reflect the impact of increased water management practices that occurred around 1940, and the disruption of the natural hydroperiod of the Everglades and its delivery of freshwater to Florida Bay. The seagrass record from RB19B and MB1, as represented by the relative abundance of epiphytic taxa such as *M. circularis* (Brasier, 1975), indicates abundance and density similar to present-day conditions but also displays the same high frequency and amplitude fluctuations exhibited in the salinity record.

The results from micropaleontologic analyses of sediment cores from the southern Florida region provide critical data useful to both the data managers and the land- and water-use managers. The results show that conditions (salinity and seagrass) in Florida Bay were different before significant human alteration (pre-1900) of southern Florida. Several key events appear to have had profound effects on the Florida Bay ecosystem: the construction of the Flagler Railway in the early 1900s, which restricted tidal exchange with the back reef, and the water management practices (canal and levee systems) enacted in the mid-1900s through to the present, which altered the natural freshwater delivery to Florida Bay. The results provide land- and water-use managers with target conditions for their restoration goals and data to aid in the selection of critical sites to monitor the results of implemented restoration plans. Modelers can use this information to test their models through hindcasting and to predict the impact of specific restoration actions.

ACKNOWLEDGMENTS: I wish to thank G. Lynn Brewster-Wingard, Debra A. Willard, Peter-N Webb, Ronald E. Martin, and an anonymous reviewer for their helpful comments on this manuscript. This work was supported by the U.S. Geological Survey and the South Florida Water Management District.

References

Bock, W. D., 1971, *A Handbook of the Benthonic Foraminifera of Florida Bay and Adjacent Waters, Miami Geol. Soc. Mem.* (W. D. Bock, W. W. Hay, J. I. Jones, G. W. Lynts, S. L. Smith, and R. C. Wright, eds.), Miami Geological Society, Miami, 72 pp.

Boesch, D. F., Armstrong, N. E., D'Elia, C. F., Maynard, N. G., Paerl, H. W., and Williams, S. L., 1993, Deterioration of the Florida Bay ecosystem: An evaluation of the scientific evidence. Unpublished Report to the Interagency Working Group on Florida Bay (September 1993).

Boyer, J. N. and Jones, R. D., 1996, The Florida Bay water quality monitoring program: Assessing status and trends (1989–1995), Florida Bay Science Conference, University of Florida, Key Largo, pp. 19–21.

Brasier, M. D., 1975, Ecology of recent sediment-dwelling and phytal foraminifera from the lagoons of Barbuda, West Indies, *J. Foram. Res.* **5**:42–62.

Brewster-Wingard, G. L., Ishman, S. E., Edwards, L. E., and Willard, D. A., 1996, Preliminary report on the distribution of modern fauna and flora at selected sites in north-central and northeastern Florida Bay. U. S. Geol. Survey Open-File Rept. 96–732, p. 34.

Brewster-Wingard, G. L., Ishman, S. E., Willard, D. A., Edwards, L. E., and Holmes, C. W., 1997. Preliminary paleontologic report on cores 19A and 19B, from Russell Bank, Everglades National Park, Florida Bay, U. S. Geol. Survey Open-File Rept. 97–460, p. 29.

Brewster-Wingard, G. L., Ishman, S. E., and Holmes, C. W., 1998. Environmental impacts on the southern Florida coastal waters: A history of change in Florida Bay, *J. Coast. Res.* **26**:162–172.

Bush, J., 1958, The foraminifera and sediments of Biscayne Bay, Florida, and their ecology. Unpublished Ph. D. thesis, University of Washington, Seattle, Washington.

Craighead, F. C. J., 1971, *The Natural Environments and Their Succession: Trees of South Florida*, Vol. 1. University of Miami Press, Coral Gables.

Ishman, S. E., Graham, I. and D'Ambrosio, J., 1997, Modern benthic foraminifer distributions in Biscayne Bay: Analogs for historical reconstructions, U.S. Geol. Survey Open-File Rept. 97–34, p. 23.

Ishman, S. E., Cronin, T. M., Brewster-Wingard, G-L., Willard, D. A., and Verardo, D. J., 1998, A record of ecosystem change, Manatee Bay, Barnes Sound, Florida, *J. Coast. Res.* **26**:125–138.

Lee, T. N. and Williams, E., 1996, Flow within Florida Bay and interaction with surrounding waters, Florida Bay Conference, University of Florida, Key Largo, pp. 48–49.

Lidz, B. H. and Rose, P. R., 1989, Diagnostic foraminiferal assemblages of Florida Bay and adjacent shallow waters: A comparison, *Bull. Mar. Sci.* **44**(1):399–418.

Loeblich, A. R., Jr. and Tappan, H., 1988, *Foraminiferal Genera and Their Classification*, Van Nostrand-Reinhold, New York.

Lynts, G. W., 1962. Distribution of recent foraminifera in upper Florida Bay and associated Sounds, *Contr. Cushman Found. Foram. Res.* **13**:127–144

Martin, R. E. and Wright, R. C., 1988, Information loss in the transition from life to death assemblages of foraminifera in back reef environments, Key Largo, Florida, *J. Paleontol.* **62**:399–410.

McPherson, B. F. and Halley, R., 1996, The south Florida environment: A region under stress, U. S. Geol. Survey Circ. 1134, p. 61.

Moore, W. E., 1957, Ecology of recent foraminifera in northern Florida Keys, *Am. Assoc. Petrol. Geol. Bull.* **41**:727–741.

Murray, J. W., 1991, *Ecology and Palaeoecology of Benthic Foraminifera*, John Wiley & Sons, New York.

Phleger, F. P., 1951. Ecology of foraminifera, northwest Gulf of Mexico: Part 1. Foraminifera distribution, *Geol. Soc. Am. Bull.* **46**:1–88.

Phleger, F. P., 1966, Living foraminifera from coastal marsh, southwestern Florida, *Soc. Geol. Mexicana Bol.* **28**:1–44.

Poag, C. W., 1978. Paired foraminiferal ecophenotypes in Gulf coast estuaries: Ecological and paleoecological implications, *Gulf Coast Assoc. Geol. Socs. Trans.* **28**:395–421.

Poag, C. W., 1981, *Ecologic Atlas of Benthic Foraminifera of the Gulf of Mexico,* Hutchinson and Ross, Stroudsburg, Penn.

Prager, E. J., Halley, R. B., and Hansen, M., 1996. Sediment transport processes and sea-floor mapping in Florida Bay, Florida Bay Conference. University of Florida, Key Largo, pp. 68–69.

Prager, E. J., Halley, R. B., and Orton, P., 1997, Sediment resuspension, surface properties, and seagrass in Florida Bay, U.S. Geol. Survey Open-File Rept. 97–385, pp. 69–70.

Robblee, M. B., Barber, T. B., Jr. Carlson, P. R., Durako, M. J., Fourqurean, J. W., Muehstein, L. M., Porter, D., Yabro, L. A., Zieman, R. T., and Zieman, J. C., 1991, Mass mortality of the tropical seagrass, *Thalassia testudinum,* in Florida Bay (USA), *Mar. Ecol. Prog. Ser.* **71**:297–299.

Rose, P. R. and Lidz, B., 1977, Diagnostic foraminiferal assemblages of shallow-water modern environments: south Florida and the Bahamas, *Sedimenta.* **VI**, 55 pp.

Schnitker, D., 1971, Distribution of foramininfera on the North Carolina continental shelf, *Tulane Studies Geol. Paleontol.* **8**:169–215.

Smith, T. J., III, Hudson, J. H., Robblee, M. B., Powell, G. V. N. and Isdale, P. J., 1989, Freshwater Flow from the Everglades to Florida Bay: A Historical Reconstruction Based on Fluorescent Banding in the Coral *Solenastrea bournoni, Bull. Mar. Sci.* **44**(1):274–282.

Swart, P. K., Healy, G. F., Dodge, G. F., Kramer, P., Hudson, P., Halley, R. B., and Robblee, M. B., 1996, The stable oxygen and carbon isotopic record from a coral growing in Florida Bay: A 160 year record of climatic and anthropogenic influence, *Palaeogeogr. Palaeoclim. Palaeoecol.* **123**:219–237.

VanArman, J., 1984. South Florida's estuaries, in: *Environments of South Florida: Present and Past II* (P. J. Gleason, ed.), Miami Geological Society, Miami.

Wanless, H. R., Cottrell, D. J., Tagett, M. G., Tedesco, L. P., and Warzeski, E. R., Jr, 1995, Origin and growth of carbonate banks in south Florida, in: *Carbonate Mud-Mounds, Int. Assoc. Sedimentologists Spec. Publ.* (C. V. Monty, D. W. J. Boscence, P. H. Bridges, and B. R. Pratt, eds.), pp. 439–473.

Chapter 17

Variation in Natural vs. Anthropogenic Eutrophication of Shelf Areas in Front of Major Rivers

G. J. VAN DER ZWAAN

1. Introduction

Rivers are major transport elements in the global nutrient cycle: dissolved and particulate matter is brought from the continents to deltaic areas, where it is spread over the shelves and beyond. Basically, most of the transported material consists of erosion products, including all essential nutrients for life. Additionally, organic remains from terrestrial life are swept into the shallow marine realm. The total process forms the basis of the marine food web, and through marine primary production the pelagic, and later on the benthic, foodweb is fueled.

Variation in a number of processes leads to variation in this natural eutrophication of marine systems. On short timescales riverine discharge plays a major role. If discharge is high the quantity of dissolved and particulate nutrients, together with sediments, is large as well. Riverine discharge is

G. J. VAN DER ZWAAN • Institute of Earth Sciences, University of Utrecht, 3508 TA Utrecht, and Department of Ecology/Biogeology, Faculty of Biology, University of Nijmegen, 6525 ED Nijmegen, The Netherlands.

Environmental Micropaleontology, Volume 15 of *Topics in Geobiology*, edited by Ronald E. Martin. Kluwer Academic/Plenum Publishers, New York, 2000.

directly connected with the amount of runoff in the drainage basin. In turn, runoff is a function of vegetation; the more vegetation the lower the runoff. Variation in precipitation is another factor that is indirectly responsible for marine eutrophication, since precipitation steers runoff and discharge in rain-fed rivers. Possibly the most important factor in this context is the quality of the bedrock in the drainage basin; if "erodibility" is high, the quantity of erosion products and nutrients is high as well (Butcher *et al.*, 1991; Schlesinger, 1997).

Upon arrival at the land–sea transition, nutrients are stripped by primary producers from the outflowing riverine plume as soon as the light intensity is high enough. In meltwater rivers such as the Po River, this leads to high spring production in the surface waters. The resulting biomass settles from the surface layers down to the benthic environment, where it is consumed and remineralized.

In most shelf areas riverine input is essential for marine life; the influence of rivers can be followed over considerable distances (e.g., Poag, 1981; Van der Zwaan and Jorissen, 1991). Normally, even a high nutrient input and the resulting high organic production can easily be absorbed by a well-ventilated shelf environment. The organic mass arriving at the seafloor can lead to (regionally limited) anoxia, particularly below organic matter accumulations. If, however, oxygen supply is cut off, high organic loads lead to extremely rapid spreading of anoxia. In most modern-day situations, the combination of increased organic load and decreased oxygen supply is related to riverine input in spring and thermal stratification in summer, respectively (e.g., Tyson and Pearson, 1991).

Sometimes natural variation is profoundly disturbed by human action. The anthropogenically increased nutrient load that rivers transport to the land–sea transition has affected the shelf areas surrounding modern industrialized countries. In this chapter we discuss the effects of raised nutrient contents on continental shelves. The riverine effects of the post-Industrial Revolution time will be contrasted with the more natural variation in stress that occurred prior to AD 1840. We focus on benthic foraminifera because these unicellular organisms are widely present and sensitive indicators of stress. The main advantage of this group is that baseline studies are easy to perform because the good fossilization potential guarantees that the modern situation can always be compared with that of the past (see also in this volume Coccion, Ch. 3; Dixit and Smol, Ch. 12; Dale, Ch. 13; Alve, Ch. 14; and Ishman, Ch. 16).

2. Eutrophication of Shelf Ecosystems

Although anoxia is an important expression of anthropogenic eutrophication, natural situations can also be characterized by extensive anoxia. For instance, all land-locked or strongly silled basins are prone to stagnation (e.g., Tyson and Pearson, 1991). Fjords and larger basins such as the Black Sea are typical examples of this. Here stratification is caused by the riverine freshwater wedge. The pycnocline that originates owing to salinity stratification is very effective and cuts off the oxygen supply more rapidly than temperature stratification. The Atlantic Bight is an example of a region affected by naturally induced anoxia under thermal stratification in spring and summer (Tyson and Pearson, 1991).

Anthropogenic eutrophication of shelves is typically related to major rivers. The reason for this is obvious: the nutrient load gathered by a river is proportional to the size, the population density, and the degree of industrialization of the drainage basin. The sources of the nutrients are mainly related to agricultural and human waste. Phosphates are especially important in this regard (e.g., Malone, 1991; Justic, 1991). The primary production generated in the neighborhood of the riverine point source provides the organic load that settles in shallow water, where the combination of lowered salinity and raised temperature leads in summer to effective stratification cutting off oxygen supply. In deeper water (>200 m) the organic supply would be too reduced upon arrival at the seafloor to induce anoxia, owing to degradation on the way down (see, e.g., Suess, 1980, Betzer *et al.*, 1988).

Most European shelf areas are affected by anthropogenically raised biological production and subsequently increased oxygen demand at the bottom. The coastal areas of the United States, in particular the Gulf of Mexico, also show signs of artificial eutrophication (Boesch and Rabelais, 1991; Rabelais *et al.*, 1991; Sen Gupta *et al.*, 1996). Furthermore, there is ample evidence that anoxia did increase considerably over the past decades owing to higher primary production caused by artificial eutrophication (e.g., Justic, 1991, Rabelais *et al.*, 1991, Sen Gupta *et al.*, 1996). However, it has to be realized that not all shelf anoxia is necessarily induced by high primary production. For instance, Malone (1991) shows that in shallow waters seasonal stratification can be sufficient to generate anoxia if an adequate organic load is already present in the sediment.

In combination with nutrients, rivers load other compounds into shelf environments. Alve (1991, 1995), Bressler and Yanko (1995), Collins *et al.* (1995), Scott *et al.* (1995), and Schafer *et al.* (1995) all report on specific elements of anthropogenic pollution. It is clear that the effects of, for instance, heavy metal increase in shallow marine systems can be considerable. Such pollution differs from eutrophication because it is almost always limited to a narrow and well-defined zone in the freshwater and estuarine part of the system. Eutrophication, however, can be shown to affect large shelf areas tracing the river plumes over hundreds of kilometers (see the discussion in Van der Zwaan and Jorissen, 1991).

All marine benthic groups are affected by the consequences of eutrophication if anoxia occurs. Stachowitz (1991) showed that in the Adriatic Sea the benthic macrofauna was completely wiped out over large areas after summer anoxia. The effects of artificial eutrophication can also be shown to be present in northwestern Europe (Pearson and Rosenberg, 1978; Pearson, 1987; Moodley *et al.*, 1993, Alve and Murray, 1997). The consequences for the meiofaunal community are less well studied than those for macrofauna, partly because observations on this scale are difficult. In an early study Josefson and Widbom (1988) demonstrated that as anoxic conditions increased, larger invertebrates disappeared rapidly. The meiofauna, in particular foraminifera, survived relatively well. This seems likely, since oxygen consumption and metabolism are tuned to bodyweight; i.e., the smaller the body weight the less the oxygen consumption per unit (Vernberg and Vernberg, 1976; Hannah *et al.*, 1994). This would leave meiofauna and microfauna as the elements that would longest endure the prevailing anoxia.

Pearson and Rosenberg (1978) and Stachowitz (1991) documented the effects of the rising redox front after the beginning of anoxic events. Once thermal stratification started, anoxia proceeded extremely rapidly and within days infaunal macrobenthos was forced to the surface. With further rise of the redox front all macrobenthics disappeared. In a laboratory study (Moodley *et al.*, 1997) we showed that the sequence of fauna disappearance is indeed that macrofaunal elements are killed off first. Within 10 days of the onset of anoxia, ostracodes and other more advanced meiofaunal elements died. It should be noted that H_2S does not necessarily play a role at this stage. In our experiments (Moodley *et al.*, 1997, 1998), the fauna remaining after 10 days consisted almost entirely of nematodes and foraminifera. Eventually the nematodes also disappeared; only one nematode species remained. Soft-walled foraminifera died next, whereas the hard-shelled (calcareous and agglutinated) foraminifera survived the longest, in considerable numbers. Ongoing research shows that only the combination of anoxia and substantial H_2S production is lethal for all meiofaunal organisms, including foraminifera, within 6 days. This tends to confirm the earlier observations of Bernhard (1993).

Field observations indicate that these laboratory experiments provide a reliable picture of the effects of anoxia. The main difference is that the lab experiments suggest that many benthic foraminiferal species are able to endure anoxia, at least for the experimental period of 80 days. In contrast, field evidence shows that long-duration anoxia leads to very impoverished assemblages. We assumed that the eutrophied clay belt below the outflowing Po plume is sterile in summer owing to the upward rising redox front (Van der Zwaan and Jorissen, 1991). Indeed, total numbers there are low and many fossil anoxic deposits of the past (sapropels) testify to the fact that long duration of anoxia leads to extermination of the fauna (e.g., Van der Zwaan, 1982; Verhallen, 1989).

Oxygen profiles taken with needle probes show that in the Adriatic Sea the system is completely oxygen controlled (Fig. 1). This is even the case during the winter, the time of best ventilation owing to the absence of thermal stratification. Oxygen contents rapidly diminish in the first 0.5 cm, and further downward to less than 0.2 ml/liter within the first centimeter. As expected, in summer this is even worse: the oxygen contents decrease rapidly over the first couple of millimeters. This is somewhat different in the more sandy substrates to the north and east of the Po River delta. Here, oxygen penetration is

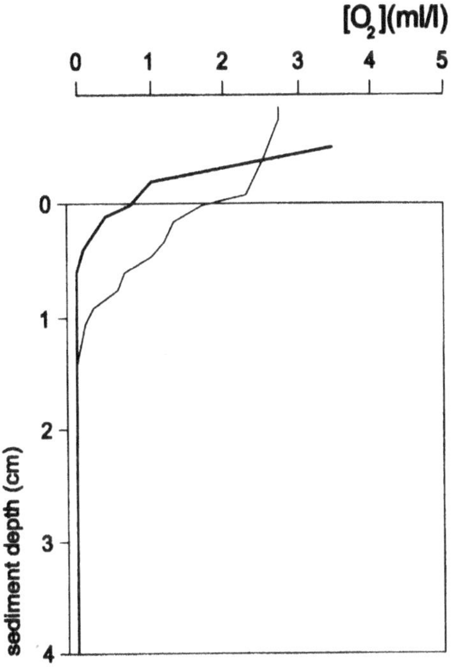

FIGURE 1. February and August 1997 oxygen concentrations in the top sediment layer at Station 108, 19 m depth, northern Adriatic Sea. February concentration (thin line) is higher and oxygen penetrates much deeper into the sediment column than in August (heavy line).

somewhat deeper although below 2 cm oxygen contents are almost nil. The foraminiferal microhabitat patterns found in the Adriatic Sea claybelt (Fig. 2) also suggest that the system is largely oxygen controlled. Only in the first centimeter do substantial numbers of benthic foraminifera occur and below that level numbers decrease rapidly. This indicates that foraminifera can survive in the dysoxic and even anaerobic zone, as is confirmed by the laboratory experiments. The standing stock, however, decreases rapidly. Whether this is caused by a reduction in the number of offspring or of reproduction frequency is not clear. Both mechanisms would lead to significantly lower standing stocks below the first centimeter level. It is even possible that below the first (aerated) centimeter level foraminifera forage and survive, and that reproduction only takes place in the fluffy, aerated top centimeter layer. To date, our information on this is too scanty.

Seasonal variation is large in river-influenced shelf systems. This is caused by a number of factors. In the case of the Po River, meltwater input is an important reason for significant increase of spring discharge. In other cases, seasonality in precipitation plays a role. Figure 3 shows the averaged seasonal pigment concentration in front of the Po (source: Ocean Pigment Project) over 1987–1993. It demonstrates that the primary production is clearly highest in the cooler winter months. This is surprising since one would expect that peaks in primary production are most pronounced during periods of highest river

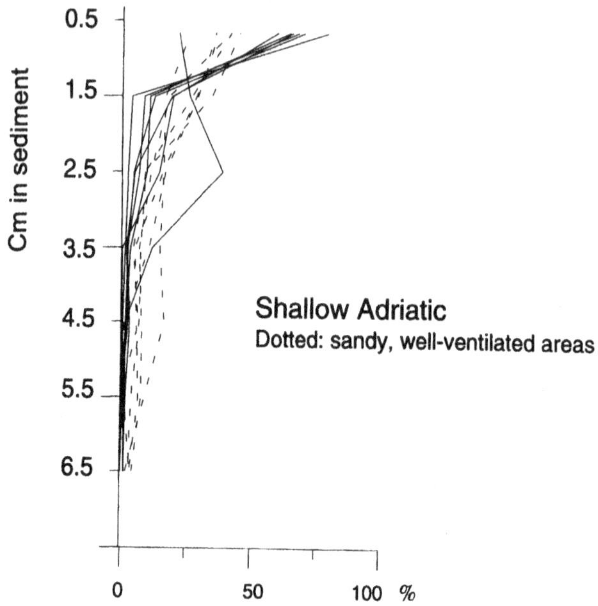

FIGURE 2. Benthic foraminiferal microhabitat patterns of 14 stations sampled in October 1991; total standing stock is taken as 100% and relative abundance is given per centimeter slice. Dotted curves are from stations in the sandy part of the northern Adriatic Sea, which was well-ventilated at the time of sampling (based on data of Jorissen *et al.*, 1992).

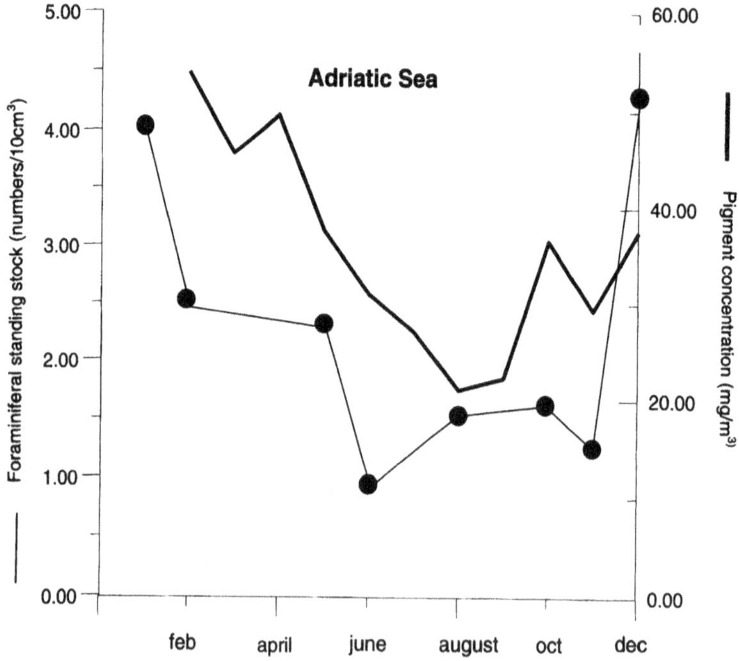

FIGURE 3. Seasonal variation in foraminiferal standing stock (data from Duijnstee *et al.*, 1997, and in preparation) and pigment data in station 108 (see Fig. 7 for location) [Pigment data from Coastal Zone Color Scan program (internet)].

discharge, i.e., in spring. It seems that the system reacts far more indirectly than expected at first sight. Apparently, increased nutrient input in spring is only partly absorbed by the marine ecosystem. This could be due to the fact that summer stratification prevents further proliferation of especially benthic fauna. Later, the accumulated nutrient load is used efficiently by flora and fauna during periods of ventilation in the cooler winter, when stratification is destroyed and the water-column is mixed. Remineralized nutrients are brought into the surface layer, which leads to increased pigment concentration. Benthic production, at least foraminiferal production, increases during these cooler months (Fig. 3), suggesting that meiofauna is limited more by oxygen and less by the abundant nutrients.

The periodic stagnation and ensuing oxygen deficiency lead to selection of specific faunal elements. Species that are able to survive the anoxic events (with an average duration of 2–4 months) or those that are able to remigrate rapidly from their refugia, dominate the assemblages. The specific foraminiferal composition of these systems, illustrated with data mostly from the Adriatic Sea, is addressed in the next sections.

3. Foraminiferal Assemblages from River-Dominated Shelf Systems

Earlier we summarized the foraminiferal biofacies patterns on shelves in front of a number of large river systems (Van der Zwaan and Jorissen, 1991). Only some of these rivers are polluted by man-raised nutrient contents. All have in common that the foraminiferal biofacies patterns display halos around the point source, where the river enters the shelf. Presumably, this reflects the influence of the river plume. This is demonstrated nicely by the patterns in front of the South American Orinoco River, which are dominated to a great extent by the riverine outflow (Fig. 4). The foraminiferal biofacies is marked over large distances by an exceptionally large input of organic matter that is mostly of natural origin. Apparently, oxygen consumption is extremely high, leading to conditions that are tolerated by relatively few species, in particular *Ammonia parkinsoniana, A. beccarii (=sarmentoi), Bolivina seminuda (=B. lowmanni),* and *Nonionella turgida.* Even at great distances from the point source biofacies patterns are still influenced. This suggests that the transport of nutrients and organic matter by currents has a considerable impact. Based on the statistical analysis that we presented (Van der Zwaan and Jorissen, 1991), one can infer a stress sequence: assemblages III, IV, and I go from presumably low-salinity environments to marine environments characterized by a high organic content. *A. parkinsoniana* and *A. beccarii* dominate the most estuarine part whereas *N. turgida* and *B. seminuda* dominate the organic-rich and presumably oxygen-deficient area in front of the rivers. *Uvigerina peregrina* is dominant in the less stressed marine environments still characterized by high organic content.

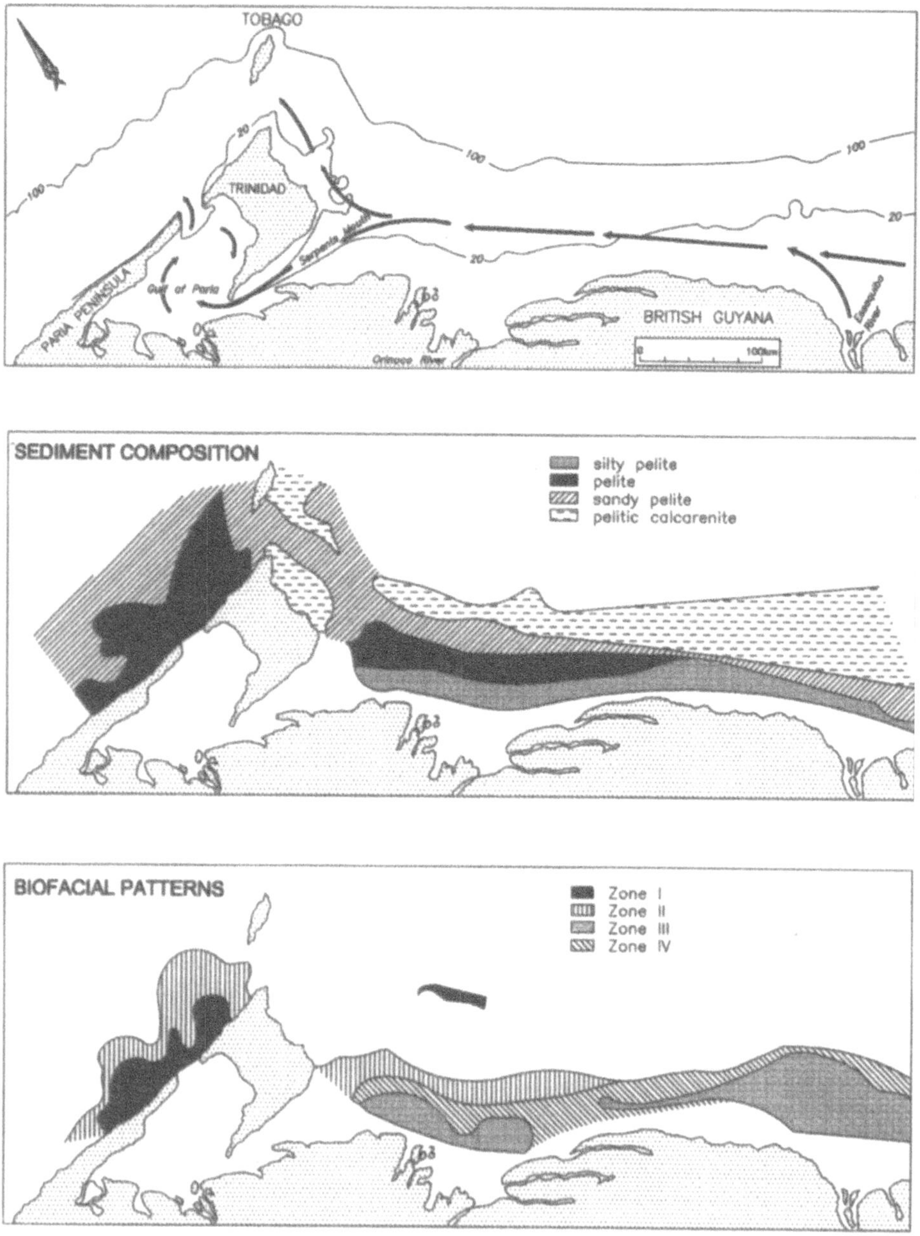

FIGURE 4. Circulation patterns, sediment composition, and biofacies patterns on the shelf bordering the Orinoco delta. For explanation see text (modified after Van der Zwaan and Jorissen, 1991).

Seiglie (1968, 1971) summarized the patterns in a large number of Caribbean sites, and found the genera *Buliminella, Nonionella, Fursenkoina* and *Florilus*, together with *Uvigerina* and *Bulimina* in deeper waters, to be dominant in organic-rich facies. Other areas with active sedimentation in front of major rivers also are characterized by relatively few species in the dead assemblage: *Bulimina, Bolivina, Fursenkoina* and *Buliminella* species in case of the Mississippi (Van der Zwaan and Jorissen, 1991). Lankford (1959, see also Parker, 1954, Phleger, 1954) found specific assemblages there, composed of *Buliminella tenuata, Nonionella turgida* (=opima), *Bolivina lowmanni,* and *Epistominella vitrea*, in the clay belt in front of the delta. The extraordinarily high abundances indicate a fertile environment.

The assemblages listed above are found on shelves where conditions are marine. In the more estuarine-dominated regions, or in the sometimes even brackish environments where sewage outfall or aquaculture dominates the patterns, the assemblages are different. Here arenaceous taxa (*Eggerella advena, E. scabra,* sometimes *Trochammina,* see Bandy *et al.*, 1964*a, b,* 1965, and compare to Collins *et al.*, 1995, Scott *et al.*, 1995, Schafer *et al.*, 1995) and *Elphidium* species (specifically *E. excavatum,* see review in Alve, 1995) occur. However, in most cases the exact reasons for the observed patterns are unclear: eutrophication, heavy metal pollution, or pollution from other sources. Alve (1995) summarized a large body of literature, and showed that in estuarine environments pollution with organic matter always coincides with raised benthic numbers. She also noted that the sites of maximum pollution are barren, but did not discuss the exact reasons for this barrenness.

Foraminiferal assemblages from the Adriatic Sea are relatively well studied. Previous studies indicate that the shelf environments in the northern Adriatic Sea are considerably affected by the presence of man (e.g., Faganelli *et al.*, 1991; Justic, 1991; Van der Zwaan and Jorissen, 1991; Barmawidjaja *et al.*, 1995). Jorissen (1987) found that the biofacies patterns in the Adriatic Sea are dominated to a great extent by the plume of the Po River. The area in front of the Po River and the clay belt that stretches southward from the point source are characterized by dead foraminiferal assemblages dominated by only three species: *Bulimina marginata, Valvulineria complanata* (=*bradyana*), and *Nonionella turgida* (Fig. 5). Similar dead assemblages are found in front of the Rhone delta (Kruit, 1955) and the Nile delta (unpublished personal observations). Earlier we concluded that organic matter accumulates especially in the belt of active clay sedimentation (Van der Zwaan and Jorissen, 1991). During summer stratification oxygen supply is low, whereas the oxygen demand remains high owing to the high organic load. We inferred that the consequent upward shift of the redox front would be tracked by various foraminiferal assemblages. The stress sequence would be *Ammonia* spp.>*Elphidium* spp.>*Textularia agglutinans*>*Nonionella turgida*>*Bulimina marginata*, the last being most tolerant to oxygen deficiency. The fully marine sequence would be somewhat different; i.e., *Nonion barleeanum*>*Cassidulina laevigata*>*Valvulineria complanata*>*Bulimina aculeata*, the latter being the most tolerant. This suggests that shelf patterns in front of rivers are influenced not

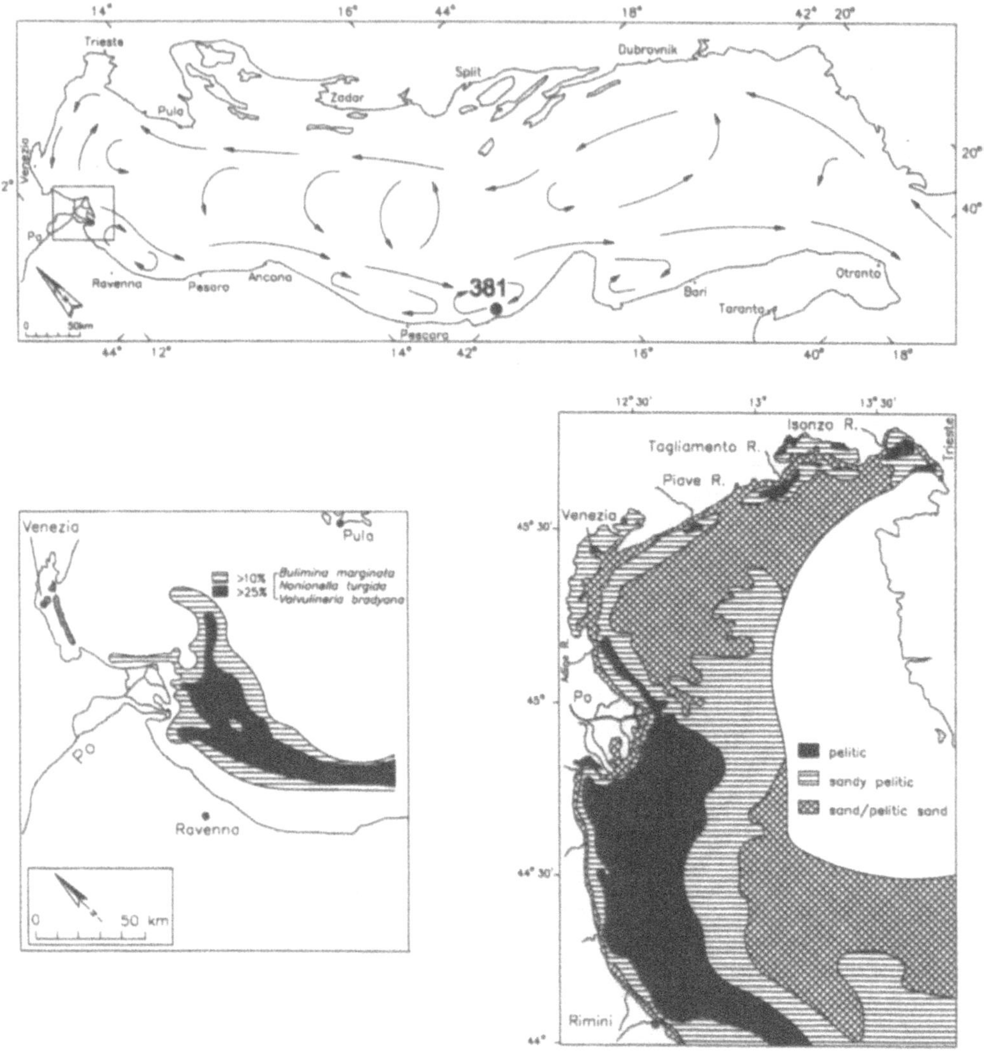

FIGURE 5. Circulation patterns, sediment distribution, and foraminiferal biofacies in front of the Po delta, Adriatic Sea. Square indicates study area as depicted in Fig. 7 (after Van der Zwaan and Jorissen, 1991).

only by oxygen changes due to the organic load, but also by salinity. Stress sequences in the estuary/delta clearly differ from those observed in the more marine localities.

The living assemblage is different from the dead one (see Hohenegger *et al.*, 1993; De Stigter *et al.*, 1998*a,b*) and appears to be seasonally variable (Barmawidjaja *et al.*, 1992; Duijnstee *et al.*, 1997; Jannink *et al.*, 1997). The differences between the living and dead assemblages can be partly ascribed to differences in the size fraction studied. Most studies of dead assemblages are in the 150-μm size fraction, whereas living assemblages are studied in the >63-μm size fraction. In the smaller size fractions, *Stainforthia fusiformis, Nonionella turgida, Hopkinsina pacifica,* and *Bolivina seminuda* are abundant. But in addition to a size fraction difference, the discrepancy between living and dead assemblages might be caused by taphonomic processes. For instance, small species such as *Nonionella turgida* and *Bolivina seminuda* are extremely fragile and thinly calcified. Preservation, especially in an acid, organic-rich, and anoxic facies, could be poor although we lack quantitative data in this respect.

Barmawidjaja *et al.* (1992) characterized a group of (living) calcareous species as potentially infaunal. The taxa (*Bolivina spathulata, B. dilatata, Bulimina marginata, Epistominella exigua,* and *Hopkinsina pacifica*) appear to adopt positions at varying depths in the sediment column throughout the year, depending on the ventilation. These species, in addition to *Nonionella turgida, Stainforthia fusiformis, Bolivina seminuda,* and *B. striatula,* are highly abundant in the sediment surface layers in front of the Po River. They are apparently tolerant to low oxygen since they proliferate in the highly fertile but presumably low-oxygen belt in October, just after seasonal stratification (Jorissen *et al.*, 1992).

4. Indicator Species of Organic Pollution in Shelf Areas

The impact of riverine input on shelves is keenly felt by marine organisms through changes in bottom and pore water oxygen contents. Moodley and Hess (1992), Alve and Bernhard (1995), and Moodley *et al.* (1997, 1998) showed experimentally that the gradual development of anaerobic conditions considerably affects the benthic foraminiferal assemblages. Alve and Bernhard (1995) concluded that *Stainforthia fusiformis* and *Nonionella* form subsurface maxima under aerobic conditions. These species migrated upward when oxygen contents diminished, apparently tracking oxygen gradients. *Bulimina marginata* also proved to be an oxygen-tolerant species. More detailed experiments (Boom and Rutten, 1997; Moodley *et al.*, 1997, 1998; Rutten and Boom, 1997) suggest that there is some differentiation among these species under gradually developing oxygen deficiency. *Nonionella turgida* appears to be the most resistant taxon, followed by others that are increasingly more affected by anaerobic conditions. Although it is difficult to establish it exactly, the experimental sequence from most to least tolerant seems to be: *N. turgida* > *Bolivina seminuda* > *Hopkinsina pacifica* > *S. fusiformis* > *B. marginata* (compare Moodley *et al.*, 1997). The pattern of *N. turgida* is representative in this respect (Fig. 6); even under rather prolonged anaerobic conditions the species seems to reproduce and increase in number.

Measurements in the field indicate that even outside of the clay belt below the river plume oxygen penetration in the sediment is not deep: in most cases

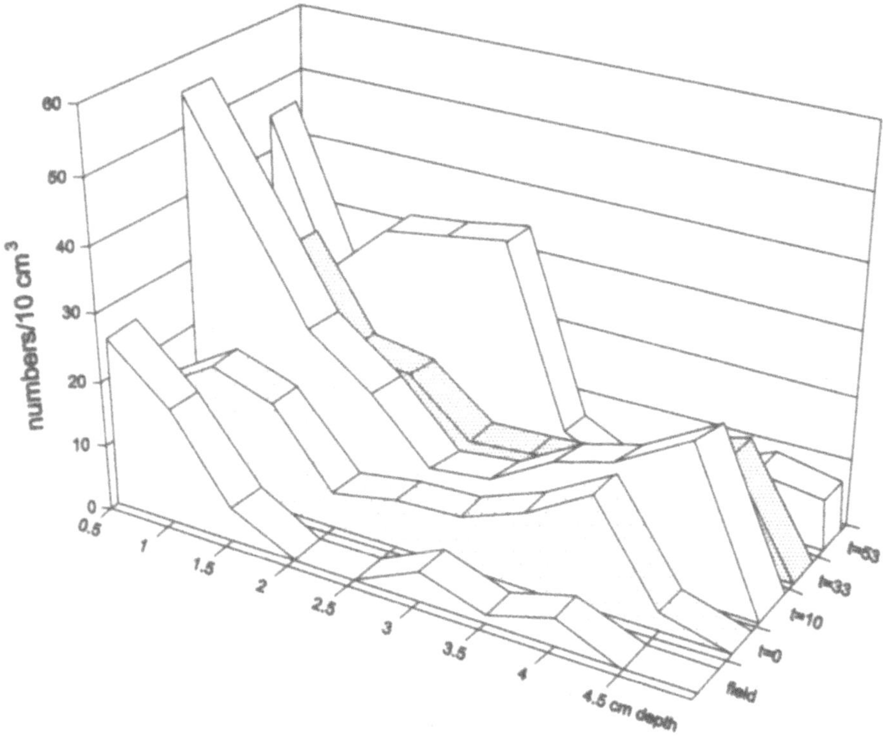

FIGURE 6. Microhabitat pattern of *Nonionella turgida* under experimental anoxic conditions. Anoxia started at day 0 (= first day) and patterns are shown at sampling moments on day 0, day 10, day 33, and day 53. Apparently *Nonionella* reproduced between day 0 and day 10. Total population flourished and tolerated anoxia well; microhabitat patterns show a shift to slightly shallower habitation, which may be due to the onset of sulfide-producing conditions (based on data from Boom and Rutten (1997), and Moodley *et al.* (1997).

the oxic–dysoxic front resides at depth less than 2 cm. Below this level species occur that are as tolerant of low oxygen conditions as the different group of species that proliferate under the river plume. Therefore, it seems likely that a factor other than oxygen alone induces the high abundances of the river plume species. It is logical to assume that food is such a factor since the river-plume is enriched in nutrients and organic matter. The nutrients lead to enhanced primary production, which, together with the organic matter derived from the continent, lead to an increase of the organic load in the sediments below the river plume. Benthic foraminifera feed on this organic matter and on the high-bacterial standing stocks remineralizing the organic mass. Clearly, taxa such as *Nonionella* and *Bolivina* must have an opportunistic edge over other, equally oxygen-tolerant, ones. This edge could involve at least two properties. Experiments show that species such as *Nonionella* and *Bolivina*, as well as *Bulimina*, reproduce rapidly. For instance, in the context of 56-day experiments *Nonionella* successfully increases its population by reproduction (see Fig. 6). Other opportunistic behavior might consist of active and rapid movement. Ernst *et al.* (1998) suggest that a species such as *Leptohalysis scottii* moves rapidly and is very active, which could be the basis of its successful competition against slow-moving, deeper-living species. Similarly, laboratory observations (unpublished results) show that *Nonionella* is very active even below the measured oxic–dysoxic interface.

5. Natural vs. Anthropogenic Signal

Foraminifera are among the few organisms that allow us to compare present and past in great detail. Owing to their abundance and high fossilization potential, benthic foraminifera are suitable for carrying out baseline studies to assess the impact of anthropogenic pollution on natural environments. Examples of these are given by Scott *et al.* (1995) and Barmawidjaja *et al.* (1995). Here we will enlarge the dataset discussed by Barmawidjaja *et al.* with data from two additional cores, jointly covering the period between the present and AD 1350, thus the past 640 years. In this period, western Europe, and particularly northern Italy, underwent considerable changes. Many of these were connected to the steady increase in the population and the concomitant increase in industrial and agricultural waste. Ciabatti (1966), Nelson (1970), and Gandolfi *et al.* (1982) summarize data which together give a clear impression of the changes that took place in the Po delta. In Fig. 7 the coastline is pictured from pre-AD 1500 until recent times. It can be seen that the main outlet shifted regularly, initially due to natural causes. In the period 1599–1604 Venetian rulers began to interfere with natural processes, mainly because they feared that the sediment load would silt up their lagoons and waterways. Owing to the construction of canals and dikes, in combination with delta progradation, the coastline changed periodically as did the position of the main point source.

The lithology and datings of the cores are described in detail elsewhere. Core 108, which is located close to core 101 in front of the Po delta (Fig. 7), was discussed by Barmawidjaja *et al.* (1995). Core 381 is at a great distance from these two cores, southeast of Pescara (see Fig. 5). The lithology and dating of core 101 has been discussed by Puscaric *et al.* (1990). The age constraints on the top part are very tight, using ^{137}Ce and ^{210}Pb. The downcore dating was done by extrapolating the average sedimentation rate as derived from the upper, well-dated, part of the core. This leads to an age of AD 1371 for the lowest sample and AD 1989 for the topmost sample. Obviously, if sedimentation rates varied considerably this casts some doubt on the accuracy of the datings of the lower part of the core. Until we have better datings at our disposal, this issue cannot be resolved. The age constraints on core 381 are provided by five ^{14}C-AMS datings, which are discussed by Jorissen (1988). On the basis of the average sedimentation rates, ages were derived for the samples. The oldest one studied (19–20 cm below the core top) has an age of AD 1535, the top can be regarded as equal to AD 1968.

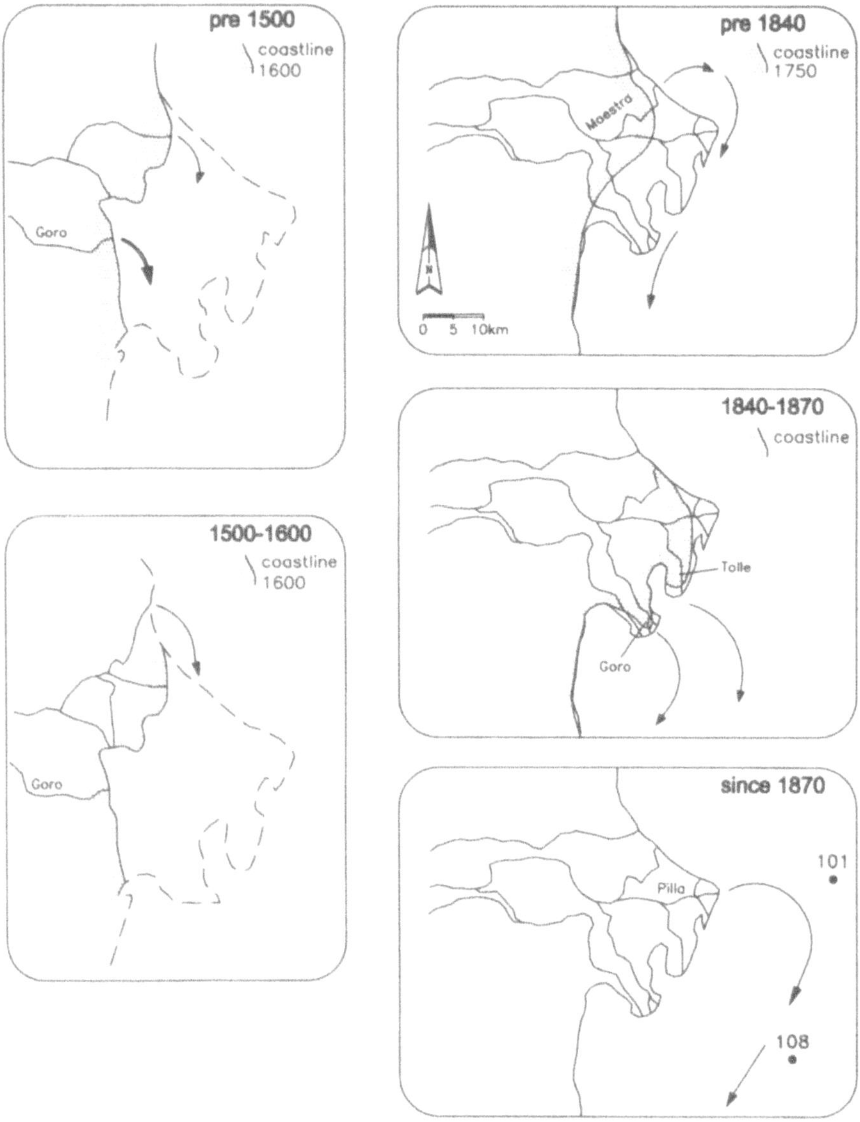

FIGURE 7. Evolution of coastal area of Po delta since medieval times. Location of stations 101 and 108 is indicated. Coastline pattern based on Ciabatti (1966), Nelson (1970), and Gandolfi *et al.*, (1982).

All cores were sampled at every centimeter, resulting in 19 samples for core 381 and 38 samples for core 101. Faunal data were discussed by Van Dijk (1995). The faunal patterns in 57 samples from core 108 were quantified; the results were given by Barmawidjaja *et al.* (1995).

All three cores, even the one far from the Po outlet, are characterized by decreasing quantities of epiphytes in the younger parts. *Elphidium* species (mostly small species but in particular *E. granosum* and *E. poeyanum*), *Neoconorbina terquemi, Buccella granulata, Asterigerinata adriatica,* and *Reussella spinulosa* (the last only in core 381) all become less abundant. The species that compensate for this decrease differ for the three cores. Immediately in front of the Po delta (core 108), *Epistominella exigua* and *Bulimina marginata* increase, followed later by *Hopkinsina pacifica* and *Bolivina seminuda*, and still later by *Stainforthia fusiformis*. In core 101, located more to the northern margin of the Po plume, *Pseudoeponides falsobeccarii* compensates for the decrease of the epiphytes. *B. marginata* also becomes more abundant but remains a minor constituent, just as *Nonionella turgida*. Even farther away from the point source (core 381), *B. marginata* and *Valvulineria complanata* become especially abundant. We interpret this overall pattern of increasing abundances of mud dwellers at the expense of epiphytes as having been caused by a significant decline in the shallow water vegetation. The marine vegetated biotope is very sensitive to pollution and is under great stress at many Mediterranean localities (Marba *et al.*, 1996; Short and Wyllie-Echevarria, 1996).

Figure 8 shows a dendrogram based on the total data of the three cores which gives an impression of the ecological grouping of the most abundant species. Species dominating the younger parts of the cores, all indicative of stress, cluster nicely together. To avoid detail, only the first or second

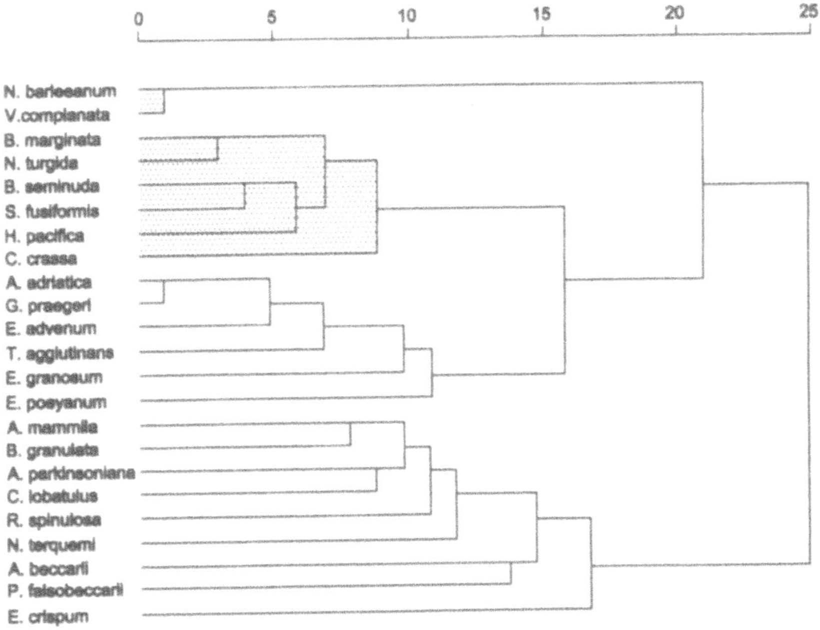

FIGURE 8. Dendrogram showing ecological grouping of the most abundant species in cores 101, 108, and 381. Analysis based on 114 samples. The group of opportunistic and stress-tolerant species, abundant under anoxic conditions, is indicated by shading.

principal component of the multivariate analyses of the three cores are displayed in Fig. 9. These axes, all dominated by *Bulimina marginata, Nonionella turgida,* or *Valvulineria complanata,* are taken to reflect the history of eutrophication. According to Barmawidjaja *et al.* (1995) the pattern in core 108 reflects major waterworks at about AD 1840 and AD 1870, resulting in increased sediment supply. From about AD 1880 an increasing eutrophication, which further increased at about AD 1930, can be observed. In Barmawidjaja *et al.* (1995) we concluded that these trends reflect the use of artificial fertilizer and an increasing amount of waste from the Po valley. The concomitant change in primary production and subsequent organic load in the clay belt below the Po plume caused increasing oxygen stress. The area affected by anoxia has enlarged over the past decades and is still growing (see Justic, 1991). The strong shift toward higher stress is also visible in core 101. Here it

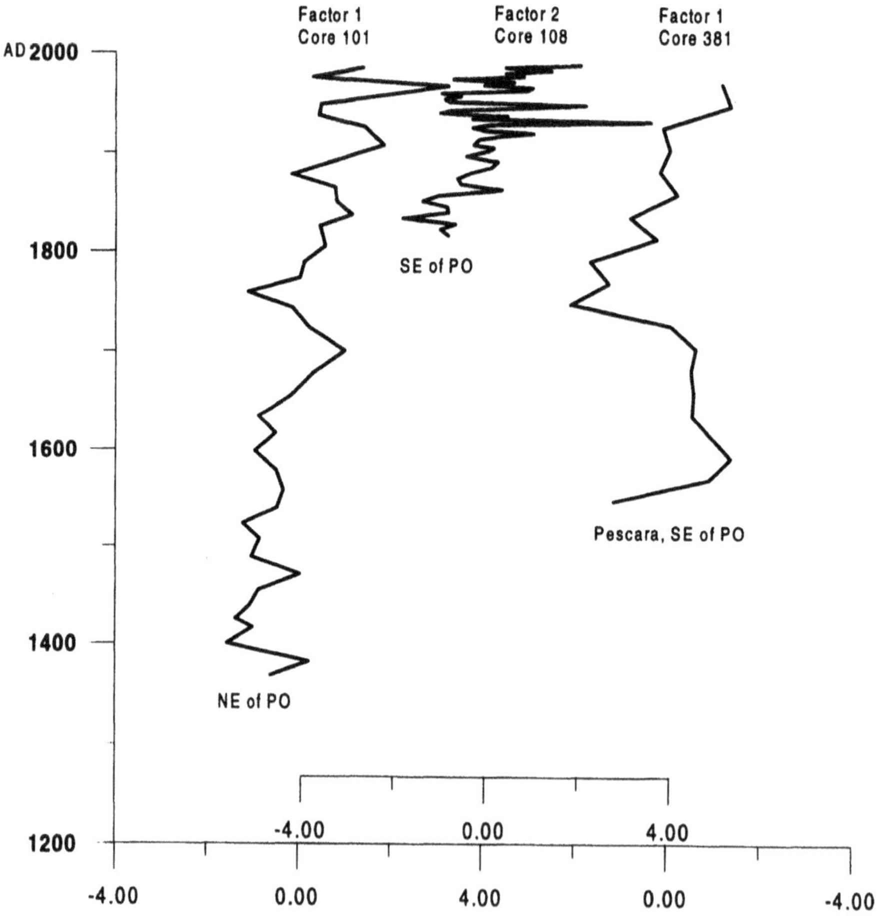

FIGURE 9. Plot of scores of first or second axes loaded by *Bulimina marginata, Valvulineria complanata,* and *Nonionella turgida* for cores 101, 108, and 381, respectively. Factors indicate degree of eutrophication-induced stress through time. For explanation see text.

becomes clear that the increase in stress and decrease in epiphytes/vegetation starts much earlier. The first signs of stress can be seen at the level of about AD 1600. In view of the historical record it is highly likely that this reflects the intensified deforestation and urbanization of the delta since that time.

Core 381, located far from the delta, displays a distinct variation through time. The multivariate analysis indicates a period of increased fertility and stress (from about AD 1600–1700, with a peak stress at about AD 1650), followed by a less stressed period with a minimum at about AD 1750. Although the youngest part of the core again seems to reflect the modern increase in stress, it seems unlikely that the older period of stress can be attributed to human action. To our knowledge there are no data to support such substantial human intervention so far away from the delta at about that time. It is much more likely that we are dealing here with the natural variability of the river-shelf system. Probably only the newest part (from about AD 1880, as the correlation to core 108 suggests) was affected by anthropogenic eutrophication. If so, it is striking that the oldest part of the pattern of core 381 seems very similar to the one displayed by isotope records on the Northern Hemisphere. These show a warm period around AD 1650, followed by the cool (second phase of the Little Ice Age) Maunder period around about AD 1750–1780 (compare Crowley and Norton, 1991). If true, it could imply that at this distance from the delta regional and more long-term variations are reflected by the faunal record. At this stage we are inclined to speculate that this is indeed the case, and that the pattern in core 381 is caused by extensive runoff during the warm intervals between AD 1600 and 1700 and from AD 1840 onward.

The examples discussed suggest that shelf systems are extremely sensitive to changes in riverine input, even without anthropogenic intervention. In systems where seasonal stratification occurs, shelves are readily affected by the organic load. The observed magnitude of the natural variability in core 381 is substantial and probably able to generate even anoxic events. In view of the fact that all three cores display a similar pattern of increase in stress from about AD 1880 onward, it seems likely that certainly from that time on human action adds substantially to variability. To get a really detailed insight into this matter much more data are needed than we have now at hand. However, this chapter does show that baseline studies using benthic foraminifera are worthwhile, since the effects of preanthropogenic levels of organic loads and industrial ones can be assessed. Such studies are likely to be successful since benthic foraminifera are extremely sensitive to the organic pollution itself as well as to the ensuing effects on the oxygen contents in bottom and pore waters.

ACKNOWLEDGMENTS: I thank Frans Jorissen for many discussions. Cindy Looy and Esther van Dijk performed part of the faunal analyses. Tom van Hinte made the drawings. Discussions with Maryke den Dulk, Ivo Duijnstee, Sander Ernst, and Natasja Jannink are acknowledged. The comments made by two reviewers and Tanja Kouwenhoven greatly improved the manuscript.

References

Alve, E., 1991. Benthic foraminifera in sediment cores reflecting heavy metal pollution in Sorfjord, western Norway, *J. Foram. Res.* **21**(1):1–19.

Alve, E., 1995, Benthic foraminiferal response to estuarine pollution: A review, *J. Foram. Res.* **25**:190–203.

Alve, E. B., and Bernhard, J. M., 1995. Vertical migratory response of benthic foraminifera to controlled oxygen concentrations in an experimental mesocosm, *Mar. Ecol. Prog. Ser.* **116**:137–151.

Alve, E. and Murray J. W., 1997. High benthic fertility and taphonomy of foraminifera: A case study of the Skagerrak, North Sea, *Mar. Micropaleontol.* **31**:3–4.

Bandy, O. L., Ingle, J. C., and Resig, J. M., 1964*a*, Foraminiferal trends, Laguna Beach outfall area, California, *Limnol. Oceanogr.* **9**:112–123.

Bandy, O. L., Ingle, J. C., and Resig, J. M., 1964*b*, Foraminifera, Los Angelos County outfall, California, *Limnol. Oceanogr.* **9**:124–137.

Bandy, O. L., Ingle, J. C., and Resig, J. M., 1965, Foraminiferal trends, Hyperion outfall, California, *Limnol Oceanogr.* **10**:314–332.

Barmawidjaja, D. J., Jorissen, F. J., Puskaric, S., and Van der Zwaan, G. J., 1992, Microhabitat selection by benthic foraminifera in the northern Adriatic Sea, *J. Foram. Res.* **22**(4):297–317.

Barmawidjaja, D. M., van der Zwaan, G. J., Jorissen, F. J., and Puskaric, S., 1995, 150 years of eutrophication in the northern Adriatic Sea: Evidence from a benthic foraminiferal record, *Mar. Geol.* **122**(4):367–384.

Bernhard, J. M., 1993, Experimental and field evidence of Antarctic foraminiferal tolerance to anoxia and hydrogen sulfide, *Mar. Micropaleontol.* **20**:3–4.

Betzer, P. R., Showers, W. J., Laws, E. A., Winn, C. D., DiTullio, G. R., and Kroopnick, P. M., 1988, Primary productivity and particle fluxes on a transect of the equator at 153 W in the Pacific Ocean, *Deep-Sea Res.* **31**:1–11.

Boesch, D. F., and Rabelais, N. N., 1991, Effects of hypoxia on continental shelf benthos: Comparisons between the New York Bight and the Northern Gulf of Mexico, in: *Modern and Ancient Continental Shelf Anoxia* (R. V. Tyson and T. H. Pearson (eds.), Geological Society, Spec. Publ. No. 58, pp. 27–34.

Boom, R. C. E., and Rutten, G. M. W., 1997. Laboratory experiments on the vertical migratory activity of benthic foraminifera in soft sediment. MSc Thesis, University of Nijmegen, 61 pp.

Bressler, V. and Yanko, V., 1995, Chemical ecology: A new approach to the study of living benthic epiphytic foraminifera, *J. Foram. Res.* **25**:267–279.

Butcher, G. S., Charlson, R. J., Orians, G., H., and Wolfe, G. V., 1991, *Global Biogeochemical Cycles*, Academic Press, New York.

Ciabatti, M., 1966, Richerche sull'evoluzione del delta padano, *Gir. Geol.* (ser. 2) **34**:1–26.

Collins, E. S., Scott, D. B., Gayes, P. T. and Medioli, F. S., 1995, Foraminifera in Minjah Bay and North Inlet Marshes, South Carolina: Relationship to local pollution sources, *J. Foram. Res.* **25**:212–223.

Crowley, T. J., and Norton, G. R., 1991, *Paleoclimatology*, Oxford Monographs on Geology and Geophysics, Vol. 16, 339 pp.

De Stigter, H. C., Jorissen, F. J., and Van der Zwaan, G. J., 1998*a*. Bathymetric distribution and microhabitat partitioning of live (Rose Bengal stained) benthic foraminifera along a shelf to deep sea transect in the southern Adriatic Sea, *J. Foram. Res.* **28**, 40–65.

De Stigter, H. C., Van der Zwaan, G. J., and Langone, L., 1998*b*, Differential rate of benthic foraminiferal test production in surface and subsurface sediment habitats in the southern Adriatic Sea, *Palaeogeogr. Palaeoclimatol. Palaeoecol.* **149**:67–68.

Duijnstee, I. A. P., Jannink, N. T., Ernst, S., and Van der Zwaan, G. J., 1997, Seasonality, trophic variability and microhabitat structure: Preliminary field data and experimental studies. *Proc. Xth Congr.* on *Protozoology* (Australia), p. 79.

Ernst, S. R., Duijnstee, I. A. P., Jannink, N. T., and Van der Zwaan, G. J., 1998, Experimental study on microhabitat preferences and mobility of benthic foraminifera, with special reference to *Leptohalysis scottii*, in *Proc. IVth Workshop Agglutinated Foraminifers* (in press).

Faganelli, J., Pezdic, J., Ogorelec, B., Herndl, G. J., and Dolenec, T., 1991, The role of sedimentary biogeochemistry in the formation of hypoxia in shallow coastal waters (Gulf of Trieste, northern Adriatic), in: *Modern and Ancient Continental Shelf Anoxia* (R. V. Tyson and T. H. Pearson (eds.), Geological Society, Spec. Publ. No. 58, pp. 107–118.

Gandolfi, G., Mordenti, A. and Paganelli, L., 1982, Composition and longshore dispersal of sands from the Po and Adige rivers since pre-Etruscan age, *J. Sed. Petrol.* **52**:797–805.

Hannah, F., Rogerson, A., and Laybourn-Parry, J., 1994, Respiration rates and biovolumes of common benthic foraminifera (Protozoa), *J. Mar. Biol. Assoc. UK* **74**:301–312.

Hohenegger, H., Piller, W. E., and Baal, C., 1993, Horizontal and vertical spatial microdistribution of foraminifers in the shallow subtidal Gulf of Trieste, northern Adriatic Sea, *J. Foram. Res.* **23**:79–101.

Jannink, N. T., Duijnstee, I. A. P., Almog-Labin, A., Herut, B., and Van der Zwaan, G. J., 1997, Seasonality, trophic variability and microhabitat structure: The reliability of benthic foraminifera as paleoecological tools, in *Proc. Xth Int. Congr. Protozoology* (*Australia*), p. 115.

Jorissen, F. J., 1987, The distribution of benthic foraminifera in the Adriatic Sea, *Mar. Micropaleontol.* **12**(1):21–48.

Jorissen, F. J., 1988, Benthic foraminifera from the Adriatic Sea: Principles of phenotypic variation, *Utrecht Micropaleontol. Bull.* **37**:1–174.

Jorissen, F. J., Barmawidjaja, D. M., Puskaric, S., and van der Zwaan, G. J., 1992, Vertical distribution of benthic foraminifera in the northern Adriatic Sea: The relation with the organic flux, *Mar. Micropaleontol.* **19**:1–2.

Josefson, A. B., and Widbom, B., 1988. Differential response of benthic macrofauna and meiofauna to hypoxia in the Gullmar Fjord basin, *Mar. Biol.* **100**:31–40.

Justic, D., 1991. Hypoxic conditions in the northern Adriatic Sea: Historical development and ecological significance, in *Modern and Ancient Continental Shelf Anoxia* (R. V. Tyson and T. H. Pearson (eds.), Geological Society, Spec. Publ. No. 58, pp. 95–106.

Kruit, C., 1955, Sediments of the Rhone delta: Grain size and macrofauna. Verhandelingen Koninklijk Mijnbouwkundig Genootschap, Geologische Serie, 15:357–514.

Lankford, R. R., 1959, Distribution and ecology of foraminifera from east Mississippi delta margin, *Bull. Am. Assoc. Petrol. Geol.* **43**:2068–2099.

Malone, T. C., 1991, River flow, phytoplankton production and oxygen depletion in Chesapeake Bay, in: *Modern and Ancient Continental Shelf Anoxia* (R. V. Tyson and T. H. Pearson (eds.), Geological Society, Spec. Publ. No. 58, pp. 83–94.

Marba, N, Duarte, C. M., Cebrian, M. E., Olsesen, B., and Sand-Jensen, K., 1996, Growth and population dynamics of *Poseidonia oceanica* on the Spanish Mediterranean coast: Elucidating seagrass decline, *Mar. Ecol. Prog. Ser.* **137**:203–213.

Moodley, L., and Hess, C., 1992, Tolerance of infaunal benthic foraminifera from low and high oxygen concentrations, *Biol. Bull.* **183**:94–98.

Moodley, L., Troelstra, S., and Van Weering, T., 1993, Benthic foraminiferal response to environmental change in the Skagerrak, northeastern North Sea, *Sarsia* **78**:129–139.

Moodley, L., Van der Zwaan, G. J., Herman, P. M. J., Kempers, A. J., and Van Bruegel, P., 1997, Differential response of benthic meiofauna to anoxia with special reference to Foraminifera (Protista: Sarcodina), *Mar. Ecol. Prog. Ser.* **158**:151–163.

Moodley, L., Van der Zwaan, G. J., Rutten, G. M. W., Boom, R. C. E., and Kempers, L., 1998, Subsurface activity of benthic foraminifera in relation to pore water oxygen content: Laboratory experiments, *Mar. Micropaleontol.* **34**, 91–106.

Nelson, B. W., 1970, Hydrography, sediment dispersal and recent historical development of the Po River Delta, Italy, *SEPM Spec. Publ.* **15**:152–184.

Parker, F. L., 1954, Distribution of the foraminifera in the northeastern Gulf of Mexico, *Bull. Mus. Compar. Zool.* **111**:453–588.

Pearson, T. H., 1987, Benthic ecology in an accumulated sludge disposal site, in: *Biological Processes and Waters in the Ocean* (J. M. Capuzzo and D. A. Kester (eds.), R. E. Krieger, Malabar, Fla. pp. 195–200.

Pearson, T. H., and Rosenberg, R., 1978, Macrobenthic succession in relation to organic enrichment an pollution of the marine environment, *Oceanogr. Mar. Biol. Ann. Rev.* **16**:229–311.

Phleger, F. B., 1954, Ecology of foraminifera, Northwest Gulf of Mexico: Part 1, *Mem. Geol. Soc. Am.* **46**:1–88.

Poag, C. W., 1981, *Ecologic Atlas of Benthic Foraminifera of the Gulf of Mexico*, Marine Science International, Woods Hole, Mass., 174 pp.

Puscaric, S., Berger, G. W., and Jorissen, F. J., 1990, Successive appearance of subfossil phytoplankton species of the northern Adriatic Sea and its relation to the increased eutrophication pressure, *Est. Coast. Shelf Res.* **31**:177–187.

Rabelais, N. N., Turner, R. E., Wiseman, W. I., and Boesch, D. F., 1991, A brief summary of hypoxia on the northern Gulf of Mexico continental shelf, in: *Modern and Ancient Continental Shelf Anoxia* (R. V. Tyson and T. H. Pearson (eds.), Geological Society, Spec. Publ. No. 58, pp. 107–118.

Rutten, G. M. W., and Boom, R. C. E., 1997. Experimental evidence of northern Adriatic Sea benthic foraminiferal tolerance to long-term anoxia. MSc thesis, University of Nijmegen, 60 pp.

Schafer, C. T., Winters, G. V., Scott, D. B., Pocklington, P., and Cole, F. E., 1995, Survey of living foraminifera and polychaete populations at some Canadian aquaculture sites: Potential for impact mapping and monitoring, *J. Foram. Res.* **25**:236–259.

Schlesinger, W. H., 1997, *Biogeochemistry*, Academic Press, New York, 443 pp.

Scott, D. B., Schafer, C. T., and Younger, D. C., 1995, Temporal variations of benthic foraminiferal assemblages under or near aquaculture operations: Documentation of impact history, *J. Foram. Res.* **25**:224–235.

Seiglie, G. A., 1968, Foraminiferal assemblages as indicators of high organic carbon content in the sediments and polluted waters, *Am. Assoc. Petrol. Geol. Bull.* **52**:2231–2241.

Seiglie, G. A., 1971. A preliminary note on the relationships between foraminifers and pollution in two Puerto Rican bays. *Carib. J. Sci.* **11**:93–98.

Sen Gupta, B. K., Turner, R. E., and Rabalais, N., 1996, Seasonal oxygen depletion in continental-shelf waters of Louisiana: Historical record of benthic foraminifers, *Geology* **24**(3):227–230.

Short, F. T., and Wyllie–Echevarria, S., 1996, Natural and human-induced disturbances of seagrasses, *Environ. Conserv.* **23**:17–27.

Stachowitz, M., 1991. Anoxia in the northern Adriatic Sea: Rapid death, slow recovery, in: *Modern and Ancient Continental Shelf Anoxia* (R. V. Tyson and T. H. Pearson, eds.), Geological Society, Spec. Publ. No. 58, pp. 119–130.

Suess, E., 1980. Particulate organic carbon flux in the oceans: Surface productivity and oxygen utilisation, *Nature* **288**:260–263.

Tyson, R. V., and Pearson, T. H., 1991. *Modern and Ancient Continental Shelf Anoxia*, Geological Society, Spec. Publ. No. 58, 470 pp.

Van der Zwaan, G. J., 1982, Paleoecology of Late Miocene foraminifera, *Utrecht Micropalentology Bulletin, Vol. 25*, 202 pp.

Van der Zwaan, G., and Jorissen, F., 1991, Biofacial patterns in river-induced shelf anoxia, in: *Modern and Ancient Continental Shelf Anoxia* (R. V. Tyson and T. H. Pearson, eds.), Geological Society Spec. Publ., No. 58, pp. 65–82.

Van Dijk, E., 1995, A 700-year record of anthropogenic influence in the northern Adriatic Sea. MSc thesis, Nijmegen University, 48 pp.

Verhallen, P. J. J. M., 1989, Late Pliocene to early Pleistocene Mediterranean mud-dwelling foraminifers: Influence of a changing marine environment on community structure and evolution, *Utrecht Micropaleontology Bulletin, Vol. 40*, 219 pp.

Vernberg, F. J., and Vernberg, W. B., 1976, *Functional Adaptation of Marine Organisms*, Academic Press, New York, 347 pp.

V

Aquifers and Engineering

Chapter 18

Establishing a Hydrostratigraphic Framework Using Palynology

An Example from the Savannah River Site, South Carolina

ROBERT S. VAN PELT, DONALD W. ENGELHARDT, RAYMOND A. CHRISTOPHER, and JOYCE LUCAS-CLARK

1. Introduction

The Savannah River Site (SRS) occupies 310 km^2 within Aiken, Barnwell, and Allendale counties in southwestern South Carolina, U.S.A. (Fig. 1). Bedrock (Paleozoic metamorphic and Triassic clastic rock) and overlying Coastal plain

ROBERT S. VAN PELT • Bechtel Savannah River Inc., Environmental Restoration Engineering & Technology, Savannah River Site, Aiken, South Carolina 29808. DONALD W. ENGELHARDT • Earth Sciences Resources Institute, University of South Carolina, Columbia, South Carolina 29801. RAYMOND A. CHRISTOPHER • Department of Geological Sciences, Clemson University, Clemson, South Carolina 29634. JOYCE LUCAS-CLARK • Clark Geological Services, Fremont, California 94536.

Environmental Micropaleontology, Volume 15 of *Topics in Geobiology*, edited by Ronald E. Martin. Kluwer Academic/Plenum Publishers, New York, 2000.

sediments (Upper Cretaceous through Holocene unconsolidated sediments) constitute the hydrologic system beneath the SRS and surrounding areas (Fig. 2). Direction of groundwater flow in the aquifers at SRS is primarily toward the Savannah River and its tributaries. Aquifers in the Cretaceous sediments are the primary source of drinking water for the SRS and surrounding communities.

The primary mission of the SRS is to produce nuclear materials for national defense. In the 1980s concern about environmental issues precipitated studies of the hydrogeology of the SRS. Since 1990, the mission has expanded to include environmental restoration following more than 40 years of waste-producing activities. Numerous waste sites at SRS are currently undergoing environmental assessments and/or remediation under state and federal regulations.

An essential part of the environmental assessments is the characterization of the subsurface hydrogeology. Hydrogeological characterization involves establishing the hydrologic and geologic conditions (including the hydrostratigraphy), and incorporating this information into groundwater flow and contaminant transport models. Results from the hydrogeology and groundwater models are then applied in the selection of remediation strategies for a

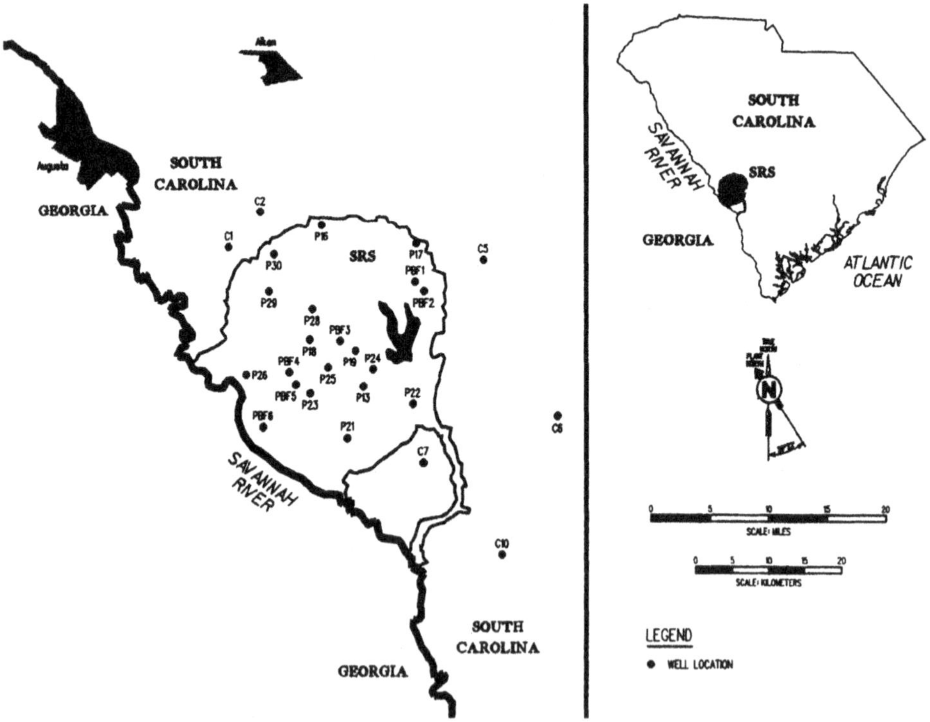

FIGURE 1. Location of the study area, Savannah River Site (SRS). Palynological data have been obtained for all or part of the wells shown on the map (modified from Fallaw and Price, 1995).

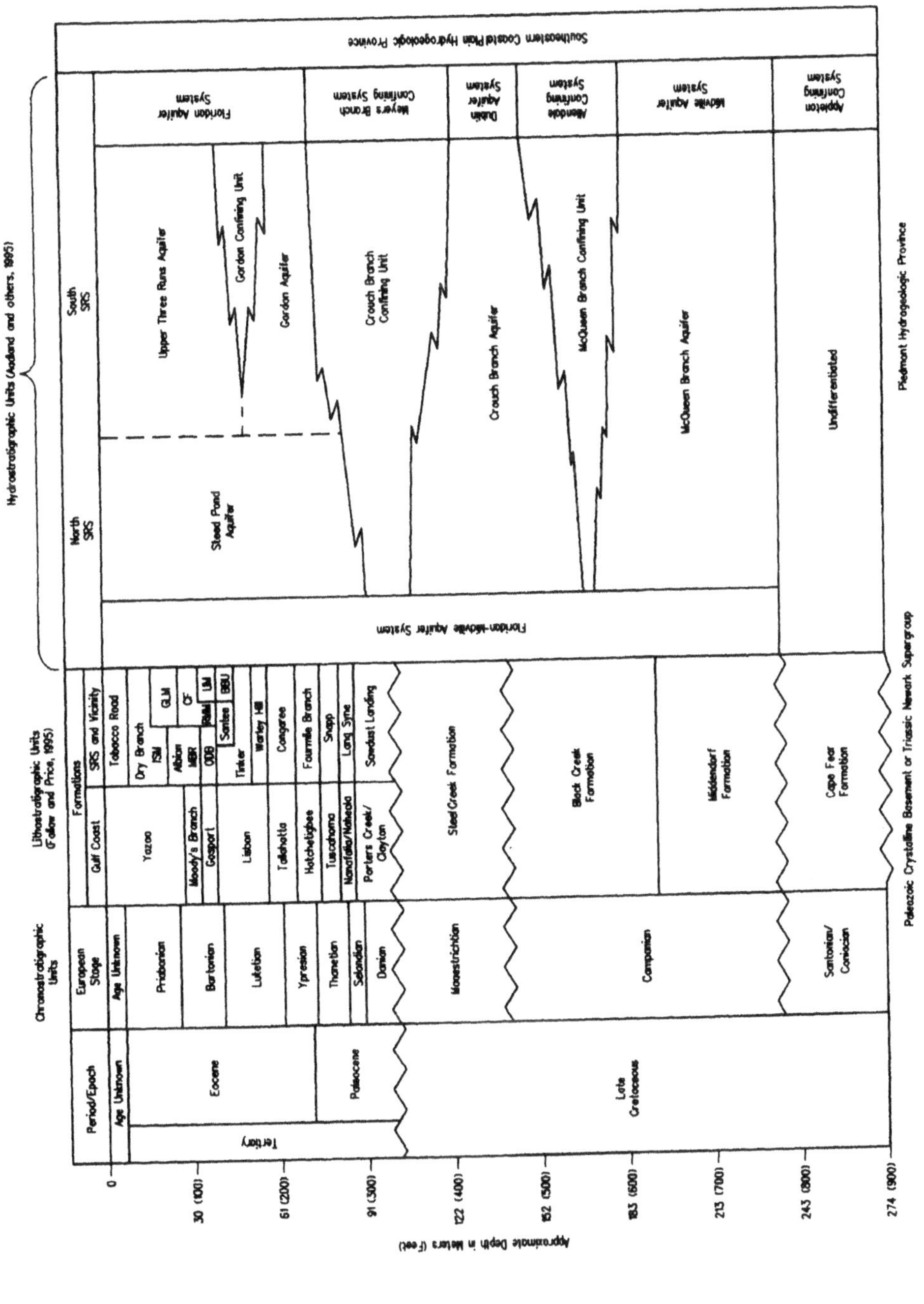

FIGURE 2. Comparison of chronostratigraphic, lithostratigraphic, and hydrostratigraphic units in the SRS region: IRM = Irwinton Sand member; CF = Clinchfield formation; RMM = Riggins Mill member; BBU = Blue Bluff unit; UM = Utley member; GLM = Griffins Landing member; ODB = Orangeburg district bed.

waste site. At the SRS, palynology has played a critical role in establishing the hydrogeologic framework, including the identification and characterization of hydrogeologic units and the relationships among them.

Differentiating among the sedimentary units that comprise the hydraulic system at SRS can be difficult when based solely on lithology. Most of the units consist of sediments of near-shore and deltaic origin, regardless of age, such that similar depositional origins mask lithostratigraphic distinctions. Many sedimentary units are lenticular, comprise several depositional facies, and transgress time. Palynology has been applied to the various sedimentary units for age determination and paleoenvironmental interpretation.

Palynology also plays an important role in the detection of groundwater preferential pathways, such as facies changes and structural elements (e.g., faults). Prediction of such pathways is essential for developing accurate models of groundwater contaminant flow and transport. The subsurface investigation at the SRS represents one of the frontier applications of palynology to hydrogeology and environmental restoration.

2. Previous Palynological Work at the Savannah River Site

As early as the middle of the 1970s, the need for accurate interpretation of the hydrostratigraphy beneath the SRS became apparent to those concerned with potential contamination problems (Cahil, 1982; Faye and Prowell, 1982; Marine and Root, 1975). Palynology has been used from the middle of the 1960s to the present as a basis for revising the formal stratigraphy (Siple 1967; Prowell *et al.*, 1985*a*, Gohn and Campbell, 1992; Nystrom *et al.*, 1991; Fallaw and Price, 1992, 1995). For example, the Sawdust Landing Formation (hydrostratigraphically equivalent to the middle part of the Crouch Branch Confining Unit in the northern part of the SRS; Fig. 2) was determined to be Paleocene rather than Cretaceous in age by dinoflagellate and pollen analysis (Lucas-Clark, 1992*a*). Subsequently, this confining unit was separated into three distinct unconformity-bound units based on palynological age interpretations (Aadland *et al.*, 1992). In general, age interpretations based on palynology allow for more accurate correlation with other stratigraphic units in the southeastern coastal plain, in what has come to be considered the Floridan Aquifer System. The revised lithostratigraphy and study of groundwater transport led to a complete formal hydrostratigrapic nomenclature for the SRS and surrounding vicinity (Aadland, 1990; Aadland and Bledsoe, 1992; Aadland *et al.*, 1992, 1995) (Fig. 2).

Studies in the middle of the 1980s focused on Tertiary carbonate units, which were considered as conduits for contaminants and were difficult to correlate owing to their discontinuous, lenticular, reeflike nature. Although palynomorphs are usually sparse in the carbonates themselves, they are abundant in the interbedded and encapsulating clays and mud. Determining the age and identifying reliable palynological data in these otherwise unfossiliferous fine-grained detrital units was determined to be more critical to the correlation of the carbonates than the fossils in the carbonates themselves.

More recent emphasis on environmental restoration has required compliance with state and federal regulations [e.g., Resource Conservation and Recovery Act (RCRA)/ Comprehensive Environmental Response Compensation and Liability Act (CERCLA)], which require a complete subsurface hydrogeologic/geologic characterization of waste sites, such as landfills, settling basins, and storage tank areas (Van Pelt *et al.*, 1993; Lucas-Clark and Van Pelt, 1993; Van Pelt and Lucas-Clark, 1996). In nearly all cases, palynology has been utilized to aid in identifying hydrostratigraphic units and improving and confirming well log correlation.

Certain hydrostratigraphic units, specifically confining units, have characteristic dinoflagellate and pollen assemblages that can provide age and paleoenvironment interpretations. The interpretations allow for accurate correlation as well as prediction of pinchouts and leaks in the confining units. Some confining units, such as the lower part of the Crouch Branch Confining Unit, were deposited under marine conditions and represent a major transgressive sequence. Other units, such as the late Paleocene upper part of the Crouch Branch Confining Unit, are characterized by dinoflagellates that are abundant in deltaic and possibly brackish water environments. Overall, shelfal dinoflagellate assemblages appear to dominate confining units that are more reliable as continuous barriers to hydrologic flow except where breached by unconformities; shallow water (e.g., estuarine, lagoonal) dinoflagellate assemblages generally comprise confining units that are not reliable barriers to vertical groundwater flow. This relationship is illustrated by the upper part of the Crouch Branch Confining Unit, which grades into coarser-grained nonmarine sediments on the southeastern side of the Savannah River and is absent on the river's northwest side. This conclusion was particularly significant to the Trans-River Flow study conducted in the middle of the 1990s by the United States Geological Survey and the Georgia State Geological Survey. The focus of the Trans-River Flow study was to determine the possibility of contaminated groundwater migrating from the SRS under the Savannah River and into the aquifers in Georgia (Clarke *et al.*,1994; Leeth and Nagle, 1996; Falls *et al.*, 1997; Huddlestun and Summerour, 1996; and Clarke and West, 1997).

3. Savannah River Site Hydrogeology and Palynological Stratigraphy

The hydrology beneath the SRS and surrounding area is controlled by the pre-Cretaceous Paleozoic metamorphic and Triassic clastic rocks and overlying Coastal Plain sediments. Poor water quality and low permeability prevent usage of ground water from within the pre-Cretaceous units. Overlying these units, however, the Coastal Plain sediments comprise a multilayer hydrologic system in which retarding clay and marl beds are interlayered with sands and limestone beds that transmit water more readily (Aadland and Bledsoe, 1992). In ascending stratigraphic order, the hydrologic system is divided into four aquifers (McQueen Branch, Crouch Branch, Gordon, and Upper Three Runs), separated from one another by intervening confining units (Fig. 3). The Gordon and Upper Three Runs Aquifers coalesce in the northern part of the area to form the Steed Pond Aquifer (Lewis and Aadland, 1992).

The relation between the palynology and the hydrogeology is presented below. Selected palynomorphs that are useful in hydrostratigraphic correlation are shown in Figs. 4 through 8. The hydrogeology is presented in ascending order.

3.1. Appleton Confining System

Separating the aquifers of the Coastal Plain sequence from basement is the Appleton Confining System, which consists of saprolite overlying crystalline basement rock and the sedimentary unit referred to as the Cape Fear Formation. The Cape Fear Formation is mainly nonmarine and dated by pollen and spores as middle Turonian to Santonian (Late Cretaceous) in age (Lucas-Clark, 1992*a*). The Appleton Confining System/Cape Fear Formation is characterized by the presence of a distinctive pollen type that has been referred to as aff. *Porocolpopollenites* spp. by Doyle and Robbins (1977) (Fig. 6a). In addition, the unit contains diverse and abundant representatives of the genus *Complexiopollis*, including *C. abditus* Tschudy 1973 (Fig. 6b), as well as other described and undescribed forms. All species of *Complexiopollis* that occur in the unit are of the "newer" variety as characterized by Christopher (1979). Representatives of the genus *Pseudoplicapollis* (Fig. 6c) also occur throughout the unit, but they are not so abundant as are the *Complexiopollis* forms. Other genera that are present in assemblages from the Appleton Confining System/Cape Fear Formation are *Praecursipollis* (Fig. 6d), *Osculapollis* (Fig. 6e), *Trudopollis* (Fig. 6f), *Plicapollis* (Fig. 6g), *Labrapollis* (Fig. 6h), and numerous undescribed forms.

3.2. McQueen Branch Aquifer

The McQueen Branch Aquifer overlies the Appleton Confining System and is the lower of two Cretaceous aquifers at SRS. This aquifer represents the principal source for domestic water use at SRS. Domestic wells in these sands commonly yield more than 1000 gal/min of high-quality water.

The McQueen Branch Aquifer corresponds to the sands that are referred to as the Middendorf Formation as well as these in the lower part of the Black Creek Formation at SRS. The Middendorf Formation is considered nonmarine

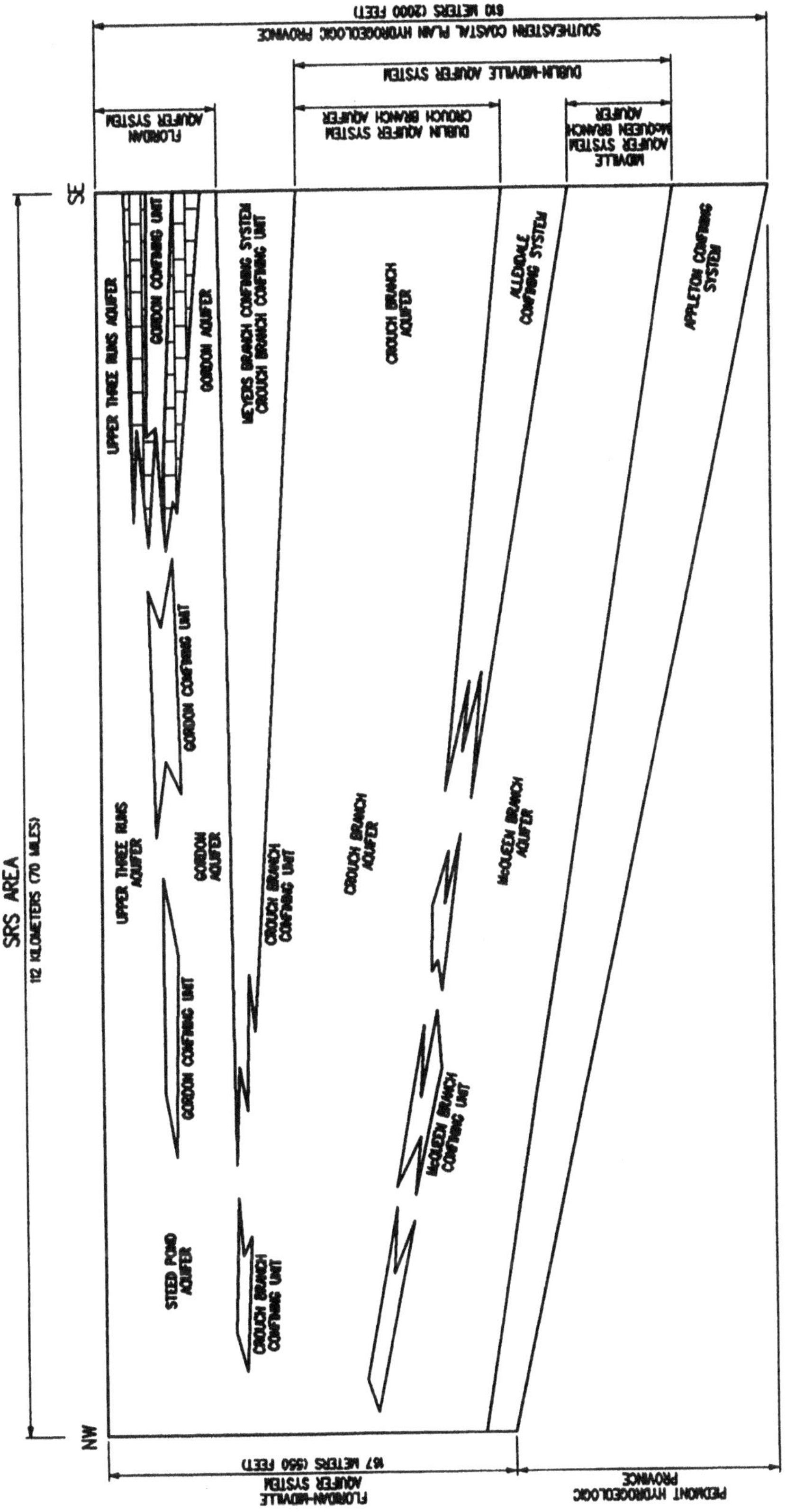

FIGURE 3. Cross section of the Upper Cretaceous and Tertiary hydrostratigraphic sequence beneath the SRS (modified from Aadland *et al.*, 1995).

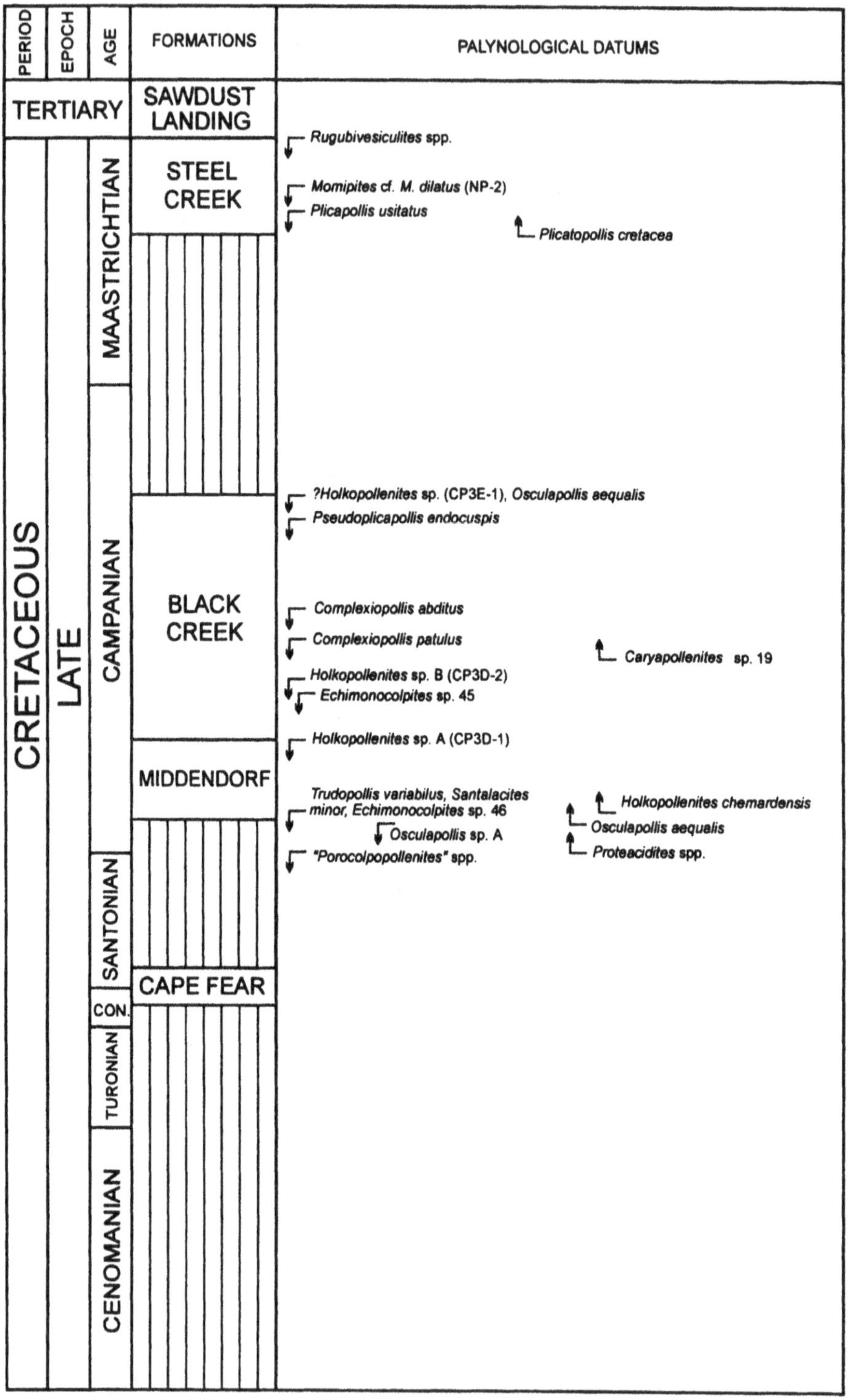

FIGURE 4. Selected Cretaceous palynomorph datums. The relationship of formational boundaries to stage boundaries is modified following Berggren *et al.*, 1996.

FIGURE 5A. Selected Tertiary palynomorph (first appearance) datums. The relationship of formational boundaries to stage boundaries is modified following Berggren *et al.*, 1996: IRM = Irwinton Sand Member; CF = Clinchfield Formation; RMM = Riggins Mill Member; BBU = Blue Bluff Unit; UM = Utley Member; GLM = Griffins Landing Member; ODB = Orangeburg District Bed.

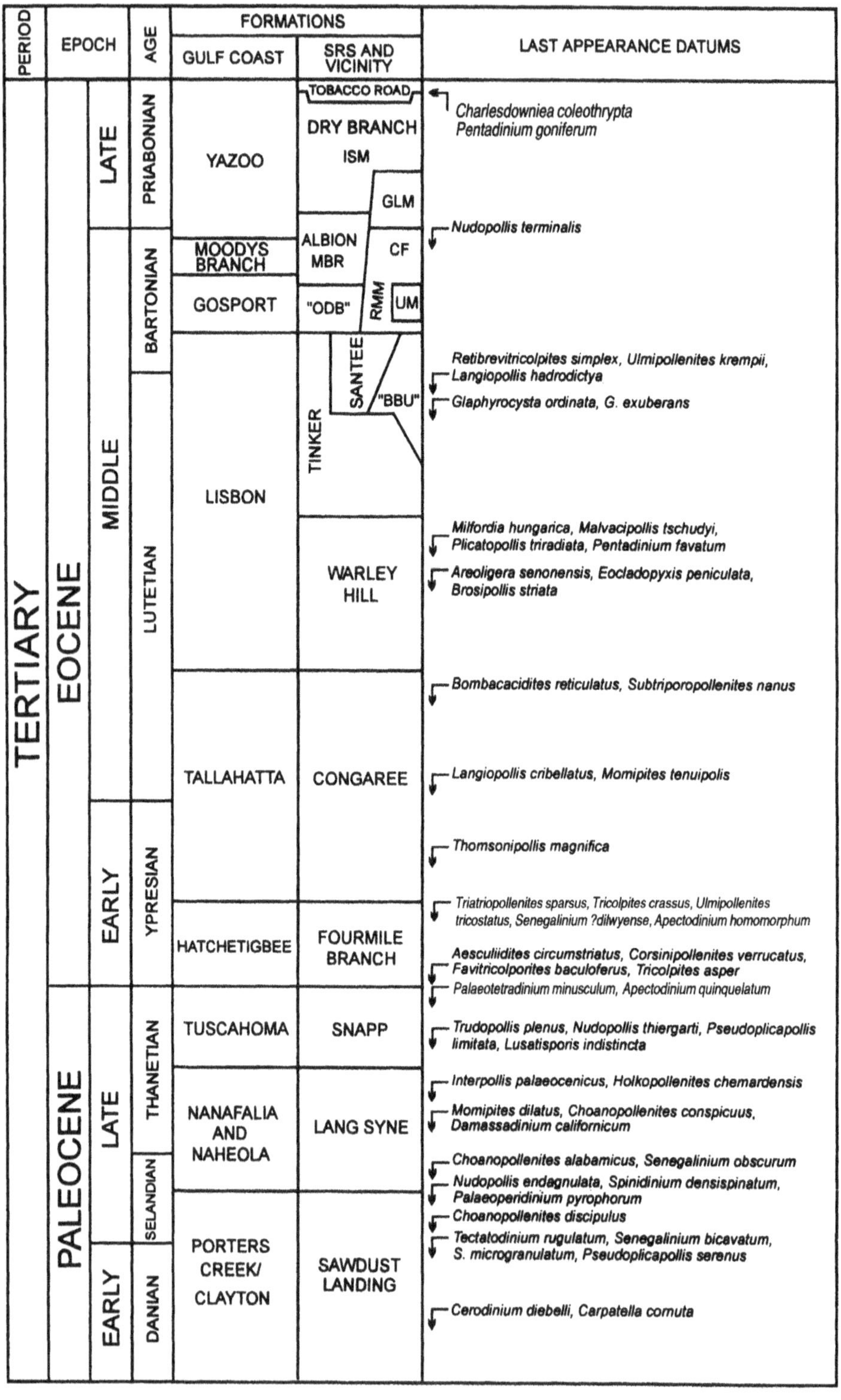

FIGURE 5B. Selected Tertiary palynomorph (last appearance) datums. The relationship of formational boundaries to stage boundaries is modified following Berggren *et al.*, 1996: IRM = Irwinton Sand Member; CF = Clinchfield Formation; RMM = Riggins Mill Member; BBU = Blue Bluff Unit; UM = Utley Member; GLM = Griffins Landing Member; ODB = Orangeburg District Bed.

to shallow marine in origin, and correlates with the lower (but not basal) part of the Tar Heel Formation in North Carolina. Dinoflagellates and pollen in the shallow marine facies have yielded mainly early Campanian (Late Cretaceous) ages. In most marine samples from the Middendorf Formation, dinoflagellates are sparse and assemblages consist mainly of spiny cysts and a few peridiniacean genera such as *Subtilisphaera* Jain and Millepied, *Spinidinium* Cookson and Eisenack, and *Palaeocystodinium* Alberti. Pollen assemblages from units mapped as the Middendorf Formation at SRS are distinctly different from those in the underlying Appleton Confining System/Cape Fear Formation, which reflects the unconformity that separates these units. Occurring in the Middendorf assemblages are *Plicapollis usitatus* Tschudy 1975 (Fig. 6i), representatives of the genus *Proteacidites* (Fig. 6j), two undescribed species with monocolpate apertures and echinate surface ornamentation and a variety of undescribed taxa. In addition, the lowest stratigraphic occurrence of *Holkopollenites chemardensis* Fairchild in Stover *et al.*, 1966 (Fig. 6m,n) occurs within this unit (Fig. 4). The pollen data suggest an early Campanian age for the unit.

The lower part of the Black Creek Formation is marine to nonmarine in origin, and contains pollen assemblages similar to those in the underlying Middendorf Formation. Dinoflagellates in the shallow marine facies have yielded mainly Campanian ages. In most of the marine samples, dinoflagellates are abundant, and assemblages consist of abundant small peridiniacean species of *Subtilisphaera, Spinidinium, Palaeohystrichophora* Deflandre, and larger peridiniacean species of *Andalusiella* (Riegel) Riegel and Sarjeant, *Xenascus* Cookson and Eisenack, *Palaeocystodinium* Alberti, *Cerodinium* Vozzennikova, and *Trithyrodinium* Drugg. Gonyaulacacean dinoflagellates other than ceratioid types are sparse (Lucas-Clark, 1992*a*). Age diagnostic dinoflagellate species in the McQueen Branch Aquifer/Black Creek Formation include *Andalusiella spicata* (May) Lentin and Williams, *Xenascus sarjeantii* (Corradini) Stover and Evitt, and *Palaeohystrichophora infusorioides* Deflandre.

3.3. McQueen Branch Confining Unit

The McQueen Branch Confining Unit separates the two Cretaceous (McQueen Branch and Crouch Branch) aquifers. The unit thins and pinches out in the vicinity of the well cluster P 19 (see Fig. 1) near the center of the SRS (where the two Cretaceous aquifers are in communication), but it persists throughout the remainder of the site region. The unit corresponds to the clays of the middle part of the Black Creek Formation.

Dinoflagellates from this unit have yielded mainly late Campanian ages. Characteristics of the dinoflagellate assemblages vary. Some assemblages are dominated by spiny cysts; others by *Areoligera coronata* (Wetzel) Lejeune-Carpenter/*senonensis*; others by small peridiniacean cysts; and others by *Xenascus ceratioides* (Deflandre) Lentin and Williams/*sarjeantii* complex.

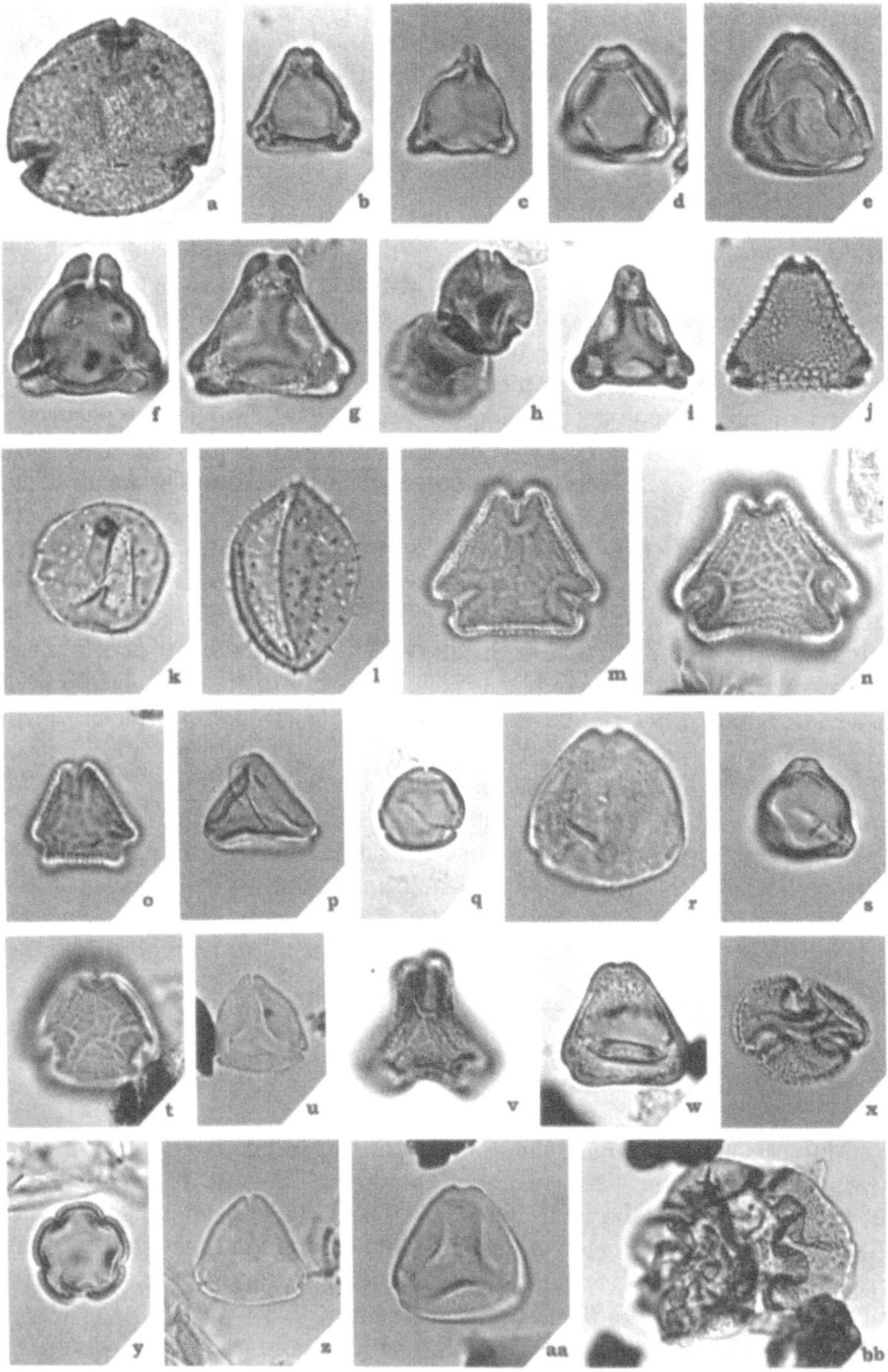

FIGURE 6. Selected Cretaceous pollen used in hydrostratigraphic correlation. All figures are 800×, except BB, which is 400× (reduced 25% for reproduction): (a) *Porocolpopollenites* sp.

Useful species of dinoflagellates include *Dinogymnium euclaense* Cookson and Eisenack, *D. digitus* (Deflandre) Evitt *et al.*, and *Palaeohystrichophora infusorioides* var. A and B of Aurisano (1989). Resedimentation of early Cretaceous dinoflagellates is not uncommon (Lucas-Clark, 1992*a*).

Pollen assemblages from the McQueen Branch Confining Unit include a variety of triporate forms that can be assigned to the genera *Momipites* (Fig. 6q,z), *Caryapollenites* (Fig. 6r), and *Betulaceoipollenites* (Fig. 6s) In addition, representatives of the genus *Holkopollenites* (Fig. 6v) are diverse and, at times, common elements in these assemblages. A late Campanian age is suggested by the pollen data (Fig. 4).

3.4. Crouch Branch Aquifer

The Crouch Branch Aquifer represents the upper of the two Cretaceous aquifers at SRS. It consists of the sand layers at the uppermost part of the Black Creek Formation and the greater part of the overlying Steel Creek Formation. Dinoflagellates from the fine-grained marine portions of this aquifer suggest ages of late Campanian and Maastrichtian. Unfortunately, the clean sands and highly altered kaolinitic clays comprising the upper part of this aquifer often do not contain well-preserved fossil material. However, generic level identification of dinoflagellates such as *Dinogymnium* spp. and species of *Rugubivesiculites* (Fig. 6bb) pollen are useful in distinguishing Cretaceous from overlying Tertiary units (Lucas-Clark and Van Pelt, 1993). Diagnostic dinoflagellate species include: *Dinogymnium acuminatum* Evitt *et al.*, *Palaeocystodinium benjaminii* Drugg, and *Cerodinium striatum* (Drugg) Lentin and Williams.

Pollen data from the Crouch Branch Aquifer suggest the presence of an unconformity within the unit; this unconformity corresponds to the contact between the Black Creek and Steel Creek Formations and represents the uppermost part of the Campanian and most of the lower Maastrichtian. *Osculapollis aequalis* Tschudy 1975 (Fig. 6w) is restricted to the units below the unconformity, as are several morphotypes of the genus *Holkopollenites* (Fig. 6o,v). Restricted to the units above the unconformity are *Libopollis jarzenii* (Fig. 6x) and a thin-walled species of *Baculostephanocolpites* (MPH-1

(*sensu* Doyle, 1969); (b) *Complexiopollis abditus* Tschudy 1973; (c) *Pseudoplicapollis longiannulata* Christopher 1979; (d) *Praecursipollis plebus* Tschudy 1975; (e) *Osculapollis* sp.; (f) *Trudopollis* sp.; (g) *Plicapollis* sp.; (h) *Labrapollis* sp.; (i) *Plicapollis usitatus* Tschudy 1975; (j) Proteacidites sp.; (k) *Echimonocolpites* sp. 45 (C1-45); (l) Echimonocolpites sp. 46 (C1-46); (m,n) *Holkopollenites chemardensis* Fairchild in Stover, Elsik and Fairchild 1966; (o) *Holkopollenites* sp. A (=CP3d-1 of Wolfe, 1976); (p) *Santalacites minor* Christopher 1979; (q) *Momipites* sp.; (r) *Caryapollenites* sp.; (s) *Betulaceoipollenites* sp. (=NO-3 of Wolfe, 1976); (t) *?Holkopollenites* sp. (CP3E-1 of Wolfe, 1976); (u) *Pseudoplicapollis endocuspis* Tschudy 1975; (v) *Holkopollenites* sp. B (=CP3D-2 of Wolfe, 1976); (w) *Osculapollis aequalis* Tschudy 1975; (x) *Libopollis jarzenii* Farabee, Daghlian, Canright and Oftedahl 1984; (y) *Baculostephanocolpites* sp. B (=MPH-2 of Wolfe, 1976); (z) *Momipites dilatus* group (=NP-2 of Wolfe, 1976); (aa) *Plicatopollis cretacea* Frederiksen and Christopher 1979; (bb) *Rugubivesiculites* sp.

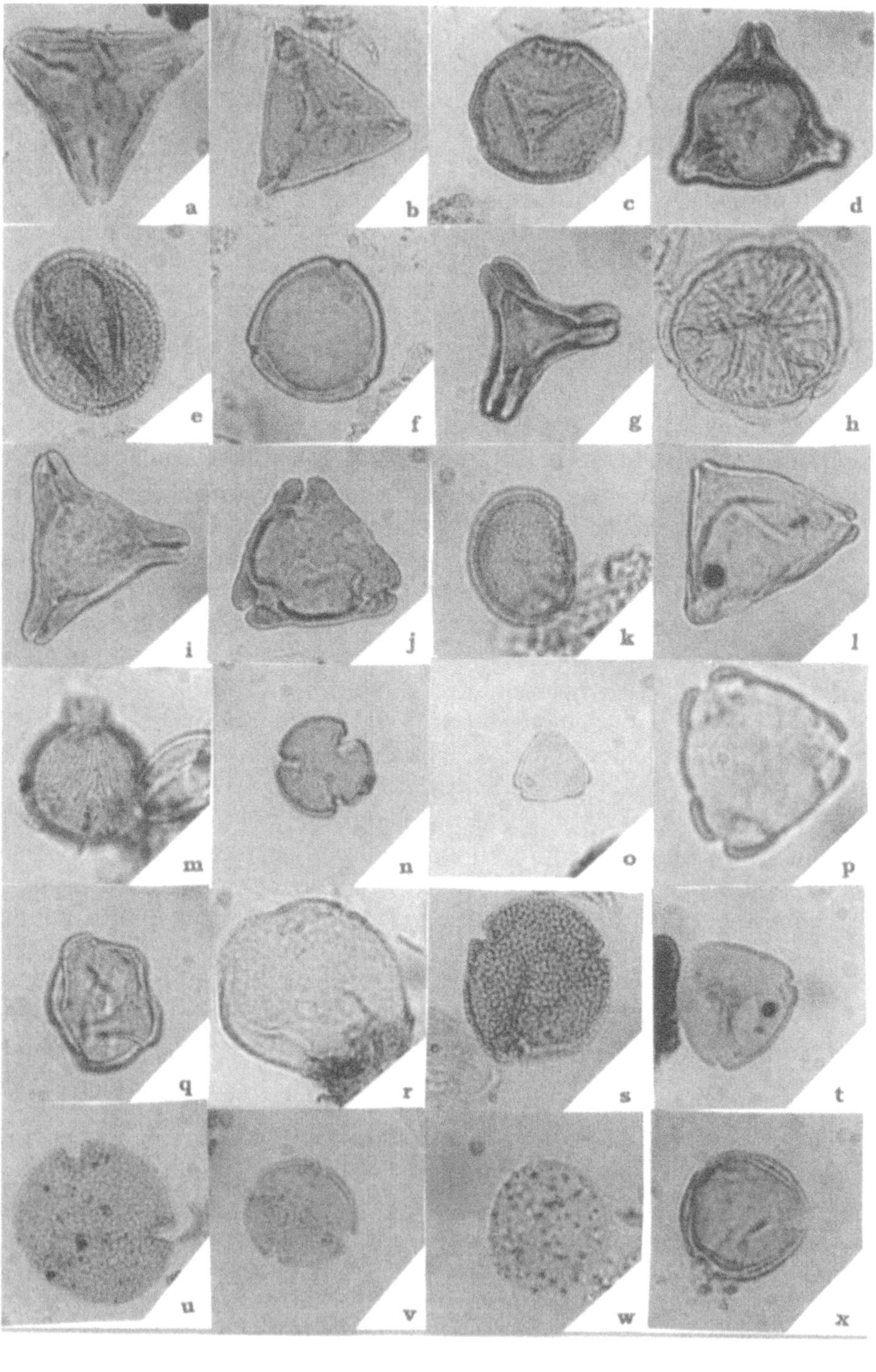

FIGURE 7. Selected Tertiary pollen used in hydrostratigraphic correlation. All figures are 625 ×, except for (p), which is 1500 × (reduced 25% for reproduction): (a) *Boehlensipollis holii* Krutzsch

of Wolfe, 1976) (Fig. 6y). In the SRS and surrounding vicinity, the Steel Creek Formation is primarily nonmarine to marginal marine, whereas the underlying Black Creek Formation is near shore to open marine in origin.

3.5. Crouch Branch Confining Unit

The confining system that overlies the Crouch Branch Aquifer represents the principal confining unit for the SRS and vicinity. This confining unit prevents migration of contaminated water from the Steed Pond and Gordon Aquifers of Tertiary age into the underlying Cretaceous Crouch Branch Aquifer. However, in some areas within the SRS, the confining unit is discontinuous, resulting in local communication of the Tertiary and Cretaceous aquifers. It is thus critical to identify the confining unit where it is present, and important to understand its relation to other units.

The Crouch Branch Confining Unit corresponds to lower and upper Paleocene and lower Eocene lithologic units, which correspond to the Lang Syne, Sawdust Landing, the Snapp, and the Fourmile Branch Formations. The base of the Crouch Branch Confining Unit/Sawdust Landing Formation is characterized by the first appearance datum (FAD) of several Danian (Paleocene) dinoflagellate species: *Carpatella cornuta* Grigorovich (Fig. 8a), *Damassadinium californicum* (Drugg) Fensome *et al.* (Fig. 8r) *Spiniferites septatus* (Cookson and Eisenack) McLean, *Senegalinium ?dilwynense* (Cookson and Eisenack) Stover and Evitt (Fig. 8b), and *Spinidinium pulchrum* (Bensen) Lentin and Williams.

The upper part of the Crouch Branch Confining Unit (equivalent to the upper part of the Lang Syne, Snapp, and Fourmile Branch Formations) is characterized by dinoflagellate assemblages that are dominated by one or, at most, a few species. These assemblages may comprise small peridiniacean species along with *Areoligera* Lejeune-Carpenter spp. and/or *Cordosphaeridium* Eisenack spp. *Apectodinium homomorphum* (Deflandre and Cookson) Lentin and Williams (Fig. 8h), *A. quinquelatum* (Williams and Downie) Lentin

---

1962; (b) *Pseudoplicapollis limitata* Frederiksen 1978; (c) *Thomsonipollis magnificus* (Pflug in Thomson and Pflug) Krutzsch 1960; (d) *Nudopollis terminalis* (Thomson and Pflug in Thomson and Pflug, 1953); (e) *Favitricolporites baculoferous* (Thomson and Pflug) Pflug 1953; (f) *Subtriporopollenites nanus* (Thomson and Pflug in Thomson and Pflug, 1953); (g) *Choanopollenites discipulus* Tschudy 1973; (h) *Lusatisporis indistincta* Frederiksen 1979; (i) *Nudopollis thiergartii* (Thomson and Pflug) Pflug 1953; (j) *Trudopollis plenus* Tschudy 1975; (k) *Tricolpites asper* Frederiksen 1978; (l) *Piolencipollis endocuspoides* Frederiksen 1979; (m) *Brosipollis striata* Frederiksen 1988; (n) *Tricolpites crassus* Frederiksen 1979; (o) *Pseudoplicapollis serenus* Tschudy 1975; (p) *Pseudoplicapollis serenus* Tschudy 1975 1500 ×; (q) *Lymingtonia* cf. *L. rhetor* Erdtman 1960 in Frederiksen 1988; (r) *Milfordia hungarica* (Kedves, 1965) Krutzsch and Vanhoorne 1977; (s) *Intratriporopollenites stavensis* Frederiksen 1980; (t) *Plicatopollis triradiata* (Nichols 1973) Frederiksen and Christopher 1978; (u) *Bombacacidites reticulatus* Krutzsch 1961; (v) *Quadrapollenites vagus* Stover in Stover *et al.*, 1966; (w) *Malvacipollis tschudyii* (Frederiksen 1973) Frederiksen 1980; (x) *Retibrevitricolpites simplex* Frederiksen 1988.

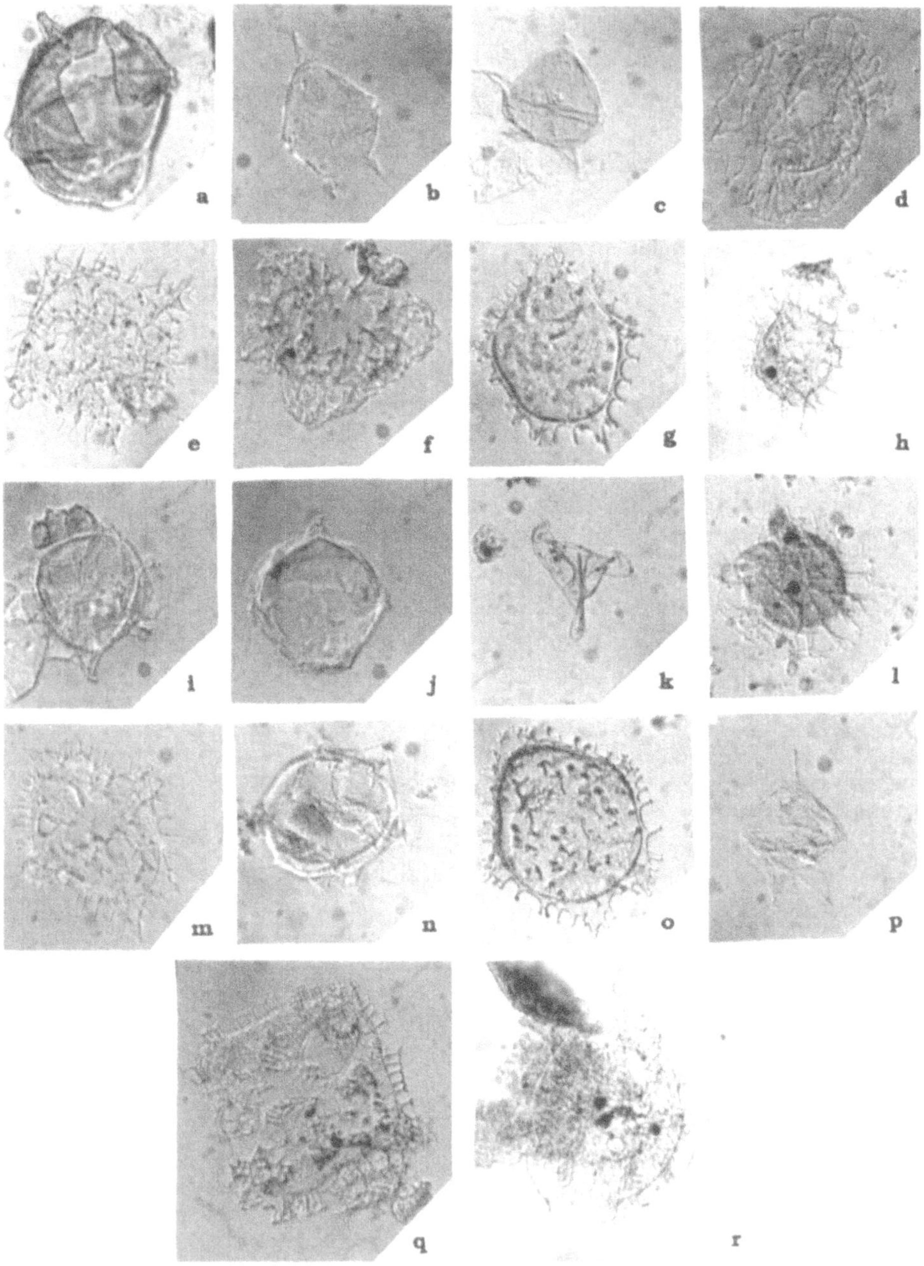

FIGURE 8. Selected Tertiary dinoflagellates used in hydrostratigraphic correlation. All figures are 500×. (reduced 25% for reproduction): (a) *Carpatella cornuta* Grigorovich 1969; (b) *Senegalinium ?dilwynense* (Cookson and Eisenack) Stover and Evitt 1978; (c) *Senegalinium bicavatum* Jain and Millepied 1973; (d) *Adnatosphaeridium multispinosum* Williams and Downie 1966; (e) *Wetzeliella articulata* Eisenack 1938; (f) *Glaphyrocysta ?vicina* (Eaton) Stover and Evitt 1978; (g) *Dracodinium* sp.; (h) *Apectodinium homomorphum* (Deflandre and Cookson) Harland 1979; (i)

and Williams, and *Eocladopyxis peniculata* Morgenroth make their first stratigraphic appearance in the upper part of the confining unit (Edwards, 1990; Lucas-Clark, 1992*b*).

There are several pollen species that have their bases or first appearances (FAD) in the Crouch Branch Confining Unit/Sawdust Landing Formation (Fig. 5a,b). These include *Choanopollenites conspicuus* (Groot and Groot 1962) Tschudy 1973, *Choanopollenites discipulus* Tschudy 1973 (Fig.7g), *Favitricolporites baculoferus* (Pflug in Thompson and Pflug 1953) Srivastrava 1972 (Fig. 7e), *Nudopollis terminalis* (Pflug and Thompson in Thompson and Pflug, 1953) Pflug 1953 (Fig. 7d), *Nudopollis thiergarti* (Thompson and Pflug 1953) Pflug 1953 (Fig. 7i), *Momipites dilatus* Fairchild in Stover *et al.* (1966), *Pseudoplicapollis limitata* Frederiksen 1978, *Tricolpites asper* Frederiksen 1978, and *Trudopollis plenus* Tschudy 1975 (Fig. 7j). All of these taxa have tops or last appearances (LADs) higher in the stratigraphic section but the assemblage is characteristic of the early Paleocene age. The range top of *Pseudoplicapollis serenus* Tschudy 1975 (Fig. 7o,p) that is within the Sawdust Landing Formation is considered by Frederiksen (1991) to mark the top of the Danian or early Paleocene.

In the upper part of the Crouch Branch Confining Unit, which includes the upper Lang Syne and Snapp Formations, there are several pollen and spore species that have their last appearances. These include taxa that had their first appearances in the lower part of the Crouch Branch confining unit: *Choanopollenites conspicuus, Favitricolporites baculoferus* (Fig. 7e), *Interpollis paleocenicus* (Elsik 1968) Frederiksen 1980, *Lusatisporis indistincta* Frederiksen 1979 (Fig. 7h), *Momipites dilatus* (Fig. 6z), *Nudopollis thiergartii* (Fig. 7i), *Pseudoplicapollis limitata* (Fig. 7b), *Tricolpites asper* (Fig. 7k), and *Trudopollis plenus* (Fig. 7j). *Holkopollenites chemardensis* Fairchild in Stover *et al.* (1966); Fig. 6m,n), which first appears in the Upper Cretaceous has a top in this interval. *Corsinipollenites? verrucatus* Frederiksen 1988 has a first and last appearance within the Snapp Formation. Frederiksen (1998) reports this species from the lower Eocene. Taxa such as *Subtriporopollenites nanus* (Pflug and Thompson in Thompson and Pflug 1953) Frederiksen 1980 (Fig. 7f), *Pseudolaesopollis ventosus* (Potonie' 1931) Frederiksen 1979, and *Milfordia hungarica* (Kedves 1965) Krutzsch and Vanhoorne 1977 (Fig. 7r) have their first appearances or bases within the Snapp Formation or equivalents in the southeastern United States (Frederiksen 1980*a*,*b*, 1988, 1998).

Pentadinium goniferum Edwards 1982; (j) *Samlandia chlamydophora* Eisenack 1954; (k) Palaeotetradinium minusculum (Alberti) Stover and Evitt 1978; (l) *Hafniasphaera septata* (Cookson and Eisenack) Hansen 1977; (m) *Wetzeliella* cf. *W. articulata* Eisenack 1938; (n) *Pentadinium laticinctum* Gerlach 1961; (o) *Wetzeliella* sp.; (p) *Lentinia* sp.; (q) *Charlesdowniea variabilus* (Bujak) Lentin and Vozzhennikova 1989; (r) *Damassadinium californicum* (Drugg) Fensome et al., 1993.

3.6. Gordon/Steed Pond Aquifer and Gordon Confining Unit

The middle-to-upper Eocene aquifer (Steed Pond/Gordon) and confining unit (Gordon) that overlies the Crouch Branch Confining Unit is a complex set of strata consisting, in part, of carbonates (limestone) that are lenticular and discontinuous, as well as alternating clay and sand layers that act as confining units. Understanding the potential complexity of groundwater flow requires considerable accuracy in the identification and correlation of the lithologic units.

Lithologic units that correspond to the Steed Pond and equivalent aquifers include the Congaree Formation, Warley Hill Formation, Santee Formation, Clinchfield Formation, Irwinton Sand and Griffins Landing Members of the Dry Branch Formation, and the Tobacco Road Formation. Sediments comprising the Tobacco Road Formation are consistently barren of organic-walled microfossils. The other lithologic units have yielded dinoflagellate and pollen assemblages, usually from the clay- and silt-rich intervals.

Dinoflagellates useful in correlating the complex of middle-to-upper Eocene units include: *Pentadinium favatum* Edwards, *P. goniferum* Edwards (Fig. 8i), *Glaphrocysta* Stover and Evitt spp. (Fig. 8f), *Wetzelliella articulata* Eisenack group (Fig. 8e), *Membranophoridium aspinatum* Gerlach, *M. bilobatum* Michoux, and *Charlesdowniea variabilis* (Bujak) Lentin and Vozzhennikova (Fig. 8q).

Throughout most of the SRS, confining conditions (Gordon Confining Unit) exist to separate the Steed Pond Aquifer into two distinct aquifer zones, the Upper Three Runs and underlying Gordon Aquifer zones. The Gordon Confining Unit consists of sandy clays to clayey sands, which represent parts of the Warley Hill and Tinker Formations, and the Blue Bluff Unit. Dinoflagellates first appearing within this confining unit include *Pentadinium laticinctum* Gerlach (Fig. 8n), *Cordosphaeridium cantharellus* (Brosius) Gocht, and *Charlesdowniea coleothrypta* (Williams and Downie) Lentin and Vozzhennikova. *Pentadinium favatum* last appears within the upper part of this confining unit.

There are not as many extinctions or tops for pollen taxa in the middle and late Eocene Gordon Confining Unit as in the Crouch Branch Confining Unit. The most distinctive species are *Bombacacidites* aff. *B. reticulatus* Krutzsch 1961, *Subtriporopollenites nanus* (Fig. 7f), *Malvacipollis tschudyi* (Frederiksen 1973) Frederiksen 1980 (Fig. 7w), *Brosipollis striata* Frederiksen 1988 (Fig. 7m), *Milfordia* hungarica (Fig. 7r), *Lanagiopollis hadrodictya* Frederiksen 1988, *Plicatopollis triradiata* (Nichols 1973) Frederiksen and Christopher 1978 (Fig. 7t), *Ulmipollenites krempii* (Anderson 1960) Frederiksen 1979, and *Retibrevitricolpites simplex* Frederiksen 1988 (Fig. 7x). *Dicolpopollis* spp. and *Proxapertites* spp. are reported by Tschudy (1973*a*,*b*) and Frederiksen (1988) to have their last appearances in the middle and late Eocene.

In addition to the tops observed in the Gordon Confining Unit several species have first occurrences within this unit. *Retibrevitricolpites simplex* (Fig. 7x) first appears in the Congaree Formation. Other species that have bases in the Congaree and Warley Hill and Santee Formations are *Juglans nigripites* Wodehouse 1933, *Ulmipollenites undulosus* Wolff 1934, *Lymingtonia* cf. *L. rhetor* Erdtman 1960 (Fig. 7q) in Frederiksen (1988), *Symplocos? jacksoniana* Traverse 1955, *Gothanipollis cockfieldensis* Engelhardt 1964, and *Boehlensipollis hohlii* Krutzsch 1962 (Fig. 7a). *Graminidites* spp. is reported by Tschudy (1973) and Frederiksen (1988) to have a first appearance in the late Eocene.

Many of the first appearance datums or bases (FADs) and last appearance datums (LADs) or tops are shown in Figs. 5a,b, which summarize the results of observations of palynomorph taxa in the cores from the Savannah River Site and published data. The principal publications on pollen and spores that were utilized in the study include: Elsik, 1974; Elsik and Dilcher, 1974; Frederiksen, 1978, 1979, 1980*a*,*b*, 1988, 1991, 1998, in press, Frederiksen and Christopher, 1978; Tschudy 1973*a*, 1975.

4. Conclusions

Dinoflagellate and pollen biostratigraphy has been applied successfully as a geologic tool for helping establish the subsurface hydrogeologic framework beneath the SRS. Because of the rapid vertical and lateral changes in updip coastal plain setting, dinoflagellate, and pollen and spore biostratigraphy, and to some extent dinoflagellate paleoenvironmental interpretations, have been critical in the correlation of aquifers and confining units and in the identification of facies changes and structural elements (e.g., Pen Branch Fault, Aadland *et al.*, 1995). Recognition of these features is important as they serve as preferential pathways for groundwater movement and contaminant transport.

ACKNOWLEDGMENTS: We thank Lucy E. Edwards and Norman O. Frederiksen for helpful discussions on particular aspects of these palynofloras. Support for this project was provided by Department of Energy under contract No. DE-AC09-96-SR18500. We gratefully acknowledge Bechtel Savannah River Incorporated for permission to publish this information.

References

Aadland, R. K., 1990, Hydrogeologic Characterization of the Cretaceous-Tertiary coastal plain sequence at the Savannah River Site, South Carolina, in: *Savannah River Region: Transition between the Gulf and Atlantic Coastal Plains, Proc. 2nd Bald Head Island Conference on Coastal Plains Geology* (V. A. Zullo, W. B. Harris, and V. Price, eds.), University of North Carolina at Wilmington, pp. 57–60.

Aadland, R. K., and Bledsoe, H. W., 1992, Hydrogeologic classification of the Cretaceous-Tertiary coastal plain sequence at the Savannah River Site, *Geol. Soc. Am. Prog. Abstr.* **24**(2)1.

Aadland, R. K., Thayer, P. A., and Smits, A. D., 1992, The hydrogeological atlas of west-central South Carolina, Westinghouse Savannah River Company, Westinghouse Savannah River Company Technical Report 90-987.

Aadland, R. K., Gellici, J. A., and Thayer, P. A., 1995, Hydrogeologic framework of west-central South Carolina, South Carolina Department of Natural Resources, Water Resources Division Report No. 5, 200 p.

Aurisano, R. W., 1989, Upper Cretaceous dinoflagellate biostratigraphy of the subsurface Atlantic coastal plain of New Jersey and Delaware, U.S.A., *Palynology* **13**:143–180.

Berggren, W. A., Kent, D. V., Swisher, C. C. III, and Aubry, M.-P., 1996, A revised Cenozoic geochronology and chronostratigraphy, in: *Geochronology, Time Scales and Global Stratigraphic Correlation*: Society of Economic Paleontologists and Mineralogists, *Spec. Publ.* No. 54 W. A. Berggren, D. V. Kent, M.-P. Aubry, and Jan Hardenbol, eds.), pp. 129–212.

Cahil, J. M., 1982, Hydrology of the low-level radioactive solid-waste burial site and vicinity near Barnwell, South Carolina, U.S. Geological Survey Open File Report 82-863, 101 p.

Christopher, R. A., 1979, Normapolles and triporate pollen assemblages from the Raritan and Magothy formations (Upper Cretceous) of New Jersey: *Palynology* **3**:73–121.

Clarke, J. S., and West, C. T., 1997, Groundwater levels, predevelopment groundwater flow, and stream-aquifer relations in the vicinity of the Savannah River Site, Georgia and South Carolina, United States Geological Survey, Water-Resources Investigations Report 97-4197, 120 p.

Clarke, J. S., Falls, W. F., Edwards, L. E., Frederiksen, N. O., Bybell, L. M., Gibson, T. G., and Litwin, R. J., 1994, Geologic, hydrologic, and water quality data for a multi-aquifer system in coastal plain sediments near Millers Pond, Burke County, Georgia, 1992-93, Georgia Department of Natural Resources Information Circular 96, 34 p.

Doyle, J. A., and Robbins, E. I., 1977, Angiosperm pollen zonation of the continental Cretaceous of the Atlantic coastal plain and its application to deep wells in the Salisbury embayment: *Palynology* **1**:43–78.

Edwards, L. E., 1990, Dinocysts from the Lower Tertiary units in the Savannah River area, South Carolina and Georgia, in: *Savannah River Region: Transition between the Gulf and Atlantic Coastal Plains, Proc. 2nd Bald Head Island Conference on Coastal Plains Geology* (V. A. Zullo, W. B. Harris, and V. Price, eds.), pp. 89–90.

Elsik, W. C., 1974, Characteristic Eocene palynomorphs in the Gulf Coast, U.S.A., *Palaeontographica, Abt. B* **149**:90–111.

Elsik, W. C., and Dilcher, D. L., 1974, Palynology and age of clays exposed in Lawrence clay pit, Henry County, Tennessee, *Palaeontographica, Abt. B* **146**:65–87.

Engelhardt, D. W., 1964, A new species of Gothanipollis Krutzsch from the Cockfield Formation (middle Eocene) of Mississippi: *Poll. Spores* **6**(2):597–600.

Erdtman, G., 1960, On three new genera from the Lower Headon Beds, Berkshire, *Bot. Not.* **113**:46–48.

Fallaw, W. C., and Price, V., 1992, Outline of stratigraphy at the Savannah River Site, in: *Carolina Geological Society Field Trip Guide Book, Nov. 13-15, 1992, Geological Investigations of the Central Savannah River Area, South Carolina and Georgia* (W. C. Fallaw and V. Price, eds.), U.S. Department of Energy and S.C. Geological Survey, pp. CGS-92-B-11-1-33.

Fallaw, W. C., and Price, V., 1995, Stratigraphy of the Savannah River Site and vicinity, *Southeast. Geol.* **35**(1):21–58.

Falls, W. F., Baum, J. S., and Prowell, D. C., 1997, Physical stratigraphy and hydrostratigraphy of Upper Cretaceous and Paleocene sediments, Burke and Screven Counties, Georgia, *Southeast. Geol.* **36**(4):153–176.

Faye, R. E., and Prowell, D. C., 1982, Effects of Late Cretaceous and Cenozoic faulting on the geology and hydrology of the Coastal Plain near the Savannah River, Georgia, and South Carolina, U.S. Geological Survey Open File Report 82-156.

Frederiksen, N. O., 1978, New Paleogene pollen species from the Gulf Coast and Atlantic coastal plains, *U.S. Geol. Surv. J. Res.* **6**:691–696.

Frederiksen, N. O., 1979, Paleogene sporomorph biostratigraphy, northeastern Virginia, *Palynology* **3**:129–167.

Frederiksen, N. O., 1980*a*, Sporomorphs from the Jackson group (upper Eocene) and adjacent strata of Mississippi and western Alabama, U.S. Geol. Survey Professional Paper 1084, 75 p.

Frederiksen, N. O., 1980*b*; Paleogene sporomorphs from South Carolina and quantitative correlations with the Gulf Coast, *Palynology* **4**:125–179.

Frederiksen, N. O., 1988, Sporomorph biostratigraphy, floral changes, and paleoclimatology, Eocene and earliest Oligocene of the eastern Gulf Coast, U.S. Geol. Survey Professional Paper 1448, 68 p.

Frederiksen, N. O., 1991, Midwayan (Paleocene) pollen correlations in the eastern United States, *Micropaleontology* **37**(2):101–123.

Frederiksen, N. O., 1998, Upper Paleocene and lowermost Eocene angiosperm pollen biostratigraphy of the eastern Gulf Coast and Virginia, *Micropaleontology* **44**(1):45–68.

Frederiksen, N. O., In Press, Lower Tertiary pollen biostratigraphy of testholes in Georgia near the Savannah River Site, U.S. Geol. Survey Professional Paper.

Frederiksen, N. O., and Christopher, R. A., 1978, Taxonomy and biostratigraphy of Late Cretaceous and Paleogene triatriate pollen from South Carolina, *Palynology* **2**:113–145.

Gohn, G. S., and Campbell, B. G., 1992, Recent revisions to the stratigraphy of subsurface Cretaceous sediments in the Charleston, South Carolina area, *S. Carolina Geol.* **34**:25–38.

Huddlestun, P. F., and Summerour, J. H., 1996, The lithostratigraphic framework of the uppermost Cretaceous and lower Tertiary of Eastern Burke County, Georgia, *Georgia Geol. Surv. Bull.* No. 127, 94 p.

Krutzsch, W., 1960, Ueber Thomsonipollis magnificus (Th. and Pf. 1953) n.f. gen. n. comb. und Bemerkungen zur regionalen Verbreitung einiger Pollengruppen im aelteren Palaeogen, *Freib. Fors.* **C86**:54–65.

Krutzsch, W., 1961, Beitrag zur Sporenpalaeontologie der praaoberoligozaenen kontinentalen und marinen Tertiaerablagerungen Brandenburgs, *Ber. Geol. Geds. DDR* **5**:290–343.

Krutzsch, W., 1962, Stratigraphisch bzw. botanisch wichtige neue Sporen -und Pollenformen aus dem deutschen Tertiaer, *Geologie* **11**(3):265–307.

Krutzsch, W., and Vanhoorne, R., 1977, Die Pollenflora von Epinois und Loksbergen in Belgien, *Palaeontographica, Abt. B* **163**:1–110.

Lewis, S., and Aadland, R. K., 1992, Hydrogeologic setting of the A/M area: Framework for groundwater transport, Westinghouse Savannah River Company Technical Report 92-355, 80 p.

Leeth, D. C., and Nagle, D. D., 1996, Shallow subsurface geology of part of the Savannah River alluvial valley in the upper coastal plain of Georgia and South Carolina, *Southeast. Geol.* **36**(1):1–14.

Lucas-Clark, J., 1992*a*, Problems in Cretaceous palynostratigraphy (dinoflagellates and pollen) of the Savannah River Site area, in: *Savannah River Region: Transition between the Gulf and Atlantic Coastal Plains, Proc. 2nd Bald Head Island Conference on Coastal Plains Geology* (V. A. Zullo, W. B. Harris, and V. Price, eds.), p. 74.

Lucas-Clark, J., 1992*b*, Some characteristic fossil dinoflagellate cysts of Eocene strata, Savannah River Site, South Carolina, in: *Geological Investigations of the Central Savannah River Area, South Carolina and Georgia, Carolina Geological Society Field Trip Guidebook 1992* (W. Fallaw and V. Price, eds.), pp. 58–60.

Lucas-Clark, J., and Van Pelt, R. S., 1993, Application of dinoflagellate biostratigraphy to environmental restoration activities at the Savannah River Site, South Carolina, U.S.A. (abstr.) *Prog. Abstr. 5th Int. Conf. Modern and Fossil Dinoflagellates, Zeist,* The Netherlands, April 1993.

Marine, L. W., and Root, K. R., 1975, A ground water model of the Tuscaloosa aquifer at the Savannah River Plant, in: Savannah River Laboratory Environmental Transport and Effects Research Annual Report (T. V. Crawford, ed.), USERDA Report DP-1374, E.I. duPont de Nemours & Co., Savannah River Laboratory, Aiken, South Carolina.

Martini, E., 1971, Standard Tertiary and Quaternary calcareous nannoplankton zonation, in: *Proc. II Planktonic Conf. Rome,* 1970 (A. Farinacci, ed.), Ed. Tecnoscienza, pp. 739–785.

Nystrom, P. G., Jr., Willoughby, R. H., and Price, L. K., 1991, The Cretaceous and Tertiary stratigraphy of the South Carolina upper Coastal Plain, in: *Geology of the Carolinas* (J. W. Horton, Jr. and V. A. Zullo, eds.), University of Tennessee Press, Knoxville, pp. 221–240.

Prowell, D. C., Christopher, R. A., Edwards, L. E., Bybell, L. M., and Gill, H. E., 1985*a*, Geologic section of the updip coastal plain from central Georgia to western South Carolina, U.S. Geological Survey Miscellaneous Field Studies, Map MF-1737.

Prowell, D. C., Edwards, L. E., and Frederiksen, N. O., 1985*b*, The Ellenton formation in South Carolina: A revised age designation from Cretaceous to Paleocene, *U.S. Geol. Surv. Bull.* **160-A**:63–69.

Siple, G. E., 1967, Geology and ground water of the Savannah River plant and vicinity, South Carolina. U.S. Geological Survey Water-Supply Paper 1841, 113 p.

Sissingh, W., 1977, Biostratigraphy of Cretaceous calcareous nannoplankton, *Geol. Mijnbow* **56**(1):37–65.

Srivastava, S. K., 1972, Some spores and pollen from the Paleocene Oak Hill member of the Naheola formation, Alabama (U.S.A.), *Rev. Palaeobot. Palyn.* **14**:217–285.

Stover, L. E., Elsik, W. C., and Fairchild, W. W., 1966, New genera and species of Early Tertiary palynomorphs from the Gulf Coast, Kansas University *Paleontolog. Contr. Paper* 5, p. 10.

Thomson, P. W., and Pflug, H., 1953, Pollen and sporen des mitteleuropaEischen TertiaErs *Palaeontographica, Abt. B* **94**:1–138.

Traverse, A. F., Jr., 1955, Pollen analysis of the Brandon lignite of Vermont, U.S. Bureau of Mines Rept. Inv. 5151, p. 107.

Tschudy, R. H., 1973*a*, Stratigraphic distribution of significant Eocene palynomorphs of the Mississippi embayment, U.S. Geological Survey Professional Paper 743-B, p. 24.

Tschudy, R. H., 1973*b*, *Complexiopollis* pollen lineage in Mississippi embayment rocks: U.S. Geological Survey Professional Paper 743-C, p. C1-C15.

Tschudy, R. H., 1975, Normapolles pollen from the Mississippi embayment: U.S. Geological Survey Professional Paper 865, 42 p.

Van Pelt, R. S., and Lucas-Clark, J., 1996, The Application of Palynology to Environmental Restoration Activities: An Example from the Savannah River Site, South Carolina (Abstr.), *9th Int. Palynological Cong. Prog. Abstr.* June 1996, p. 82.

Van Pelt, R. S., Waanders, G. L., Lucas-Clark, J., and Harris, M. K., 1993, The application of palynology to waste-site characterization: An example from the Savannah River Site, South Carolina, (Abstr.) *Am. Assoc. Stratigraphic Palynologists 1993 Ann. Meet. Prog. Abstr.*, p. 82.

Wodehouse, R. P., 1933, Tertiary pollen: II. The oil shales of the Eocene Green River formation, *Torrey Bot. Club Bull.* **60**:479–524.

Wolff, H., 1934, Mikrofossilien des pliocaenen Humodils der Grube Freigericht bei Dettingen a.M. und Vergleich mit alteren Schichten des Tertiaers sowie posttertiaeren Ablagerungen, *Preuss. Geol. Landesanstalt, Inst. Palaeobot. Petrogr. Brenn. Arb.* **5**: 55–86.

Wolfe, J. A., 1976, Stratigraphic distribution of some pollen types from the Campanian and lower Maastrichtian rocks (Upper Cretaceous) of the Middle Alantic states, U.S. Geological Survey Professional Paper 977, p. 1–18.

Chapter 19

Construction of the Thames Barrier

An Application of Micropaleontology to the Solution of an Environmental Problem

MALCOLM B. HART

Introduction

London is one of the world's most important capital cities, having expanded over 2000 years since being founded by the Romans (as Londinium) during the time of their occupation of Britain. The original settlement was located on a gravel bank very close to the present City of London. This relatively high location afforded protection from the occasional high tides that, even then, had begun to encroach on the flat marshland that bordered the nearby river.

MALCOLM B. HART • Department of Geological Sciences and Plymouth Environmental Research Center, University of Plymouth, Plymouth PL4 8AA, United Kingdom.

Environmental Micropaleontology, Volume 15 of *Topics in Geobiology*, edited by Ronald E. Martin. Kluwer Academic/Plenum Publishers, New York, 2000.

Throughout almost two millennia flooding was accepted as a regular occurrence, but it was only in 1928, when a tide of 5.8 m was recorded, that those in charge of London became concerned. The devastating flooding of 1953, which caused significant loss of life in both the United Kingdom and The Netherlands, was the final straw and the Thames Barrier and its associated flood defenses can be seen as a direct result of that disaster.

In the eventual site investigation prior to the construction of the Barrier, micropaleontology was an important factor in the planning process. The Barrier, which eventually cost £440,000,000 (at 1984 prices), was first raised "in anger" on October 31, 1982, some 10 years after the site investigation work. The micropaleontological work accounted for approximately 1% of the total site investigation cost which, in itself, was less than 0.0025% of the final budget.

1.1. Development of the River Thames

The River Thames developed during the Pleistocene (Rose and Allen, 1977) and, with sea level some 120 m lower than the present day, flowed eastward to a confluence with a proto-Rhine (Fig. 1) in the area of the present southern North Sea Basin. This vast river then flowed southwestward through the Dover Straits and out over the continental shelf.

At the end of the last glacial episode (the Devensian cold stage) the Flandrian transgression began (Anderton *et al.*, 1979) and sea level rose to just below present-day levels forming the wide Thames estuary. In the immediate hinterland the river ran through low marshes in a wide, open valley. Compared to its modern counterpart the river was much smaller, nontidal, and unaffected by the lower mean sea level.

It was on a gravel bank close to this nontidal river that the Romans founded Londinium. As relative sea level continued to rise slowly in the outer estuary occasional "surge" tides began to be recorded in the vicinity of the developing settlement. As the city developed as a trading center the first London Bridge was built of timber. The result of this construction was the development of the port downstream of the bridge, thereby avoiding the effort of demasting ships to allow them to pass upstream. Tidal berths developed on both sides of the river, especially on the southern side (site of the later Surrey Docks) next to an expanse of marshland. These docks were protected from increased river flow or the occasional storm surge by a series of earth banks. These banks probably made things worse as they deprived the marshes of sediment that would normally have been deposited by subsequent flooding events. The marshland was, therefore, deprived of the sediment needed to maintain its level at that of the river.

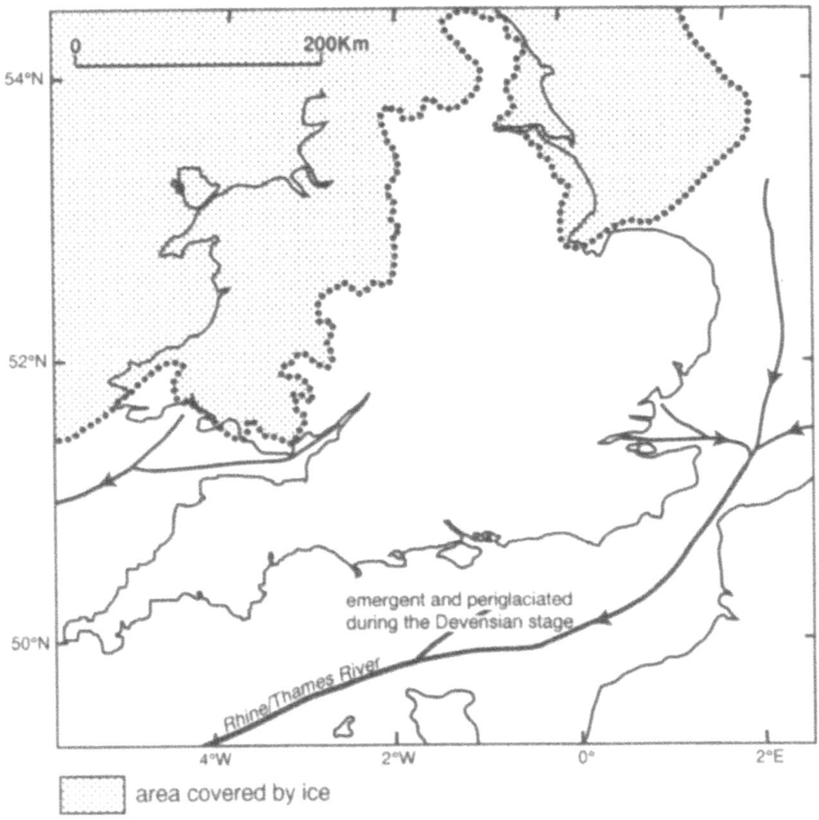

FIGURE 1. Pleistocene paleogeography of the Thames estuary and southern North Sea basin (adapted from Graham and Straw, 1992).

1.2. History of Flooding

In Roman times no normal, astronomically controlled tides reached London, but there were occasional records of high sea levels causing flooding of the area near the river. Records are intermittent but there were clearly major floods in 1099 (see *Anglo-Saxon Chronicle*) and 1236 (recorded by Matthew Paris in *Chronica Majora*). This latter event occurred in early November and appears to have been the coincidence of a spring tide and a wind-generated surge. It reportedly lasted over a period of five successive tides (60–70 hr). Flooding events are also noted by John Stow in *Chronicles of England* and Samuel Pepys in his diaries (see Gilbert and Horner, 1984).

It was during the 12th century that the initial wooden London Bridge was replaced by one built of stone. This bridge had very short spans supported by massive piers. The narrow openings of the bridge seemed appropriate at the time, but as the tidal range increased, the velocity of the water passing through the restricted openings increased alarmingly. This eroded the riverbed so badly that the piers were undermined. Construction of buildings on the bridge made things worse and the bridge became unsafe; hence the old nursery rhyme "London Bridge is falling down." Engineers trying to protect the bridge laid flagstones on the riverbed, constricting the flow even further.

With this bridge in place all shipping berthed downstream and the Port of London was developed by the construction of nontidal basins. These expanded during the 18th and 19th centuries as the population rose from 985,000 in 1801 to 2,995,000 in 1901 (Gilbert and Horner, 1984).

2. Geological Problems

The rapid development of London meant that areas prone to flooding were used for both housing and industry. Occasional catastrophic flooding events were largely ignored. Archaeological evidence of Roman riverside usage indicates that the level of the Thames has risen by about 4.6 m in 2000 years (approximately). This rate of sea level rise is, clearly, not the product of sea-level rise alone and must be a combination of both the postglacial sea-level rise and the sinking of the London area. In 1881 a "surge" tide, well above normal levels (Fig. 2), should have been a warning that there was a third variable: the weather.

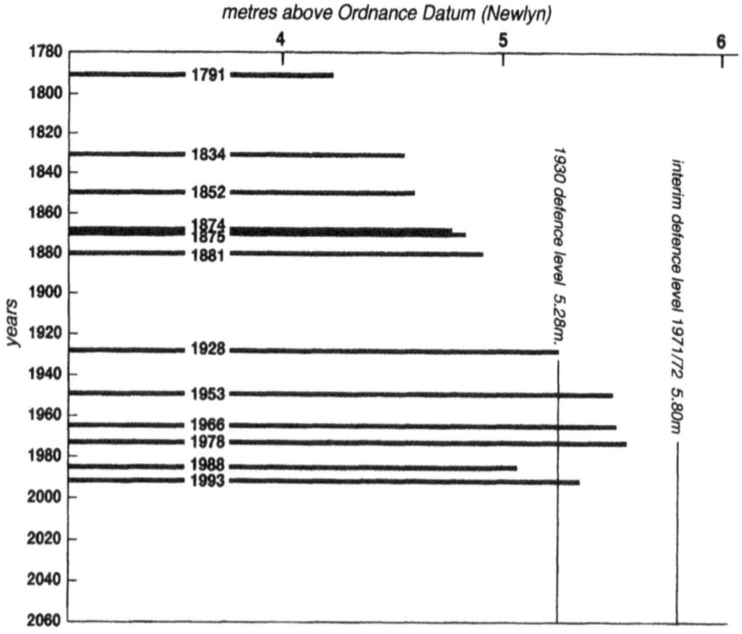

FIGURE 2. Rising levels of tidal maxima at London Bridge (adapted from Horner, 1987, by addition of new data supplied by the Environment Agency).

The 1881 event drew attention to the problems of a relative rise in sea level in the London area but it was the floods of 1928, and the associated loss of life in the cellars and basements of Chelsea, that caused a public outcry. The 1928 event, exceeding the previous (1881) record by 0.3 m, prompted the government to act. The Liverpool Tidal Institute and the Meteorological Office were asked to investigate the cause of the flooding and, in 1929, Doodson and Dines identified the "storm surge effect" as the third variable. With the publication of this research the three causes of London's problems had been identified; global sea-level rise (=eustatic change), subsidence of the area, and storm surges.

2.1. Eustatic Sea-Level Change

The term "eustacy" has been attributed to Suess (in *Das Antlitz der Erde*, as *eustatische Bewegungen*, which was written in 1888 and later translated into French in 1921) and become accepted as any type of absolute sea-level change, except those generated by local or regional influences. Eustacy was thus defined as worldwide, simultaneous change in sea level. With the discovery of variations in geoid relief (Morner, 1976) this concept became redundant as sea-level change could no longer be claimed to be worldwide or synchronous. Morner (1987) defines eustacy as any absolute sea level change regardless of cause. The variables are, thus:

1. Tectonoeustatic (ocean basin volume changes)
2. Glacioeustatic (ocean water distributional changes)
3. Geoidaleustatic (geoid relief deformation or change)

Glacioeustatic changes, following the Devensian glacial episode, created the Flandrian transgression (Fig. 3) and this separated the United Kingdom from mainland Europe. Should global warming become a proven reality then this may translate into a further rise in sea level.

2.2. Subsidence of the London Area

The theory of plate tectonics suggests that the surface of the Earth is composed of thin, rigid, semielastic plates of crustal material that "float" upon the upper portion of the mantle. Loaded by ice, plates are downwarped and then, when the ice is removed by melting, the process is reversed. Although the London area was periglacial during the Devensian (Fig. 1), northern Britain along with much of northern Europe was covered in ice and is now in a rebound phase. Southern England is, effectively, being tilted downward by this uplift in the north (Fig. 4).

Additional sinking of the London area is caused by the extraction of groundwater from strata below the city (Wilson and Grace, 1942; Poland and Davis, 1969). The resulting compaction of the clay (and chalk) beneath the city

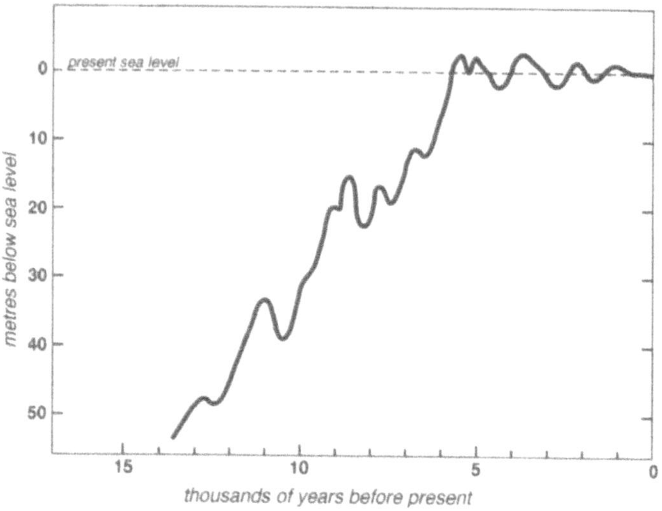

FIGURE 3. The Flandrian sea-level rise (after Eddles and Hart, 1989).

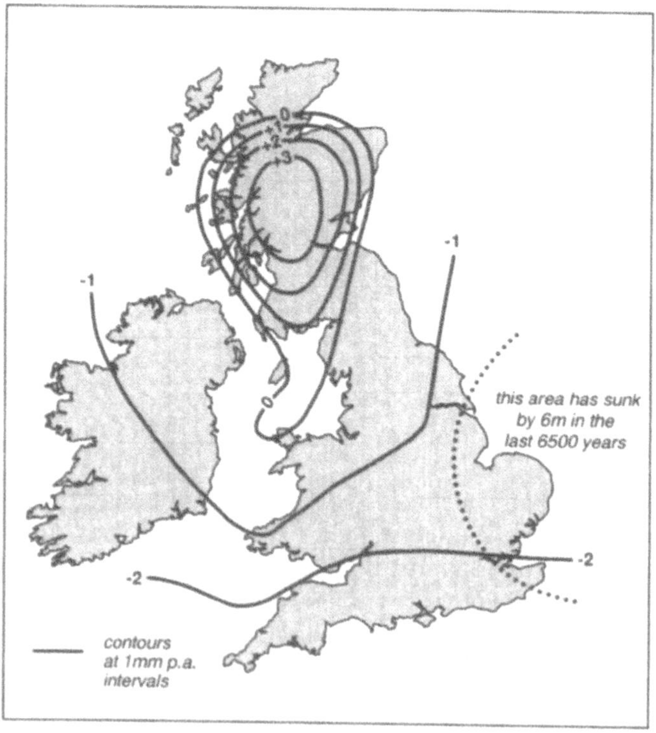

FIGURE 4. The tilting of the United Kingdom as a result of glacial unloading and isostatic readjustment (after Dunning *et al.*, 1978).

(over an area of some 295 km^2) may account for an additonal settlement of 0.2 m. While the loading of the clays beneath the city is irreversible, it is also unlikely that any recharge of groundwater levels would reverse the drying/compaction process.

2.3. Storm Surges

The low-pressure weather systems that habitually cross the British Isles arise as a result of the interaction of oceanic circulation and atmospheric phenomena. The interaction of the warm Gulf Stream and the cold Labrador Current off the coast of eastern Canada causes a drop in air pressure and as cyclonic winds blow into the centers of these "depressions" they intensify as they move eastward across the North Atlantic Ocean. Low pressure in the atmosphere allows the sea surface to rise and beneath an average depression a swell 0.3 m high, covering an area of ca. 2590 km in diameter (Ermington, 1989), may be formed. As this swell mounts the continental shelf, the shallowing magnifies the sea-surface rise.

The usual tracks of depressions are to the northeast, into the Norwegian Sea, but occasionally they track across the United Kingdom. If the depression is very deep, with an associated area of high pressure to the west of the United Kingdom, a northerly gale (winds of 63–74 km/hr) can blow southward, moving the dome of water into the southern North Sea.

This is precisely what happened in January/February 1953 (Fig. 5) as a depression swung southward across the North Sea. The coastline of England and the Low Countries funneled the waters into the constricted shape of the southern North Sea and this added to the height of the "surge." The strong northerly winds also banked the water up against these coastlines. In the estuaries, the narrowing "V-shaped" profiles of the rivers further exacerbated the problem. The disaster caused by the 1953 floods generated a number of investigations (Waverley Report, 1954; Bondi Report, 1967). The risks to London were starkly presented by Sir Herman Bondi, who concluded that the financial cost of a catastrophic flood could be around £5,000,000,000 (1966 prices) and that no responsible government could not act to prevent it. The Greater London Council (GLC) was invited to investigate the matter.

3. The Site Investigation Process

Prior to the main investigation (in 1971/1972) a number of Thames tidal control structures had been proposed, all of which attempted to protect the areas at risk (Fig. 6). Designs involving barriers (or other forms of defensive structures) downstream from London (Fig. 7) were, generally, ruled out as being too costly and/or disruptive to shipping. One proposal to build a barrage from Margate to Clacton, almost identical to The Netherlands Ooste Scheldte

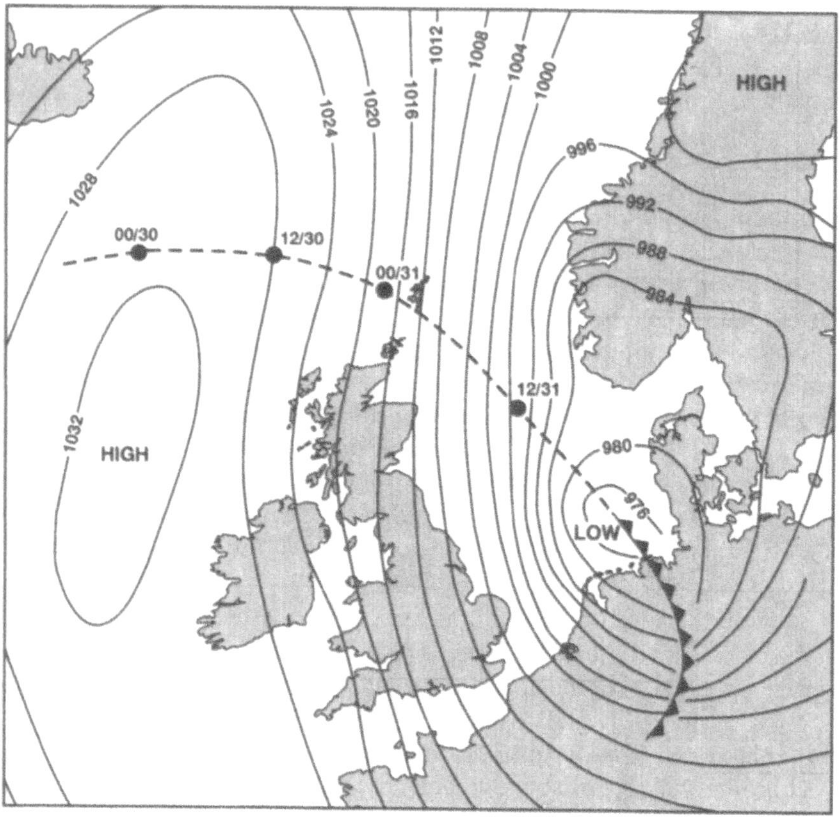

FIGURE 5. Weather map for February 1, 1953 (after Heaps, 1977). The center of the depression is located at 12-hr intervals from midnight on January 29/30 through to its position at midnight on January 31. Barometric pressures are given in millibars.

project, was discounted on grounds of cost. It was recognized during these investigations that a barrier located in the Woolwich area would also involve coastal defense works downstream in order to protect the area bordering the lower estuary. A barrier at Long Reach would cost more but would involve much less coastal protection work.

The Thames' riverbed is largely underlain by alluvial sediments, including sands, silts, and gravels. The construction of a massive barrier would place a considerable load on this substrate and the bearing strength of these unconsolidated modern sediments was clearly insufficient. It became clear that solid foundations, with low compressibility and high shear strength, were necessary and this consideration began to influence the site selection process. The geological succession below the superficial sediments provides only one such rock: the chalk (of Late Cretaceous age).

The potential site was, therefore, limited to locations where chalk was seen in outcrop or where reasonably unweathered chalk would be encoun-

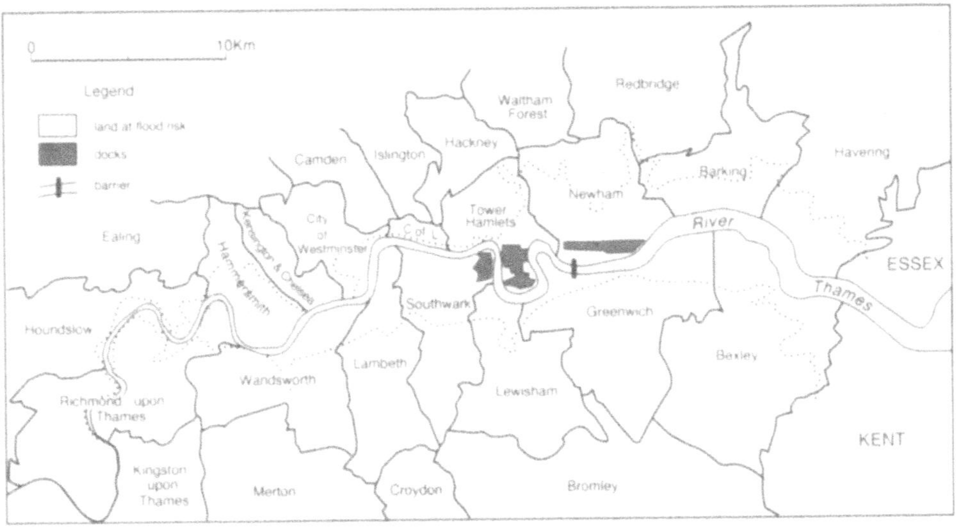

FIGURE 6. Areas at risk of flooding prior to barrier construction (after Freeman, 1993). The final location of the barrier is indicated as are the commercial docks that were in operation during the 1960s.

tered by the foundations. Only two sites were available: near the village of Chalk (now subsumed into the eastern suburbs of Gravesend) in Kent and Woolwich Reach. The site near Chalk is located immediately downstream of the (then) developing Tilbury Docks and as such was deemed unsuitable as *either* of the locations would result in closure of all the docks just east of the City of London (Fig. 6) and it was planned for Tilbury to become the main port for London. The site at Woolwich was a considerable distance upstream and,

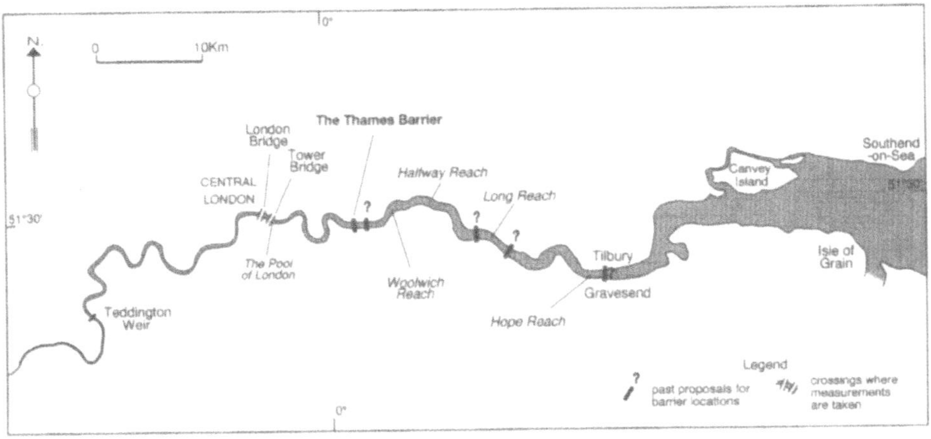

FIGURE 7. Potential barrier sites (after Freeman, 1993).

as a result, would require a much smaller structure. Construction of the barrier in that area would, however, necessitate a much more extensive program of "bank-raising" downstream—including another (quite small) barrier on the Barking Creek (see later discussion).

With Woolwich Reach appearing to be the favored location, consulting engineers were employed to investigate the site in detail and this involved a complete geological assessment of the potential location of the barrier.

3.1. Preliminary Site Investigation

The geology of the London area is well known, having been mapped on several occasions by the British Geological Survey in the previous 100 years. Figure 8 is a map of the area reinvestigated for the barrier works and shows the location of the proposed site, together with the faults that were known to cross it. It was soon clear that a number of important questions required answers:

1. Did major faults cross the actual barrier site?
2. Did minor faults cross the actual barrier site?

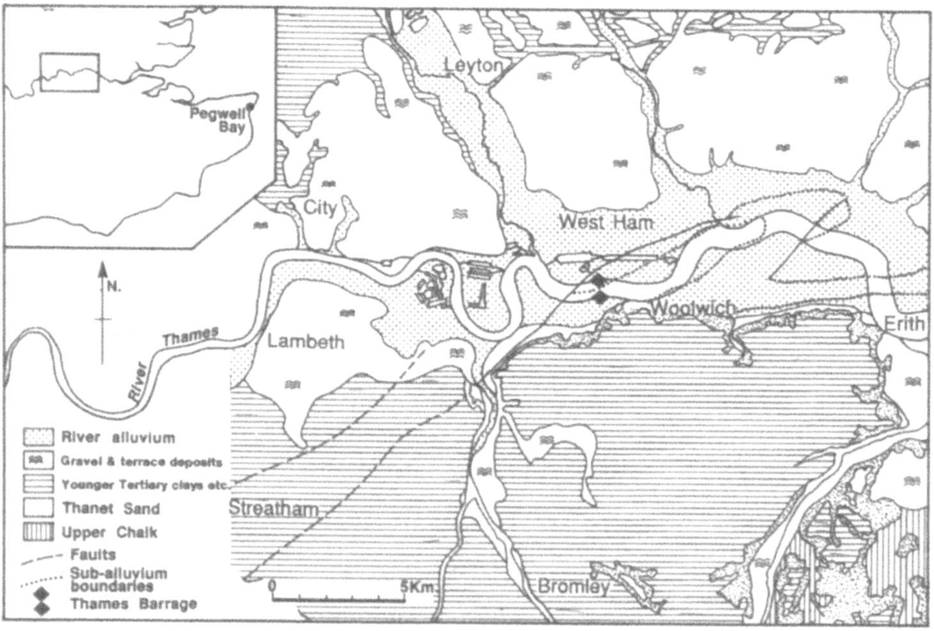

FIGURE 8. Outline geological map of the Thames Barrier site. The area of the map is indicated on the inset map of southeast England. Pegwell Bay (Kent) is the nearest outcrop of the chalk/Thanet Sands succession and this was used as a reference section during the investigations (after Hart, 1993).

3. Was the chalk between the suites of faults disturbed in any way?
4. Was the stratigraphical succession (and, potentially, the engineering properties of the chalk) variable across the site?

As the investigation was to involve quite narrow diameter boreholes (4 inch or 10.2 cm) drilled through fairly uniform white chalk Prof. P. G. Fookes (one of the consulting engineers on the project) realized that micropaleontology could be used to provide the detailed correlation that would be required. D. J. Carter (then at Imperial College, University of London) was taken on as a consultant, and I assisted him (I was then at the University of Newcastle-on-Tyne). The drilling, sampling, and micropaleontological work was undertaken in 1971 and early 1972.

At an early stage in the site investigation process a decision was taken not to use extensive geophysical surveys or undertake geophysical logging of boreholes. I am not aware of the exact reasons for these decisions, but it must be remembered that this investigation was undertaken in 1971/1972 and the various technologies that would be used today were not available at that time. Wimpey Laboratories Ltd carried out a limited water-borne acoustic survey of the river channel to provide information on subriver bed features. Two systems with energy at different frequencies were used—described as "pinger" (transducer source) and "boomer" (mechanical percussive source). The pinger proved more successful, but only in the southern half of the river. The data were used to confirm the faults and folds predicted by the micropaleontological results (see later discussion).

3.2. Micropaleontological Problems of Working on a Small Site

Most geologists and micropaleontologists are trained in interregional and interbasinal correlation. In such cases it is normal for a few taxa, common to both areas, to be relied upon for the creation of an appropriate stratigraphy. Using planktonic Foraminifera in the Upper Cretaceous, for instance, Robaszynski and Caron (1995) developed a viable zonation that is applicable extensively across several continents in what is commonly described as the "Tethyan Realm."

In engineering geology, however, the site may be less than 1 km across and the accuracy of correlation demanded by the contractor is usually far beyond that which is either possible or affordable (effort/accuracy = time = money). In prestige construction programs such as the Thames Barrier and the Channel Tunnel (Hart, 1993) sampling and placing of boreholes is usually adequate for the task but that is certainly not the case in most investigations. In every case it must be understood that the contractor does not want the succession fitted into a global stratigraphy but *demands* that one provide a precise and understandable (to them) local correlation. This can be difficult! While extinctions and first appearances of taxa provide such guidelines to the international biostratigrapher it is impossible to guarantee that such an event is recorded anywhere on the site, or that it can be relied upon as being synchronous on such a local scale (bearing in mind the sampling interval). Figure 9 attempts to show how the problem manifests itself in a series of imaginary boreholes.

In any preliminary research the published literature will almost certainly be consulted and, while this can be useful, it can also be misleading. Published, academically oriented research will often depend on a large number of closely spaced samples, with resampling undertaken at key boundaries. Taxonomic splitting, often using biometrics, may well be beyond what is possible with either the material or the time available. In many cases it is probably safer to use an informal, "in-house" generated taxonomy that is appropriate for the job in hand. Although quite obvious, it is probably worth repeating that the contractor, especially in specific, one-off, engineering

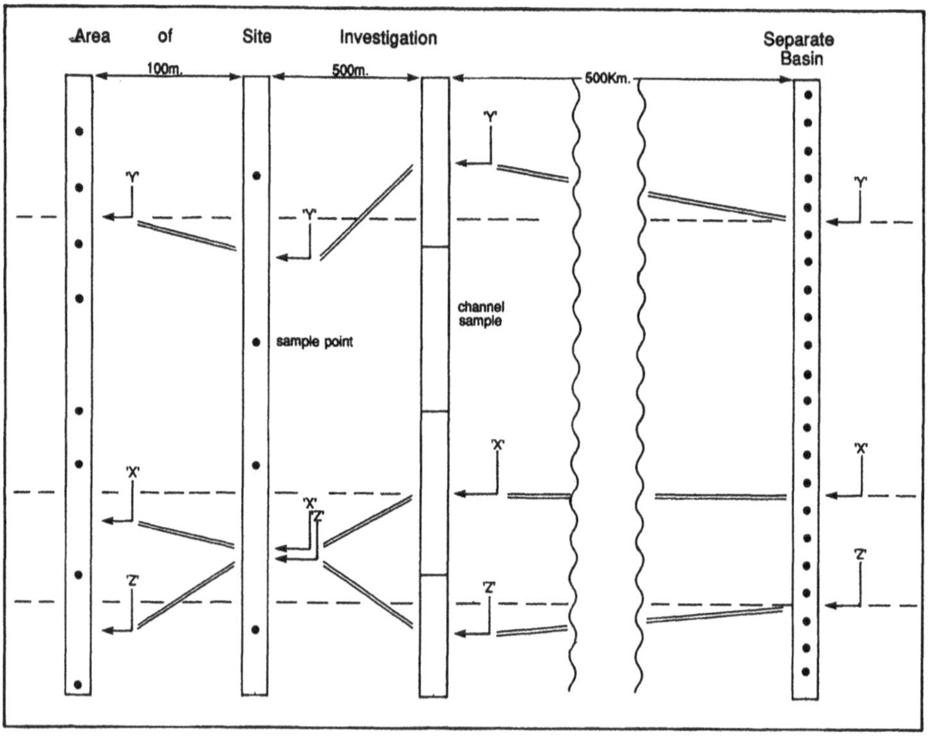

FIGURE 9. Theoretical model of the correlation problems that arise in engineering work. Bioevents "X," "Y," and "Z" in the boreholes show how the samples from the engineering site are almost more difficult to correlate than to use one borehole and a site in a separate basin. The variable sample spacing generates one type of problem while the use of "channel" samples means that any bioevent can only be attributed to the midpoint of the sample (after Hart, 1993). In engineering geology channel samples are often the result of extruding material from a U100 and "bagging" the whole of the sample. The only way to collect a valid micropaleontological sample from such material is to take a piece of every fragment of rock (to sample the whole interval) and then plot any fossil occurrences at the midpoint of the sampled interval.

investigations, has not the slightest interest in the fauna being employed by the consultant or in its wider significance.

Overuse of specific names of the fossils used and the detailed use of "zones" (e.g., the *Rotalipora cushmani* Taxon Range Zone) may worry the contractor, who probably feels that he/she is paying for some rather expensive academic indulgence. It is often much better to say, for example, that the extinction of *Rotalipora* sp. identifies the top of the "green zone." The engineers can then, relaxed by the thought that they have not paid *too* much for some useful data, draw a green line across their site plan. In subsequent diagrams and discussion the use of titles such as the "Blue Zone," "Red Zone" and "Yellow Zone" will be noticed!

3.3. Micropaleontological Investigation

As all the pier foundations were designed to penetrate the alluvium and the Thanet Sand Formation no detailed work on these units was undertaken as a part of the investigation. Sixty-eight boreholes were drilled to investigate the Upper Chalk. A number of these were subjected to *in-situ* engineering testing while all the boreholes were sampled for both laboratory engineering testing and the micropaleontological investigation. Many of the borehole sites were begun as percussion cores, hammering in 100-mm diameter tubes (U100's), and then continued downward as continuously cored rotary boreholes with variable recovery rates. Small samples were taken out of these cores as 4-mm-diameter plugs (U4's). An undisturbed sample (as the "U" denotes) can be regarded as one that is removed from its natural condition without disturbing its structure, density, porosity, moisture content, or stress condition. The standard sampling tube for obtaining samples from cohesive soils is referred to as the U100. This has a diameter of 100 mm, a length of approximately 450 mm, and walls 1.2 mm thick. On withdrawl from the borehole the sample is sealed into the tube with paraffin wax and end caps screwed on: it is later extruded and sampled.

There were 49 over-water borings, 3 of which were done from staging connected to the south bank. The remainder were done from barges. There were 19 land boreholes, 2 of which were much longer and used as reference points on each bank. The boreholes followed a pattern dictated by the final structure of the barrier. All the chalk samples yielded abundant microfaunas and most of the chalk cores showed very little drilling disturbance. The main cause of coring failure and/or disruption was an occasional small flint (chert) jammed inside the drill-bit. D. J. Carter and I logged and sampled all the cores laid out on the floor of a large warehouse adjacent to the site.

The engineering properties of chalk vary little acoss a relatively small site (Carter and Mallard, 1974) provided that there is no great variation in stratigraphical level of the material and that the chalk is *in-situ* and not deeply weathered or frost-shattered. The micropaleontological investigation had, therefore, two clear objectives:

1. The production of a correlation sufficiently detailed to permit construction of geological sections across the site that would show the founding level of all piers and the position of any faults/folds that might leak or affect the stability of the barrier when constructed.
2. The location of boundaries between the undisturbed *in-situ* chalk and any overlying disturbed, soliflucted, and/or frost-shattered chalk.

Although the foraminiferal fauna of the Lower Chalk (approximately Cenomanian) had been studied in detail for the site investigations for the Channel Tunnel (Carter, 1966; Hart, 1993) and in subseqent research (Carter and Hart, 1977*a*; Jefferies, 1961, 1963), this was not the case for the Middle Chalk (Turonian–Coniacian) or the Upper Chalk (Coniacian–Maastrichtian). Some of the faunas had been described by Williams-Mitchell (1948), Barnard and Banner (1953), Barr (1962, 1966), Hart (1970), and Owen (1970), although there was, no comprehensive study of the microfauna to be encountered on the site at the time of the investigation. Much basic research was, therefore, required in order to determine the overall succession and to apply it to the problems in hand. Many of the taxa were left in open nomenclature or given "best fit" names, even though the authors knew that in any normal taxonomic study they would clearly be invalid.

The sampling and processing procedures have been described by Carter and Hart (1977*b*). In most of the borehole cores, samples were taken every 1.5 m, but where lithological variation was observed in the opened cores sampling intervals were reduced to 1.0 m or less. Samples from the cores were taken very carefully, after extrusion, always removing the disturbed material from around the edge of the core prior to sampling. The majority of samples could be crushed easily under water in a pestle and mortar and then sieved on a large-diameter 200-mesh (75-μm) sieve.

Borehole 25 (Fig. 10), located on the south bank of the river, was used as a reference section and studied first of all. Three grain-size-fractions were studied (>500 μm, 500–250 μm, and 250 – 75 μm) but counts were only done on the middle-size fraction as it contained the majority of the diagnostic fauna. The other size fractions were only scanned for certain taxa that were, eventually, identified as having some stratigraphical value [e.g., *Loxostomum eleyi* (Cushman)]. As only a minority of the species encountered were described in the available literature, all species—except those immediately identifiable—were given an approximate generic assignment and an informal "letter" designation (e.g., *Gaudryina* sp.B) or, in some cases, a number. Certain species known to have long ranges or taxonomic problems associated with them were not counted individually, but in groups. Seventy-one species of benthonic and eight species of planktonic Foraminifera were identified and over 250,000 individuals counted.

Standard distribution charts for all the boreholes were produced; an example of these is shown in Fig. 10. As work progressed not all the 79 taxa were plotted on such charts as it was clear that only certain taxa had any value relevant to the problems of the site. Different methods of presentation and

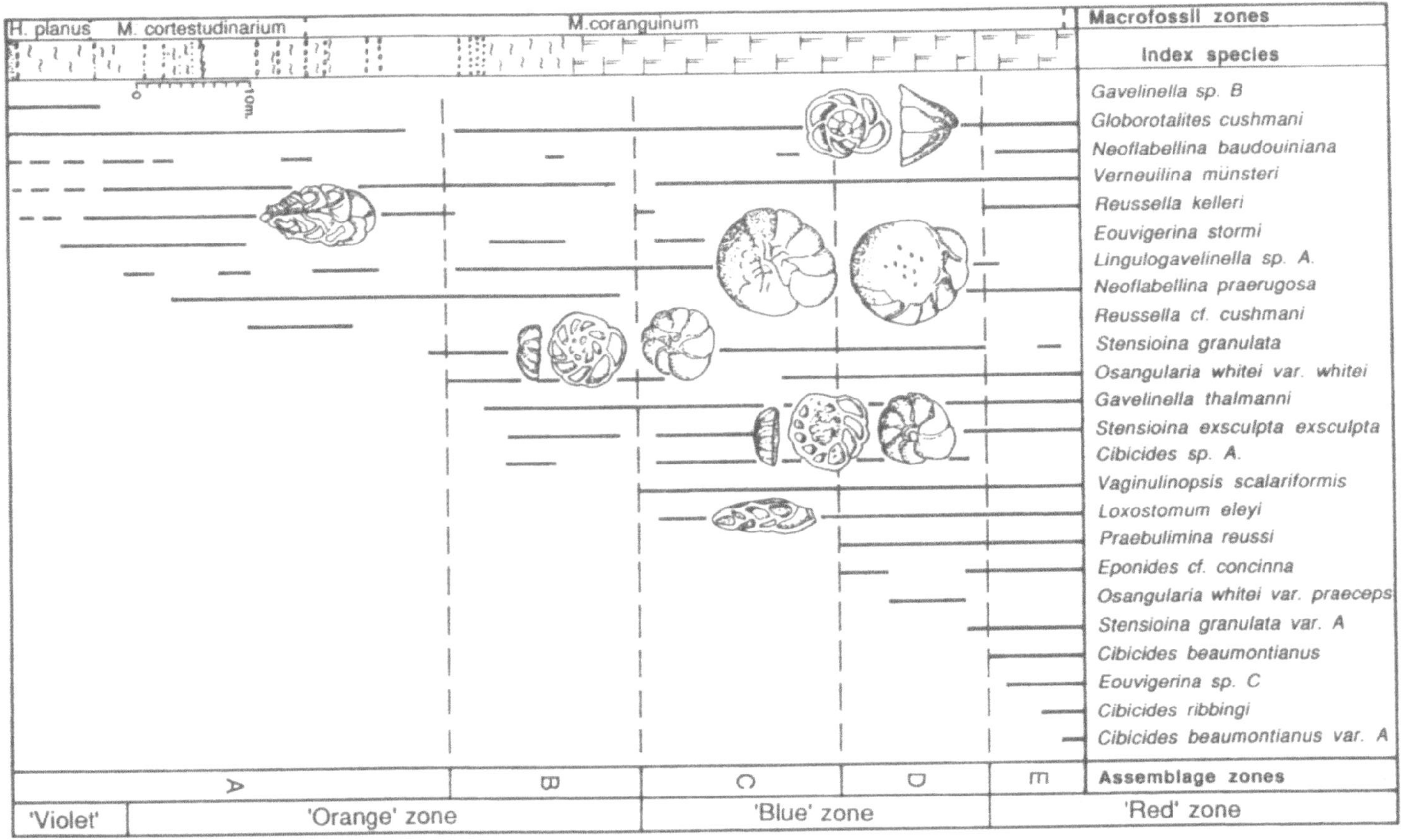

FIGURE 10. Foraminiferal analysis of borehole 25, located on the south bank of the River Thames. The informal zonation used in the site investigation is compared with the zonation (A–E) established by Bailey (1978) following a comparison with other successions in the southeast of England. The "color" zones are also used in Figs. 11 and 12 (after Hart, 1993).

grouping of species were developed and employed in different ways during the work. The following plots were found to be useful:

1. A simple range chart showing the total range of stratigraphically significant taxa.
2. A distribution chart for the 500 to 250-μm-grain-size fraction showing the percentage abundance of species and/or groups of species within one benthonic Superfamily (e.g., Cassidulinacea *sensu* Loeblich and Tappan, 1964) plotted as a cumulative histogram (Fig. 11, Graph A).
3. A distribution chart for the 500 to 250-μm-grain-size fraction showing the percentage abundance of species groups of planktonic foraminifera in the total assemblage, expressed as a cumulative percentage graph (Fig. 11, Graph B).
4. A distribution chart for the 500 to 250-μm-grain-size fraction showing the percentage abundance of the four superfamilies (*sensu* Loeblich and Tappan, 1964) represented in the benthonic fauna expressed as a cumulative percentage graph [Fig. 11, Graph C (Buliminacea only)].

The foraminiferal zonation was based on (1) and (2). The recognition of intrazonal subdivisions was based on (2), (3), and (4). Foraminiferal zones were erected for borehole 25 and the overlapping borehole 19 (Fig. 11) and then applied across the site (Fig. 12). To avoid the use of complex foraminiferal nomenclature, and to placate the engineers, these were given the names of colors (which were later plotted on the geological sections across the site—as in Fig. 12). In 1971/1972 the relationship of these zones to the then accepted macrofossil zonation in the lower part of the Upper Chalk was unclear, but our subsequent work (Bailey, 1978, Bailey and Hart, 1979, and Hart *et al.*, 1989) clarified the situation.

Microfossil zones are based on the vertical ranges, or overlap of vertical ranges, of species. Any species at the beginning or near the end of its vertical range tends to be rare and, therefore, irregularly distributed laterally (now called the Signor-Lipps effect for species at the end of their range); for this reason species can appear and disappear at slightly different levels in closely spaced boreholes (e.g., *Stensioeina exsculpta exsculpta* (Reuss) appears before *Loxostomum eleyi* (Cushman) upsection in borehole 25, but the order of appearance is reversed in borehole 52 (see Fig. 11). For this reason the positions of zonal boundaries can vary slightly from borehole to borehole and correlations based on them are not always completely accurate on a local scale. The Red/Blue zonal boundary, however, is sharp and easily recognizable. The microfaunal change marking it occurs within the stratigraphical thickness of 0.01 m and the accuracy with which it is located in the boreholes depends solely on the sampling interval (see Fig. 9). The change downward from the Red to the Blue zone involves the sudden appearance of the very striking *Lingulogavelinella* sp. A [later named *Lingulogavelinella* sp. cf. *vombensis* (Brotzen)]. Subsequent work (Bailey, 1978, Robaszynski *et al.*, 1980, and Hart *et al.*, 1989) has shown the value of this taxon in correlation, although it is

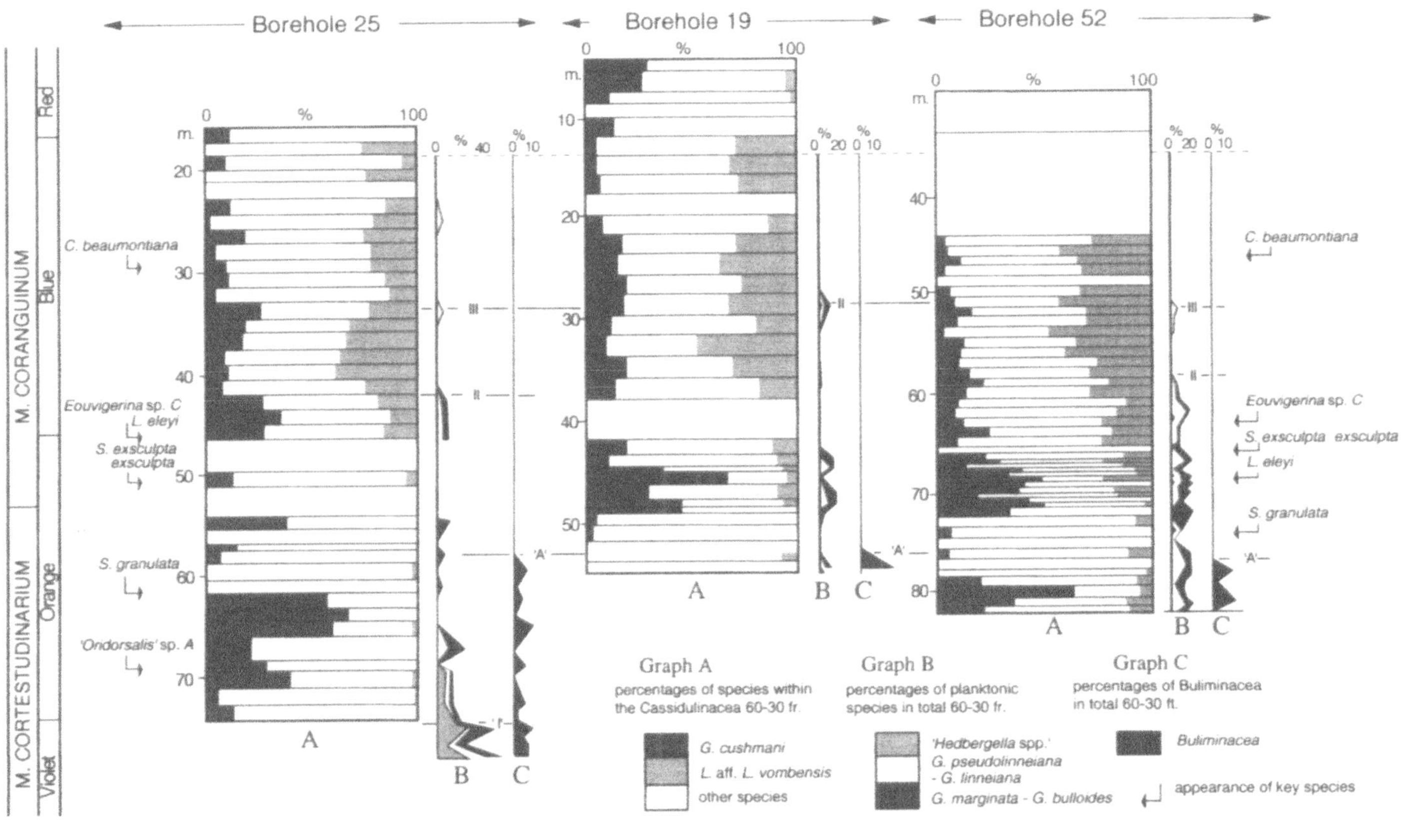

FIGURE 11. Raw micropaleontological data from boreholes 25, 19, and 52 (see locations on Fig. 12). The 60–30 fraction is the 500- to 250-μm-size fraction, which was used as the standard in the project. This provides both ease of microscope work and, in the chalk of the United Kingdom, provides the bulk of adult specimens of most species represented in the samples. Adding the smaller, or larger, size fractions gave no better resolution. Graphs A, B, and C were used to effect a correlation. As demonstrated in Fig. 9, one can see that in boreholes 25 and 52 the first appearances of *Loxostomum eleyi* and *Stensioeina exsculpta exsculpta* are reversed. It is also clear that the correlation of lines "A" ll and lll provide a good correlation as does the sharp reduction in the abundance of *Lingulogavelinella* sp. cf. *vombensis* (after Hart, 1993).

now accepted as not being the *vombensis* of the original author (a Maastrichtian species from Sweden).

Correlations based on zonal boundaries were supplemented by others based on different criteria.

3.3.1. Fluctuations in the Abundance of Benthonic Species

Many long-ranging benthonic species fluctuate in abundance through the drilled section. These fluctuations are particularly well marked in the Superfamily Cassidulinacea (*sensu* Loeblich and Tappan, 1964) and are best demonstrated using cumulative percentage histograms of species or groups of species within the superfamily in the 500 to 250-μm-grain-size fraction. The cumulative histograms for boreholes 19, 25, and 52 are compared in Fig. 11, Graph A). The overall distribution patterns are governed largely by those of two constituent species, *Globorotalites cushmani* Goel [including transitional individuals with the younger *G. micheliniana* (d'Orbigny)] and *Lingulogavelinella* sp. cf. *vombensis* (Brotzen).

Similar histograms were prepared for 18 boreholes. Differences were found to occur both as the distances between the boreholes and between samples in the same boreholes increase. The former reflects increasing difference induced by lateral changes in environment between assemblages living on the sediment substrate, and the latter shows increasing distortion of the true distribution pattern caused by increasingly incomplete representation. Correlation lines based on the connection of features of these intrinsically unstable patterns must be treated with considerable reserve and considered as guides to correlation rather than as firm ties.

3.3.2. Planktonic:Benthonic Ratio

Planktonic foraminifera in general evolved rapidly, but in these assemblages they show little differentiation and the same taxa are sporadically distributed throughout the succession. However, studies of those in borehole 25 reveal changes in their abundance at various levels. Graphs for boreholes 19, 25, and 52 are also shown in Fig. 11. Three features are easy to recognize: the sudden fall in planktonic percentages up-section at the Orange/Violet boundary (level 1 in the figure); the sudden disappearance of planktonic species up-section in the lower part of the Blue zone (level 11 in the figure); and their sudden, temporary, reappearance in fair abundance near its middle (level 111 in the figure). The unequivocal identification of these and other features depends on the availability of long, tightly sampled sequences from the boreholes to be correlated. In the majority of the boreholes from engineering works such good recovery is rarely realized.

3.3.3. Percentage Abundance of Superfamilies

When the 500–250-μm grain size fraction of benthonic assemblages from borehole 25 were separated into their constituent superfamilies (*sensu* Loeblich and Tappan, 1964), the overall pattern was not sufficiently distinctive to serve as a basis for correlation. However, one superfamily (Buliminacea—Loeblich and Tappan, 1964), which is well represented in the <250-μm grain size fraction, contains a few species that attain large size in some parts of the succession and, therefore, also appear in the 500–250-μm grain size fraction. These large-sized specimens appear sporadically in the lower part of the Violet zone, increasing in abundance upward, and may constitute up to 15% of the 500–250-μm grain size fraction in the lower part of the Orange zone. About two-thirds of the way up the Orange zone these large specimens abruptly disappear at a level thought to represent a slight depositional break. This horizon (level A), which might approximate the *Micraster coranguinum/ Micraster cortestudinarium* zonal boundary, is sharply defined and a good datum: its positions in boreholes 19, 25, and 52 is indicated in Fig. 11. Unfortunately, because of its low position in the succession, it is absent from the majority of boreholes.

It is generally accepted on evidence from Kent and Surrey that the Thanet Sand Formation, often with the "Bull Head Bed" (a flint conglomerate) at its base, overlies chalk that has been very gently folded and tilted prior to erosion and the deposition of the overlying strata. Profiles of identifiable horizons within the chalk on site demonstrate that although the erosion plane cuts only one macrofossil zone (the *M. coranguinum zone*), this had been thrown into a series of shallow flexures before beveling occurred. The river alluvium and flood plain gravels were shown to rest on the Thanet Sand Formation only to the north of borehole 12; to the south of it they rest directly on chalk. Post-Thanetian structures were investigated by plotting a stratum contour map of the sub-Thanet erosion surface (top of chalk) based on triangulation of site borehole levels. We discussed this fully (Carter and Hart, 1977*b*) identifying an en-echelon suite of minor northwest–southwest faults (of Miocene age?) across the area (Fig. 12).

In order to determine the nature of any pre-Thanetian structures, we (Carter and Hart, 1977*b*) used the Red/Blue zonal boundary for another contouring exercise (Fig. 13), which identified a series of minor folds that must postdate deposition of the Red zone and be pre-Thanetian. They probably relate to inversion of the chalk at the close of the Cretaceous (the so-called Laramide movements).

Failure to differentiate frost-shattered and/or soliflucted materials from firm *in-situ* bedrock can lead both to stratigraphical miscorrelations in site investigations and major difficulties during construction. These deposits are difficult to recognize, particularly in poorly sampled successions, since the one often passes imperceptibly into the other. Both can pass downward or laterally into *in-situ* chalk and even in surface sections the placing of boundaries is largely subjective. However, solifluction chalk that has shifted appreciably can often be recognized micropaleontologically since it contains fine-grained exotic material (e.g., Tertiary sand grains, glauconite, vegetable material, small snail shells) introduced during movement. In the U4's and U100's the recognition of frost-shattering is extremely difficult as secondary fracturing is often produced as the tubes are driven into the succession. When

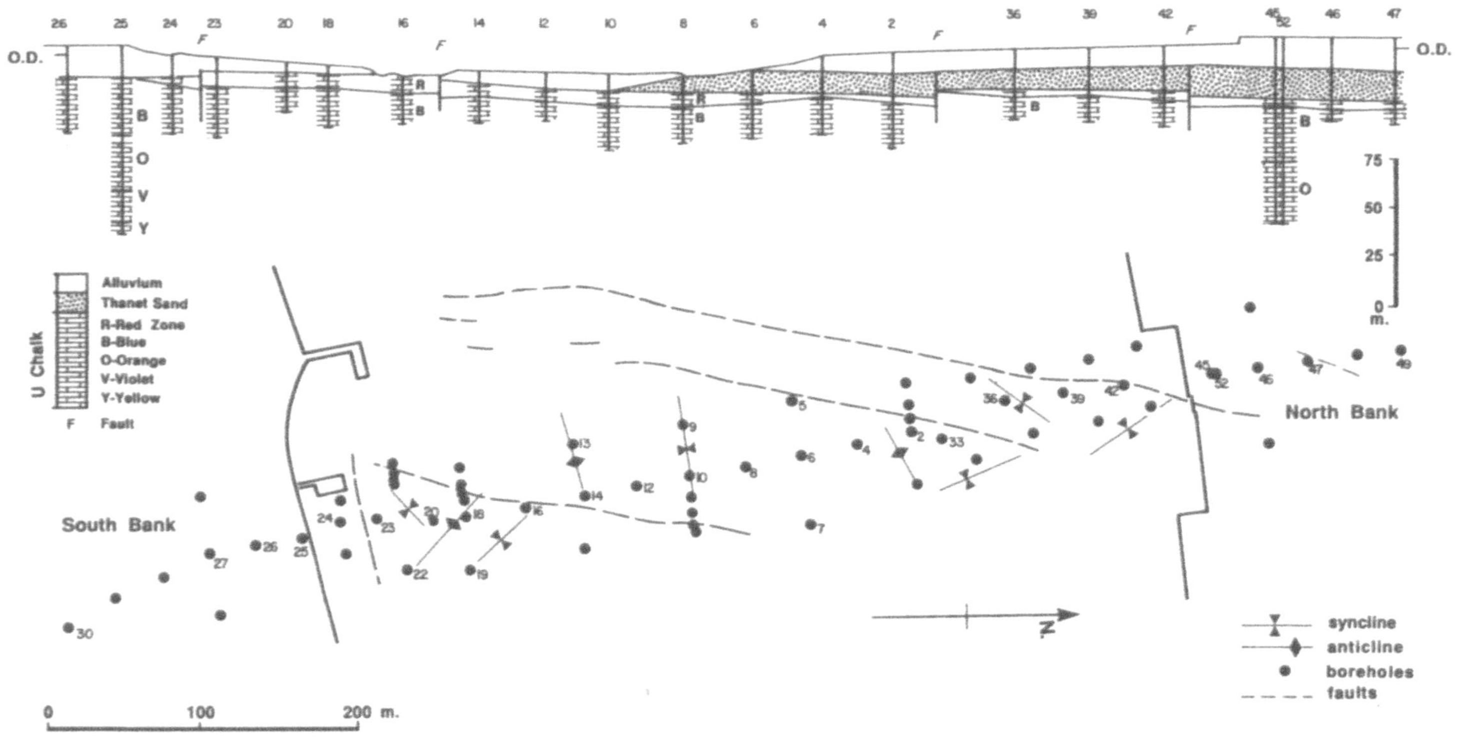

FIGURE 12. Location of the boreholes used in the investigation. Boreholes 25 and 52 provided the two main reference sections. The edge of the south bank of the river is located between boreholes 24 and 25 while the edge of the north bank is between 42 and 45. It must be stressed that the faults and folds mapped using the paleontology are very small structures and it was decided that they posed no threat to the construction. The Thanet sand formation is seen to pinch out in midriver, leaving the chalk exposed below the thin cover of alluvium (after Hart, 1993).

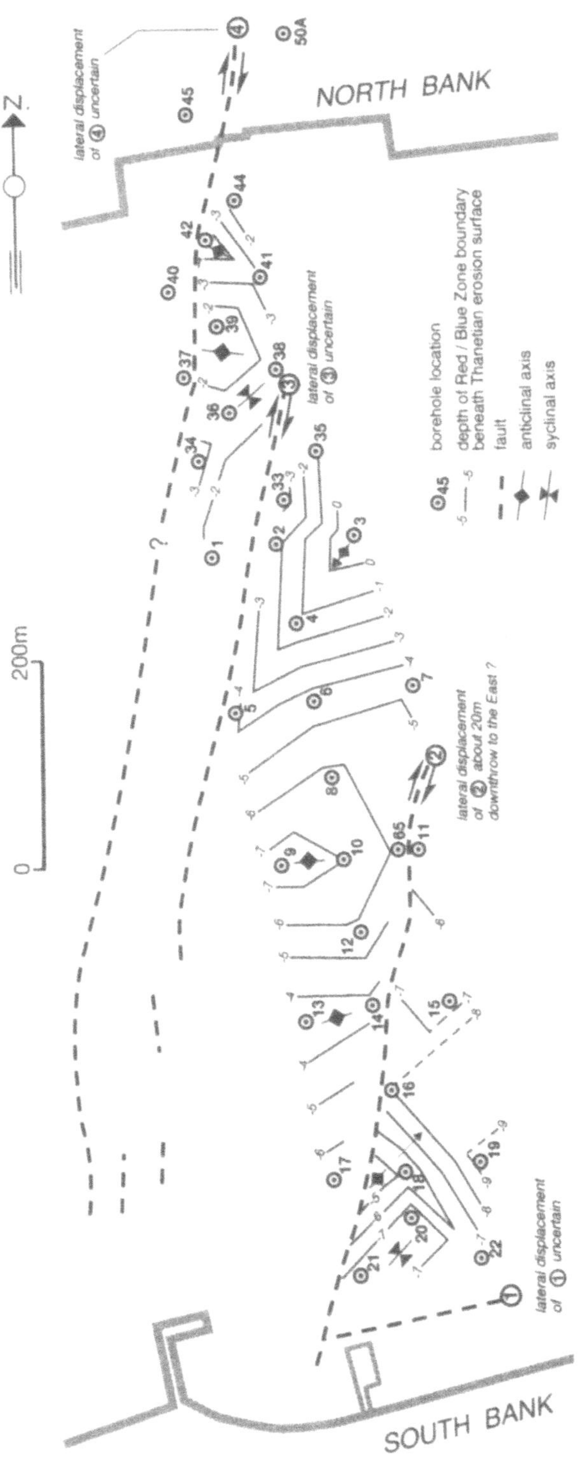

FIGURE 13. Stratum contour map showing the depth of the Red/Blue zone boundary below the top of the chalk. This is a section of part of the map shown in Fig. 12; its position can be located by means of the borehole sites, especially those located along the eventual midline of the barrier (boreholes 20, 18, 16, 14, etc.).

the chalk is examined closely the underlying structure tends to be obscured by a network of secondary fractures, but from a distance of about 2 m *in-situ* material gives an impression of general continuity of bedding. This can be enhanced by studying the cores in the dark and using a portable UV lamp, the purple fluorescence of the chalk often being interrupted by faint yellow lines (uranium in phosphates) that represent very minor, bedding-parallel, pauses in sedimentation. The top meter of unmoved, frost-shattered, chalk may show solution-rounding of fragments and contain foreign materials from above. U100's of *in-situ* chalk taken immediately below uncased and unconsolidated cover are usually contaminated by caving; such contamination was minimized by ensuring that all samples for micropaleontological analysis were taken from the center of all cores.

No typical solifluction chalks were encountered on site, and frost-shattered chalk was encountered only south of borehole 10, where the Thanet Sand Formation is missing. This suggests removal of the Thanet Sand Formation prior to the most severe Plio–Pleistocene climatic deterioration. It is not known if the late-glacial flood plain gravels, which overlie the broken chalk and Thanet Sand Formation, were in place at that time. Presumably shattering could have occurred through thin, saturated cover. However, the occurrence of undisturbed late-glacial gravels overlying frost-affected chalk passing laterally into solifluction deposits at Swanscombe, near Northfleet (Kent), indicate that the gravels were deposited later.

Frost-damaged chalk extends down from the Red zone into the upper part of the Blue zone, and this information was added to the geological sections of the site (Table 1).

TABLE 1. Levels to Which Frost-Shattering Was Noticed or Determined in Boreholes

Borehole No.	Frost-shattered chalk seen to:	*In-situ* chalk seen at:
12		10.00 m (?)
13	10.00 m (top)	10.00 m (bottom)
15	7.00 m (top)	7.00 m (top)
17		8.40 m (?)
18	5.00 m (top)	5.00 m (bottom)
19	4.75 m	6.75 m
20	4.50 m	7.00 m
21	14.00 m	16.00 m
22	12.00 m (top)	12.00 m (bottom)
25	23.45 m	24.90 m
26	21.50 m (?)	23.75 m
28	15.25 m (?)	26.80 m
29	15.25 m	18.25 m
30	15.10 m (?)	18.75 m (?)
31	17.72 m (?)	20.02 m (?)
64	9.00 m (top)	9.00 m (bottom)
67	5.00 m	7.00 m

The total site investigation cost approximately £100,000 (at 1971/1972 prices) and showed that the site was underlain by strata of adequate bearing capacity for the foundations of the piers and abutments of the barrier. The micropaleontological (and limited geophysical) studies established the existence of small-scale faulting (and folding), which was not considered to have significantly affected the engineering properties of the materials. It was also concluded that there was little risk of damaging movements from these faults or from the larger faults known to be present in the region. The investigations also indicated that deep-frost-shattering and major solution features were not present, or at least posed no risk to the structure.

4. Construction of the Barrier

The design of the barrier (chosen from 41 proposals) is compact, attractive, practical, and environmentally sensitive. Charles Draper was the engineer whose idea for radial gates on sills evolved from the "gas-tap" principle.

The design requirement stipulated that:

1. Four main navigation openings 61.0 m wide were to be provided.
2. Two subsidiary navigation openings of 31.0 m were also to be provided.
3. Piers to be kept as narrow as possible, consistent with their function.
4. Structures in the bed of the river in the navigation openings were to be at ruling depth to avoid restriction of navigation.
5. Adequate overhead clearance was to be provided in the navigation openings for the passage of shipping.
6. The reduction of the cross section of the waterway by the permanent works was not to exceeed 25% and by the temporary works not more than 30%.
7. Time required to close all gates was not to exceed 1 hr.
8. The structure was to be designed to withstand an extreme surge differential of 9.9 m and a reverse head differential of 6.1 m.

Initially the Port of London authorities demanded a 135-m main opening, with two supporting 61-m passages. The closure of the upstream dock facilities, together with evidence that vessels up to 20,000 tonnes could negotiate the 61-m passage below Tower Bridge with little difficulty, allowed this to be revised to the provision of four 61-m openings. As all the main openings were to be identical this allowed for a major cost saving in the construction phase. The final plan of the structure is shown in Fig. 14 and the operation of the gates is shown in Fig. 15.

When dangerous conditions arise, the Meteorological Office Storm Surge Warning Service notifies the barrier operators—the Environment Agency. This allows the barrier gates to be rotated upright to stop the flow upstream in 30 min. Notification of the closure is communicated to shipping by the Port

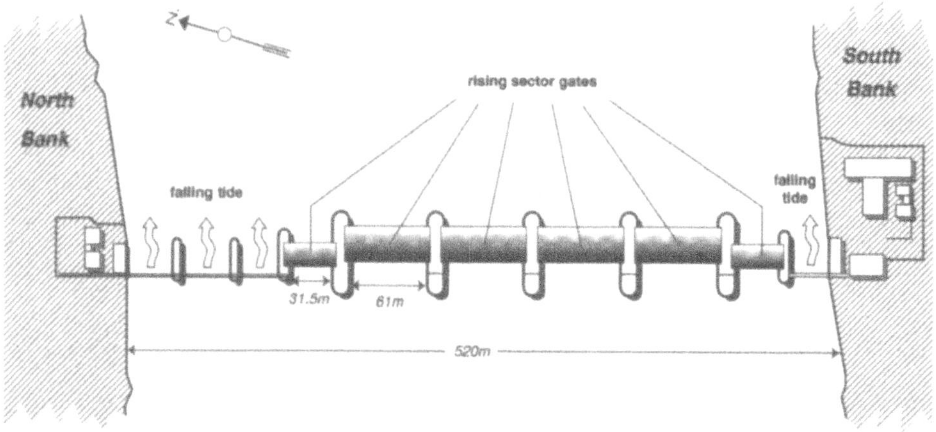

FIGURE 14. Plan of the final barrier construction (courtesy of Environment Agency). The four large and two small rising sector gates (see Fig. 15 for operation) are indicated in the center of the river. The remaining four openings are closed by falling radial gates (see Gilbert and Horner, 1984, for details).

of London Authority. Although not used at the present time, the gates can be rotated a further 20°, generating an additional 1.1 m of protection. This was supposed to extend the life of the barrier nearly to the year 2200 by predictions made in the 1970s. If global warming is a reality then this estimate may be at risk.

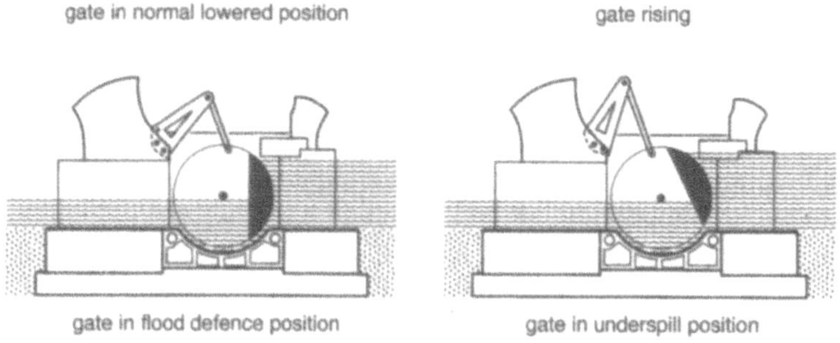

FIGURE 15. Diagrams explaining the operation of the barrier (courtesy of Environment Agency). The gate in the underspill position (tilted 20° beyond the vertical) gives the extra height explained in the text.

FIGURE 16. Photograph of the final barrier (kindly provided by the Environment Agency).

The Thames Barrier, currently valued at over £1,000 million, took 8 years to build and was officially inaugurated by Her Majesty the Queen in 1984. Half-a-million tonnes of concrete were used in the foundations on which the superstructures were constructed. When raised from their submerged position (resting inside the sills) each of the four main gates are as high as a five-storey building. Each of the gates, and the arms that rotate them, weigh 3700 tonnes. The hydraulic power packs (sheltered by the stainless steel shells) are electrically driven, using three alternative supplies, routed via each of the river banks and, should these options become unavailable, from three on-site power generators. The Thames Barrier in its final form is shown in Fig. 16.

In order that the water backed up by the raised barrier does not flood the lower reaches of the River Thames 32 km of flood defenses were built downstream, with bank levels some 2 m higher than had previously existed. Prior to construction, in 1971/1972, 102 km of "interim defenses" were constructed (see Fig. 17). These included the 60-m-high Barking Barrier, which has a drop gate that is held well above water level when not in use, thereby allowing access to Barking Creek for some commercial shipping. Upstream

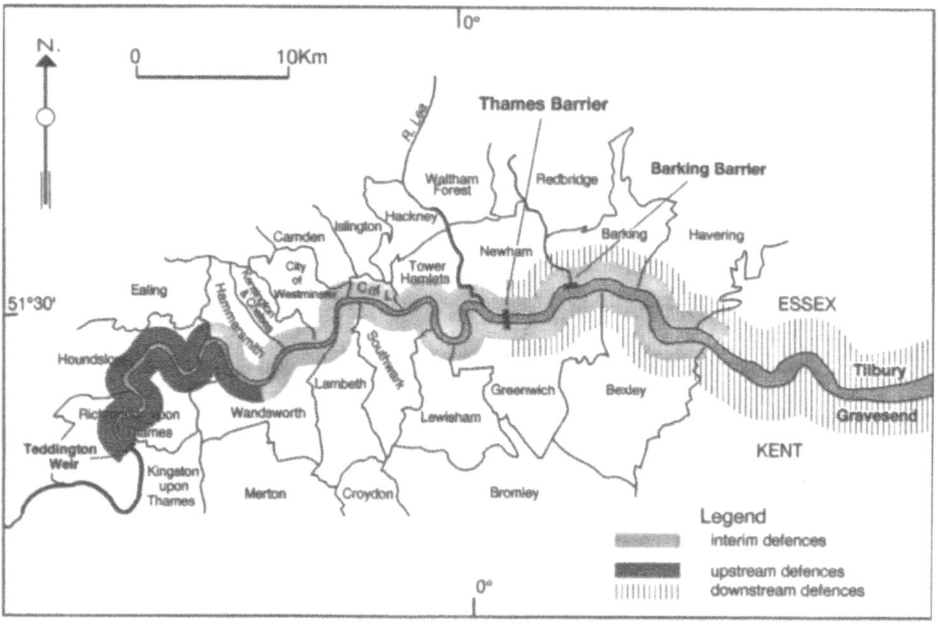

FIGURE 17. Map of interim, upstream, and downstream flood protection measures (courtesy of Environment Agency). Teddington weir is the upstream limit of tidal activity and also marks the end of the upstream defenses. The downstream defenses involved 32 km of works, raising the bank levels 2 m above their previous levels. The interim defenses were constructed during 1971/1972 (at the same time as the site investigation for the barrier) and involved 102 km of bank-raising works. This included the construction of the Barking Barrier, which has a drop gate that is held out of the water when not in use, to allow uninterrupted passage by commercial shipping using Barking Creek.

defenses were also constructed in the areas around Wandsworth, Richmond upon Thames, and Hounslow. These protect vulnerable areas from flooding that might occur when high river flows (due to rainfall) coincide with periods of barrier closure.

5. The Future

The Thames Barrier and associated flood defenses were designed to protect London from flooding until at least the year 2030. There is no reason to doubt that this target date will be achieved, especially if one uses the increased height available by rotating the gates into the 20° to the vertical position (Fig. 15). To do this on a regular basis would require additional downstream and upstream bank-raising works (see Fig. 17).

To the present day the barrier has been closed just over 200 times. Of these closures 27 have been to protect London from tidal flooding and 3 have been to prevent river flooding. All the rest, with one exception, have been for maintenance, experiments, or tests. The one exception (the 87th closure) was on August 20, 1989, when the *Marchioness* (a pleasure craft) was sunk in a collision with the *Bowbelle*. This resulted in a significant loss of life and the barrier was raised to allow the salvage and rescue work to be completed.

There is, however, a major concern over global warming and the possible impact that this might have on sea-level change during the next millennium. This is well beyond the predictions that were available during the design phase of the work on the barrier. Estimates of this change vary significantly (see Table 2) and not all authors allow for all the variables (e.g., thermal expansion of the worlds' oceans, subsidence of oceanic crust as a result of the increased loading by water). As indicated in Table 2 the predicted changes,

TABLE 2. Estimates of Changes in Global Sea Level (after Hart, A. B., 1997)

Authors	Best estimate (cm)	Range (cm)	To (year)
Hoffman *et al.* (1983)		56–345	2100
PRB (1985)		10–160	2100
Hoffman *et al.* (1986)		58–367	2100
Robin (1986)	80[a]	25–165[a]	2080
Thomas (1986)	100	60–230	2100
Jaeger (1988)	30	−2–51	2025
Raper *et al.* (1990)	21[b]	5–44[b]	2030
van der Veen (1988)[c]		28–66	2085
Oerlemans (1989)	20	0–40	2025
Clayton (1990)	164	26–365	2030
Pugh (1990)[d]	110	90–170	2100
Warrick and Oerlemans (1990) (based on IPCC, 1990)	66	30–110	2100
IPCC "BAU" Scenario (1990)	47	18–109	2100
Church *et al.* (1991)[e]	35	15–70	2050
Wigley and Raper (1992)		15–90[f]	2100
(based on IPCC, 1992)		(22–115)[g]	2100
Wigley and Raper (1993)	46[h]	3–124[i]	2100
Woodworth (1993)	61		2087
Titus and Narrayanan (1995)[j]	34	5–77[k]	2100
IPCC (1996)[a,b,c,l]	49	20–86	2100

[a] Estimated from global sea level and temperature change from 1880–1980 and global warming of 3.5 ± 2.0°C for 1980–2080.
[b] Internally consistent synthesis of components.
[c] For a global warming of 2–4°C.
[d] Surface air temperatures are assumed to increase linearly until 2050, to an average value of 3° higher than at present, and then to remain constant.
[e] Assumes rapid warming of 3°C by 2050 for best guess scenarios.
[f] Best guess with a temperature change of 1.7–3.8°C.
[g] Base case forcing with a temperature change of 2.1–5.0°C.
[h] For IPCC (1992) Policy Scenario B, best estimate model parameters.
[i] For IPCC (1992) forcing scenarios A and C with high and low model parameters, respectively.
[j] Incorporates subjective probability distributions for model parameter values based on expert opinion.
[k] Represents 90% confidence level.
[l] For the IPCC IS92a forcing scenario, using a climate sensitivity of 2.5°C for the mid projection and 1.5° and 4.5° for the low and high projections respectively. (See also Raper *et al.*, 1996).

which range from a drop of 2 cm to a rise of 350 cm, would clearly have an impact on the long-term viability of the barrier.

In 1988, the United Nations and the World Meteorological Organization jointly set up the *Intergovernmental Panel on Climate Change* (IPCC). The aim of this group is to provide an authoritative international consensus of scientific opinion on climate-related issues—including global warming. The various IPCC scenarios consider such things as:

- World population estimates
- Economic growth
- GNP growth
- Energy supplies, demands, and changes in usage
- CFC, HCFC, CO_2, CH_4, N_2O, and other halocarbon emissions
- Levels of deforestation and/or replanting
- Cumulative net carbon emissions
- Cumulative net fossil carbon emissions

A number of authors have varied the IPCC variables and reached independent conclusions on the nature and extent of the resulting sea-level rise. A total melt of all ice caps and sheets would result in a sea-level rise of 75–80 m, but even the worst estimates indicate that this is unlikely in the next 1000 years. The most accepted predictions would appear to indicate that by 2050 there will probably need to be some changes in the operation of the barrier and a raising of some of the flood defenses. As this is well within the original estimates of the life of the barrier this would appear to vindicate the planning phase of the project, which had that date as its initial target.

ACKNOWLEDGMENTS: I wish to acknowledge that the work described here was a team effort involving a large number of specialists. Particular thanks go to David Carter and Prof. Peter Fookes. The Environment Agency kindly provided the photograph used in Fig. 16, as well as data on barrier operation and flood defenses. The Agency also provided details of all barrier closures since the time of construction. Mr. J. Abraham is thanked for producing the final versions of the diagrams.

References

Anderton, R., Bridges, P. R., Leeder, M. R. and, Sellwood, B. W, 1979, *A Dynamic Stratigraphy of the British Isles*, Unwin Hyman, London, 301 pp.

Bailey, H. W., 1978, A foraminiferal biostratigraphy of the Lower Senonian of Southern England. Unpublished Ph.D. Thesis, CNAA/Plymouth Polytechnic.

Bailey, H. W., and Hart, M. B., 1979, The correlation of the early Senonian in Western Europe using Foraminiferida, *Aspekte des kreide Europas, IUGS, Ser. A.* **6**:159–169.

Barnard, T., and Banner, F. T., 1953. Arenaceous foraminifera from the Upper Cretaceous of England, *Quart. J. Geol. Soc. Lond.* **109**:173–216.

Barr, F. T., 1962, Upper Cretaceous planktonic foraminifera from the Isle of Wight, England, *Palaeontology* **5**:552–580.
Barr, F. T., 1966, The foraminiferal genus *Bolivinoides* from the Upper Cretaceous of the British Isles, *Palaeontology* **9**:220–243.
Bondi, H., 1967, London flood barrier, Unpublished Report to the Ministry of Housing and Local Government.
Carter, D. J., 1966, Channel Tunnel Study Group Site Investigation: Appendix 3 (Micropalaeontology Report), pp. MI-1–MA-14; Appendix 4 (Figures).
Carter, D. J., and Hart, M. B., 1977*a*. Aspects of Mid-Cretaceous stratigraphical micropalaeontology, *Bull. Brit. Mus., Nat. Hist. (Geol.)* **29**:1–135.
Carter, D. J., and Hart, M. B., 1977*b*, Micropalaeontological investigations for the site of the Thames Barrier, London. *Quart. J. Eng. Geol.* **10**:321–338.
Carter, P. G., and Mallard, D. J., 1974, A study of the strength, compressibility and density trends within the Chalk of South East England. *Quart. J. Eng. Geol.* **7**:43–55.
Church, J. A., Godfrey, J. S., Jackett, D. R., and McDougall, T. J., 1991, A model of sea level rise caused by ocean thermal expansion, *J. Clim.* **4**:438–456.
Clayton, K. M., 1990, Sea-level rise and coastal defences in the UK, *Quart. J. Eng. Geol.* **23**:283–287.
Doodson, A. T., and Dines, J. S., 1929, Report on Thames floods and meteorological conditions associated with high tides in the Thames. *Geophys. Mem.* **47**:1–39.
Dunning, F. W., Mercer, I. F., Owen, M. P., Roberts, R. H., and Lambert, J. L. M., 1978, *Britain Before Man*. H.M.S.O., London, 36 pp.
Eddles, A. P., and Hart, M. B., 1989, Late Quaternary Foraminiferida from Plymouth Sound: Preliminary investigation. *Proc. Ussher Soc.* **7**:168–171.
Ermington, S., 1989, Barrier protects London from floods. *Wat. Eng. Mgmt.* **October**: 8–10.
Freeman, P., 1993, An assessment of the effectiveness of the Thames Barrier in response to global mean sea level rise, Unpublished M.Sc. Thesis, University of Plymouth.
Gilbert, S., and Horner, R. W., 1984, *The Thames Barrier*, Thomas Telford, London, 182 pp.
Graham, C. C., and Straw, A., 1992, Quaternary, in: *Atlas of Palaeogeography and Lithofacies* (J. C. W. Cope, J. K. Ingham, and P. F. Rawson, eds.), Geological Society, London, pp. 149–153.
Hart, A. B., 1997, A risk assessment of the inland area around Charmouth, West Dorset. Unpublished M.Sc. Thesis, University of Portsmouth.
Hart, M. B., 1970, The distribution of the Foraminifera in the Albian and Cenomanian of SW England. Unpublished Ph.D. Thesis, University of London.
Hart, M. B., 1993, Engineering and economic geology, in: *Applied Micropalaeontology* (D. G. Jenkins, ed.), Kluwer Academic Publishers, Dordrecht, pp. 225–262.
Hart, M. B., Bailey, H. W., Crittenden, S., Fletcher, B. N., Price, R. J., and Swiecicki, A., 1989, Cretaceous, in: *Stratigraphical Atlas of Fossil Foraminifera 2nd Ed.* (D. G. Jenkins and J. W. Murray, eds.), Ellis Horwood, Chichester, pp. 273–371.
Heaps, N. S., 1977, Development of storm surge models at Bidston. Internal Report No. 51, Institute of Oceanography.
H.M.S.O., 1954, Report of the departmental committee on coastal flooding. Cmd 9165. H.M.S.O., London [the Waverly Report].
Hoffman, J. S., Keyes, D., and Titus, J. G., 1983, Projecting future sea level rise: Methodology, estimates to the year 2000, and research needs. United States Government Printing Office #055-000-0236-3, Washington, DC.
Hoffman, J. S., Wells, J. B., and Titus, J. G., 1986, Future global warming and sea level rise, in: *Iceland Coastal and River Symposium* (G. Sigbjarnason, ed.), National Energy Authority, Reykjavik, pp. 245–266.
Horner, R. W., 1987, The Thames Barrier. *Proc. Instn. Civil Eng.* **134** (November):752–760.
IPCC, 1990, *Climate Change*: The IPCC Scientific Assessment (J. R. Houghton, G. T. Jenkins, and J. J. Ephraums, eds.), Cambridge University Press, Cambridge, 364 pp.
IPCC, 1992, *Climate Change 1992: The Supplementary Report to the IPCC Scientific Assessment* (J. T. Houghton, B. A. Callander, and S. K. Varney, eds.), Cambridge University Press, Cambridge, 200 pp.

IPCC, 1996*a*, *Climate Change 1995: The Science of Climate Change*, Contribution of Working Group I to the Second Assessment Report of the IPCC (J. T. Houghton, L. G. Meira Filho, B. A. Callander, N. Harris, A. Kattenberg, K. Maskell, and J. A. Lakeman, eds.), Cambridge University Press, Cambridge, 572 pp.

IPCC, 1996*b*, *Climate Change 1995: Impacts, Adaptions and Mitigation of Climate Change*. Contribution of Working Group II to the Second assessment Report of the IPCC (R. T. Watson, M. C. Zinyowera, and R. H. Moss, eds.), Cambridge University Press, Cambridge.

IPCC, 1996*c*, *Climate Change 1995: Economic and Social Dimensions of Climate Change*. Contribution of Working Group III to the Second Assessment Report of the IPCC (J. P. Bruce, H. Lee, and E. F. Haites, eds.), Cambridge University Press, Cambridge.

Jaeger, J., 1988, Developing policies for responding to climate change: A summary of the discussions and recommendations of the workshops held in Villach (1987) and Bellagio (1987). WMO/TD-No. 225.

Jefferies, R. P. S., 1961, The palaeoecology of the *Actinocamax plenus* Subzone (lowest Turonian) in the Anglo–Paris Basin, *Palaeontology* **4**:609–674.

Jefferies, R. P. S., 1963, The stratigraphy of the *Actinocamax plenus* Subzone (Lowest Turonian) in the Anglo–Paris Basin, *Proc. Geol. Assoc.* **74**:1–33.

Loeblich, A. R., and Tappan, H., 1964, Protista 2. Sarcodina, chiefly "Thecamoebians" and Foraminiferida, in: *Treatise on Invertebrate Paleontology*, Part C (R. C. Moore, ed.), University of Kansas Press, Kansas, pp. 1:1–510; 2:511–590.

Morner, N.-A., 1976, Eustasy and geoid changes, *J. Geol.* **84**:123–151.

Morner, N.-A. 1987, Models of global sea-level changes, in: *Sea Level Changes* (M. J. Tooley and I. Shennan, eds.), Institute of British Geographers, Spec. Publ. Ser. No. 20, Blackwell, Oxford.

Oerlemans, J., 1989, A projection of future sea level, *Clim. Change* **15**:151–174.

Owen, M., 1970, Turonian foraminifera from southern England, Unpublished Ph.D. Thesis, University of London.

Poland, J. F., and Davis, G. H., 1969, Land subsidence due to withdrawals of fluids, *Rev. Eng. Geol.* **2**:187–269.

PRB (Polar Research Board), 1985, Glaciers, ice sheets and sea level: Effect of a CO_2-induced climate change. Report of a Workshop held in Seattle, Washington, September 13–15, 1984, United States DOE/ER/60235-1.

Pugh, D. T., 1990, Is there a sea level problem? *Proc. Instn. Civil Eng.* **88**:347–366.

Raper, S. C. B., Warrick, R. A., and Wigley, T. M. L., 1990, Global sea level rise: past and future, in: *Proc. SCOPE Workshop in Rising Sea Level and Subsiding Coastal Areas, Bancock 1988* (J. D. Milliman, ed.), John Wiley & Sons, Chichester.

Robaszynski, F., Amedro, F., Foucher, J. C., Gaspard, D., Magniex-Jannin, F., Manivit, H., and Sornay, J., 1980, Synthese Biostratigraphique de l'Aptien au Santonien du Boulonnais a partir de sept groupes paléontologiques: Foraminifères, Nannoplancton, Dinoflagelles et Macrofaunes *Rev. Micropaleontol.* **22**:195–321.

Robaszynski, F., and Caron, M., 1995, Foraminifères planctoniques du Cretace: commentaire de la zonation Europe-Mediterranee, *Bull. Soc. Geol. France* **166**:681–692.

Robin, G. de Q., 1986, Changing sea level, in: *The Greenhouse Effect, Climatic Change and Ecosystems* (B. Bolin, B. R. Doos, J. Jaeger, and R. A. Warrick, eds.), John Wiley & Sons, Chichester, pp. 323–359.

Rose, J., and Allen, P., 1977, Middle Pleistocene stratigraphy in Southeast Suffolk, *J. Geol. Soc. Lond.* **133**:83–102.

Thomas, R. H., 1986, Future sea level rise and its early detection by satellite remote sensing, in: *Effects of Changing Stratospheric Ozone and Global Climate. Vol. 4, Sea Level Rise* J. G. Titus, ed.).

Titus, J. G., and Narrayanan, V., 1995, The probability of sea level rise, US Environmental Protection Agency, Office of Policy, Planning and Evaluation, Climate Change Division. EPA230-R-95-008, EPA, Washington D.C., 186 pp.

van der Veen, C. J., 1988, Projecting future sea level, *Survs. Geophys.* **9**:389–418.

Warrick, R. A., and Oerlemans, J., 1990, Sea level rise, in: *Climate Change: The IPCC Scientific Assessment* (J. T. Houghton, G. J. Jenkins, and J. J. Ephraums, eds.), Cambridge University Press, Cambridge, pp. 257–281.

Wigley, T. M. L., and Raper, S. C. B., 1992, Implications for climate and sea level of revised IPCC emissions scenarios, *Nature* **357**:293–300.

Wigley, T. M. L., and Raper, S. C. B., 1993, Future changes in global mean temperature and sea level, in: *Climate and Sea Level Change: Observations, Projections and Implications* (R. A. Warrick, E. M. Barrow, and T. M. L. Wigley, eds.), Cambridge University Press, Cambridge.

Williams-Mitchell, E., 1948, The zonal value of Foraminifera in the chalk of England, *Proc. Geol. Assoc.* **59**:91–112.

Wilson, G., and Grace, H., 1942, The settlement of London due to underdrainage of the London Clay, *J. Instn. Civil Engrs.* **19**:100–127.

Woodworth, P. L., 1993, Sea level changes, in: *Climate and Sea Level Change: Observations, Projections and Implications* (R. A. Warrick, E. M. Barrow, and T. M. L. Wigley, eds.), Cambridge University Press, Cambridge.

Index

GPSR Compliance

The European Union's (EU) General Product Safety Regulation (GPSR) is a set of rules that requires consumer products to be safe and our obligations to ensure this.

If you have any concerns about our products, you can contact us on ProductSafety@springernature.com

In case Publisher is established outside the EU, the EU authorized representative is:

Springer Nature Customer Service Center GmbH
Europaplatz 3
69115 Heidelberg, Germany

Batch number: 10761599

Printed by Printforce, the Netherlands